Lattice Coding for Signals and Networks

A Structured Coding Approach to Quantization, Modulation and Multi-user Information Theory

Unifying information theory and digital communication through the language of lattice codes, this book provides a detailed overview for students, researchers and industry practitioners.

It covers classical work by leading researchers in the field of lattice codes and complementary work on dithered quantization and infinite constellations, and then introduces the more recent results on "algebraic binning" for side-information problems, and linear/lattice codes for networks. It shows how high-dimensional lattice codes can close the gap to the optimal information theoretic solution, including the characterization of error exponents.

The solutions presented are based on lattice codes, and are therefore close to practical implementations, with many advanced setups and techniques, such as shaping, entropy-coding, side-information and multi-terminal systems. Moreover, some of the network setups shown demonstrate how lattice codes are potentially more efficient than traditional random coding solutions, for instance when generalizing the framework to Gaussian networks.

Ram Zamir is a Professor at the Department of Electrical Engineering Systems at Tel-Aviv University, Israel. He consults in the areas of radar and communications (DSL and WiFi), and is the Chief Scientist of Celeno Communications.

Lattice Coding for Signals and Networks

A Structured Coding Approach to Quantization,
Modulation and Multi-user Information Theory

RAM ZAMIR

Tel Aviv University

with contributions by

BOBAK NAZER

Boston University

and

YUVAL KOCHMAN

The Hebrew University of Jerusalem

and with illustrations by

Ilai Bistritz

CAMBRIDGE
UNIVERSITY PRESS

CAMBRIDGE
UNIVERSITY PRESS

University Printing House, Cambridge CB2 8BS, United Kingdom

Cambridge University Press is part of the University of Cambridge.

It furthers the University's mission by disseminating knowledge in the pursuit of education, learning and research at the highest international levels of excellence.

www.cambridge.org
Information on this title: www.cambridge.org/9780521766982

© Cambridge University Press 2014

First published 2014

Printed in the United Kingdom by Clays, St Ives plc

A catalogue record for this publication is available from the British Library

Library of Congress Cataloguing in Publication data
Zamir, Ram.
Lattice coding for signals and networks : a structured coding approach to quantization, modulation, and multiuser information theory / Ram Zamir, Tel Aviv University.
    pages   cm
Includes bibliographical references and index.
ISBN 978-0-521-76698-2 (hardback)
1. Coding theory.   2. Signal processing – Mathematics.   3. Lattice theory.   I. Title.
TK5102.92.Z357   2014
003'.54 – dc23   2014006008

ISBN 978-0-521-76698-2 Hardback

To my parents Eti and Sasson Zamir

Contents

Preface

Digital communication and information theory talk about the same problem from very different aspects. Lattice codes provide a framework to tell their mutual story. They suggest a common view of source and channel coding, and new tools for the analysis of information network problems.

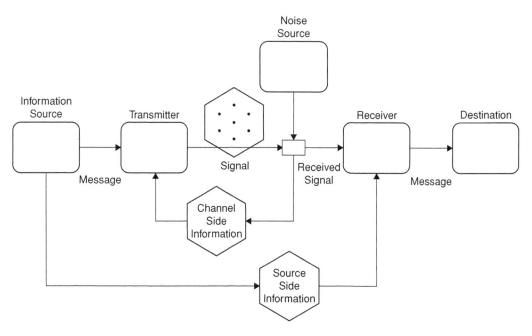

This book makes the language of quantization and modulation more accessible to the hard core information theorist. For him or her, lattices serve as a bridge from the high dimension of Shannon's theory to that of digital communication techniques. At the same time, lattices provide a useful tool for the communication engineer, whose scope is usually limited to the low – sometimes even one or two – dimensions of practical modulation schemes (e.g., QAM or PCM). She or he can "see," through the lattice framework, how signals and noise interact as the dimension increases, for example, when modulation is combined with coding. Surprisingly for both disciplines, the generalization of the lattice framework to "Gaussian networks" is not only very natural, but in some cases is more powerful than the traditional techniques.

This book is beneficial to the "Gaussian-oriented" information theorist, who wishes to become familiar with network information theory from a constructive viewpoint (as opposed to the more abstract random-coding/random-binning approach). And it is a useful tool for the communication practitioner in the industry, who prefers a "geometric" and "signal-processing oriented" viewpoint of information theory in general, and multi-user problems in particular. The algebraic coding theorist can celebrate the variety of new applications for lattice codes found in the book. The control theorist, who wishes to add communication constraints into the system, will find the linear-additive model of dithered lattice quantization useful. Other readers, like those having a background in signal processing or computer networks, can find potential challenges in the relations to linear estimation and network coding.

Ram Zamir
Tel Aviv
March 2014

Acknowledgements

In the spring of 1989 I took a data compression course with Meir Feder at Tel Aviv University, and read Jacob Ziv's 1985 article "On universal quantization." Ziv showed that the redundancy of a randomized (dithered) uniform scalar quantizer is always bounded by ≈ 0.754 bit; he also stated, without a proof, that if the scalar quantizer is replaced by a "good" high-dimensional lattice quantizer, then this universal bound can be reduced to half a bit – provided that Gersho's conjecture is true. In my final project – while learning about Gersho's conjecture and verifying Ziv's statement – I fell in love with the world of lattice codes.

Many people with whom I have collaborated since then have contributed to the material presented in this book. The roots of the book are in my MSc and PhD research – under Meir's supervision – about entropy-coded dithered lattice quantization (ECDQ) of signals. Tamas Linder offered rigorous proofs for some of the more technical results (and became my colleague and coauthor for many years). Two other staff members in the EE department in Tel Aviv University – Gregory Poltyrev and Simon Litsyn – while contributing from their wide knowledge about lattices, opened the way to the later applications of dithered lattice codes to signaling over the AWGN channel.

Toby Berger – my post-doctoral mentor at Cornell University in the years 1994–1995 – introduced me to the fascinating world of multi-terminal source coding. This became the first instance where randomized lattice codes were applied in network information theory. A year later, Shlomo Shamai and Sergio Verdú, with whom I communicated about systematic lossy source-channel codes, inspired me to introduce the idea of nested lattices for the Wyner–Ziv (source coding with side information) problem. This idea, which started as a toy example for a more practical systematic source-channel code, grew later into a general framework for "algebraic binning" for information networks.

Uri Erez took my first advanced information theory course in the spring of 1997; in his final project he developed an interesting technique for using channel-state information at the transmitter. His PhD research then became a fusion center of many ideas in lattice coding for noisy channels: following a pointer given to us by Shlomo Shamai to the Costa problem, Uri came up with the innovative idea of lattice pre-coding for the "dirty-paper" channel (a channel with interference known at the transmitter), using dither and Wiener estimation. Simon Litsyn helped in showing the existence of lattices which are "good for almost everything," which turned out to be a crucial element in the asymptotic optimality of nested lattice based coding schemes for general network problems. Dave Forney provided insightful comments about Uri's work, and – after noticing that the zero-interference case resolves an open question about lattice decoding of Voronoi codes – summarized his interpretations under the multiple-meaning title "Shannon meets Wiener" (2002).

Emin Martinian and Greg Wornell contributed the idea of lattice codes with variable partition (for source coding with distortion side information at the encoder) during my Sabbatical at MIT in 2002–2003.

The work in my research group during the years 2003–2010 revealed two new exciting aspects of lattice codes. Yuval Kochman developed the modulo-lattice modulation technique for joint source-channel coding, and in particular, for bandwidth conversion (an idea proposed earlier in Zvi Reznic's PhD work). Tal Philosof discovered (during his PhD research with Uri Erez and myself) that lattice codes are stronger than random codes for the "doubly dirty" multiple-access channel.

Although the material had been there for quite a few years, it took some courage and encouragement to initiate this book project. The idea was thrown into the air during my visit at Andy Loeliger and Amos Lapidoth's groups at ETH, in the summer of 2008, and suggested again by Jan Østergaard during my visit at Aalborg University a couple of months later. Dave Forney gave me important comments and suggestions in the early stages of the writing, and I thank him for that. Tom Cover, whose book with Joy Thomas was a source of inspiration for many years, was kind enough to give me a few writing-style tips during my visit at Stanford in the summer of 2009.

Our research students in Tel Aviv University provided enormous help during the writing of this book. The chapter about lattice error exponents grew from extensive discussions with Amir Ingber. Sergey Tridenski and Arie Yeredor made specific contributions to the section on error exponents for Voronoi codebooks. Or Ordentlich and Uri Erez helped me shape the material about the existence of good lattices and nested lattices. My thanks are due to Yuval Domb, Eli Haim, Anatoly Khina, Adam Mashiach, Nir and Michal Palgy, Nir Weinberger and Yair Yona for many fruitful discussions; and to the students who participated in my "Lattices in information theory" course in the fall of 2011 for the valuable feedback.

Special thanks are due to my programming assistant Ilai Bistritz, whose good advice went much beyond the numerical work, graphs and illustrations that he contributed to this book.

During the work I received help and good advice from Ofer Amrani, Benny Appelbaum, Joseph Boutrus, Shosh Brosh-Weitz, Robert Calderbank (who gave me his class notes on coded modulation), Avner Dor, Moran and Tal Gariby, Michael Gastpar, Bo'az Klartag, Frank Kschischang, Stella Achtenberg, Tamas Linder, Bobak Nazer, Jan Østergaard, Dan Rephaeli, Kenneth Rose, Yaron Shany, Anelia Somekh-Baruch and Alex Vardy. (And I have surely missed here some important people who helped along the way.) Comments on early drafts of the book were kindly provided by Ling Cong, Adam Mashiach, Jan Østergaard, Danilo Silva, Shlomo Shamai and Yaron Shani, who also provided references and pointers.

This writing project could last forever without the constant attention and professional advice of my editors at Cambridge University Press, Phil Meyler, Sarah Marsh and Mia Balashova.

I was extremely happy when Bobak Nazer agreed to join me in writing the chapter about Gaussian networks; his deep understanding of the subject and clear writing style took this part of the book to a much higher level. Also the chapter about modulo-lattice modulation greatly benefitted from the collaboration in writing with Yuval Kochman.

Last but not least, I could not have survived these four long years of writing without the infinite love and patience of my wife Ariella and three children Kessem, Shoni and Itamar.

Notation

Lattices

Λ	lattice
G	generating matrix (columns are basis vectors)
$\det(\Lambda)$	lattice determinant
$\mathcal{P}_0$	fundamental cell
$\mathcal{V}_0, \mathcal{V}_\lambda$	fundamental Voronoi cell, Voronoi cell of lattice point λ
$V(\Lambda)$	cell volume
$\gamma(\Lambda)$	point density
$\bmod \Lambda, \bmod_{\mathcal{P}_0} \Lambda, \mathbf{x}/\Lambda$	modulo-lattice operations
$Q_\Lambda(\cdot)$	lattice quantizer
$Q_\Lambda^{(NN)}(\cdot)$	nearest-neighbor lattice quantizer
$d_{\min}$	minimum distance
$N_\Lambda(d)$	number of lattice points at distance d from the origin
$N_\Lambda(d_{\min})$	kissing number of Λ
$r_{\text{pack}}(\Lambda), r_{\text{cov}}(\Lambda)$	packing radius, covering radius
$r_{\text{eff}}(\Lambda)$	effective radius
$\rho_{\text{pack}}(\Lambda), \rho_{\text{cov}}(\Lambda)$	packing and covering efficiencies
$\sigma^2(\Lambda)$	second moment
$G(\Lambda)$	normalized second moment (NSM)
$\Gamma_q(\Lambda), \Gamma_s(\Lambda)$	vector-quantizer granular gain, shaping gain
$P_e(\Lambda, \sigma^2)$	error probability (in the presence of AWGN)
$\mu(\Lambda, \sigma^2)$	volume to noise ratio (VNR)
$\mu(\Lambda, P_e)$	normalized volume to noise ratio (NVNR)
$\mu_{\text{matched}}(\Lambda, \mathbf{Z}, P_e)$	noise-matched NVNR
$\mu_{\text{euclid}}(\Lambda, \mathbf{Z}, P_e)$	Euclidean (mismatched) NVNR
$\mu_{\text{mix}}(\Lambda_1, \Lambda_2, P_e, \alpha),$	mixture-noise NVNR
$\quad \mu_{\text{mix}}(\Lambda_1, \Lambda_2, P_e, \xi)$	
$\Gamma_c(\Lambda, P_e)$	coding gain (relative to cubic lattice)
$\mathbf{U}, \mathbf{U}_{\text{eq}}$	dither, equivalent dithered quantization noise
R_{ECDQ}	entropy rate of lattice quantizer
$R_\infty(\Lambda)$	rate per unit volume

$\mathbb{L}$	Minkowski–Hlawka–Siegel (MHS) ensemble
$N_{\mathcal{S}}(\Lambda)$	number of non-zero lattice points in $\mathcal{S}$
J	nesting matrix
$\Gamma = \Gamma(\Lambda_1, \Lambda_2)$	nesting ratio
Λ_1/Λ_2	quotient group, relative cosets
$\mathcal{C}_{\Lambda_1, \mathcal{P}_0(\Lambda_2)}$, $\mathcal{C}_{\Lambda_1, \mathcal{V}_0(\Lambda_2)}$	lattice-shaped codebook, Voronoi codebook
$\mathcal{C}_{\mathbf{u},\Lambda_1,\mathcal{P}_0}$	dithered codebook
$R(\Lambda_1/\Lambda_2)$	codebook rate [bit per dimension]
$\mathbf{Z}_{\mathrm{eq}}$	equivalent noise in mod Λ channel

Information theory

$H(X)$	regular entropy (of random variable X)
$H_B(p)$	binary entropy
$h(X)$	differential entropy
$I(X;Y)$	mutual information (between a pair of random variables)
$P_E(X)$	entropy power of a random variable $(P_E(X) = 2^{2h(X)}/2\pi e)$
C	channel capacity
C_∞	capacity per unit volume (Poltyrev's capacity)
$C^{(d)}$, $C^{(\mathrm{euclid\text{-}th})}$	mismatched capacities
$R(D)$	rate-distortion function
$\mathcal{A}_\epsilon^{(n)}$	typical set
$\mathcal{C}$	codebook
$r_{\mathrm{noise}} = \sqrt{n\sigma^2}$	typical AWGN radius

General

x, y	scalar variables
$\mathbf{x}, \mathbf{y}$	vector variables
X, Y	random variables
$\mathbf{X}, \mathbf{Y}$	random vectors (column form)
$\mathbf{X}^t$	$\mathbf{X}$ transpose (row form)
$\mathrm{Var}(\mathbf{X})$	average variance per dimension
$\mathbb{R}^n$	Euclidean space
$\mathbb{Z} = \{0, \pm1, \pm2, \ldots\}$	integers
$\mathbb{Z}_q = \{0, 1, \ldots, q - 1\}$	modulo-q group
$N(\mu, \sigma^2)$	Gaussian distribution with mean μ and variance σ^2
$\mathcal{B}_r$	a ball of radius r centered about the origin
$\mathcal{B}(\mathbf{x}, r)$	a ball of radius r centered at $\mathbf{x}$
V_n	volume of a unit-radius n-dimensional ball

Vol(S)	volume of a set S
$\dot{=}$	equality to the first order in the exponent
$\otimes$	binary convolution ($p \otimes q = p(1 - q) + q(1 - p)$)
$[x]^+$	maximum between x and zero.

Abbreviations

AWGN	additive white-Gaussian noise
BPSK	binary phase-shift keying
BSC	binary-symmetric channel
BSS	binary-symmetric source
ECDQ	entropy-coded dithered quantization
MAC	multiple-access channel
ML	maximum likelihood
MSE	mean-squared error
NN	nearest-neighbor
NSM	normalized second moment
NVNR	normalized volume to noise ratio
PAM/QAM	pulse/quadrature-amplitude modulation
PAPR	peak to average power ratio
SER	symbol error rate
SNR	signal to noise ratio
VNR	volume to noise ratio
VNER	volume to noise entropy power ratio
VRDM	variable-rate dithered modulation

1 Introduction

Roughly speaking, a lattice is a periodic arrangement of points in the n-dimensional Euclidean space.[1] It reflects the "geometry of numbers" – in the words of the late nineteenth century mathematician Hermann Minkowski. Except for the one-dimensional case (where all lattices are equivalent up to scaling), there are infinitely many shapes of lattices in each dimension. Some of them are better than others.

Good lattices form effective structures for various geometric and coding problems. Crystallographers look for symmetries in three-dimensional lattices, and relate them to the physical properties of common crystals. A mathematician's classical problem is to pack high-dimensional spheres – or cover space with such spheres – where their centers form a lattice. The communication engineer and the information theorist are interested in using lattices for quantization and modulation, i.e., as a means for lossy compression (source coding) and noise immunity (channel coding). Although these problems seem different, they are in fact closely related.

The effectiveness of good lattices – as well as the complexity of describing or using them for coding – increases with the spatial dimension. Such lattices tend to be "perfect" in all aspects as the dimension goes to infinity. But what does "goodness" mean in dimensions 2, 3, 4, . . .?

In two dimensions, the *hexagonal lattice* is famous for the honeycomb shape of its Voronoi cells. The centers of the billiard (pool) balls in Figure 1.1 fall on a hexagonal lattice, which forms the tightest packing in two dimensions. The same hexagonal lattice defines a configuration for deploying cellular base stations that maximizes the coverage area per base station.

Interestingly, however, for higher dimensions the problems of packing and covering are *not* equivalent. In Figure 1.2, the centers of the oranges fall on the face-centered cubic (FCC) lattice, which is the best known sphere packing in three dimensions. In contrast, the best deployment of cellular base stations in a skyscraper (which maximizes their *three-dimensional* coverage) is over a body-centered cubic (BCC) lattice, illustrated in Figure 1.3.

---

[1] See the Wikipedia disambiguation page for other meanings of the word "lattice": in art and design, music, math and science.

Figure 1.1 Billiard (pool) balls packed in a triangle, for an initial game position.

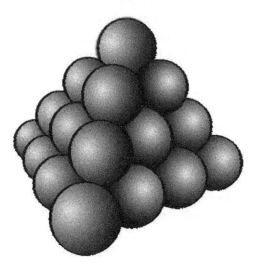

Figure 1.2 Packing oranges in a pile: each row is half-diameter shifted with respect to the previous row to reduce the unused volume. Similarly, each layer is staggered to fill the holes in the layer below it. The centers of the oranges form a lattice known as a face-centered cubic (FCC) lattice.

Which is the "best" lattice in each dimension is a question we shall not address; issues of efficient design and coding complexity of lattices are not at the focus of this book either. Instead, we characterize the performance of a lattice code by its thickness (relative excess coverage) and density (relative packed volume), and by the more communication-oriented figures of merit of normalized second moment (NSM) for quantization, and normalized volume to noise ratio (NVNR) for modulation. We define these quantities in detail in Chapter 3, and use them in Chapters 4–9 to evaluate lattice codes for the basic *point-to-point* source and channel coding problems. As we shall see, high-dimensional lattice codes can close

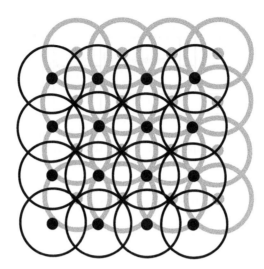

Figure 1.3 Three-dimensional sphere covering with a BCC lattice, describing the best deployment of cellular base stations in a skyscraper. The solid line shows even layers; the gray line shows odd layers. Compare the staggering pattern with that of the pile of oranges in Figure 1.2.

the gap to the information theoretic limits of communication: the capacity and rate-distortion function, quantities introduced by Shannon in his seminal 1948 paper [240], and further refined during the 1950s and 1960s.

The 1970s and 1980s saw the blooming of network information theory. Remarkably, some of the fundamental network problems were successfully solved using Shannon's information measures and *random coding* techniques, now with the additional variant of random binning. Simple examples of such network setups are *side-information* problems: the Slepian–Wolf and Wyner–Ziv source coding problem, and the Gelfand–Pinsker "dirty-paper" channel coding problem. The lattice framework provides a *structured coding* solution for these problems, based on a nested pair of lattices. This nested lattice configuration calls for new composite figures of merit: one component lattice should be a good channel code (have a low NVNR), while the other component lattice should be a good quantizer (have a low NSM). For joint source-channel coding problems, lattices with a good NSM-NVNR product are desired. We shall develop these notions in Chapters 10 and 11.

The curious reader may still wonder why we need a book about lattices in information theory. After all, Shannon's probabilistic measures and random coding techniques characterize well the limits of capacity (channel coding) and compression (source coding), and they also allow the study of source and channel networks [53, 64]. From the practical world side, communication theory provides ways to combine modulation with "algebraic" codes and approach the Shannon limits.

All this is true, yet between the theoretical and the constructive points of view something gets lost. Both the probabilistic and the algebraic approaches somewhat

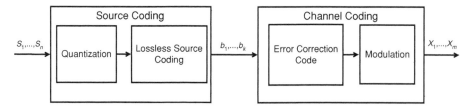

Figure 1.4 Source coding followed by channel coding. For an analog source and channel, the combined system maps a point in $\mathbb{R}^n$ (a source vector) to a point in $\mathbb{R}^m$ (a channel input vector). The ratio m/n is known as the "bandwidth-expansion factor."

hide the interplay between analog signals like sound or noise (created by nature) and digital modulation signals (created by man). Lattices are discrete entities in the analog world, and as such they bridge nicely the gap between the two worlds. At large dimensions, good lattices mimic the behavior of Shannon's random codes. For small dimensions, they represent an elegant combination of modulation and digital coding. As a whole, lattices provide a unified framework to study communication and information theory in an insightful and inspiring way.

Recent developments in the area of network information theory (mostly from the 2000s) have added a new chapter to the story of lattice codes. In some setups, structured codes are potentially *performance-wise better* than the traditional random coding schemes! And as Chapter 12 shows, the natural candidates to achieve the benefit of structure in Gaussian networks are, again, lattice codes.

1.1   Source and channel coding

Let us describe briefly how lattices fit into the framework of digital communication and classical information theory.

By Shannon's *separation principle*, transmission of an information source over a noisy channel is split into two stages: *source coding*, where the source is mapped into bits, and *channel coding*, where the digital representation of the source is mapped into a channel input signal. These two stages, which we describe in detail below, are illustrated in Figure 1.4.

The *source coding* (or compression) problem deals with compact digital representation of source signals. In *lossless* compression, our goal is to *remove redundancy* due to asymmetry in the frequency of appearance of source values, or to "memory" in the source. In this case, the source signal is available already in a digital form, say, as a sequence of binary symbols. And the task is to map n "redundant" source bits $\mathbf{s} = s_1, \ldots, s_n$ into $k = k(\mathbf{s})$ code bits, where $k < n$. [2]

---

[2] We would like k to be smaller than n for most source vectors (or for the most likely ones) in order to compress; but not too small, so the mapping would be invertible for (almost) all source vectors, for lossless reproduction.

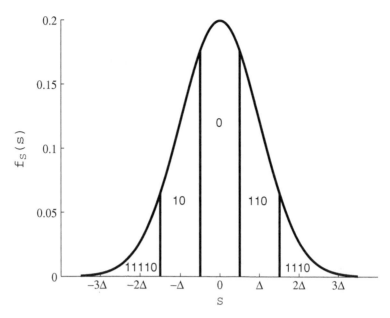

Figure 1.5 Scalar uniform quantization of a Gaussian source, followed by variable-length coding, i.e., $n = 1$ and k is varying. Each quantization level represents a range of source values.

In *lossy* compression, the source is usually *continuous* in nature: an analog representation of speech, sound, picture or video signal. Digitizing an analog signal consists first of converting it into a *discrete* form (both in time and in amplitude), and then coding it in the discrete alphabet domain. In discrete time the source is again a vector $\mathbf{s} = s_1, \ldots, s_n$, representing n consecutive source samples. After the vector $\mathbf{s}$ is encoded into a k-bit codeword, it is decoded and reconstructed as $\hat{\mathbf{s}} = \hat{s}_1, \ldots, \hat{s}_n$. The overall operation of mapping $\mathbf{s}$ to $\hat{\mathbf{s}}$ is called *quantization*, and the image (for a fixed k, the set of all 2^k possible reconstruction vectors $\hat{\mathbf{s}}$ in $\mathbb{R}^n$) is the quantization *codebook*.

A *lattice quantizer* codebook consists of points from an n-dimensional lattice. The codebook can be a truncated version (of size 2^k) of the lattice, or the whole lattice (with a variable codeword length $k = k(\hat{\mathbf{s}})$). We would like to make the bit rate $R = k/n$ (or the average coding rate $R = \bar{k}/n$) as *small* as possible, subject to a constraint on the reconstruction fidelity. Figure 1.5 shows the case of a scalar ($n = 1$) lattice quantizer with a variable code length $k(\hat{\mathbf{s}})$.

Channel coding deals with transmitting or storing information over a noisy channel or on a storage device. Our goal here is to *add redundancy* to the transmitted signal, to make it distinguishable from the noise. The channel input alphabet may be *discrete*, say, binary. In this case, transmission amounts to mapping k bits of information into n "redundant" code bits, where $n > k$.

The most common communication links are, however, over *continuous* media: telephone lines, cables or radio waves. The *baseband* channel representation is in

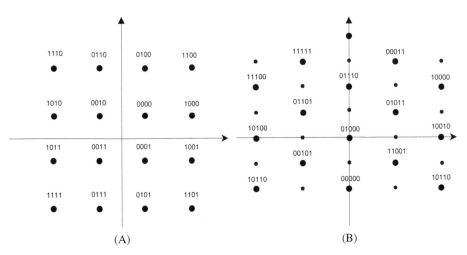

Figure 1.6 Two-dimensional finite lattice constellations, consisting of 16 points ($k = 4$). (A) A simple square constellation, representing uncoded quadrature-amplitude modulation (QAM); here $n' = k = 4$. (B) A hexagonal lattice constellation, represented as a mapping of redundant binary vectors of length $n' = 5$ into a rectangular constellation.

discrete time, so the channel input is a vector $\mathbf{x} = x_1, \ldots, x_n$. Coding over such a channel turns out to be in many ways the *dual* of encoding an analog source. It consists of two stages: an *error-correction coding* stage, where redundancy is added in the discrete alphabet domain (e.g., by converting k information bits to $n' > k$ code bits); and a *modulation* stage, where the digital codeword is mapped into the vector $\mathbf{x}$. The overall encoder mapping is thus of a k-bit information vector into a point in $\mathbb{R}^n$ (representing n consecutive channel inputs). The set of all 2^k possible input vectors $\mathbf{x}$ is called a codebook or a *constellation*.

A *lattice constellation* is a truncated version (of size 2^k) of an n-dimensional lattice. We would like to make the coding rate $R = k/n$ – which is now the (usually fixed) number of transmitted information bits per channel input – as *large* as possible, subject to a constraint on the probability of decoding error. See two examples of two-dimensional lattice constellations in Figure 1.6.

One benefit of the lattice coding framework that we can immediately recognize is that coding and modulation (or quantization) are combined as a *single entity*; a lattice code directly maps digital information (say, an index) into a vector in $\mathbb{R}^n$, and vice versa.

1.2   The information theoretic view

Information theory characterizes the ultimate performance limits of source and channel coding, as the code block length n goes to infinity.

In the channel coding case, the coding rate R is upper bounded (for a vanishing error probability) by the *Shannon capacity* C of the channel. The quantity C (associated with a memoryless channel with a transition distribution $p(y|x)$) is calculated by maximizing the mutual information (a functional of $p(x)$ and $p(y|x)$) over the input distribution $p(x)$. The maximizing input distribution $p^*(x)$ is used to prove the achievability of C: a set of $\approx 2^{nC}$ codewords is generated *randomly* and independently with an i.i.d. distribution $p^*(x)$; a *random coding* argument is then used to show that based on the channel output, the decoder can guess the correct transmitted codeword with a high probability as $n \to \infty$.

We see that *à la* Shannon, good codewords look like *realizations of random noise*. In the case of a binary-symmetric channel, the code generating noise consists of equally likely 0/1 bits. In the quadratic-Gaussian case, the code should be generated by a *white-Gaussian noise* (WGN).

Rate-distortion theory uses similar ideas to establish the ultimate performance limits of lossy source coding [18]. The Shannon *rate-distortion function $R(D)$* lower bounds the coding rate R of any lossy compression scheme with distortion level of at most D (under some given distortion measure). And similarly to the channel coding case, computation of $R(D)$ induces an optimal reconstruction distribution, which is used to generate a good random codebook: independent realizations of a Bernoulli(1/2) sequence compose the codewords for a binary-symmetric source under Hamming distortion, while independent realizations of WGN compose the codewords for a white-Gaussian source under mean-squared distortion.

The fact that good codewords look like white noise is intriguing. Intuitively, one would expect the symbols of a codeword to be *dependent*, to distinguish them from the channel noise. This has made the random coding idea, on the one hand, a source of inspiration for many since Shannon presented his landmark theory in 1948. On the other hand, it sets a challenge for finding more *structured* ways to approach the information theoretic limits, ways in which the dependence between the code symbols is more explicit. Can noise be realized in a structured way?

1.3   Structured codes

The Hamming code – mentioned already in Shannon's 1948 paper – was the early bird of the structured coding approach. It was followed by the breakthrough of algebraic coding theory in the 1950s and 1960s [21]. The implication was that, in fact, a good collection of random-like bits can be constructed as an additive group in the binary modulo-2 space. These *linear codes* take various forms, such as Reed–Muller, BCH and, more recently, LDPC, turbo and polar codes, and they also have extensions to non-binary (Reed–Solomon) codes and convolutional (trellis) codes.

Common to all these codes is that for a random message, the resulting n-length codeword is indeed roughly uniformly distributed over the n-dimensional binary space. That is, each code bit takes the values 0 and 1 with equal probability; furthermore, small subsets of code bits are roughly independent.

The extension of this concept to continuous signals is however less obvious: can a code mimic Gaussian noise in a structured way? A first step towards this goal is provided by Shannon's *asymptotic equipartition property* (AEP). In a high dimension n, the *typical set* of WGN of variance σ^2 is a spherical shell of radius $\approx \sqrt{n\sigma^2}$. Thus, the codewords of a good code are roughly uniformly distributed over such a spherical shell.

The concept of *geometrically uniform codes* (GUC) [86] suggests a deterministic characterization for a "uniform-looking" code: every codeword should have the same *distance spectrum* to its neighboring codewords. This concept captures the desired property of a good Euclidean code, in both the block and the convolutional (*trellis*) coding frameworks.

Due to their periodic and linear structure, lattices are natural candidates for *unbounded* GUCs. For example, the commonly used QAM constellation shown in Figure 1.6(A) is a truncated version of the *square lattice*, while the more "random-like" set of two-dimensional codewords shown in Figure 1.6(B) is a truncated version of the *hexagonal lattice*. Moreover, the code designer can shape the borders of these constellations to be more round, for example, by truncating them into a circle or into a coarser hexagonal cell. And as the dimension gets high, lattices which are truncated into a "good" coarse lattice cell become closer to a randomly generated Gaussian codebook.

1.4   Preview

We shall get to the exciting applications mentioned earlier after building up some necessary background. The book starts by introducing lattices in Chapter 2, and the notions of lattice goodness in Chapter 3. Chapter 4 introduces two central players in our framework: dithering, which is a means to randomize a lattice code, and Wiener estimation, which is a means to reduce the quantization or channel noise. The importance of these techniques will be revealed gradually throughout the book.

Equipped with these notions and techniques, we consider variable-rate ("entropy-coded") dithered quantization (ECDQ) using an *unbounded* lattice in Chapter 5. In particular, we shall see how the NSM characterizes the redundancy of the ECDQ above Shannon's rate-distortion function. The reader who is interested primarily in channel coding may skip Chapter 5, and continue directly to modulation with an unbounded lattice constellation in Chapter 6.[3] This chapter shows how the NVNR determines the gap from capacity of a lattice constellation. It also describes variable-rate dithered modulation, which is the channel coding counterpart of ECDQ.

---

[3] Sections which are optional reading for the flow of the book are denoted by an asterisk.

Before moving to more advanced coding setups, we stop to examine the existence of asymptotically good lattices in Chapter 7. In Chapter 8 we define nested lattices, and *finite* Voronoi-shaped codebooks taken from a lattice. These notions form in Chapter 9 the basis for Voronoi modulation, which achieves the capacity of a power-constrained AWGN channel, and for Voronoi quantization, which achieves the quadratic-Gaussian rate-distortion function. In both these solutions, dither and Wiener estimation play crucial roles.

A small step takes us from the point-to-point communication setups above to side-information problems in Chapter 10. We shall construct lattice code solutions for the Wyner–Ziv problem (source coding with side information at the decoder) and the "dirty-paper" problem (channel coding with side information at the encoder). These lattice coding schemes serve as building blocks for common multi-terminal communication problems: encoding of distributed sources and broadcast channels. Before moving to more general networks, we examine in Chapter 11 a lattice-based joint source-channel coding technique, called modulo-lattice modulation (MLM). A combination of MLM and prediction leads to "analog matching" of sources and channels with mismatched spectra, and to "bandwidth conversion." Chapter 12 extends the discussion on multi-terminal problems to general Gaussian networks. There we shall see that when side information is distributed among several nodes of the network, lattice codes are not only attractive complexity-wise, but sometimes they have better performance than traditional random coding and binning techniques.

Chapter 13 complements the discussion of asymptotically good lattice codes in Chapter 7 by examining their error exponents. As for capacity, good lattice codes turn out to be optimal also in terms of this more refined aspect.

Information theory is not a critical prerequisite for reading this book, but (starting from Chapter 5) we use information measures, such as entropy, mutual information and capacity, to assess system performance. To keep the book self-contained, the Appendix includes elementary background in information theory, as well as some other complementary material.

As mentioned above, dithering and Wiener estimation are central concepts in the lattice coding framework. The question of where and in what sense they are necessary will follow our discussion throughout the book.

What's *not* in the book?

The writer has the freedom to focus on his favorite subject. Naturally (in the case of this writer) the book takes an information theoretic flavor, with less emphasis on coding theoretic aspects. For algebra of lattices, and for specific constructions of lattices and coded-modulation schemes from error-correcting codes, the reader is referred to the comprehensive book of Conway and Sloane [49], and to the excellent class notes of Forney [81] and Calderbank [28].

Encoding and decoding complexity is a topic of theoretical as well as practical importance, although traditionally neglected by information theory. A good introduction to the subject can be found in the survey paper of Agrell *et al.* [3]. The vast literature on MIMO communication contains numerous publications about the design of linear coded-modulation schemes and efficient lattice decoding algorithms.

In the fight between a timely manuscript and time of publication, some topics which are natural to the spirit of the book were left out. One such topic is the extension to *colored*-Gaussian sources and channels; see, for example, [211, 288, 291]. Another topic is the emerging area of lattice wiretap codes; see, for example, the survey paper by Liang *et al.* [156] and other recent work [118, 168]. Hopefully these topics will find their way to a later edition of the book.

Finally, since the late 1990s lattice-based cryptography has been a major area of research in computer science. Its connection to lattice codes for communication is yet to be explored; see the book by Micciancio and Goldwasser [186], and the survey by Micciancio and Regev [188].

2 Lattices

The simplest lattice is the one-dimensional grid $\{\ldots, -2\Delta, -\Delta, 0, \Delta, 2\Delta, \ldots\}$. In one dimension, all lattices are equivalent up to scaling. To make life more interesting – and to obtain better geometric properties – we must consider multi-dimensional lattices.

This chapter presents n-dimensional lattices and important ideas associated with lattice codes that are used throughout the book. We take a geometric and, for some asymptotic results, probabilistic viewpoint. The algebraic aspects of lattices – although crucial for their implementation at a low complexity – are secondary for our purposes, and will not be treated in this book.

We restrict our attention to communication problems in which the lattice code is selected by the system designer. Thus, we rely on the existence of lattices with certain "good" properties, and on algorithms for encoding and decoding them at a reasonable complexity. [1]

We start with the basic definitions of a lattice and lattice partition.

2.1 Representation

A lattice is a regular array in the Euclidean space. Mathematically, it is a *discrete sub-group* of $\mathbb{R}^n$: a set of points which is *closed under reflection and real addition*. The set is *discrete* in the sense that the distance between any two points is greater than some positive number. If a point λ is in the lattice then so is its reflection $-\lambda$, and if two points λ_1 and λ_2 are in the lattice then so is their vector sum $\lambda_1 + \lambda_2$. Thus, the origin (the point $\mathbf{0}$) is always in the lattice because it is the sum of λ and $-\lambda$.

---

[1] The situation is different when the lattice is selected by nature or at random. For example, in digital communication (e.g., QAM) over a fading MIMO channel, the physical multi-path channel behaves like a random matrix which creates an equivalent lattice constellation at the receiver. In cryptography, a "hard-to-break" lattice is created by a random number generator.

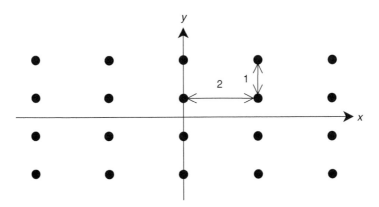

Figure 2.1 The two-dimensional grid $\{(2i, j) : i, j \in \{0, \pm 1, \pm 2, \ldots\}\}$ contains all points in the plane whose y-coordinate is an integer and whose x-coordinate is an integer multiple of 2.

Furthermore, the lattice is a countably infinite set: it must contain all integer multiples $\pm 2\lambda, \pm 3\lambda, \pm 4\lambda, \ldots$ of any lattice point λ, as well as all integer linear combinations $\lambda_1 \pm \lambda_2, \lambda_1 \pm 2\lambda_2, \ldots, 3\lambda_1 \pm 2\lambda_2, \ldots$, of any two lattice points λ_1 and λ_2, etc.

We can obtain simple multi-dimensional lattices by taking the Cartesian product of scalar lattices, like the two-dimensional grid shown in Figure 2.1. Such a simple grid, however, would not allow us to obtain the efficient arrangements of oranges and cellular base stations shown in Figures 1.2 and 1.3. Our next step is to define a lattice in a more general and constructive way.

The linearity property of the lattice reminds us of a linear vector space. It is only in the latter that any real-valued coefficients, and not just integer multiples, are possible. This analogy calls for a definition of a lattice in terms of a *basis*.

Definition 2.1.1 *A non-degenerate n-dimensional lattice Λ is defined by a set of n linearly independent basis (column) vectors $g_1, \ldots, g_n$ in $\mathbb{R}^n$. The lattice Λ is composed of all integral combinations of the basis vectors, i.e.,*

$$\Lambda = \left\{ \lambda = \sum_{k=1}^{n} i_k g_k : i_k \in \mathbb{Z} \right\}$$

$$= \left\{ \lambda = G \cdot \mathbf{i} : \mathbf{i} \in \mathbb{Z}^n \right\}, \tag{2.1}$$

where $\mathbb{Z} = \{0, \pm 1, \pm 2, \ldots\}$ is the set of integers, $\mathbf{i} = (i_1, \ldots, i_n)^t$ is an n-dimensional integer (column) vector, and the $n \times n$ generator matrix G is given by

$$G = [\, g_1 \,|\, g_2 \,|\, \cdots \,|\, g_n \,].$$

The resulting lattice is denoted $\Lambda(G)$.

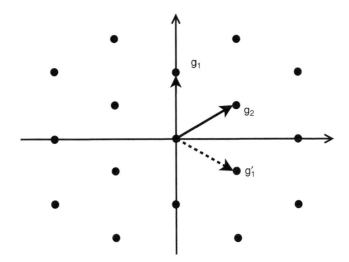

Figure 2.2 The hexagonal lattice is generated by the basis vectors $\mathbf{g}_1 = (0, 2)^T$ and $\mathbf{g}_2 = (\sqrt{3}, 1)^T$. But it can also be generated by the pair $\mathbf{g}_1' = (\sqrt{3}, -1)^T$ and the same $\mathbf{g}_2$, or by the pair $\mathbf{g}_1'' = (\sqrt{3}, 1)^T$ and $\mathbf{g}_2'' = (-\sqrt{3}, 1)^T$. Clearly this lattice cannot be written as a Cartesian product of two scalar lattices. Nevertheless, we can construct it by alternating between two staggered horizontal scalar lattices, one for the even rows and one, half-step shifted, for the odd rows.

Figure 2.2 shows the famous two-dimensional hexagonal lattice – denoted as A_2. The reason why it is called "hexagonal" will become clear in the next section.

We shall soon discuss the degenerate case, where the number of basis vectors in G is less than the dimension n, or the basis vectors are linearly dependent. When G is an *identity matrix*, we get the *integer lattice*, $\Lambda = \mathbb{Z}^n$, also called the *cubic lattice* or "$\mathbb{Z}$ lattice." Any lattice can be viewed as a linear transformation, by the generator matrix, of the integer lattice:

$$\Lambda = G \cdot \mathbb{Z}^n, \tag{2.2}$$

which is simply another way of writing (2.1).

However, the generator matrix is not unique for a given lattice. A lattice is invariant to a *unimodular transformation* of its basis.

Proposition 2.1.1 (Change of basis) *A matrix G' generates the same lattice as G, i.e., $\Lambda(G') = \Lambda(G)$, if and only if*

$$G' = G \cdot T = [\, G\mathbf{t}_1 \mid G\mathbf{t}_2 \mid \ldots \mid G\mathbf{t}_n \,] \tag{2.3}$$

for some unimodular matrix $T = [\, \mathbf{t}_1 \mid \mathbf{t}_2 \mid \ldots \mid \mathbf{t}_n \,]$, *i.e., an integer matrix with a unit absolute determinant,* $\det(T) = \pm 1$.

Proof If T satisfies the condition, then each column of G' is an integer combination of the columns of G, i.e., $\mathbf{g}_j' = G\mathbf{t}_j = \sum_{i=1}^n t_{ij}\mathbf{g}_i$. Thus, by Definition 2.1.1, $\Lambda(G')$

is contained in $\Lambda(G)$. Conversely, since $\det(T) = \pm 1$, the inverse matrix T^{-1} is a (unit-determinant) integer matrix too (by Cramer's rule for matrix inversion), so $\Lambda(G')$ also contains $\Lambda(G)$. Hence, $\Lambda(G')$ and $\Lambda(G)$ must be identical. To prove the "only if" part, note that since the basis vectors are linearly independent, T must be integer valued otherwise $\Lambda(TG)$ will contain points outside $\Lambda(G)$. The same argument shows that if $|\det(T)|$ is greater than 1, then $|\det(T^{-1})|$ is smaller than 1, hence $\Lambda(T^{-1}G')$ contains points outside $\Lambda(G')$. Thus $\Lambda(G') = \Lambda(G)$ implies $|\det(T)| = 1$. $\square$

A by-product of Proposition 2.1.1 is that all (square) generator matrices of a lattice have the same absolute determinant: $\det(G') = \det(GT) = \det(G)\det(T) = \pm \det(G)$. Thus, the absolute value of the determinant of the generator matrix is an invariant property of the lattice.

Definition 2.1.2 (Lattice determinant) [2] *The lattice determinant $\det(\Lambda)$ is defined as the absolute determinant of its generator matrix $|\det(G)|$.*

Due to the linear independence of the basis vectors, the matrix G is non-singular, thus $\det(\Lambda) > 0$.

As we saw in Figure 2.1, a simple way to construct high-dimensional lattices is by taking Cartesian products of lower-dimensional lattices.

Definition 2.1.3 (Cartesian product) *The Cartesian product of two lattices Λ_1 and Λ_2 of dimensions n_1 and n_2 is an $n = n_1 + n_2$ dimensional lattice:*

$$\Lambda_1 \times \Lambda_2 = \left\{ \begin{pmatrix} \mathbf{x} \\ \mathbf{y} \end{pmatrix} : \mathbf{x} \in \Lambda_1, \mathbf{y} \in \Lambda_2 \right\}. \tag{2.4}$$

The generator matrix of the product lattice is a block-diagonal matrix

$$G = \begin{pmatrix} G_1 & 0 \\ 0 & G_2 \end{pmatrix} \tag{2.5}$$

with the component generator matrices on its diagonal, hence its determinant is the product of the component determinants $\det(\Lambda_1 \times \Lambda_2) = \det(\Lambda_1) \cdot \det(\Lambda_2)$.

Equivalent dimension We expect that under a "natural" goodness measure, the product lattice $\Lambda \times \cdots \times \Lambda$ is as good as its component lattice Λ; hence, both lattices have the same *equivalent dimension*.

2.1.1   Characterization of lattice bases

Does the invertible generator matrix form (2.1) describe the most general arrangement of points satisfying the linearity property at the beginning of the section?

---

[2] In the literature (e.g., [49]) $\det(\Lambda)$ is sometimes defined as $\det^2(G)$, which is also the determinant of the Gram matrix $G^t G$.

Degenerate and dense lattices A lattice in $\mathbb{R}^n$ may have *less* than n basis vectors (or the basis vectors may be linearly dependent). In this case, the lattice is contained in a linear sub-space of $\mathbb{R}^n$, and is called *degenerate*. For example, the basis vectors $[1, -1, 0]^t$ and $[0, 1, -1]^t$ span a two-dimensional hexagonal lattice, which is embedded in some tilted plane in $\mathbb{R}^3$. See Problem 2.1.

On the other hand, we never need more than n basis vectors to generate a lattice in $\mathbb{R}^n$: if some of the vectors are linearly dependent then, either a smaller basis for the lattice can be found, or the set generated by (2.1) is *dense* (non-discrete) and therefore cannot be considered as a lattice.

Example 2.1.1 (Dense lattice) *In one dimension, $G = (1, \sqrt{2})$ generates a* dense *set; that is, integer combinations of 1 and $\sqrt{2}$ can arbitrarily approach any point in* $\mathbb{R}$. [3]

Example 2.1.2 (Extended basis) *A basis is not necessarily a subset of an extended basis. In one dimension, the points 9 and 10 span the entire $\mathbb{Z}$ lattice, but none of them can span it alone. In two dimensions, the three points (1,2), (2,1) and (2,2) span the entire $\mathbb{Z}^2$ lattice, but neither pair does.*

Primitive points A lattice point λ is called *primitive* if it is the shortest lattice point in its direction, i.e., $\alpha\lambda$ is *not* in Λ, for all $0 < \alpha < 1$. Basis vectors are necessarily primitive, but the opposite is not true.

Example 2.1.3 (Checkerboard lattice) *A set of n linearly independent primitive vectors does* not *necessarily form a basis for a lattice. Consider as an example the n-dimensional "checkerboard" lattice, which consists of all the all-even and all-odd vectors in $\mathbb{R}^n$, i.e., the union of $2\mathbb{Z}^n$ and $[1, \ldots, 1] + 2\mathbb{Z}^n$. Figure 2.3 shows the two-dimensional case. In three dimensions, this is exactly the BCC lattice of Figure 1.3. A simple basis for this lattice consists of the all-one vector $[1, \ldots, 1]$, plus any $n - 1$ vectors from the set of n elementary even vectors $[2, 0, \ldots, 0], \ldots, [0, \ldots, 0, 2]$. Note that the elementary even vectors are primitive, independent of each other, and, for $n > 4$, shorter than the all-one vector. However, they cannot span odd vectors; hence, without the all-one vector they do not form a basis for the checkerboard lattice. (See for comparison the definition of the D_n lattice in Example 2.4.2.)*

Good basis for a given lattice Since the basis is not unique, we may ask which basis is "best" for a given lattice. The answer is, however, not precise. A common rule of thumb for a good basis is that

- the basis vectors $g_1, \ldots, g_n$ are the shortest possible,
- the basis vectors are nearly orthogonal.

---

[3] *Quasicrystals can be modeled using a basis with more than three vectors in $\mathbb{R}^3$ [158].*

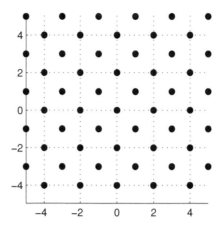

Figure 2.3 Checkerboard lattice in two dimensions.

The first criterion guarantees numerical stability, while the second is useful for reducing the *complexity* of searching for the *closest lattice point* to a given point in space – a problem which is at the heart of the coding and decoding of lattice codes. It nevertheless turns out that the two criteria are closely related by the Hadamard inequality [53]:

$$\det(\Lambda) = |\det(G)| \leq \prod_{i=1}^{n} \|\boldsymbol{g}_i\| \tag{2.6}$$

with equality if and only if the basis vectors are *orthogonal*. Thus, a short basis also tends to be close to orthogonal. The *LLL algorithm* [154] reduces a given basis into a new and usually shorter basis, which satisfies a certain "near-orthogonality" criterion.

Interestingly, the *n shortest* lattice vectors do *not* necessarily form a basis. Take, for example, the checkerboard lattice (Example 2.1.3): the elementary even vectors do not form a basis for that lattice, although for dimension $n > 4$ they are the shortest (in particular, shorter than the all-one vector, whose length is $\sqrt{n}$).

2.1.2   Cosets

The final point we should discuss before the end of this section is that of a lattice shift, or *coset*, defined as

$$\Lambda_{\mathbf{x}} = \mathbf{x} + \Lambda = \{\mathbf{x} + \lambda : \lambda \in \Lambda\}. \tag{2.7}$$

A coset is a discrete set of points such that the difference vector between every pair of points belongs to the lattice. However, the coset itself is, in general, *not* a lattice, as it is not closed under reflection and addition; in particular, it does not contain the origin.

Clearly, the union of $\Lambda_{\mathbf{x}}$ over all shifts $\mathbf{x}$ covers the entire space $\mathbb{R}^n$. But this union contains many overlaps. A natural question to ask then is: what is the *minimal set of shifts S* such that

$$\bigcup_{\mathbf{x} \in S} \Lambda_{\mathbf{x}} = \mathbb{R}^n ? \tag{2.8}$$

This question leads us to the subject of *lattice partition*.

2.2 Partition

A lattice induces a division of the Euclidean space into *congruent cells*. Like the lattice representation, this division is not unique; there are many ways to partition space with respect to a given lattice Λ.

From a geometric viewpoint, the most important division is the *Voronoi* partition, which uses a *nearest-neighbor* (NN) rule. Let $\| \cdot \|$ denote some norm, for example, Euclidean distance. The distance of a point $\mathbf{x}$ in $\mathbb{R}^n$ from Λ is defined as

$$\|\mathbf{x} - \Lambda\| \triangleq \min_{\lambda \in \Lambda} \|\mathbf{x} - \lambda\|. \tag{2.9}$$

The *nearest-neighbor quantizer* $Q_{\Lambda}^{(NN)}(\cdot)$ maps $\mathbf{x}$ to its closest lattice point:

$$Q_{\Lambda}^{(NN)}(\mathbf{x}) = \arg\min_{\lambda \in \Lambda} \|\mathbf{x} - \lambda\|, \tag{2.10}$$

and the *Voronoi cell $\mathcal{V}_{\lambda}$* is the set of all points which are quantized to λ:

$$\mathcal{V}_{\lambda} = \{\mathbf{x} : Q_{\Lambda}^{(NN)}(\mathbf{x}) = \lambda\}. \tag{2.11}$$

The breaking of ties in (2.10) is carried out in a systematic manner, so that the resulting Voronoi cells $\{\mathcal{V}_{\lambda}, \lambda \in \Lambda\}$ are congruent.

If not stated otherwise, the Voronoi partition refers to using the Euclidean norm in (2.9) and (2.10). In this case, the Voronoi cell $\mathcal{V}_{\lambda}$ is a convex polytope, which – like the lattice – is symmetric about the origin. See Problem 2.2. Each face of $\mathcal{V}_{\lambda}$ is determined by a hyperplane, crossing orthogonally to the line connecting λ to one of its neighbors. These neighbors are then called *face-determining points*.

Example 2.2.1 (Honeycomb) *The Voronoi partition of the lattice of Figure* 2.2 *(with $G = \begin{pmatrix} 0 & \sqrt{3} \\ 2 & 1 \end{pmatrix}$) divides the plane into equilateral hexagonal cells with edge length $2/\sqrt{3}$, as shown in Figure 2.4(A). A possible "tie breaking" rule, which keeps the cells congruent, is that each cell contains three out of its six edges and two out of its six corners, with the* same *orientation for all cells.* [4]

---

[4] Any systematic association of *half* the (non-corner) boundary points to each cell would keep the cells congruent. This is because each of these points is on the border of two cells, while each corner point is on the border of three cells. For n-dimensional cells, boundary points are classified into n types of

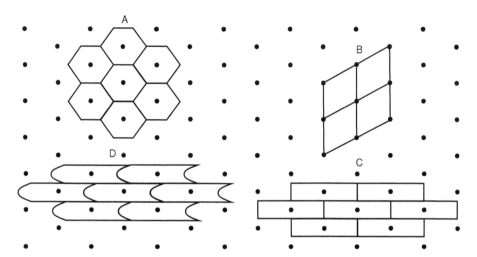

Figure 2.4 Hexagonal lattice with four possible partitions: (A) Voronoi partition; (B) parallelepiped partition; (C) "brick wall" partition generated by successive quantization (first quantize the y-component then, conditioned on that, quantize the x-component); (D) a general (non-polytope) fundamental cell.

The *fundamental* Voronoi cell $\mathcal{V}_0$ is the Voronoi cell associated with the origin ($\lambda = \mathbf{0}$). Due to the periodic nature of the lattice, all the Voronoi cells are shifted versions (by the lattice points) of $\mathcal{V}_0$. Hence, any point in space can be uniquely expressed as the sum of a lattice point and a point in the fundamental Voronoi cell.

As mentioned previously, we do not have to use the Euclidean distance in (2.10). A periodic partition will result by using *any* function of the difference $\mathbf{x} - \lambda$; an example comparing the ℓ_2 and ℓ_4 norms is shown in Figure 2.5. An alternative definition for a general lattice-based partition, which does not rely explicitly on a distance measure, is based on the notion of a *fundamental cell*.

We say that a collection of sets $\{S_i\}$ *covers* the Euclidean space if any point in space is in one of the sets, i.e., $\cup_i S_i = \mathbb{R}^n$. We say that the sets are *packed* in the Euclidean space if no point in space belongs to more than one set, i.e., $S_i \cap S_j = \emptyset$ for all $i \neq j$. Finally, if the sets both cover $\mathbb{R}^n$ *and* are packed in $\mathbb{R}^n$, then $\{S_i\}$ is a *partition* of $\mathbb{R}^n$.

Definition 2.2.1 (Fundamental cell, lattice partition) *A fundamental cell $\mathcal{P}_0$ of a lattice Λ is a bounded set, which, when shifted by the lattice points, generates a partition $\mathcal{P} = \{\mathcal{P}_\lambda\}$ of $\mathbb{R}^n$. That is,*

(i) each cell $\mathcal{P}_\lambda$ is a shift of $\mathcal{P}_0$ by a lattice point $\lambda \in \Lambda$

$$\mathcal{P}_\lambda = \mathcal{P}_0 + \lambda = \{\mathbf{x} : (\mathbf{x} - \lambda) \in \mathcal{P}_0\};$$

---

k-dimensional edges, for $k = 0, 1, \ldots, n - 1$. Although the boundary has zero volume, its association to the cell is critical for lattice codebooks (see Chapter 9).

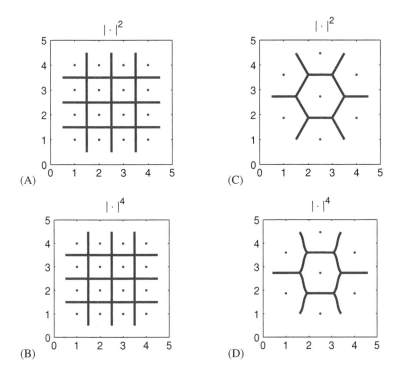

Figure 2.5 Examples of lattices and lattice partitions: (A) the $\mathbb{Z}^2$ lattice with a Euclidean Voronoi partition; (B) the $\mathbb{Z}^2$ lattice with a fourth-power norm-based Voronoi partition; (C) a hexagonal lattice with a Euclidean Voronoi partition; (D) a hexagonal lattice with a fourth-power norm-based Voronoi partition.

(ii) the cells do not intersect, $\mathcal{P}_\lambda \cap \mathcal{P}_{\lambda'} = \emptyset$ for all $\lambda' \neq \lambda$; and
(iii) the union of the cells covers the whole space, $\bigcup_{\lambda \in \Lambda} \mathcal{P}_\lambda = \mathbb{R}^n$.

It is convenient to think of a fundamental cell as a connected region, although the definition does not require that.

Definition 2.2.1 implies that given a lattice Λ and a fundamental cell $\mathcal{P}_0$, any point $\mathbf{x}$ in space can be *uniquely* expressed as a sum

$$\mathbf{x} = \lambda + \mathbf{x}_e, \quad \text{where } \lambda \in \Lambda \text{ and } \mathbf{x}_e \in \mathcal{P}_0. \quad (2.12)$$

We may think of λ in (2.12) as the *quantization* of $\mathbf{x}$ to the lattice Λ,

$$\lambda = Q_\Lambda(\mathbf{x}), \quad (2.13)$$

and of $\mathbf{x}_e$ in (2.12) as the *quantization error*. This extends the notion of a nearest-neighbor quantizer (2.10) with Voronoi partition (2.11), to the case of a general fundamental cell $\mathcal{P}_0$ inducing a lattice partition $\mathcal{P} = \Lambda + \mathcal{P}_0$.

The Voronoi partition generated by the nearest-neighbor quantizer (2.10) clearly satisfies the properties in Definition 2.2.1 (provided that ties are broken in a systematic manner). The simplest lattice partition is, however, a *parallelepiped partition* generated by some lattice basis $g_1, \ldots, g_n$. Here $\mathcal{P}_0$ is the *fundamental parallelepiped*, consisting of all points which are linear combinations of the basis vectors with coefficients between zero and one:

$$\mathcal{P}_0 = \left\{ \mathbf{x} = \sum_{i=1}^{n} \alpha_i g_i : 0 \leq \alpha_1, \ldots, \alpha_n < 1 \right\} \tag{2.14}$$

$$= G \cdot \text{Unit Cube}, \tag{2.15}$$

where Unit Cube $= \{\mathbf{x} : 0 \leq x_i < 1, i = 1, \ldots, n\}$. Note that the unit cube is the parallelepiped partition of the $\mathbb{Z}$-lattice. [5] See Figure 2.4(B).

Since the lattice has more than one basis, its parallelepiped partition is not unique. Moreover, a shift or reflection of a fundamental cell is another fundamental cell, which generates another partition of the lattice. Interestingly, though, it follows from a simple "volume preservation" argument that the volume of a cell is the same under *any* lattice partition. And, as we shall see later, all lattice partitions are, in fact, *equivalent* in several senses; for example, any fundamental cell is a complete set of coset shifts in (2.8).

Proposition 2.2.1 (Cell volume) *The cell volume*

$$V = \text{Vol}(\mathcal{P}_0) = \int_{\mathcal{P}_0} d\mathbf{x} \tag{2.16}$$

is independent of the lattice partition $\mathcal{P}$*, and it is equal to the lattice determinant of Definition* 2.1.2

$$V = \det(\Lambda) = |\det(G)| \overset{\Delta}{=} V(\Lambda). \tag{2.17}$$

Proof Consider first the parallelepiped partition (2.14) induced by the generator matrix G. By a change of variables $\mathbf{x} = G\mathbf{x}'$, and using (2.15), we have

$$V = \int_{\mathcal{P}_0} d\mathbf{x} = |\det(G)| \int_{\text{Unit Cube}} d\mathbf{x}' = |\det(G)|. \tag{2.18}$$

Next, for a general partition, consider the cells contained in a large cube B. Since the cells have a finite diameter, the volume of the fractional cells at the boundary of the cube B becomes negligible when B is sufficiently large. Thus, if there are $N(B)$ lattice points inside B, then the cell volume is roughly

$$V \approx \frac{\text{Vol}(B)}{N(B)}, \tag{2.19}$$

independent of the shape of the cells, and this approximation becomes exact when the edge length of B, and hence also $N(B)$, go to infinity. □

---

[5] To see that the parallelepiped cell $\mathcal{P}_0$ in (2.14) satisfies the conditions of Definition 2.2.1, note that (i) the difference between any two points in $\mathcal{P}_0$ is *not* a lattice point, and (ii) every point outside $\mathcal{P}_0$ can be written as a sum of a point in $\mathcal{P}_0$ and a lattice point. See Lemma 2.3.2 in the next section.

In the following section (see Corollary 2.3.1) we shall see an alternative (more direct) proof for the second part of the proof above, showing that the cell volume is partition invariant.

The approximation in (2.19) holds, in fact, for any body which is large compared to the cells, i.e., the number of lattice points $N(S)$ in a large body S is approximately $\text{Vol}(S)/V(\Lambda)$. We thus define the lattice *point density* as the reciprocal of the cell volume:

$$\gamma(\Lambda) = \frac{1}{V(\Lambda)}, \tag{2.20}$$

measured in *points per unit volume*.

2.3 Equivalent cells and coset leaders

An even stronger notion of equivalence between partitions holds: all the fundamental cells of a lattice are identical modulo a fixed partition. More explicitly, any fundamental cell can be decomposed into pieces and rearranged (via lattice shifts) to form another fundamental cell. Although this may seem to be a geometric property, it is, in fact, a consequence of the lattice being a sub-group of the Euclidean space.

Definition 2.3.1 (Mod $\mathcal{P}_0$, Mod Λ) *For a given lattice partition $\mathcal{P}$ with a fundamental cell $\mathcal{P}_0$, the* modulo fundamental cell *operation is defined as*

$$\mathbf{x} \bmod \mathcal{P}_0 = \mathbf{x}_e = \mathbf{x} - Q_\Lambda(\mathbf{x}), \tag{2.21}$$

where $Q_\Lambda(\mathbf{x})$ and $\mathbf{x}_e$ are the quantization and quantization error (2.12), *respectively, induced by the partition $\mathcal{P}$. We shall call this a* modulo-lattice *operation – and use the notation $\mathbf{x} \bmod \Lambda$, or $\mathbf{x}/\Lambda$ – whenever there is no ambiguity about the assumed partition of Λ.*

Proposition 2.3.1 (Modulo laws) *The modulo-lattice operation satisfies the* shift-invariance *property*

$$(\mathbf{x} + \lambda) \bmod \Lambda = \mathbf{x} \bmod \Lambda, \quad \forall \lambda \in \Lambda, \tag{2.22a}$$

and the distributive *law,*

$$(\mathbf{x} \bmod \Lambda + \mathbf{y}) \bmod \Lambda = (\mathbf{x} + \mathbf{y}) \bmod \Lambda. \tag{2.22b}$$

Proof If $\mathbf{x} = \lambda' + \mathbf{x}_e$ (with $\lambda' \in \Lambda$ and $\mathbf{x}_e \in \mathcal{P}_0$) is the unique decomposition (2.12) of $\mathbf{x}$ with respect to a partition $\mathcal{P}$, then $\mathbf{x} + \lambda = (\lambda + \lambda') + \mathbf{x}_e$ must be the unique decomposition of $\mathbf{x} + \lambda$ with respect to $\mathcal{P}$, i.e., both $\mathbf{x}$ and $\mathbf{x} + \lambda$ have the same quantization error $\mathbf{x}_e$, which proves the shift-invariance property. The distributive law now follows because the inner modulo operation in (2.22b) amounts to shifting

$\mathbf{x}$ by some lattice point λ. Thus $(\mathbf{x} \bmod \Lambda + \mathbf{y}) \bmod \Lambda = (\mathbf{x} - \lambda + \mathbf{y}) \bmod \Lambda = (\mathbf{x} + \mathbf{y}) \bmod \Lambda$. $\qquad\square$

When considering different fundamental cells of the lattice in the modulo-lattice (or modulo fundamental cell) operation (2.21), the following simple relations hold.

Lemma 2.3.1 (Modulo different cells) *The result of the modulo-lattice operation is the same for any fundamental cell of the lattice, up to a shift by a lattice point:*

$$(\mathbf{x} \bmod \mathcal{P}_0) - (\mathbf{x} \bmod \mathcal{Q}_0) \in \Lambda, \tag{2.23a}$$

where $\mathcal{P}_0$ and $\mathcal{Q}_0$ are fundamental cells of Λ. Iterating modulo-lattice operations is dominated by the last fundamental cell:

$$(\mathbf{x} \bmod \mathcal{Q}_0) \bmod \mathcal{P}_0 = \mathbf{x} \bmod \mathcal{P}_0. \tag{2.23b}$$

Finally, all fundamental cells of Λ are equal up to a modulo-lattice operation:

$$\mathcal{Q}_0 \bmod \mathcal{P}_0 = \mathcal{P}_0. \tag{2.23c}$$

Proof See Problem 2.5. $\qquad\square$

The following proposition provides a more direct relation between the "pieces" created by taking one cell modulo another cell.

Proposition 2.3.2 (Breaking a cell into pieces) *For partitions $\mathcal{P}$ and $\mathcal{Q}$ with fundamental cells $\mathcal{P}_0$ and $\mathcal{Q}_0$, respectively, let*

$$\mathcal{A}_\lambda = \mathcal{Q}_0 \cap \mathcal{P}_\lambda \quad and \quad \mathcal{B}_\lambda = \mathcal{P}_0 \cap \mathcal{Q}_\lambda, \quad for\ \lambda \in \Lambda. \tag{2.24}$$

Then,

$$\mathcal{B}_\lambda = \mathcal{A}_{-\lambda} + \lambda. \tag{2.25}$$

See the example of a hexagonal lattice in Figure 2.6, where $\mathcal{P}_0$ is a hexagon and $\mathcal{Q}_0$ is a parallelogram.

Proof By Definition 2.2.1 of lattice partition,

$$\mathcal{A}_\lambda = \mathcal{Q}_0 \cap (\lambda + \mathcal{P}_0) \quad and \quad \mathcal{B}_\lambda = \mathcal{P}_0 \cap (\lambda + \mathcal{Q}_0). \tag{2.26}$$

Intersecting with a shift is the same as first shifting backward, then intersecting, then shifting forward. Thus,

$$\mathcal{B}_\lambda = [(-\lambda + \mathcal{P}_0) \cap \mathcal{Q}_0] + \lambda = \mathcal{A}_{-\lambda} + \lambda. \tag{2.27}$$

$$\square$$

Since the "pieces" in (2.25) are disjoint, and are identical up to a shift, Proposition 2.3.2 implies that

$$\mathrm{Vol}(\mathcal{P}_0) = \mathrm{Vol}(\dot{\cup}_{\lambda \in \Lambda}\ \mathcal{B}_\lambda) = \mathrm{Vol}(\dot{\cup}_{\lambda \in \Lambda}\ \mathcal{A}_{-\lambda}) = \mathrm{Vol}(\mathcal{Q}_0), \tag{2.28}$$

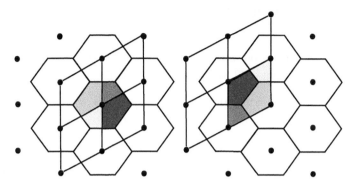

Figure 2.6 The basic Voronoi cell $\mathcal{V}_0$ modulo the basic parallelepiped cell $\mathcal{P}_0$: the intersections of $\mathcal{V}_0$ with the $\mathcal{P}_\lambda$ cells can be rearranged (via shifts by lattice points) to generate $\mathcal{P}_0$.

where $\dot{\cup}$ denotes the *disjoint union*. We have therefore obtained an alternative proof for the second part of Proposition 2.2.1.

Corollary 2.3.1 (Volume invariance property) *Every fundamental cell has the same volume.*

2.3.1   Minimal set of coset shifts

We now return to the notion of a *lattice coset* (2.7), and to our question regarding a minimal set of coset shifts which covers $\mathbb{R}^n$. This turns out to be related to the properties of a valid fundamental cell.

Observe that if $\mathcal{P}_0$ is a fundamental cell of the lattice Λ, then the difference vector between any two points in $\mathcal{P}_0$ *cannot* be a lattice point, i.e.,

$$\mathbf{x} - \mathbf{x}' \notin \Lambda, \quad \text{for all} \ \ \mathbf{x}, \mathbf{x}' \in \mathcal{P}_0. \tag{2.29}$$

Otherwise, i.e., if $\mathbf{x} - \mathbf{x}' = \lambda \in \Lambda$ for some $\mathbf{x}, \mathbf{x}' \in \mathcal{P}_0$, then $\mathbf{x} = \mathbf{x}' + \lambda$ is in the cell $\mathcal{P}_\lambda = \lambda + \mathcal{P}_0$, i.e., the point $\mathbf{x}$ is both in $\mathcal{P}_0$ and in $\mathcal{P}_\lambda$, contradicting property (ii) of Definition 2.2.1. We say that S is a *maximal set* satisfying condition (2.29) if it is impossible to add any other point to S without violating the condition.

Lemma 2.3.2 (Fundamental cell property 1) *A set $\mathcal{P}_0$ is a fundamental cell of Λ if and only if it is a maximal set satisfying condition (2.29).*

Proof The "only if" part (necessity) follows from the discussion above, i.e., condition (2.29) for a set S is necessary (and sufficient) so that the sets $\{\lambda + S, \ \lambda \in \Lambda\}$ do not intersect (property (ii) of Definition 2.2.1). The "if" part (sufficiency) follows because for a *maximal* set S, any point $\mathbf{x}'$ outside S must be equal to $\mathbf{x} + \lambda$ for some pair $\mathbf{x} \in S$ and $\lambda \in \Lambda$ (or otherwise $\mathbf{x}'$ could be included in S without violating (2.29)), thus also property (iii) of Definition 2.2.1 is satisfied: $\cup_{\lambda \in \Lambda} (\lambda + S) = \mathbb{R}^n$.   □

Returning to our question, we say that S is a *minimal set of coset shifts* covering $\mathbb{R}^n$ if the union in (2.8) is *disjoint*, i.e., $\Lambda_x \cap \Lambda_{x'} = \emptyset$ for any $x, x' \in S, x \neq x'$.

Lemma 2.3.3 (Fundamental cell property 2) *A set $\mathcal{P}_0$ is a fundamental cell of the lattice Λ if and only if it is a minimal set of coset shifts covering $\mathbb{R}^n$, i.e.,*

$$\dot{\bigcup}_{x \in \mathcal{P}_0} \Lambda_x = \mathbb{R}^n, \tag{2.30}$$

where $\dot\cup$ denotes a disjoint union.

Proof See Problem 2.6.   □

In group theory, each coset is seen as an *equivalence class* of elements, and the set of all cosets is called a *quotient group*, denoted $\mathbb{R}^n / \Lambda$. Lemma 2.3.3 implies that each coset in $\mathbb{R}^n / \Lambda$ has a unique intersection point with any given fundamental cell. The intersection point can thus "represent" the coset.

Definition 2.3.2 (Coset representative) *Given a fundamental cell $\mathcal{P}_0$, the representative of a coset Λ_x is the (unique) intersection point of the coset and the fundamental cell, i.e.,*

$$\text{coset representative of } \Lambda_x = \Lambda_x \cap \mathcal{P}_0.$$

Since the points of the coset Λ_x are shifts of x by the lattice points, the modulo law (2.22a) implies that they all collapse to the representative when reduced modulo $\mathcal{P}_0$, i.e.,

$$\text{coset representative of } \Lambda_x = \Lambda_x \bmod_{\mathcal{P}_0} \Lambda = x \bmod_{\mathcal{P}_0} \Lambda. \tag{2.31}$$

Example 2.3.1 (Coset leader) *Since a modulo-Voronoi cell operation does not increase the length of a vector, the coset representative in the fundamental Voronoi cell is the* shortest vector *in the coset, called the* coset leader*:*

$$\text{coset leader of } \Lambda_x = \arg\min_{y \in \Lambda_x} \|y\| \tag{2.32}$$

with ties broken according to the Voronoi partition rule (2.10); see Problem 2.7. It follows that the Voronoi cell $\mathcal{V}_0$ is the set of all coset leaders.

We have considered two alternative necessary and sufficient conditions for a set to be a fundamental cell of the lattice: a maximal set of non-lattice differences (Lemma 2.3.2), and a minimal covering set of coset shifts (Lemma 2.3.3). By the latter condition, any fundamental cell $\mathcal{P}_0$ is a *complete set* of coset representatives, and the quotient group (set of all cosets) can be expressed as

$$\mathbb{R}^n / \Lambda = \{\Lambda_x : x \in \mathcal{P}_0\}. \tag{2.33}$$

While the choice of a fundamental cell $\mathcal{P}_0$ determines the result of the modulo operation in Definition 2.3.1, it is irrelevant for the question of whether two points are *equal modulo the lattice*. That is,

$$\mathbf{x} = \mathbf{y} \bmod \Lambda \qquad (2.34)$$

means that $\mathbf{x}$ and $\mathbf{y}$ belong to the same coset of Λ. The equality, if true, holds regardless of the choice of fundamental cell for Λ.

2.4   Transformation and tiling

Suppose that a lattice Λ' is generated by a linear transformation of another lattice Λ, i.e., $\Lambda' = T \cdot \Lambda$, for some full rank square matrix T. By (2.2),

$$\Lambda' = T \cdot G \cdot \mathbb{Z}^n, \qquad (2.35)$$

where G is a basis of Λ. Thus

$$G' = T \cdot G = [T\mathbf{g}_1 \mid T\mathbf{g}_2 \mid \ldots \mid T\mathbf{g}_n] \qquad (2.36)$$

is a basis of Λ', i.e., each basis vector is transformed by the matrix T. (Note the distinction from right-multiplication when changing a basis for the same lattice (2.3).) In particular, by Proposition 2.2.1, the cell volume of Λ' is given by

$$V(\Lambda') = |\det(T)| \cdot V(\Lambda) \qquad (2.37)$$

for example, for a scalar transformation, $V(t\Lambda) = |t|^n V(\Lambda)$.

It is easy to see that if $\mathcal{P}_0$ is a fundamental cell of Λ, then $T \cdot \mathcal{P}_0$ is a fundamental cell of $T \cdot \Lambda$. In particular, any parallelepiped partition of $\Lambda' = T \cdot \Lambda$ is just a linear transformation of some parallelepiped partition of Λ, i.e., $\mathcal{P}_0(T \cdot \Lambda) = T \cdot \mathcal{P}_0(\Lambda)$. Interestingly, however, this is not always the case for Voronoi partition, i.e., in general,

$$\mathcal{V}_0(T \cdot \Lambda) \neq T \cdot \mathcal{V}_0(\Lambda). \qquad (2.38)$$

Example 2.4.1 (Transformation and Voronoi partition) *The hexagonal lattice A_2 can be written as $G \cdot \mathbb{Z}^2$, with G as given in Example 2.2.1. The Voronoi cells of $\mathbb{Z}^2$ are squares. The linear transformation of these squares results in partition (B) (the rhombuses) in Figure 2.4, whose fundamental cell is*

$$\{\mathbf{x} = G \cdot \alpha : -0.5 \leq \alpha_1, \alpha_2 \leq 0.5\}. \qquad (2.39)$$

This is a centralized version of the parallelepiped *fundamental cell of A_2. In contrast, the fundamental Voronoi cell of A_2 is* hexagonal *(partition (A) in Figure 2.4). Another example, of vertical scaling of the hexagonal lattice and its effect on the partition, is shown in Figure 2.7.*

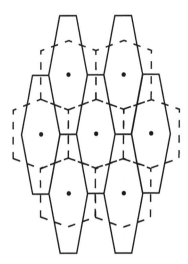

Figure 2.7 Vertical stretching of the hexagonal lattice and its Voronoi partition: partition followed by stretch (solid line); stretch followed by partition (dashed line).

The reason for (2.38) is that, in general, $\|\mathbf{x} - \lambda\| \leq \|\mathbf{x} - \lambda'\|$ does not imply $\|T(\mathbf{x} - \lambda)\| \leq \|T(\mathbf{x} - \lambda')\|$, unless T is proportional to some measure-preserving transformation (i.e., for square norm, if and only if $TT^t \propto I$); thus, the nearest lattice point may change after the transformation.

Closest lattice-point search complexity

Many communication problems – like combined equalization and detection over filter channels, or predictive quantization of correlated sources – would be extremely simple to solve if we could interchange the order of linear transformation and nearest-neighbor (Voronoi) partition. Unfortunately, as illustrated in Figure 2.7, the cells at the transform domain do not reflect minimum Euclidean association at the lattice code domain.

The same problem underlies the complexity of searching for the closest lattice point to a given vector $\mathbf{x}$ in $\mathbb{R}^n$. This search is at the heart of lattice quantization under a mean-squared error criterion, or maximum likelihood decoding of a lattice constellation in the presence of AWGN. If the lattice is cubic, i.e., $\Lambda = a\mathbb{Z}^n$, then the closest vector $Q_\Lambda^{(NN)}(\mathbf{x})$ can be obtained simply via coordinate-by-coordinate quantization. In the general lattice $\Lambda(G)$ case, a naive approach would be to transform the search to a cubic domain by the inverse generator matrix, quantize coordinate-by-coordinate, and transform back to the lattice domain, i.e.,

$$\hat{\mathbf{x}} = G \cdot Q_{\mathbb{Z}^n}(G^{-1}\mathbf{x}). \tag{2.40}$$

However, as explained above, $\hat{\mathbf{x}} \neq Q_\Lambda^{(NN)}(\mathbf{x})$, unless the matrix G is orthogonal, or $\mathbf{x}$ is "very close" to a lattice point; see [3, 46]. A sufficient condition for "closeness" is that $\mathbf{x}$ is inside the ball inscribed in the centralized fundamental parallelepiped associated with G; see (2.39) and the discussion about a "good basis" in Section 2.1.1.

2.4.1 Scaling laws

Recall the unique decomposition (2.12) of a point $\mathbf{x} \in \mathbb{R}^n$ into a lattice point $\lambda = Q_\Lambda(\mathbf{x})$ and a quantization error $\mathbf{x}_e = \mathbf{x} \bmod \Lambda$ (see Definition 2.3.1). Both the quantizer $Q_\Lambda(\mathbf{x})$ and the modulo operation are defined with respect to some fundamental cell $\mathcal{P}_0$.

In the context of an invertible linear transformation T, this decomposition satisfies the following identities:

$$Q_\Lambda(T\mathbf{x}) = T \cdot Q_{T^{-1} \cdot \Lambda}(\mathbf{x}) \tag{2.41}$$

and

$$[T\mathbf{x}] \bmod {}_{\mathcal{P}_0}\Lambda = T \cdot [\mathbf{x} \bmod {}_{T^{-1} \cdot \mathcal{P}_0} T^{-1} \cdot \Lambda]. \tag{2.42}$$

See Problem 2.8. Specifically, for a scalar transformation,

$$Q_\Lambda(\alpha\mathbf{x}) = \alpha \, Q_{\frac{\Lambda}{\alpha}}(\mathbf{x}) \tag{2.43}$$

and

$$[\alpha\mathbf{x}] \bmod \Lambda = \alpha \left[\mathbf{x} \bmod \frac{\Lambda}{\alpha}\right]. \tag{2.44}$$

Thus, scaling a vector before quantization is equivalent to quantizing the non-scaled vector by a scaled version of the lattice.

2.4.2 Similarity

A scaled orthogonal transformation can enlarge, shrink or rotate a lattice, but it does not change its *shape*. (It can also reflect it, but a lattice is anyway invariant to reflection.)

Definition 2.4.1 (Similarity) *We say that two lattices Λ_1 and Λ_2 have the same shape, are* equivalent *or* similar*, if*

$$\Lambda_2 = a \cdot T \cdot \Lambda_1 \tag{2.45}$$

for some positive scalar a and orthonormal matrix T (i.e., $TT^t = I$).

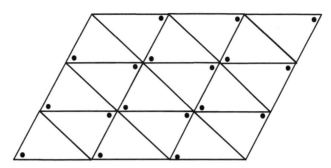

Figure 2.8 Tiling the plane with triangles.

In view of the basis-transformation (2.36) and basis-change (2.3) rules, two matrices G_1 and G_2 generate similar lattices if

$$G_2 = a \cdot T \cdot G_1 \cdot \tilde{T} \tag{2.46}$$

for some positive scalar a, orthonormal matrix T and unimodular matrix $\tilde{T}$.

Example 2.4.2 (D_n lattice) *The D_n lattice consists of all integer vectors in $\mathbb{R}^n$ with an even element sum:*

$$D_n = \{\mathbf{z} \in \mathbb{Z}^n : \sum_{i=1}^{n} z_i \text{ is even}\}. \tag{2.47}$$

For $n = 2$ it coincides with the n-checkerboard lattice of Example 2.1.3 (consisting of all the all-even and all-odd vectors in $\mathbb{R}^n$), while for higher dimensions it strictly contains it. For $n = 3$, both lattices are similar to each other, and are similar to the BCC lattice of Figure 1.3. For more than four dimensions, however, D_n and the n-checkerboard are not *similar. See Problems 2.9 and 2.12.*

2.4.3　Tiling

Lattice partition is a special case of tiling space with a *space-filling body*. In the general tiling case, the space is partitioned into congruent cells, where, in addition to translation as in the case of lattice partition, we also allow *rotation* of the fundamental cell. For example, as shown in Figure 2.8, we can tile $\mathbb{R}^2$ with triangles, though a triangle cannot be a fundamental cell of any lattice.

An interesting question is, then: does this extra degree of freedom of rotation allow us to obtain "better" arrangements of points in space? To answer this and many other questions, we need a measure for lattice goodness. This is the topic of Chapter 3.

2.5   Algebraic constructions

Lattices are tightly connected to number theory, combinatorics and algebraic codes over finite fields. Although this is not the focus of the book, it is beneficial to provide some background on these relations.

The matrix $G^t G$ – the product of the generator matrix and its transpose – is known as the *Gram matrix*. The diagonal elements of the Gram matrix are the square lengths $\|g_i\|^2$ of the n lattice basis vectors, while the off-diagonal elements are their $n(n-1)$ inner products $\langle g_i, g_j \rangle$. Thus, the Gram matrix is invariant to rotation or reflection of the lattice (but not to a change of basis). If the elements of the Gram matrix are integers – not necessarily meaning that the elements of G are integers – then the lattice is called *integral*. For example, the hexagonal lattice of Figure 2.2 is integral; the length of its two basis vectors is equal to 2, while their inner product is equal to 1. If, in addition, the determinant of $G^t G$ is equal to 1, then the lattice is called *unimodular*. The only unimodular lattice up to dimension 8 is the $\mathbb{Z}$-lattice. At dimension 8 we find the first non-trivial unimodular lattice, known as E_8, or the *Gosset lattice*, which has a unique algebraic nature.

Integer-valued lattices, like E_8 and the D_n lattice of Example 2.4.2, can be constructed by "lifting" a *linear q-ary code* into the Euclidean space. We describe this construction – known as "*construction A*" – below, and start with the binary alphabet ($q = 2$) case. Note that a general lattice – like the hexagonal, BCC and FCC – whose generator matrix elements have irrational ratios, cannot be constructed this way. See [49] for more examples and generalizations.

2.5.1   Linear codes and construction A

An (n, k, d) linear binary code $\mathcal{C}$ maps k information bits into binary codewords of length n. There are $M = 2^k$ codewords $\mathbf{c}_1, \ldots, \mathbf{c}_M$, where $\mathbf{c}_m = (c_{m1}, \ldots, c_{mn})$, each satisfying a set of $(n - k)$ independent linear constraints: $h_{i1} \cdot c_{m1} \oplus \ldots \oplus h_{in} \cdot c_{mn} = 0$, for $i = 1, \ldots, n - k$, where $h_{ij} \in \{0, 1\}$ and $\oplus$ denotes the exclusive-or (XOR) operation. These constraints can be written compactly in terms of an $(n - k) \times n$ *parity-check matrix* H:

$$H \cdot \mathbf{c}_m = \mathbf{0}, \tag{2.48}$$

where all sums and products are modulo 2. The parameter $d = d_{\min}(\mathcal{C})$ is the *minimum distance* of the code, i.e., the minimum Hamming weight (number of ones) over all non-zero codewords.

Construction A is a method for generating a lattice by "lifting" a linear binary code $\mathcal{C}$ to the Euclidean space. We use $\mathbf{x} \bmod 2 = (x_1 \bmod 2, \ldots, x_n \bmod 2)$ to denote a modulo-2 reduction of each of the components of $\mathbf{x} \in \mathbb{R}^n$. (This is equivalent to taking $\mathbf{x}$ modulo the cubic lattice $2\mathbb{Z}^n$.)

Definition 2.5.1 (Modulo-2 lattice) *The set of all integer vectors whose modulo-2 reduction belongs to C forms a lattice,*

$$\Lambda_C = \{\mathbf{x} \in \mathbb{Z}^n : \mathbf{x} \bmod 2 \in C\}, \tag{2.49}$$

called a modulo-2 lattice. Equivalently, $\Lambda_C = 2\mathbb{Z}^n + C$.

It is easy to verify that Λ_C is closed under reflection and addition, and hence it is indeed a lattice. [6] It is also easy to verify that the *minimum Euclidean distance* between any two points in Λ_C is $d_{\min}(\Lambda_C) = \min\{\sqrt{d}, 2\}$.

Example 2.5.1 *If $H = (1, 1)^t$, then $C = \{(0, 0)^t, (1, 1)^t\}$, and Λ_C is the D_2 ("checkerboard") lattice, consisting of all integer coordinates with even sum (Figure 2.3).*

Another useful notion associated with binary codes is the *syndrome*. For a general vector $\mathbf{x}$ in $\{0, 1\}^n$, the syndrome is a vector in $\{0, 1\}^{n-k}$, defined as

$$\mathbf{s} = H \cdot \mathbf{x}. \tag{2.50}$$

The code C is thus the set of vectors in $\{0, 1\}^n$ with a *zero* syndrome. Each non-zero syndrome $\mathbf{s}$ in $\{0, 1\}^{n-k}$ defines a k-dimensional affine sub-space $C_\mathbf{s}$ of $\{0, 1\}^n$, consisting of all vectors $\mathbf{x}$ satisfying (2.50). This sub-space is a *coset* of C, i.e., $C_\mathbf{s} = \mathbf{v} \oplus C$ for some vector $\mathbf{v}$ in $\{0, 1\}^n$. If we lift $C_\mathbf{s}$ to $\mathbb{R}^n$ as in (2.49) (i.e., take $\{\mathbf{x} \in \mathbb{Z}^n : \mathbf{x} \bmod 2 \in C_\mathbf{s}\}$), then we obtain an integer coset of the lattice Λ_C. And, vice versa, each integer coset of Λ_C corresponds to a unique syndrome $\mathbf{s}$. Thus, Λ_C has (at most) 2^{n-k} different integer cosets.

Linear q-ary code The concept of modulo-2 lattices extends to larger alphabets. For an integer $q \geq 2$, let

$$\mathbb{Z}_q = \{0, 1, \ldots, q - 1\} \tag{2.51}$$

denote the modulo-q group, i.e., a q-ary alphabet with modulo-q addition. Let

$$C = \{\mathbf{c}_0, \ldots, \mathbf{c}_{M-1}\} \tag{2.52}$$

denote a code over $\mathbb{Z}_q^n$, i.e., each codeword $\mathbf{c}_i$ is a vector of n elements from the alphabet $\mathbb{Z}_q$. We say that the code is *linear*, or that it is a *sub-group* of $\mathbb{Z}_q^n$, if the modulo-q sum of every two codewords (including a codeword with itself) is inside the code:

$$\mathbf{c}_i \oplus \mathbf{c}_j \in C, \quad \text{for any } \mathbf{c}_i, \mathbf{c}_j \in C, \tag{2.53}$$

where $\oplus$ denotes modulo-q addition. Since adding a number to itself q times modulo-q gives zero, the zero codeword $\mathbf{c}_0 = \mathbf{0}$ is always inside the code.

---

[6] If $\mathbf{x} \bmod 2 = \mathbf{c} \in C$, then, by the properties of the modulo operation, $(-\mathbf{x}) \bmod 2 = (-(\mathbf{x} \bmod 2)) \bmod 2 = (-\mathbf{c}) \bmod 2 = \mathbf{c}$. Thus, $-\mathbf{x}$ belongs to Λ as well. Closeness under real addition follows similarly.

A linear q-ary code can be written in terms of an $n \times k$ generator matrix G with elements in $\mathbb{Z}_q$, for some $1 \leq k < n$:

$$\mathcal{C} = \{\mathbf{x} = G \cdot \mathbf{w} : \mathbf{w} \in \mathbb{Z}_q^k\}, \tag{2.54}$$

where $\mathbf{w}$ runs over all k-dimensional column vectors with elements from $\mathbb{Z}_q$, and the product of the matrix G and the vector $\mathbf{w}$ is defined modulo q.[7] Thus, the code $\mathcal{C}$ is spanned by the k column vectors of the matrix G.

Prime alphabet Since $\mathbf{w}$ runs over a set of size q^k, the resulting code size (the number of distinct codewords) M is at most q^k, and it is equal to q^k if every $\mathbf{w}$ generates a distinct codeword $\mathbf{c}_m$. This condition holds if q is a *prime number* p and the columns of G are *linearly independent* within the group $\mathbb{Z}_p$.[8] In this case, both the parity-check (2.48) and generator-matrix (2.54) representations are equivalent, provided the rows of the parity-check matrix H span the *null space* of the columns of the generator matrix G. Furthermore, $M = p^k$, so the *code rate*, in bit per dimension, is given by

$$R \triangleq \frac{1}{n} \log_2(M) = \frac{k}{n} \log_2 p. \tag{2.55}$$

Similarly to the binary case, we can use construction A to lift a linear q-ary code $\mathcal{C}$ to the Euclidean space and form a modulo-q lattice.

Definition 2.5.2 (Modulo-q lattice) *A q-ary linear code $\mathcal{C}$ induces a modulo-q lattice:*

$$\Lambda_{\mathcal{C}} = \{\mathbf{x} \in \mathbb{Z}^n : \mathbf{x} \bmod q \in \mathcal{C}\}. \tag{2.56}$$

Equivalently, $\Lambda_{\mathcal{C}}$ can be written as $\Lambda_{\mathcal{C}} = q\mathbb{Z}^n + \mathcal{C}$.

Thus, $\Lambda_{\mathcal{C}}$ consists of replications of $\mathcal{C}$ by shifting it with points from $q\mathbb{Z}^n$. Figure 2.9 shows a linear code $\mathcal{C}$ in $\mathbb{Z}_{11}^2$ (i.e., $q = 11$, $k = 1$, $n = 2$) lifted to $\mathbb{R}^2$.

Proposition 2.5.1 (Properties of a modulo-q lattice)

(a) $\Lambda_{\mathcal{C}}$ *is a lattice.*

(b) $\Lambda_{\mathcal{C}}$ *contains and is contained in cubic lattices: $q\mathbb{Z}^n \subset \Lambda_{\mathcal{C}} \subset \mathbb{Z}^n$.*

(c) $\Lambda_{\mathcal{C}}$ *is non-degenerate (even if $\mathcal{C}$ contains only a single (zero) codeword).*

(d) *The cell volume is equal to the number of integer cosets, and is given by $V(\Lambda_{\mathcal{C}}) = |\mathbb{Z}^n / \Lambda_{\mathcal{C}}| = q^n / M$, where M is the size of the code $\mathcal{C}$.*

---

[7] This should not be confused with multiplication over the finite field $GF(q)$, for $q = p^m$.

[8] Linear independence (i.e., $\mathbf{w} = \mathbf{0}$ is the unique solution for $G\mathbf{w} = \mathbf{0}$) is meaningful for a prime alphabet p, because the product (modulo p) of two elements in $\mathbb{Z}_p$ is zero if and only if at least one of them is zero. This is not the case for a non-prime group size. For example, taking $q = 4$ and $n = 2$, the linear code $\mathcal{C} = \{(0, 0), (2, 2)\}$ in $\mathbb{Z}_4$ is generated by the 2×1 matrix $G = [2, 2]^t$, and it contains only two codewords, i.e., less than $q^k = 4^1 = 4$ which is equal to the alphabet size, so the "effective rank" of G is less than 1.

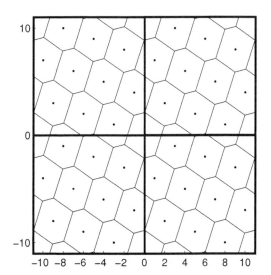

Figure 2.9 Construction A of a two-dimensional lattice. Here $q = 11$, and a generator matrix of the underlying linear code is $G = [2, 3]^t$, so C consists of the codewords $(0, 0), (2, 3), (4, 6), (6, 9), (8, 1), (10, 4), (1, 7), (3, 10), (5, 2), (7, 5)$ and $(9, 8)$. Alternatively, we can take a systematic matrix $G = [1, 7]^t$, so, by (2.59), the lattice is spanned by the vectors $[1, 7]^t$ and $[0, 11]^t$.

(e) *For a prime alphabet p, and if C is generated by a full-rank $n \times k$ matrix (2.54), then $M = p^k$, and*

$$V(\Lambda_C) = p^{n-k}. \tag{2.57}$$

(f) *The minimum distance between any two points in Λ_C is at most q; for $q = 2$ (a modulo-2 lattice), $d_{\min}(\Lambda_C) = \min\{2, \sqrt{d}\}$, where d is the minimum Hamming distance of the generating binary code C.*

(g) *Λ_C is spanned by the extended $n \times (n + k)$ generator matrix*

$$G_{\Lambda_C} = [G \mid q I_n], \tag{2.58}$$

where G is the generator matrix of C and I_n is the $n \times n$ identity matrix. See Example 2.1.2.

(h) *The extended matrix (2.58) can be reduced to a standard $n \times n$ generator matrix for Λ_C, provided that C has a systematic representation. Specifically, suppose a generator matrix of the form $G = [I_k | P^t]^t$, where I_k denotes the $k \times k$ identity matrix and P is an $(n - k) \times k$ matrix. That is, each codeword $\mathbf{c} = G\mathbf{w}$ consists of the information vector $\mathbf{w}$ itself (the "systematic part"), concatenated with*

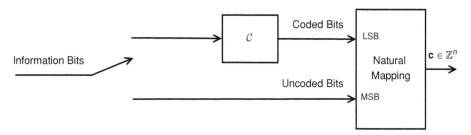

Figure 2.10 Coded modulation: construction A defines a natural mapping from information bits to lattice points in the Euclidean space. Trellis-coded modulation corresponds to the case where $\mathcal{C}$ is a convolutional code.

$P\mathbf{w}$ *(the "parity symbols"). Then,*

$$G_{\Lambda_C} = \begin{bmatrix} I_k & | & 0 \\ P & | & qI_{n-k} \end{bmatrix} \tag{2.59}$$

is an $n \times n$ generator matrix for Λ_C.

Example 2.5.2 (From Hamming code to E_8 lattice) *The E_8 lattice can be obtained by construction A from a simple binary linear code known as the (8,4,4) Hamming code (i.e., $q = 2$, $n = 8$, $k = 4$ and $d_{\min} = 4$). A systematic basis for this code is given by*

$$G^t = \begin{pmatrix} 1 & 0 & 0 & 0 & 0 & 1 & 1 & 1 \\ 0 & 1 & 0 & 0 & 1 & 0 & 1 & 1 \\ 0 & 0 & 1 & 0 & 1 & 1 & 0 & 1 \\ 0 & 0 & 0 & 1 & 1 & 1 & 1 & 0 \end{pmatrix} \tag{2.60}$$

$$\underbrace{}_{I_4} \ \underbrace{}_{P}$$

so by the reduced form (2.59),

$$G_{\Lambda_C} = \begin{bmatrix} I_4 & | & 0 \\ P & | & 2I_4 \end{bmatrix} \tag{2.61}$$

is a generator matrix for E_8.

2.5.2 Coded modulation and trellis codes

In many communication systems, the channel symbols belong to some fixed low-dimensional (say, k-dimensional) finite constellation. To increase the coding gain, the system encodes together many (say, m) such symbols. The resulting set of possible transmitted vectors is equivalent to a high-dimensional ($n = mk$) constellation. If the underlying constellation is a scalar lattice ($k = 1$), then we obtain a construction A lattice $\Lambda \in \mathbb{R}^n$. See Figure 2.10.

A special case of interest is when the encoder takes the form of a *convolutional code*. In this case, the equivalent dimension of the resulting lattice tends to infinity.

A lattice of infinite dimension is a discrete set of sequences $\{\ldots, x_{-1}, x_0, x_1, \ldots\}$ which is closed under reflection and addition. It can be described by taking the dimension of the generating matrix in Definition 2.1.1 to infinity; this amounts to a convolution between an integer sequence $\ldots, i_{-1}, i_0, i_1, \ldots$ and a real-valued time-varying filter $g_{n,k}$:

$$\Lambda_\infty = \left\{ \left\{ x_n = \sum_k g_{n,k} \cdot i_{n-k} \right\}_{n=-\infty}^{\infty} : i_k \in \mathbb{Z} \text{ for all } k \right\}. \tag{2.62a}$$

Alternatively, an integer-valued infinite-dimensional lattice can be generated via construction A (Definition 2.5.2), i.e., by lifting a q-ary *convolutional code* $\mathcal{C}_{\text{conv}}$ to the Euclidean space (in a convolutional code, the filter coefficients $g_{n,k}$ and the information symbols i_k in (2.62a) belong to some finite alphabet $\mathbb{Z}_q$, and all operations are taken modulo q):

$$\Lambda_\infty = \left\{ \{x_n\} \in \mathbb{Z}^\infty : \{x_n \bmod q\} \in \mathcal{C}_{\text{conv}} \right\}. \tag{2.62b}$$

In practice, it is desirable to use a finite, time-invariant filter $g_1, \ldots, g_K$ (or a set of such filters in parallel). Hence, the generation rule (2.62b) has a finite number (q^K) of *states*. It can then be described using a *trellis diagram*, with one axis playing the role of a state space, and the other the role of time. The resulting transformation, from a digital sequence (of bits or q-ary symbols) to a real-valued sequence x_n, is known as *trellis-coded modulation* (for the construction A of (2.62b)), or as a *signal code* (for the real convolution of (2.62a)). See [30, 93, 236].

Under appropriate conditions, a trellis code is invariant under translation, implying some kind of uniformity in space.

Definition 2.5.3 (Geometrically uniform code) *A code $\mathcal{C}$ in Euclidean space is geometrically uniform if for any two codewords $\mathbf{c}, \mathbf{c}' \in \mathcal{C}$, there exists an isometry T (a distance preserving transformation, i.e., a transformation consisting only of translation, rotation and reflection) that maps $\mathbf{c}$ to $\mathbf{c}'$ while leaving the code invariant, i.e., $T(\mathbf{c}) = \mathbf{c}'$ and $T(\mathcal{C}) = \mathcal{C}$.*

We see that lattices as well as unbounded linear trellis codes (2.62) are geometrically uniform (with the transformation T consisting only of translation). Most of the discussion in this book applies to such codes.

Summary of Chapter 2

Lattice $\Lambda(G) = G\mathbb{Z}^n = \{i_1 g_1 + \cdots + i_n g_n : i_1, \ldots, i_n \text{ are integers}\}$.

Equivalent bases $\Lambda(GT) = \Lambda(G)$ if and only if T is a unimodular matrix.

Fundamental cell $\Lambda + \mathcal{P}_0$ covers $\mathbb{R}^n$ without overlap.

Cell volume $V(\mathcal{P}_0) = V(\Lambda) = |\det(G)|$ for any fundamental cell $\mathcal{P}_0$.

Lattice quantizer $Q_\Lambda(\mathbf{x}) = \lambda$ if $\mathbf{x}$ is in $\lambda + \mathcal{P}_0$.

Modulo-lattice operation $\mathbf{x} = Q_\Lambda(\mathbf{x}) + [\mathbf{x} \bmod \Lambda]$.

Distributive law $[[\mathbf{x}] \bmod \Lambda + \mathbf{y}] \bmod \Lambda = [\mathbf{x} + \mathbf{y}] \bmod \Lambda$.

Cosets $\mathbf{x} + \Lambda$ has a unique representative in each fundamental cell $\mathcal{P}_0$.

Fundamental Voronoi cell $\mathcal{V}_0 = \{\mathbf{x} : \|\mathbf{x}\| \le \|\mathbf{x} - \lambda\|$ for all $\lambda \in \Lambda\}$. $\mathcal{V}_0$ is the set of minimum-length coset representatives, where ties are broken systematically.

Nearest-neighbor quantizer

$$Q_\Lambda^{(NN)}(\mathbf{x}) = \arg\min_{\lambda \in \Lambda} \|\mathbf{x} - \lambda\|.$$

Similarity $\Lambda(aTG)$ is similar to $\Lambda(G)$, if T is an orthogonal matrix.

Construction A If $\mathcal{C}$ is a linear q-ary code, then $\Lambda_\mathcal{C} = q\mathbb{Z}^n + \mathcal{C}$ is a modulo-q lattice.

Problems

P.2.1 (Degenerate lattices, too small basis) Show that the two basis vectors $[1, -1, 0]$ and $[0, 1, -1]$ generate a hexagonal lattice A_2 in $\mathbb{R}^3$.

P.2.2 (Voronoi cell convexity) Use the triangle inequality to prove the convexity of the cells for Euclidean Voronoi partition.

P.2.3 (Irregular fundamental cell) Show that if $\{\mathcal{P}_{0,i}\}$ is a partition of a fundamental cell $\mathcal{P}_0$ of Λ, and $\lambda_i \in \Lambda$, then the union $\cup_i(\lambda_i + \mathcal{P}_{0,i})$ is another fundamental cell of Λ.

P.2.4 (Volume formula) Prove that the approximation $V \approx \frac{\text{Vol}(B)}{N(B)}$ becomes tight for a large B, in the proof of Proposition 2.2.1. Hint: use the fact that the diameter d of the cells is finite.

P.2.5 (Modulo laws) Prove Lemma 2.3.1, using the decompositions (2.13) of $\mathbf{x}$ relative to $\mathcal{P}$ and to $\mathcal{Q}$.

P.2.6 (Covering $\mathbb{R}^n$ with cosets) Prove Lemma 2.3.3. Guidance: show that the disjoint union (2.30) can be written equivalently as $\cup_{\lambda \in \Lambda}(\lambda + \mathcal{P}_0) = \mathbb{R}^n$.

P.2.7 (Modulo reduces length) Prove that modulo-Voronoi reduction cannot increase the length of a vector, i.e., $\|\mathbf{x} \bmod_{\mathcal{V}_0} A_2\| \le \|\mathbf{x}\|$. Give an example of a non-Voronoi cell where the modulo operation increases the length.

P.2.8 (Scaling laws) Prove the transformation and scaling laws (2.42)–(2.41).

P.2.9 (Similarity) Show that the D_2 lattice of Example 2.4.2 coincides with the two-dimensional checkerboard lattice of Example 2.1.3. Show that D_3 and the three-dimensional checkerboard lattice are similar. Guidance: find an orthogonal transformation from the basis vectors $(2, 0, 0), (0, 2, 0), (1, 1, 1)$ of the checkerboard lattice in $\mathbb{R}^3$, to a scaled version of the basis vectors $(2, 0, 0), (1, 1, 0), (1, 0, 1)$ of the D_3 lattice.

P.2.10 (Construction A properties) Prove properties (a)–(g) of construction A, Proposition 2.5.1.

P.2.11 (Generating matrix of a modulo-q lattice) Prove formula (2.59) for the generating matrix of a modulo-q lattice, based on the systematic form of the generating matrix of the code. (Guidance: show (A) that for any integer vector $\mathbf{i}$, the real multiplication $G_{\Lambda_C} \cdot \mathbf{i}$ followed by a modulo-q operation is a point in C, and (B) (the reverse statement) if $\mathbf{x}$ mod q is in C (i.e., $\mathbf{x}$ is a shift of a point in C by a point in $q\mathbb{Z}^n$), then $\mathbf{x}$ can be written as $G_{\Lambda_C} \cdot \mathbf{i}$ for some integer vector $\mathbf{i}$.)

P.2.12 (Construction A for D_n) Show that D_n is a modulo-2 lattice, generated by a single parity-check binary code $C = \{\mathbf{c} \in \mathbb{Z}_2^n : c_1 \oplus \ldots \oplus c_n = 0\}$. Use that to show that

$$
G = \begin{pmatrix}
1 & 0 & & \ldots & 0 & -1 \\
0 & 1 & 0 & \ldots & & 0 \\
& & & & & \\
\vdots & & & \ddots & & \vdots \\
0 & & & \ldots & 1 & 0 \\
1 & & & \ldots & 1 & 1
\end{pmatrix}
$$

is a generator matrix for D_n. Conclude that D_n is not similar to the n-checkerboard for more than four dimensions. (Hint: observe that D_n is spanned by n of its shortest lattice vectors, while any basis of the n-checkerboard must contain at least one *non*-shortest lattice vector for $n > 4$.)

P.2.13 (Multi-level codes and minimum distance) Construction A of a modulo-2 lattice can be viewed as encoding of the least significant bits of the coordinates of the constellation, while leaving the most significant bits un-coded. That is, letting LSB(z) denote the least significant bit in a binary expansion of the integer $z \in \mathbb{Z}$, the lattice Λ_C consists of all integer vectors $\mathbf{z} = (z_1, \ldots, z_n)$ such that LSB($\mathbf{z}$) = (LSB(z_1), \ldots, LSB(z_n)) is a point in C. To increase the minimum distance beyond 2 (while still using *binary* codes – see Proposition 2.5.1(f)), we should also add redundancy to the higher bits in the binary expansion of $\mathbb{Z}$. Suppose that C_2 is an (n, k) linear binary code. Define a 4-ary linear code $C_4 = [C_2 \oplus 2 \cdot C_2]$ mod 4 (where the multiplication by 2 and addition are carried modulo 4), and form an

n-dimensional modulo-4 lattice $\Lambda_{\mathcal{C}_4} = \mathcal{C}_4 + 4\mathbb{Z}^n$. Show that: (i) $\mathcal{C}_2 = \mathcal{C}_4$ mod 2; (ii) $\Lambda_{\mathcal{C}_4} \subset \Lambda_{\mathcal{C}_2}$; (iii) conclude that all points $\mathbf{z}$ in $\Lambda_{\mathcal{C}_4}$ satisfy that LSB($\mathbf{z}$) is a point in $\mathcal{C}_2$, but the reverse statement is not true; (iv) conclude that $d_{\min}(\Lambda_{\mathcal{C}_4}) \geq d_{\min}(\Lambda_{\mathcal{C}_2})$, and the inequality is strict if $d_{\min}(\mathcal{C}_2) > 2$. See *construction D* in Forney *et al.* [92] and Conway and Sloane [49, chapter 8].

Interesting facts about lattices

The Nobel prize in chemistry in 2011 was awarded to the materials scientist Dan Shechtman, for his discovery in 1982 of quasicrystals. Until that time, every-one assumed that the atoms of an ordered solid material must form a three-dimensional lattice. Materials with such a periodic structure are characterized by a point-like diffraction pattern (i.e., a spatial Fourier series) which can only pos-sess a 2nd, 3rd, 4th and 6th order symmetry; in contrast, the diffraction pattern of quasicrystals may have an unrestricted (typically 5th, 8th, 10th or 12th) order symmetry [158].

The seventeenth century astronomer Johannes Kepler conjectured that the FCC lattice forms the best sphere packing in three dimensions. While Gauss showed that no other lattice arrangement is better, the perhaps harder part – of excluding non-lattice arrangements – remained open until a full (computer-aided) proof was given in 1998 by Hales.

The early twentieth century mathematician Hermann Minkowski used lattices to relate n-dimensional geometry with number theory – an area he called "the geometry of numbers" [36]. The Minkowski–Hlawka theorem (conjectured by Minkowski and proved by Hlawka in 1944) will play in Chapter 7 the role of Shannon's random coding technique for proving the existence of "good" lattice codes.

Although lattice codes are not mentioned in Shannon's work, he was certainly interested in sphere packing in high dimensions.[9] A story says that when David Slepian was about to retire from Bell Laboratories, he invited his close colleague Aaron Wyner to pick his favorite books from his office (before giving all of them to his other lab mates). Wyner hesitated to take the offer before Slepian left, and when he finally came to make his choice most books were already taken. Yet one book on the shelf caught his eye, "An introduction to the geometry of N dimensions," from 1929, by Sommerville. Opening it, Wyner found "C. E. Shannon" handwritten on the inside cover.[10]

The relation between error-correcting codes, sphere packing and lattices (called construction A in this chapter) was studied by Leech and Sloane [153], and Conway and Sloane [49, chapter 5] in the 1970s and 1980s. They were motivated by several

---

[9] Shannon's 1953 paper "The lattice theory of information" [242], refers to another definition of the word "lattice" (a partial order between sets).

[10] Thanks are due to Wyner's colleague, Toby Berger, for telling me this story.

discoveries in the mathematical literature during the 1960s, like the 2^m-dimensional Barnes–Wall lattices and the ultradense Leech lattice in 24 dimensions; see the introduction by Forney [83]. Another notable motivation was provided by Ungerboeck's invention of trellis-coded modulation by set partition in 1982 [259], and de Buda's asymptotic analysis of lattice-based codes [57, 58] in the late 1970s and 1980s. In a series of works through the 1980s and 1990s, Forney [83, 84, 85, 86, 87, 94] established tools to characterize and evaluate lattice codes, towards their implementation in digital communication.

ITU-T standard V.34 for voice-band telephone channel modems at 33.6 kbits per second uses a four-dimensional constellation selected from the D_4 lattice; see the book by Tretter [256]. In the wireless communication domain (in standards like 802.11 [WiFi] and LTE), the set of possible coded signals corresponds to a finite segment from some high-dimensional lattice. Lattice ("algebraic") codebooks are also used for data compression; one recommendation for the ITU-T 729.1 speech-coding standard uses the Gosset lattice E_8 as the codebook for code-excited linear prediction (CELP) [104, 144]. And obviously, the analog-to-digital (A/D) convertor at the interface of any signal compression scheme is a scalar lattice quantizer.

Some of the stronger public-key algorithms today use lattice-based cryptography, a concept that was initiated by Ajtai in 1996 [6]. Actual systems based on lattices were proposed immediately after Ajtai's discovery by Goldreich *et al.* [108] and Hoffstein *et al.* [121] (the NTRU algorithm was patented in the late 1990s). These systems rely on the asymmetry of coding and decoding, and on the difficulty of finding a "good" basis for a given lattice. It is easy to translate an integer vector to a lattice point and to perturb it slightly, but it is hard to find the closest lattice point to the perturbed vector – unless the decoder has a "good" basis; see the book by Micciancio and Goldwasser [186] and the survey paper by Micciancio and Regev [188].

3 Figures of merit

In digital communications, the cubic lattice plays the role of a simple uniform quantizer – for source coding, or an "uncoded" constellation – for channel coding. Better source-channel coding schemes can be associated with more complex lattices. This relation requires a definition of the notion of lattice goodness.

We shall develop two figures of merit of a lattice in the context of digital communication: (i) the *normalized second moment* – a measure of its goodness as a vector quantizer under a squared-error distortion measure, in Section 3.2, and (ii) the *volume to noise ratio* – a measure of its goodness as a (coded) constellation for the AWGN channel, in Section 3.3. Before that, in Section 3.1, we introduce two more fundamental quantities associated with lattice goodness for sphere packing and covering. As we may expect, all these quantities are invariant to scaling and rotation of the lattice.

3.1 Sphere packing and covering

Which shape minimizes the surface area among all shapes of a given volume in $\mathbb{R}^n$? Which shape minimizes the diameter or the second moment? The *iso-perimetric inequality* implies that for all these questions – and many more – the unique solution is a *ball*. Two of the most fundamental questions about balls (or spheres) are how efficiently they can be packed in the Euclidean space, and how efficiently they can cover it. Although these questions do not necessarily lead to a lattice arrangement, lattices offer the most natural and insightful sphere packings and coverings in the Euclidean space. [1]

Let $\mathcal{B}_r$ denote an n-dimensional ball of radius r centered at zero:

$$\mathcal{B}_r = \{\mathbf{x} \in \mathbb{R}^n : \|\mathbf{x}\| \leq r\}. \tag{3.1}$$

---

[1] In lattice literature, balls and spheres are usually synonymous, whereas in mathematics sphere usually refers to the surface of a ball and solid sphere usually refers to a ball.

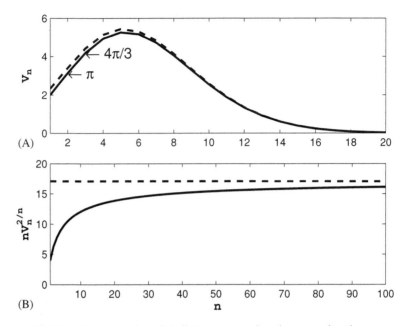

Figure 3.1 (A) The volume V_n of a unit ball $\mathcal{B}_1$, compared to the approximation $V_n \approx (2\pi e/n)^{n/2}$. (B) Convergence of $n V_n^{2/n}$ to $2\pi e \approx 17$.

Hence, $\mathcal{B}_1$ is the *unit-radius ball*, and $\mathcal{B}_r = r\mathcal{B}_1$. The volume $\mathrm{Vol}(\mathcal{B}_r) = \int_{\mathcal{B}_r} d\mathbf{x}$ of an n-dimensional ball is given by

$$\mathrm{Vol}(\mathcal{B}_r) = V_n \cdot r^n, \tag{3.2}$$

where V_n is the *volume of a unit ball*, i.e., a ball with a unit radius. That is, $V_1 = 2$, $V_2 = \pi$, $V_3 = \frac{4}{3}\pi$, and, in general,

$$V_n = \mathrm{Vol}(\mathcal{B}_1) = \frac{\pi^{n/2}}{(n/2)!}, \tag{3.3}$$

where the factorial for fractional numbers is defined via its Gamma-function extension, i.e., $(n/2)! = \Gamma(n/2 + 1) = \sqrt{\pi} \cdot \frac{1}{2} \cdot \frac{3}{2} \cdots \cdot \frac{n}{2}$ for odd n.

As Figure 3.1(A) shows, V_n increases for dimensions $1 \leq n \leq 6$, and then it decreases. As we shall see later, $V_n \approx \left(\frac{2\pi e}{n}\right)^{n/2}$ for large n up to a sub-exponential factor; see Figure 3.1(B). Thus, to avoid a *vanishing* volume, the radius r must *grow* with the dimension (at least roughly as $\sqrt{n}$). Also, noting that

$$\frac{\mathrm{Vol}(\mathcal{B}_{r(1-\epsilon)})}{\mathrm{Vol}(\mathcal{B}_r)} = (1 - \epsilon)^n \to 0, \tag{3.4}$$

as $n \to \infty$ for all $\epsilon > 0$, we conclude that most of the volume of a high-dimensional ball is concentrated near its outer shell. Thus, a ball and a spherical shell are

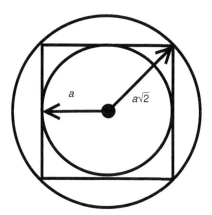

Figure 3.2 It is insightful to compare the behavior of the volume of a ball with that of a cube of "similar dimensions." The volume of an n-dimensional cube with edge length $2a$ is $(2a)^n$. A maximal ball *inscribed* in this cube has a radius a, so its volume is $V_n a^n$, which vanishes as $n \to \infty$ (it is $\approx (a\sqrt{2\pi e/n})^n$ for large n). On the other hand, a *circumscribed* sphere of this cube has a radius $a\sqrt{n}$, so its volume grows like $V_n(\sqrt{n}a)^n \approx (\sqrt{2\pi ea})^n \approx (4a)^n$. Thus, volume-wise, an n-dimensional ball is similar to the cube inscribed in it, but not to the cube containing it. This may explain why the cubic lattice is so bad for packing but quite good for covering.

asymptotically *equivalent volume-wise*. Furthermore, they are both volume-wise similar to the cube *inscribed* in them, as discussed in the caption of Figure 3.2.

3.1.1   Packing problem

Let us formalize the problem of arranging the pile of oranges in Figure 1.2. For a given lattice Λ and a radius r, the set $\Lambda + \mathcal{B}_r$ is a *packing* in Euclidean space, if for all distinct lattice points $\lambda, \lambda' \in \Lambda$, we have

$$(\lambda + \mathcal{B}_r) \cap (\lambda' + \mathcal{B}_r) = \emptyset. \tag{3.5}$$

That is, the spheres do not intersect. The *packing radius* $r_{\text{pack}}(\Lambda)$ of the lattice is defined by the largest balls the lattice can pack:

$$r_{\text{pack}}(\Lambda) = \sup\{r : \Lambda + \mathcal{B}_r \text{ is a packing}\}. \tag{3.6}$$

As Figure 3.3 shows, $r_{\text{pack}}(\Lambda)$ is the inner radius of the Voronoi cell $\mathcal{V}_0$, i.e., the radius of the largest (open) n-dimensional ball contained in $\mathcal{V}_0$.

The central ball $\mathcal{B}_{r_{\text{pack}}(\Lambda)}$ in the packing $\Lambda + \mathcal{B}_{r_{\text{pack}}(\Lambda)}$ touches each of its neighbors at one point. The number of such neighbors is called the *kissing number* of the lattice, which is also the number of *shortest lattice vectors*:[2]

$$N_\Lambda(d_{\min}) \overset{\Delta}{=} |\{\lambda \in \Lambda : \|\lambda\| = d_{\min}\}|, \tag{3.7}$$

---

[2] This is usually smaller than the kissing number of $\mathbb{R}^n$, which is the maximum number of balls that can touch another ball of the same size [49].

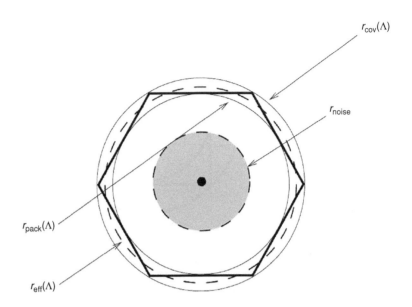

Figure 3.3 Geometric picture of $r_{\text{pack}}(\Lambda)$ and $r_{\text{eff}}(\Lambda)$ with respect to the Voronoi region, as well as the other radii defined in the text.

where $d_{\min} = d_{\min}(\Lambda)$ is the *minimum distance* of the lattice Λ:

$$d_{\min}(\Lambda) \stackrel{\Delta}{=} \min_{\lambda, \lambda' \in \Lambda, \ \lambda \neq \lambda'} \|\lambda - \lambda'\| = \min_{\lambda \in \Lambda, \ \lambda \neq 0} \|\lambda\|. \tag{3.8}$$

Note that since the touching point is on the line connecting the centers of the balls, the minimum distance is *twice* the packing radius:

$$d_{\min}(\Lambda) = 2r_{\text{pack}}(\Lambda). \tag{3.9}$$

Note also that the shortest lattice vectors are generally a subset of the *face-determining points* of the fundamental Voronoi cell, as illustrated in Figure 3.4. For the hexagonal lattice, these sets are the same; all six face-determining points are at a distance $d_{\min}(\Lambda)$ from the origin.

To assess the packing efficiency of a given lattice, we make the following definition.

Definition 3.1.1 (Effective radius) *The* effective radius *of a lattice Λ is defined as the radius of a sphere having the same volume as the lattice cells:*

$$r_{\text{eff}}(\Lambda) = \left[\frac{V(\Lambda)}{V_n} \right]^{1/n}. \tag{3.10}$$

By the properties of a ball, $r_{\text{eff}}(\Lambda) \geq r_{\text{pack}}(\Lambda)$, with equality if and only if the Voronoi cell itself is a ball. More generally, if $\text{Vol}(\mathcal{B}_r) = a^n V(\Lambda)$ (or larger), then

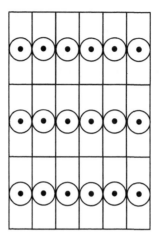

Figure 3.4 A case where the set of shortest lattice vectors is smaller than the set of the face-determining points of the fundamental Voronoi cell.

$r \geq a r_{\text{pack}}(\Lambda)$. In particular, if $\text{Vol}(\mathcal{B}_r) = 2^n V(\Lambda)$ (or larger), then by (3.9) $r \geq d_{\min}(\Lambda)$; i.e., $\mathcal{B}_r$ contains all the shortest lattice vectors. [3]

Definition 3.1.2 (Packing efficiency) *The* packing efficiency *of a lattice Λ is defined as*

$$\rho_{\text{pack}}(\Lambda) = \frac{r_{\text{pack}}(\Lambda)}{r_{\text{eff}}(\Lambda)}. \qquad (3.11)$$

The normalization by the effective radius guarantees that the packing efficiency is *invariant to scaling*, i.e., $\rho_{\text{pack}}(\alpha \Lambda) = \rho_{\text{pack}}(\Lambda)$. One use of ρ_{pack} is to express the proportion of space taken up by the spheres – a quantity known as *packing density*:

$$\Delta(\Lambda) = \frac{\text{volume of packed spheres}}{\text{volume of space}} = \frac{\text{Vol}(\mathcal{B}_{r_{\text{pack}}(\Lambda)})}{V(\Lambda)} = \rho_{\text{pack}}^n(\Lambda). \qquad (3.12)$$

When taking a *Cartesian product* of lattices, the packing radius is determined by the lattice with the *minimal* packing radius. The effective radius, on the other hand, increases roughly as the square root of the number of factors. Thus, the packing efficiency of Cartesian products generally *decreases*. See Problem 3.3.

Example 3.1.1 (Packing efficiency of $\mathbb{Z}^n$) *The packing radius of the cubic lattice $\mathbb{Z}^n$ is $1/2$ for all n. Since $V(\mathbb{Z}^n) = 1$, the effective radius is $r_{\text{eff}}(\mathbb{Z}^n) = 1/\sqrt[n]{V_n}$. Thus $\rho_{\text{pack}}(\mathbb{Z}^n) = \sqrt[n]{V_n}/2$. It follows that $\rho_{\text{pack}}(\mathbb{Z}^n) = 1, \sqrt{\pi}/2 \approx 0.886$ and*

---

[3] This argument holds not only for a "Euclidean ball," but, in fact, for any norm-based ball (i.e., any norm in (3.1)). Equivalently, this argument applies to any convex body which is symmetric about the origin. We thus obtain the *Minkowski convex-body theorem*: any zero-symmetric convex body with volume greater than $2^n V(\Lambda)$ contains at least one non-zero lattice point. See Problem 3.2.

$\sqrt[3]{\pi/6} \approx 0.806$, *for* $n = 1, 2$ *and* 3, *respectively. As for large* n, *by the asymptotic expression* $V_n \approx (2\pi e/n)^{n/2}$, *we have* $\rho_{\text{pack}}(\mathbb{Z}^n) \approx \sqrt{\pi e/2n}$. *In particular, while the effective radius grows as* $\sqrt{n}$ *(like the* diagonal *of a cubic cell), the packing radius is fixed, hence the packing efficiency decreases to zero as* $1/\sqrt{n}$. *See Figure* 3.2.

Hermite parameter

An alternative measure for the packing efficiency is the *Hermite parameter*:

$$\hbar(\Lambda) \overset{\Delta}{=} \frac{d_{\min}^2(\Lambda)}{V^{2/n}(\Lambda)}. \tag{3.13}$$

It is related to the radius-ratio measure in Definition 3.1.2, as $\hbar(\Lambda) = 4\rho_{\text{pack}}^2(\Lambda)/V_n^{2/n}$. For the cubic lattice $\hbar(\mathbb{Z}^n) = 1$, for all n. And, in general, the Hermite parameter is invariant to taking Cartesian products of the same lattice, i.e., $\hbar(\Lambda^n) = \hbar(\Lambda)$ (unlike the packing efficiency $\rho_{\text{pack}}(\Lambda)$). We shall see below, in the context of modulation in the presence of AWGN, that $\hbar(\Lambda)$ plays the role of a *nominal* coding gain.

The packing efficiency is always greater than zero and no greater than 1:

$$0 < \rho_{\text{pack}}(\Lambda) \leq 1. \tag{3.14}$$

It is exactly 1 for one dimension, where "spheres" – i.e., intervals – can be packed without holes, but then it drops strictly below 1 for all $n > 1$.

The densest lattice in the plane, i.e., the one that maximizes $\rho_{\text{pack}}(\Lambda)$ over all lattices in $\mathbb{R}^2$, is the hexagonal (A_2) lattice (Example 2.2.1), for which $\rho_{\text{pack}}(\Lambda) = \sqrt{\pi/2\sqrt{3}} \approx 0.9523$. The densest three-dimensional lattice is the *face-centered cubic* (FCC) (A_3) lattice – formed by the centers of the pile of oranges in Figure 1.2 – with $\rho_{\text{pack}}(\Lambda) = \sqrt[3]{\pi/3\sqrt{2}} \approx 0.9047$. The densest lattices are known for all dimensions up to eight, but are still unknown for most higher dimensions.

As the examples above seem to indicate, the density of the best *known* lattices tends to *decrease* with the dimension (though not monotonically). On the positive side, a theorem due to Minkowski and Hlawka [36] guarantees that in each dimension there exists a lattice whose packing efficiency is at least one-half:

$$\max_{\Lambda \in \mathbb{R}^n} \cdot \rho_{\text{pack}}(\Lambda) \geq \frac{1}{2}. \tag{3.15}$$

Thus, there is a factor $1/2$ between the upper bound (3.14), which holds for every lattice, and the lower bound (3.15), which holds for some "good" lattices. We shall return to this point in Chapter 7.

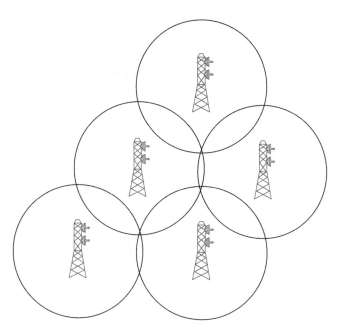

Figure 3.5 Covering space with cellular base stations.

3.1.2   Covering problem

Consider deployment of cellular base stations in a three-dimensional space, where each base station can communicate with users at a maximum range of r_{cov}, as shown in Figure 3.5. To ensure coverage, each point in space must fall inside a ball of radius r_{cov} around at least one of the base stations.

Mathematically, the associated notions for the covering problem are defined similarly to their packing counterparts. The set $\Lambda + \mathcal{B}_r$, composed of spheres centered around the lattice points, is a covering of Euclidean space if

$$\mathbb{R}^n \subseteq \Lambda + \mathcal{B}_r.$$

That is, each point in space is covered by at least one sphere. Define the covering radius of the lattice $r_{\mathrm{cov}}(\Lambda)$ by

$$r_{\mathrm{cov}}(\Lambda) = \min\{r : \ \Lambda + \mathcal{B}_r \text{ is a covering}\} \tag{3.16}$$

which is also the outer radius of the Voronoi cell, i.e., the minimum radius of a (closed) ball containing $\mathcal{V}_0$. See Figure 3.3.

Definition 3.1.3 (Covering efficiency) *The covering efficiency of a lattice is defined by*

$$\rho_{\mathrm{cov}}(\Lambda) = \frac{r_{\mathrm{cov}}(\Lambda)}{r_{\mathrm{eff}}(\Lambda)}, \tag{3.17}$$

where $r_{\mathrm{eff}}(\Lambda)$ is defined in (3.10).

Again, the normalization by the effective radius guarantees that the covering efficiency is *invariant to scaling*, i.e., $\rho_{\mathrm{cov}}(\alpha\Lambda) = \rho_{\mathrm{cov}}(\Lambda)$.

The *covering thickness* – the dual of the packing density – is the average number of spheres covering a point. It is given by the proportion of space taken up by the spheres,

$$\theta(\Lambda) = \frac{\text{volume of covering spheres}}{\text{volume of space}} = \frac{\mathrm{Vol}(\mathcal{B}_{r_\Lambda^{\mathrm{cov}}})}{V(\Lambda)} = \rho_{\mathrm{cov}}^n(\Lambda). \qquad (3.18)$$

The covering efficiency $\rho_{\mathrm{cov}}(\Lambda)$ is by definition not less than 1. As for packing, it is exactly 1 for one dimension, but then it goes above 1 for all $n > 1$. In contrast to packing, the covering efficiency does not get much worse by taking Cartesian products.

Example 3.1.2 (Covering efficiency of $\mathbb{Z}^n$) *The covering radius of the cubic lattice $\mathbb{Z}^n$, which is the distance of the cube corner to the center, is $\sqrt{n}/2$. The effective radius is $r_{\mathrm{eff}}(\mathbb{Z}^n) = 1/\sqrt[n]{V_n}$ (see Example 3.1.1). Thus $\rho_{\mathrm{cov}}(\mathbb{Z}^n) = \sqrt{n}\sqrt[n]{V_n}/2$. It follows that $\rho_{\mathrm{cov}}(\mathbb{Z}^n) = 1$, $\sqrt{2\pi}/2 \approx 1.253$ and $\sqrt{3}\sqrt[3]{\pi/6} \approx 1.396$, for $n = 1, 2$ and 3, respectively. For a large n, the approximation $V_n \approx (2\pi e/n)^{n/2}$ implies that $\rho_{\mathrm{cov}}(\mathbb{Z}^n) \approx \sqrt{\pi e/2} \approx 2.066$. That is, in contrast to its vanishing efficiency for packing, the asymptotic covering efficiency of the cubic lattice is not that bad: its covering radius is only about twice its effective radius. See Figure 3.2.*

The optimum covering efficiency is unknown for most dimensions. The thinnest lattice in the plane, i.e., the one that minimizes $\rho_{\mathrm{cov}}(\Lambda)$ over all lattices in $\mathbb{R}^2$, is, as expected, the hexagonal lattice, with $\rho_{\mathrm{cov}}(\Lambda) \approx 1.0996$. In three dimensions the thinnest lattice is surprisingly different than the densest one: it is the *body-centered cubic* (BCC) lattice – formed by the centers of the three-dimensional cellular base stations in Figure 3.5 – with $\rho_{\mathrm{cov}}(\Lambda) \approx 1.1353$. (For comparison, the covering efficiency of the FCC lattice is ≈ 1.2794.)

These examples give the impression that the covering efficiency of the best lattices becomes worse as the dimension increases. Nevertheless, a result of Rogers [230] shows that there exists a sequence of lattices Λ_n of increasing dimension n such that

$$\rho_{\mathrm{cov}}(\Lambda_n) \to 1 \quad \text{as } n \to \infty.$$

Thus, the thinnest coverings eventually improve and approach the lower bound of 1. We shall return to this point in Chapter 7.

3.2   Quantization: normalized second moment

Digital communication suggests an interesting variation on the problems of sphere packing and covering, which is more relevant for engineering applications. Vector quantization under a squared-error distortion measure is close in spirit to the sphere

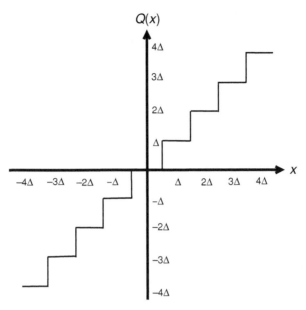

Figure 3.6 The scalar nearest-neighbor quantizer $Q_\Lambda(x)$, known also as a uniform mid-thread quantizer.

covering problem, while modulation in the presence of additive white-Gaussian noise (AWGN) is somewhat close to the sphere packing problem. As we shall see, however, the new problems exhibit different behavior as a function of the dimension.

We shall again restrict ourselves to lattice-based quantization and modulation, although more general "periodic constellations" are possible. It will be easier to start this time with the (sphere covering related) quantization problem.

For a general lattice partition $\mathcal{P}$ given in Definition 2.2.1, the lattice quantizer $Q_\Lambda : \mathbb{R}^n \to \Lambda$ maps all points inside a cell to the lattice point associated with this cell:

$$Q_\Lambda(\mathbf{x}) = \lambda, \quad \text{if } \mathbf{x} \in \mathcal{P}_\lambda \tag{3.19}$$

as we have already seen in (2.13). This becomes the nearest-neighbor quantizer (2.10) if $\mathcal{P}$ is a Voronoi partition, and it takes the familiar staircase form shown in Figure 3.6 in the scalar ($n = 1$) case.

For many sources $\mathbf{S}$ of interest, the *quantizer error vector* $Q_\Lambda(\mathbf{S}) - \mathbf{S}$ is roughly *uniformly distributed* over the quantizer fundamental cell, and we care about its mean-squared value, i.e., the *mean-squared error* (MSE).

Definition 3.2.1 (Second moment) *The second moment of a lattice is defined as the second moment per dimension of a random variable $\mathbf{U}$ which is uniformly distributed* *over the fundamental Voronoi cell $\mathcal{V}_0$:*

$$\sigma^2(\Lambda) = \frac{1}{n} E\{\|\mathbf{U}\|^2\} = \frac{1}{V(\Lambda)} \cdot \frac{1}{n} \int_{\mathcal{V}_0} \|\mathbf{x}\|^2 d\mathbf{x}, \tag{3.20}$$

where $E\{\cdot\}$ denotes expectation, and $V(\Lambda)$ is the volume of the lattice cells. Note that since the fundamental Voronoi cell is symmetric about the origin, $\mathbf{U}$ has a zero mean; hence $\sigma^2(\Lambda)$ is also the variance per dimension of $\mathbf{U}$.

A figure of merit of a lattice quantizer with respect to the MSE distortion measure is the *normalized second moment* (NSM), or the "second-moment to volume ratio."

Definition 3.2.2 (Normalized second moment)

$$G(\Lambda) = \frac{\sigma^2(\Lambda)}{V^{2/n}(\Lambda)}. \tag{3.21}$$

The minimum possible value of $G(\Lambda_n)$ over all lattices in $\mathbb{R}^n$ is denoted by G_n.

Note that $G(\Lambda)$ is a dimensionless quantity which is invariant to scaling or rotation of Λ, i.e., $G(\alpha\Lambda) = G(\Lambda)$ for all $\alpha > 0$, and $G(T\Lambda) = G(\Lambda)$ for an orthonormal transformation T.

Example 3.2.1 *The NSM of the scalar lattice quantizer under Voronoi partition is given by*

$$G_1 = \frac{1}{\Delta^3} \int_{-\Delta/2}^{\Delta/2} x^2 dx = 1/12. \tag{3.22}$$

Note that the NSM does not change if we take a Cartesian product (2.4) of the lattice, i.e.,

$$G(\Lambda \times \Lambda) = G(\Lambda). \tag{3.23}$$

In particular,

$$G(\mathbb{Z}^n) = G(\mathbb{Z}) = G_1. \tag{3.24}$$

Hence, $G_1 = 1/12 \approx 0.08333$ provides a reference level for a general lattice quantizer.

Definition 3.2.3 (Vector-quantizer gain) *The quantization granular gain of an n-dimensional lattice Λ, relative to the cubic lattice, is defined as*

$$\Gamma_q(\Lambda) = \frac{G(\mathbb{Z}^n)}{G(\Lambda)} = \frac{1/12}{G(\Lambda)}. \tag{3.25}$$

In the context of Voronoi modulation, this quantity is also known as the shaping gain, and denoted $\Gamma_s(\Lambda)$; see Chapter 9.

The NSM of the cubic lattice also provides an upper bound for G_n, the NSM of the best lattice quantizer in $\mathbb{R}^n$. On the other hand, a lower bound for G_n is provided by G_n^*, the corresponding quantity of an n-dimensional ball (i.e., the second moment per dimension of a uniform distribution over a ball normalized by the volume raised to the power of $2/n$). This quantity is monotonically decreasing with n, and approaches $\frac{1}{2\pi e} \approx 0.05855$ as n goes to infinity. We thus have

$$G_1 \geq G_n \geq G_n^* > \frac{1}{2\pi e} \qquad (3.26)$$

for all n.

Gersho's conjecture and lattice quantization

The hexagonal lattice minimizes the *average* NSM among all two-dimensional (lattice or non-lattice) quantizers [255]. This amounts to minimizing the MSE in high-resolution quantization of a uniformly distributed source (for a fixed quantizer point density). It is tempting (and analytically convenient) to assume also that in higher dimensions, the best quantizer has a periodic structure. Gersho conjectured that the best quantizer in each dimension is a *tiling* with some optimal space-filling polytope [102]. As of today, the best known quantizer up to 24 dimensions is a *lattice* quantizer [49] (i.e., we do not know of a non-lattice quantizer which is strictly better).

While packing and covering are perfect in dimension one, and become worse as the dimension increases, the sphere NSM G_n^* improves, i.e., decreases with n. This may hint at the fact that increasing the dimension may be advantageous for the quantization problem. Indeed, the NSM of the hexagonal lattice is ≈ 0.080188, which is slightly better than the cubic lattice. Moreover, there exists a sequence of lattice quantizers Λ_n of increasing dimension n, such that

$$\lim_{n \to \infty} G(\Lambda_n) = \frac{1}{2\pi e}. \qquad (3.27)$$

Thus, a sequence of "good" lattices asymptotically achieves the sphere lower bound (3.26). More on that in Chapter 7.

The operational significance of the NSM comes from classical results in high-resolution quantization theory, which express the distortion of lattice quantization in terms of G_n. It also plays a role in dithered quantization, the topic of Chapter 4, and in lattice shaping, the topic of Chapter 9.

3.3   Modulation: volume to noise ratio

An additive-noise channel is given by the input/output relation

$$\mathbf{Y} = \mathbf{X} + \mathbf{Z}, \qquad (3.28)$$

where the noise $\mathbf{Z}$ is independent of the input $\mathbf{X}$. In the AWGN channel case, $\mathbf{Z}$ is a white (i.i.d.) Gaussian noise with zero mean and variance σ^2 whose probability density function (p.d.f.) is given by

$$f_Z(\mathbf{z}) = \frac{1}{(2\pi\sigma^2)^{n/2}} e^{-\frac{\|\mathbf{z}\|^2}{2\sigma^2}}. \tag{3.29}$$

As discussed in Chapter 1, in digital communication the input $\mathbf{X}$ belongs to a codebook $\mathcal{C} = \{\mathbf{c}_i\}$, where $\mathbf{c}_i$ carries the message i. Suppose the lattice Λ is used as a codebook, so the messages are carried by the lattice points $\{\lambda_i \in \Lambda\}$.

Since the lattice is unbounded, the common notion of a *signal to noise ratio* (SNR) is not meaningful. We replace it by the following measure for the lattice *sparsity* with respect to Gaussian noise.

Definition 3.3.1 (Volume to noise ratio) *The volume to noise ratio (VNR) of an n-dimensional lattice Λ, in the presence of AWGN with variance σ^2, is defined as* [4]

$$\mu = \mu(\Lambda, \sigma^2) = \frac{V^{2/n}(\Lambda)}{\sigma^2}. \tag{3.30}$$

The VNR (3.30) still does not reflect the *performance* of a lattice as a codebook. We would like to refine this definition, to take into account the probability of *decoding error* when Λ is used as a codebook for the noisy channel (3.28).

Since the AWGN distribution (3.29) is monotonically decreasing with the *norm* of the noise $\|\mathbf{z}\|$, given a received vector $\mathbf{Y}$ it is natural to decode $\mathbf{X}$ as the *closest lattice point*:

$$\hat{\lambda}_{NN} = \arg\min_{\lambda \in \Lambda} \|\mathbf{Y} - \lambda\| = Q_\Lambda^{(NN)}(\mathbf{Y}), \tag{3.31}$$

where $Q_\Lambda^{(NN)}$ is the nearest-neighbor (Voronoi) quantizer (2.10). An error will then occur whenever $\mathbf{Y}$ falls outside the Voronoi cell $\mathcal{V}_\lambda$ of the transmitted codeword λ. Since any cell $\mathcal{V}_\lambda \in \mathcal{V}$ is a shift of the fundamental cell $\mathcal{V}_0$ by λ, an error will occur if the noise $\mathbf{Z}$ falls outside $\mathcal{V}_0$; furthermore, the *additivity* of the channel $\mathbf{Y} = \lambda + \mathbf{Z}$ implies that this event is independent of which codeword λ was transmitted.

Definition 3.3.2 (Error probability) *The error probability in nearest-neighbor decoding of the lattice Λ, in the presence of AWGN $\mathbf{Z}$ with variance σ^2, is defined as*

$$P_e(\Lambda, \sigma^2) = \Pr\{\mathbf{Z} \notin \mathcal{V}_0\}, \tag{3.32}$$

where $\mathcal{V}_0$ is the (Euclidean) fundamental Voronoi cell of Λ.

Note that $P_e(\alpha\Lambda, \sigma^2)$ increases monotonically with the noise variance σ^2, and decreases with the lattice scaling parameter α.

---

[4] In the literature the VNR is sometimes normalized by $2\pi e$, which is an asymptotic value corresponding to the Shannon capacity.

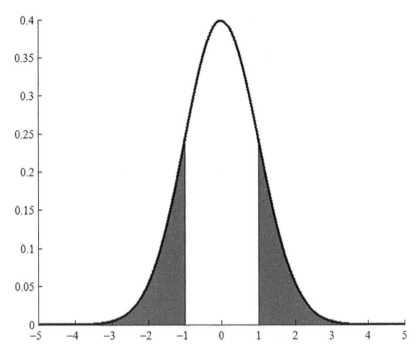

Figure 3.7 Error probability for a one-dimensional constellation.

Example 3.3.1 (Error probability of a $\mathbb{Z}$-lattice) *In the one-dimensional case, illustrated in Figure 3.7, the decoding error probability of the $\mathbb{Z}$-lattice with Voronoi partition is given by $P_e(\mathbb{Z}, \sigma^2) = 2Q(\frac{1/2}{\sigma})$, where $Q(x) = \int_x^\infty f_Z(z)dz$ is the tail probability of the normal distribution (known also as the Q-function). In n dimensions, correct decoding amounts to succeeding in n independent trials, with probability $1 - P_e(\mathbb{Z}, \sigma^2)$ each; thus*

$$P_e(\mathbb{Z}^n, \sigma^2) = 1 - \left[1 - 2Q\left(\frac{1/2}{\sigma} \right) \right]^n, \qquad (3.33)$$

which increases with n. Figure 3.8 shows this error probability as a function of the argument of the Q-function for several values of n.

Although it is hard to calculate $P_e(\Lambda, \sigma^2)$ *exactly* for general high-dimensional lattices, there are several ways to estimate it. One way is by the probability of *hyperplane crossing*, i.e., the *pairwise error probability* with respect to another lattice point. In particular, $P_e(\Lambda, \sigma^2)$ is larger than the probability that $\mathbf{Z}$ is closer to one of the shortest lattice vectors than to the origin:

$$P_e(\Lambda, \sigma^2) \geq 2Q\left(\frac{d_{\min}(\Lambda)/2}{\sigma} \right), \qquad (3.34)$$

where $d_{\min}(\Lambda)$ is the minimum distance (3.8), and the factor of two is because, as for the $\mathbb{Z}$-lattice, the shortest vectors come in symmetric pairs and their pairwise error events are disjoint. On the other hand, by the *union bound*, $P_e(\Lambda, \sigma^2)$ is *smaller*

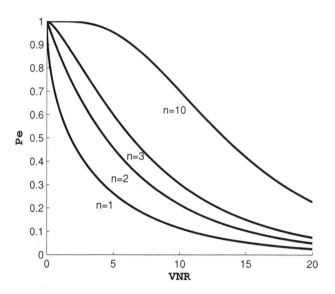

Figure 3.8 Error probability of a cubic lattice (3.33) as a function of the VNR $= 1/\sigma^2$.

than the sum of pairwise errors, where the sum is taken over all face-determining points of the fundamental Voronoi cell or, for a looser yet simpler bound, over all non-zero lattice points. [5]

Another way to estimate $P_e(\Lambda, \sigma^2)$ is by *sphere crossing*, i.e., by the probability $P_e(\mathcal{B}_r, \sigma^2)$ that the noise $\mathbf{Z}$ falls outside a ball, centered at the origin, with a radius r comparable with the lattice fundamental Voronoi cell: $P_e(\Lambda, \sigma^2)$ is smaller than $P_e(\mathcal{B}_{r_\Lambda^{\mathrm{pack}}}, \sigma^2)$, and larger than $P_e(\mathcal{B}_{r_\Lambda^{\mathrm{cov}}}, \sigma^2)$. As we shall see, it is even larger than $P_e(\mathcal{B}_{r_\Lambda^{\mathrm{eff}}}, \sigma^2)$. See more on this in Chapters 7 and 13.

By scaling the lattice Λ (or the noise variance σ^2), the error probability $P_e(\alpha\Lambda, \sigma^2)$ can take any value between zero and 1. For some target error probability $0 < \epsilon < 1$, let

$$\sigma^2(\epsilon) = \text{value of } \sigma^2 \text{ such that } P_e(\Lambda, \sigma^2) \text{ is equal to } \epsilon. \qquad (3.36)$$

We use this quantity to refine the definition of the VNR (3.30), so that it will take into account the target error probability.

---

[5] That is,

$$P_e(\Lambda, \sigma^2) \leq \sum_{d=d_{\min}}^{\infty} N_\Lambda(d)\, Q\left(\frac{d/2}{\sigma}\right) \leq \sum_{d=d_{\min}}^{\infty} N_\Lambda(d)\, e^{-d^2/8\sigma^2}, \qquad (3.35)$$

where $N_\Lambda(d)$ is the *distance spectrum*, i.e., the number of lattice points at distance d from the origin, and where the second upper bound follows from the well-known inequality $Q(x) \leq \frac{1}{2}e^{-x^2/2}$ (which is exponentially tight for large x). If $d_{\min}^2/\sigma^2 \gg 1$, then the first term dominates; so in light of (3.34), the error probability can be approximated as $e^{-d_{\min}^2/8\sigma^2}$, i.e., it is roughly determined by the minimum distance. See Section 13.5.

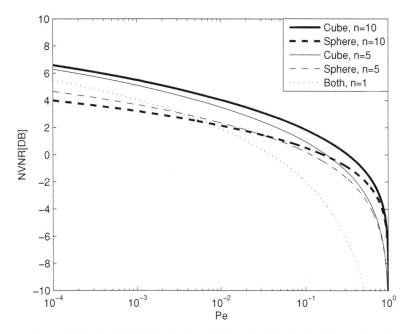

Figure 3.9 The cubic lattice NVNR in dB (i.e., $10 \log_{10}[\mu(\mathbb{Z}^n, P_e)/2\pi e]$) as a function of P_e, for dimensions $n = 1, 5$ and 10, compared to the sphere NVNR $\mu_n^*(P_e)$.

Definition 3.3.3 (Normalized volume to noise ratio) *The normalized volume to noise ratio (NVNR) of a lattice Λ, at a target error probability $0 < P_e < 1$, is defined as*

$$\mu(\Lambda, P_e) = \mu(\Lambda, \sigma^2(P_e)) = \frac{V^{2/n}(\Lambda)}{\sigma^2(P_e)}. \tag{3.37}$$

Like the NSM, the NVNR is a dimensionless number which is invariant to scaling or rotation of the lattice, i.e., $\mu(\alpha\Lambda, P_e) = \mu(\Lambda, P_e)$ for all $\alpha > 0$, and $\mu(T\Lambda, P_e) = \mu(\Lambda, P_e)$ for an orthonormal matrix T. As we can see in Figure 3.9, the NVNR decreases monotonically with P_e, and it goes from infinity to zero as P_e goes from zero to 1.

For a given target P_e, we wish to find the densest lattice, i.e., the lattice with the lowest NVNR. This would imply the largest coding rate per unit volume, as we shall see in Chapter 6. Thus, the NVNR can measure the possible performance advantage over the "uncoded" cubic lattice constellation.

Definition 3.3.4 (Coding gain) *The coding gain of a lattice Λ relative to the cubic lattice $\mathbb{Z}^n$, at some error probability P_e in the presence of AWGN, is defined as*

$$\Gamma_c(\Lambda, P_e) = \frac{\mu(\mathbb{Z}^n, P_e)}{\mu(\Lambda, P_e)}. \tag{3.38}$$

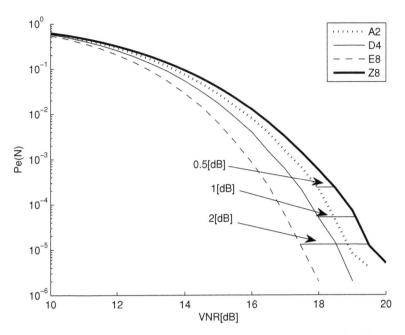

Figure 3.10 The coding gain of some common lattices: cubic ($\mathbb{Z}$), hexagonal (A_2), D_4 and E_8, at several target error probabilities. For a fair comparison, the y-axis shows the equivalent error probability at dimension 8 (i.e., we compare $\mathbb{Z}^8$, A_2^4, D_4^2 and E_8; see (3.41)).

As illustrated in Figure 3.10, the coding gain amounts to the extra noise power – compared to a cubic constellation – that the lattice Λ can tolerate, for the same point density (2.20) and error probability (3.32). Or, it can be interpreted as how much denser Λ can be for the same AWGN power σ^2 and error probability.

In the limit of small noise, the minimum distance dominates the error probability P_e; see footnote 5. Recalling the definition (3.13) of the Hermite parameter, $\hbar(\Lambda) = d_{\min}^2(\Lambda)/V^{2/n}(\Lambda)$, we thus obtain that $\mu(\Lambda, P_e) \sim -8\ln(P_e)/\hbar(\Lambda)$ (i.e., the ratio is going to one for a small error probability), so

$$\Gamma_c(\Lambda, P_e) \sim \hbar(\Lambda) \quad \text{as } P_e \to 0, \tag{3.39}$$

which is known as the *nominal coding gain* of the lattice [81]; see Problem 3.5.

Fair comparison of lattices at different dimensions When comparing the performance of lattices at different dimensions, we face the problem that the significance of the error probability in Definition 3.3.2 changes with the lattice dimension. To understand why, suppose that a k-dimensional lattice Λ_k is used m times over the n-dimensional channel (3.28), where $n = mk$. This is equivalent to a product lattice (Definition 2.1.3)

$$\Lambda_n = \underbrace{\Lambda_k \times \cdots \times \Lambda_k}_{m \text{ times}} \tag{3.40}$$

with m factors. Since correct decoding of Λ_n amounts to m successful decodings of Λ_k, the total error probability over the channel is

$$P_e(n) = 1 - (1 - P_e(k))^m \tag{3.41}$$

where $P_e(k) = P_e(\Lambda_k, \sigma^2)$ denotes the error probability of Λ_k. (The error probability of the $\mathbb{Z}^n$ lattice in Example 3.3.1 is the special case $k = 1$, $m = n$.) Clearly, $P_e(n)$ increases with m, and goes to 1 as m goes to infinity. An immediate implication is that, unlike the NSM, the NVNR is *not* invariant under a Cartesian product of the same lattice, and diverges to infinity.

For a fair comparison of lattices of different dimensions, we thus need to take a common "equivalent dimension." One way to do that for Λ_1 and Λ_2 of dimensions k_1 and k_2, respectively, is to set a target n-block error probability $P_e(n)$, at some common block length $n = m_1 k_1 = m_2 k_2$. (This is done in Figure 3.10 for $n = 8$ at $k = 1, 2, 4$ and 8.) Alternatively, we can base the comparison on the equivalent error probability per dimension, called the *symbol error rate* (SER):

$$\text{SER}(\Lambda, \sigma^2) = 1 - \sqrt[n]{1 - P_e(\Lambda, \sigma^2)}, \tag{3.42}$$

where n is the dimension of Λ. Note that if the SER of an n-dimensional lattice Λ is equal to that of a scalar constellation $\mathbb{Z}$, then $P_e(\Lambda, \sigma^2) = P_e(\mathbb{Z}^n, \sigma^2)$; i.e., the SER amounts to the error probability of a scalar constellation having the same *block* error probability.

Increasing the dimension The minimum possible value of $\mu(\Lambda, P_e)$ over all lattices in $\mathbb{R}^n$ is denoted by $\mu_n(P_e)$. What can be said about the behavior of $\mu_n(P_e)$ with n? It is clearly better (lower) than the NVNR of the cubic lattice $\mu(\mathbb{Z}^n, P_e)$, but as we saw this is a poor upper bound for a large n. From below, $\mu_n(P_e)$ is lower bounded by the NVNR $\mu_n^*(P_e)$ of an n-dimensional ball:

$$\mu_n(P_e) \geq \mu_n^*(P_e) > 2\pi e \tag{3.43}$$

for all n, where the second lower bound holds for P_e smaller than some threshold $P_e^{\text{th}} \approx 0.03$. The first inequality follows since an n-dimensional ball contains more probability mass of an AWGN vector than any other body of the same volume. As for the second inequality, we shall see in Chapter 7 that the ball NVNR is monotonically decreasing (i.e., improving) with n for $0 < P_e < P_e^{\text{th}}$, and it approaches $2\pi e$, as $n \to \infty$, for all $0 < P_e < 1$. The latter fact hints that increasing the dimension may be advantageous also for the lattice modulation problem. Indeed, there exists a sequence of "good modulation" lattices Λ_n of increasing dimension n, which approaches the sphere lower bound on the NVNR:

$$\lim_{n \to \infty} \mu(\Lambda_n, P_e) = 2\pi e, \quad \text{for all } 1 > P_e > 0. \tag{3.44}$$

Non-AWGN The noise variance, hence the VNR (3.30), is not a meaningful measure for lattice sparsity when the noise is *not* white Gaussian. Also the nearest-neighbor rule (3.31) is not optimal in the non-AWGN case. In Section 6.4 we extend the definitions of VNR, optimal decoding, error probability and NVNR for general noise distributions.

Summary of Chapter 3

Table 3.1 A table summarizing the various figures of merit ($V = V(\Lambda)$)

Goodness aspect	Representative parameter	Normalization by r_{eff}	Volume ratio	Normalization by $V^{2/n}$
Packing	$r_{\text{pack}} = \frac{d_{\min}}{2}$	$\rho_{\text{pack}} = \frac{r_{\text{pack}}}{r_{\text{eff}}}$	$\Delta = \rho_{\text{pack}}^n$	$\hbar = \frac{d_{\min}^2}{V^{2/n}}$
Covering	r_{cov}	$\rho_{\text{cov}} = \frac{r_{\text{cov}}}{r_{\text{eff}}}$	$\Theta = \rho_{\text{cov}}^n$	$-$
Quantization	$\sigma^2(\Lambda)$	$-$	$-$	$G(\Lambda) = \frac{\sigma^2(\Lambda)}{V^{2/n}}$
Modulation	$P_e(\Lambda, \sigma^2)$	$-$	$-$	$\mu(\Lambda, P_e) = \frac{V^{2/n}}{\sigma^2(P_e)}$

Problems

P.3.1 (Volume of a product of lattices) Find an expression for the effective radius of a product of two lattices, and for a product of m times the same lattice.

P.3.2 (Minkowski convex-body theorem) Prove Minkowski's convex-body theorem for a general convex zero-symmetric body (footnote 3). Hint: use the fact that a convex zero-symmetric body is a "ball" with respect to some general norm.

P.3.3 (Sphere packing in a product lattice) Prove that the packing radius of a Cartesian product of lattices $\Lambda_1 \times \Lambda_2$ is the minimal one. Find the packing efficiency of a Cartesian product of the same lattice $\Lambda \times \cdots \times \Lambda$, as a function of the number of factors m.

P.3.4 (Hermite parameter and coding gain) Prove that

$$P_e(\Lambda, \sigma^2) \geq 2Q\left(\frac{\sqrt{\hbar(\Lambda) \cdot \mu(\Lambda, \sigma^2)}}{2} \right),$$

where $\hbar(\Lambda)$ is the Hermite parameter (3.13), and $\mu(\Lambda, \sigma^2)$ is the VNR (3.30). Hint: use (3.34).

P.3.5 (Nominal coding gain) Use the bound $Q(x) \leq 1/2 \cdot e^{-x^2/2}$, and the asymptotic approximation $Q(x) \sim 1/x \cdot e^{-x^2/2}$ for large x, to prove the asymptotic approximation $\sigma^2(P_e) \sim d_{\min}^2/8 \ln(1/P_e)$ for a small P_e, which implies that the NVNR is given asymptotically by

$$\mu(\Lambda, P_e) \sim \frac{8\ln(1/P_e)}{\hbar(\Lambda)} \quad \text{as } P_e \to 0,$$

proving the asymptotic equivalence to the nominal coding gain in (3.39).

Historical notes

The packing and covering properties of n-dimensional lattices, as well as non-lattice arrangements of points in $\mathbb{R}^n$, have been explored extensively in the mathematical literature. The book of Conway and Sloane [49] is probably the best source for what is known about this topic, and how it relates to the quantization and coding problems.

Zador [282, 283] defined the coefficient of quantization efficiency under the MSE criterion (and more general rth power norm-based measures) for general (unstructured) vector quantizers in the limit of high resolution. Gersho [102] applied Zador's ideas to tessellations and lattices, and defined the normalized second moment (NSM) of the corresponding polytope. (Gersho acknowledged Sloane, his colleague at Bell Laboratories, for providing the information about lattices.) Conway and Sloane [45] examined the second moments of some well-known lattices. Their work served as a bridge between the source coding and the mathematically oriented literature about lattices. Surprisingly, it is still unknown (at least not for certain) whether the optimal quantizer in each dimension has the form of a lattice, and even not which is the optimal lattice quantizer in dimensions larger than three [248]. Tóth [49, 255] proved that the optimal two-dimensional quantizer (for a uniform source under MSE) is the hexagonal lattice quantizer. Barnes and Sloane [14] showed that the BCC is the optimal three-dimensional lattice quantizer. From a different angle, Marcellin and Fischer [178] introduced trellis-coded quantization, whose mean-squared error efficiency was further studied by Calderbank *et al.* [32].

The channel coding goodness of a lattice (as an infinite constellation in the presence of AWGN) is a more recent and evolving notion in the communication literature. The traditional approach uses the minimum distance d_{min} (for a fixed lattice cell volume) and the number of nearest neighbors $N_\Lambda(d_{min})$ to estimate the error probability, hence to assess the goodness of a lattice constellation in the presence of AWGN. Calderbank and Sloane [34], Conway and Sloane [49, chapter 3] and Forney and Wei [94] defined the coding gain of a constellation as the transmit power reduction (for the same minimum distance and coding rate) compared to the integer lattice. See the tutorial paper by Forney and Ungerboeck [93] for an extension to binary and trellis-coded signaling. However, this approach is not very accurate for rates approaching capacity; later work uses the minimum volume to noise ratio (VNR) required for a small decoding error probability; see the papers by

Poltyrev [221] and Forney *et al.* [92] (the former calls it "generalized SNR," while the latter normalize it by a factor of $2\pi e$). In this book we measure the channel coding goodness of a lattice Λ by the normalized VNR $\mu(\Lambda, P_e)$, i.e., the VNR required to meet an error probability P_e. This quantity has two advantages: (i) it is an accurate, though hard to compute, goodness measure, which is valid at all noise levels; and (ii) it is dual to the NSM (the quantization-goodness measure), and therefore it symmetrizes the characterization of schemes combining source and channel coding.

As for the quantization problem, except for the two-dimensional case (where the hexagonal lattice is optimal), it is unknown whether the best infinite constellation for an AWGN channel has the form of a lattice. The FCC lattice, for example, is known to be the best three-dimensional packing (the recently proved Kepler conjecture [49]). But this only implies asymptotic channel coding optimality in the limit of high SNR.

4 Dithering and estimation

In quantization theory, as well as in some non-linear processing systems, the term "dithering" corresponds to intentional randomization aimed at improving the perceptual effect of the quantization. Although it seems somewhat strange that adding noise can improve performance, dithering is used in practice to produce more natural sounding digital audio, while visual forms of dither reduce "blockiness" in picture coding and enable thousands of color shades on a limited 256-color display. [1]

In the context of lattice *quantization* or *shaping*, dither is an effective means of guaranteeing desired distortion or power levels, independent of the input statistics. Its role is particularly important for quantization at *low* resolution (or modulation at low SNR), where – without the dither – the quantization error is highly dependent on the quantized source (or the modulated signal is highly discrete).

In Section 3.2 we considered quantizing an "analog" source vector $\mathbf{s} = (s_1, \ldots, s_n)$ using an n-dimensional lattice Λ with some lattice partition $\mathcal{P}$. Recall that a lattice quantizer $Q_\Lambda = \{\Lambda, \mathcal{P}\}$ represents a source vector $\mathbf{s}$ that falls into the cell $\mathcal{P}_\lambda$ by the lattice point λ (3.19):

$$Q_\Lambda(\mathbf{s}) = \lambda \ \text{ if } \mathbf{s} \in \mathcal{P}_\lambda. \tag{4.1}$$

In the special case of *Voronoi partition*, λ is the *closest* lattice point to $\mathbf{s}$ (2.10).

Example 4.0.1 ($n = 1$) *The code points of the scalar lattice quantizer are located at $\{0, \pm\Delta, \pm2\Delta, \ldots\}$. Under "parallelepiped partition" (2.14) the associated cells are $\ldots, [-\Delta, 0), [0, \Delta), [\Delta, 2\Delta), \ldots$, so quantization of a point s is given by $\lfloor \frac{s}{\Delta} \rfloor \Delta$, where $\lfloor \cdot \rfloor$ denotes rounding down to the nearest integer. Under Voronoi partition, the index of the lattice point λ_i representing s is given by*

$$\arg\min_{i \in \mathbb{Z}} |s - i\Delta|,$$

---

[1] A Web article posted by an audio engineer (Google search: dither) tells that the concept of dither dates back to the 1940s, when British naval airmen discovered that the cogs and gears in the mechanical navigation systems of their airplanes would chatter and stick on the ground. Once the planes were airborne, vibration of the engines of the airplanes had a lubricating effect on these mechanical navigation systems, smoothing out their operation. This discovery led the British Navy to install small motors in the mechanical navigation systems to vibrate the cogs and gears intentionally, thus improving their performance on the ground. The "Einstein Prob-B" project in Stanford (see http://einstein.stanford.edu/highlights/hl_050605.html) uses dither to smooth out telescope vibrations.

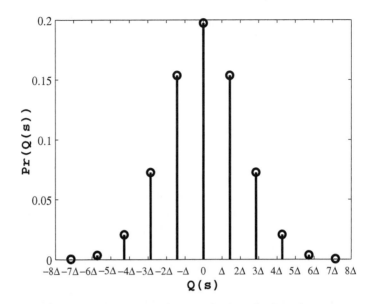

Figure 4.1 Level probabilities in scalar lattice quantization of a Gaussian source.

hence the cell edges are located at $\{\pm\Delta/2, \pm3\Delta/2, \ldots\}$. *The resulting quantizer,* $Q_\Delta(s)$, *is known as a step-Δ uniform "mid-thread" quantizer.* [2] *Figure* 4.1 *illustrates the probability distribution of* $Q_\Delta(S)$, *for a zero-mean Gaussian source S.*

The *quantization error* is defined as the difference between the input and the output [3]

$$\text{quantization error} = Q_\Lambda(\mathbf{s}) - \mathbf{s} = \lambda - \mathbf{s} = -(\mathbf{s} \bmod \Lambda), \qquad (4.2)$$

where the modulo-lattice operation was defined in (2.21), and where mod Λ is understood as mod $\mathcal{P}_0$ (see Definition 2.3.1). Clearly the quantization error depends on the source vector $\mathbf{s}$, in fact, it is a deterministic function of it. Can we make this error "independent" of the source?

Section 4.1 gives an affirmative answer, provided that we use a subtractive dither which is uniform over the quantizer cell. Section 4.2 shows that other (simpler) forms of dither result in the same answer. Section 4.3 deals with the second-order statistics of the dither; it shows that in many cases of interest, the dither can be regarded as a white noise. Sections 4.4 and 4.5 then show how to combine dithering with linear estimation in order to reduce the total mean-squared error.

---

[2] The term mid-thread follows since the input/output relation of $Q_1(s)$ takes the form of a step function, which is horizontal at the origin; see Figure 3.6. If we shift the level and threshold points by $\Delta/2$, then the quantization function becomes vertical at the origin, and the quantizer is known as *mid-rise*.

[3] The sign is opposite to that of the error $\mathbf{x}_e$ in (2.12), because of the convention in the quantization literature.

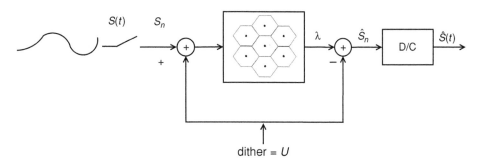

Figure 4.2 Dithered quantization.

4.1 Crypto lemma

Suppose a company sitting around a dinner table wishes to compute the sum of the ages of the participants, but without revealing the individual age of any one of them. One trick to do that (inspired by the "key" in crypto systems) is that the first participant would take a random number U and pass it on to his right neighbor, who will add to it her (true) own age, pass it on to her right neighbor, and so on. When the final number reaches the first participant he will add his own age, and then subtract U to get the age sum. If U is large enough, then we can argue that the information on the individual ages was kept secret. Or in statistical terms: without having U at hand, the number each participant gets is (almost) statistically independent of the individual ages of the preceding participants.

The idea of dithered quantization is similar; it is based on *common randomness* shared by the encoder and the decoder. Such a randomness can be obtained from the output of a *pseudo-random* number generator, whose seed is agreed in advance by the encoder and the decoder.

Definition 4.1.1 (Randomized quantization) *We say that a random vector* $\mathbf{U}$ *is "subtractive dither" if it is known at both the encoder and decoder ends, and the final reconstruction is given by*

$$\hat{\mathbf{S}} = Q_\Lambda(\mathbf{s} + \mathbf{U}) - \mathbf{U}. \tag{4.3}$$

As illustrated in Figure 4.2, the encoder adds the dither to the source prior to quantization, while the decoder subtracts the dither from the associated lattice point. The *dithered quantization error* now becomes

$$\hat{\mathbf{S}} - \mathbf{s} = Q_\Lambda(\mathbf{s} + \mathbf{U}) - \mathbf{U} - \mathbf{s}, \tag{4.4}$$

or equivalently $-((\mathbf{s} + \mathbf{U}) \bmod \Lambda)$ as in (4.2).[4]

---

[4] We shall not discuss *non*-subtractive dithered quantization, where the dither is known only at the encoder, and the reconstruction is given by $\hat{\mathbf{s}} = Q_\Lambda(\mathbf{s} + \mathbf{u})$; see [112].

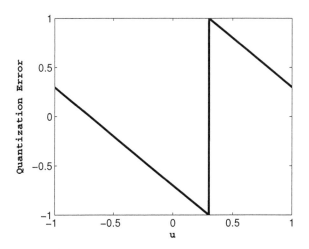

Figure 4.3 The saw-tooth function in Example 4.1.1.

Example 4.1.1 *Assume quantizing a source sample s, for some $0 \leq s < 1$, using a scalar lattice quantizer with step size $\Delta = 2$ and Voronoi partition, as in Example 4.0.1. As shown in Figure 4.3, the dithered quantization error as a function of u, for u inside the fundamental cell $-1 \leq u < 1$, has a "saw-tooth" shape (up to a minus sign):*

$$(s + u) \bmod \Delta = \begin{cases} s + u, & \text{for } -1 \leq u < 1 - s \\ s + u - 2, & \text{for } 1 - s \leq u < 1, \end{cases}$$

which spans the range $-1 \leq$ error < 1. This function has a unit slope over the entire range, except for a discontinuity at $u = 1 - s$. Thus, if the dither U is a random variable uniform over $[-1, 1)$, then so is the resulting quantization error.

This behavior is, in fact, true for dithered lattice quantization in general. As illustrated in Figure 4.4, a specific dither value amounts to a shift of the lattice partition, hence to a shift of the quantization error function. A random uniform dither makes the error uniform.

Lemma 4.1.1 (Crypto lemma) *If the dither U is uniform over the fundamental cell $\mathcal{P}_0$, i.e., with a p.d.f.*

$$f_U(\mathbf{u}) = \begin{cases} \frac{1}{V(\Lambda)}, & \mathbf{u} \in \mathcal{P}_0 \\ 0, & \mathbf{u} \notin \mathcal{P}_0, \end{cases} \tag{4.5}$$

then $(\mathbf{s} + \mathbf{U}) \bmod \Lambda$ (with mod Λ understood as mod $\mathcal{P}_0$) is uniform over $\mathcal{P}_0$, independent of $\mathbf{s}$.

Proof See the proof for a generalized dither (Lemmas 4.2.1 and 4.2.3) in the next section. □

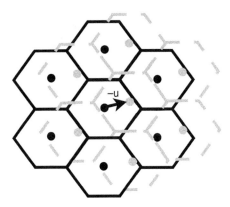

Figure 4.4 Addition and subtraction of a vector **u** before and after lattice quantization amounts to shifting the lattice quantizer $Q_\Lambda(\cdot)$ by the vector $-\mathbf{u}$.

This property is well known for finite groups (or "modulo-additive channels"): the sum of any element with a random variable uniform over the group is uniform as well. Lemma 4.1.1 ensures that the same property holds when the group is a fundamental cell of a general lattice. Informally speaking, the modulo-lattice operation is a *periodic function* in space, with the period being any fundamental lattice cell $\mathcal{P}_0$; thus, a shift of the argument **s** by a vector **U** which is uniformly distributed over that period, makes the function (i.e., the error) statistically independent of **s**, and uniform (as can be seen by setting $\mathbf{s} = 0$).

As a corollary from Lemma 4.1.1, we have the following.

Theorem 4.1.1 (Equivalent channel for dithered quantization) *For a random source* **S** *and a uniform dither* (4.5)*, the quantization error* (4.4) *is statistically independent of* **S** *and distributed as minus the dither:*

$$[Q_\Lambda(\mathbf{S} + \mathbf{U}) - \mathbf{U} - \mathbf{S}] \sim \text{Unif}(-\mathcal{P}_0). \tag{4.6}$$

A few more simple corollaries follow from Theorem 4.1.1.

1. The reconstruction $\hat{\mathbf{S}}$ (4.3) is the sum of **S** and the dithered quantization error; thus, it is equivalent to the *independent sum*

$$\hat{\mathbf{S}} \stackrel{\text{dist}}{=} \mathbf{S} - \mathbf{U} \stackrel{\text{dist}}{=} \mathbf{S} + \mathbf{U}_{\text{eq}}, \tag{4.7}$$

where $\stackrel{\text{dist}}{=}$ denotes equality in distribution, and where the *equivalent dither* is distributed as

$$\mathbf{U}_{\text{eq}} \sim \text{Unif}(-\mathcal{P}_0), \tag{4.8}$$

i.e., as $-\mathbf{U}$, as shown in Figure 4.5. If the lattice quantizer $Q_\Lambda(\cdot)$ is an NN quantizer (2.10), then the fundamental (Voronoi) cell $\mathcal{V}_0$ is symmetric about the origin, hence $\mathbf{U}_{\text{eq}} \sim \text{Unif}(\mathcal{V}_0)$.

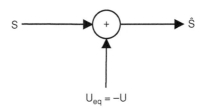

Figure 4.5 Equivalent additive-noise channel of the dithered lattice quantizer (4.3).

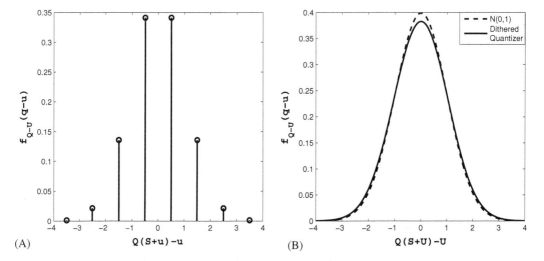

Figure 4.6 Distribution of the dithered quantizer output (4.3), for a Gaussian source $S \sim N(0, 1)$, quantized by a uniform scalar quantizer with step size $\Delta = 1$: (A) a specific dither value $U = 1/2$; (B) a random dither $U \sim \text{Unif}[0, 1)$.

2. While, for a specific dither value, the reconstruction $\hat{\mathbf{S}} = Q_\Lambda(\mathbf{S} + \mathbf{u}) - \mathbf{u}$ has a discrete distribution (Figure 4.6(A)), for a random uniform dither it is continuous (Figure 4.6(B)).

3. By Theorem 4.1.1, the MSE of the dithered quantizer

$$\sigma_Q^2 = \frac{1}{n} E \|\hat{\mathbf{S}} - \mathbf{S}\|^2 \tag{4.9}$$

is equal to the second moment of the dither (or the second moment of the fundamental cell $\mathcal{P}_0$), independent of the source $\mathbf{S}$. Under Voronoi partition this becomes the lattice second moment (3.20),

$$\frac{1}{n} E \|\mathbf{U}\|^2 = \sigma^2(\Lambda). \tag{4.10}$$

For example, for the step-Δ uniform scalar quantizer of Example 4.0.1,

$$\text{MSE} = E\{[(s + U) \bmod \Delta]^2\} = E\{U^2\} = \frac{\Delta^2}{12}. \tag{4.11}$$

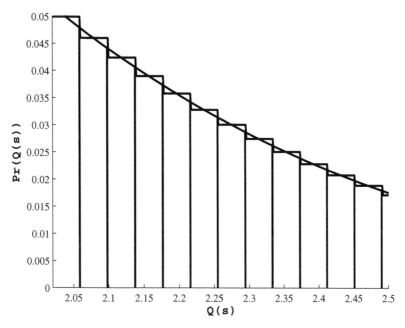

Figure 4.7 High-resolution approximation. The additive-noise model becomes a good approximation when the quantizer cells are small compared to the variations in the source p.d.f.

4.1.1 High-resolution quantization theory

High-resolution quantization (HRQ) theory assumes that the quantizer cells are small compared to the variations in the source p.d.f. This implies that the conditional distribution of the source vector **S**, given that **S** falls in some quantizer cell $\mathcal{P}_i$, is roughly *uniform* in that cell:

$$f_{\mathbf{S}|\mathbf{S} \in \mathcal{P}_i}(\mathbf{s}) \approx \mathrm{Unif}(\mathcal{P}_i), \tag{4.12}$$

as illustrated in Figure 4.7. Under this assumption, it is common to model the quantization process as adding an independent uniform noise to the source. Although, strictly speaking, this model is untrue for deterministic quantization, it is a useful tool for the analysis of systems combining quantization with (linear) signal processing.

For subtractive dithered quantization, Theorem 4.1.1 implies that the equivalent additive-noise model of Figure 4.5 is accurate at *any* resolution. The role of the dither, however, becomes less prominent as the quantizer resolution increases. For example, for any continuous source S with a p.d.f. (see [17, 165, 205]),

$$\frac{E[S \bmod \Delta]^2}{\Delta^2/12} \to 1, \quad \text{as } \Delta \to 0. \tag{4.13}$$

Thus, even without the dither in (4.11), the second moment of the quantization error approaches that of a uniform distribution.

The question of how far a given lattice-source pair is from the high-resolution quantization assumption (4.12) is discussed in Section 4.2.3 below.

Discretization of the dither

In practice the dither is a pseudo-random noise, which can only take a *finite* number of values. The continuous uniform dither $\mathbf{U}$ must therefore be replaced by some discrete approximation $\mathbf{U}_d$. How fine should the dither values be, so that the crypto lemma still – at least approximately – holds? The answer depends on the "smoothness" of the source p.d.f. Let $\mathbf{U}_d$ denote the quantization of a continuous dither $\mathbf{U} \sim \text{Unif}(\mathcal{P}_0(\Lambda))$ by some "fine" lattice Λ_f. If Λ_f satisfies the high-resolution quantization assumption (4.12) with respect to the source $\mathbf{S}$, then $\mathbf{S} + \mathbf{U}_f$ and $\mathbf{S}$ are approximately equal in distribution, where $\mathbf{U}_f = \text{Unif}(\mathcal{P}_0(\Lambda_f))$. Thus, $\mathbf{S} + \mathbf{U} = \mathbf{S} + \mathbf{U}_d + \mathbf{U}_f$ can be well approximated by $\mathbf{S} + \mathbf{U}_d$; i.e., $[\mathbf{S} + \mathbf{U}_d] \mod \Lambda$ is approximately uniform over $\mathcal{P}_0(\Lambda)$.

4.2   Generalized dither

The crypto lemma – and the consequent uniformity of the dithered quantization error (Theorem 4.1.1) – hold under more general subtractive dithering schemes. For example, since the lattice is periodic in space, shifting it by $\mathbf{u} + \lambda$ for some $\lambda \in \Lambda$ is equivalent to shifting it by $\mathbf{u}$. Thus, a dither which is uniform over any cell $\mathcal{P}_\lambda = \lambda + \mathcal{P}_0$, or even a *union* of cells $\mathcal{P}_\lambda \cup \mathcal{P}_{\lambda'}$, is equivalent to a dither which is uniform over $\mathcal{P}_0$.[5]

Definition 4.2.1 (Generalized dither) *We say that $\mathbf{U}$ is generalized dither (for a lattice Λ with a fundamental cell $\mathcal{P}_0$), if $(\mathbf{s} + \mathbf{U}) \mod \Lambda$ is uniform over $\mathcal{P}_0$, for all $\mathbf{s} \in \mathbb{R}^n$.*

We shall find three equivalent conditions for $\mathbf{U}$ to be a generalized dither. The first condition, *modulo uniformity*, implies that it is enough to verify uniformity of the dithered quantization error for the case $\mathbf{s} = 0$.

Definition 4.2.2 (Modulo-uniform dither) *We say that $\mathbf{U}$ is modulo-uniform (with respect to a lattice Λ with a fundamental cell $\mathcal{P}_0$) if $(\mathbf{U} \mod \Lambda)$ is uniform over $\mathcal{P}_0$.*

---

[5] This is in contrast to non-subtractive dither (footnote 4), which is sensitive to translation of the support of the dither.

Lemma 4.2.1 (Generalized crypto lemma I) $\mathbf{U}$ *is a generalized dither if and only if it is modulo-uniform.*

This condition holds, in particular, if $\mathbf{U}$ itself is uniform over $\mathcal{P}_0$. Hence the crypto lemma (Lemma 4.1.1) is a special case of Lemma 4.2.1. See the proof below.

4.2.1   Periodic replication

The proof of Lemma 4.2.1 is based on a relation between modulo-lattice reduction of a set and *periodic replication* of the set by the lattice or, more generally, a relation between modulo reduction of a random vector $\mathbf{X}$ and periodic replication of its density $f_{\mathbf{X}}$.

Folding and periodic replication

How do we know that a shape with area less than one can always be shifted, so that it will not intersect the integer lattice? Imagine folding the plane into the fundamental square. Since the shape area is less than one (and the folded shape can only be smaller), it must leave an uncovered hole. Now, shift the shape so that this hole falls on the corners of the fundamental square, and unfold the plane. Observation: instead of folding the plane, replicate the shape by the square lattice, and get the same projection on the fundamental square.

Periodic replication of a function $g : \mathbb{R}^n \to \mathbb{R}$ by a lattice Λ is defined as

$$g_{\mathrm{rep}\ \Lambda}(\mathbf{x}) = \sum_{\lambda \in \Lambda} g(\mathbf{x} - \lambda); \tag{4.14}$$

i.e., it is the sum of the function over the *coset* of $\mathbf{x}$. Clearly, $g_{\mathrm{rep}\ \Lambda}(\mathbf{x})$ is "Λ-periodic," i.e.,

$$g_{\mathrm{rep}\ \Lambda}(\mathbf{x} + \lambda) = g_{\mathrm{rep}\ \Lambda}(\mathbf{x}), \tag{4.15}$$

for all $\lambda \in \Lambda$.

Lemma 4.2.2 (Folded distribution) *Let $\mathbf{X}$ be a random vector with a density $f_{\mathbf{X}}$. The density of $\mathbf{X}$, after modulo-Λ reduction to a fundamental cell $\mathcal{P}_0$, is given by*

$$f_{\mathbf{X} \bmod \Lambda}(\mathbf{x}) = \begin{cases} f_{\mathbf{X}\ \mathrm{rep}\ \Lambda}(\mathbf{x}), & \mathbf{x} \in \mathcal{P}_0 \\ 0, & \mathbf{x} \notin \mathcal{P}_0, \end{cases} \tag{4.16}$$

where $f_{\mathbf{X}\ \mathrm{rep}\ \Lambda}$ is the periodic replication (4.14) of $f_{\mathbf{X}}$ by the lattice Λ.

Proof The proof follows since $(\mathbf{X} \bmod \Lambda) = \mathbf{y}$ amounts to $\mathbf{X}$ being in the coset of $\mathbf{y}$. See Appendix A.3, which follows Linder [162]. □

Note that given $\mathbf{X}$ and Λ, the choice of a fundamental cell $\mathcal{P}_0$ only affects the *support* of the distribution of $\mathbf{X}$ mod Λ in (4.16). In fact, we can view $f_{\mathbf{X} \text{ rep } \Lambda}$ as a *distribution of cosets* induced by $\mathbf{X}$; see Appendix A.3.

Lemma 4.2.2 implies an equivalent statement for the generalized crypto lemma (Lemma 4.2.1), in terms of the periodic replication of the dither p.d.f.

Lemma 4.2.3 (Generalized crypto lemma II) $\mathbf{U} \sim f_{\mathbf{U}}$ *is modulo-uniform iff the periodic replication of its density* (4.14) *is constant in space:*

$$f_{\mathbf{U} \text{ rep } \Lambda}(\mathbf{x}) = \sum_{\lambda \in \Lambda} f_{\mathbf{U}}(\mathbf{x} - \lambda) = \text{constant}, \qquad (4.17)$$

for all $\mathbf{x}$ *in* $\mathbb{R}^n$.

Proof of Lemmas 4.2.1 and 4.2.3 Lemma 4.2.2 implies that $(\mathbf{U} \text{ mod } \Lambda)$ is uniform over $\mathcal{P}_0$ if and only if condition (4.17) holds, i.e., the periodic replication of $f_{\mathbf{U}}$ is constant in space. Similarly, $(\mathbf{s} + \mathbf{U})$ mod Λ is uniform over $\mathcal{P}_0$ if and only if the periodic replication of $f_{\mathbf{s}+\mathbf{U}}$ is constant in space. But $f_{\mathbf{s}+\mathbf{U}}$ is just a shift of $f_{\mathbf{U}}$. Thus the two statements are *equivalent*, and the lemmas follow. □

The constant periodic-replication condition (4.17) is simple to check. The first two examples below are corollaries of this condition.

Example 4.2.1 (Cases of generalized dither)

- $\mathbf{U}$ *is uniform over* any fundamental cell *of* Λ *(not necessarily the one used in the modulo operation).*
- $\mathbf{U}$ *is uniform over a fundamental cell of a* sublattice $\Lambda_c \subset \Lambda$ *(Problem 4.2).*
- $\mathbf{U} = \tilde{\mathbf{U}} + \mathbf{X}$ *is the sum of uniform dither* $\tilde{\mathbf{U}}$ *with an* arbitrary *random vector* $\mathbf{X}$ *(equivalently,* $f_{\mathbf{U}} = f_{\tilde{\mathbf{U}}} * f_{\mathbf{X}}$, *where* $*$ *denotes convolution).*

The first case above can also be deduced from the *modulo equivalence* of fundamental cells (2.23c). A useful consequence is that the dither can be drawn uniformly over the *fundamental parallelepiped*. Such a dither is easy to generate; we simply draw a vector of i.i.d. components uniform on $[0, 1)$, and multiply by the generator matrix G. For a *construction A* lattice (described in Section 2.5), where $a\mathbb{Z}^n \subset \Lambda$, the second case above implies that the dither can be uniform over the cube $[0, a)^n$. Again, we can simply generate such a dither by drawing i.i.d. components uniform on $[0, a)$. The third case above amounts to "spreading" of the dither, and it follows from the crypto lemma.

4.2.2   Zeroes on dual lattice in frequency domain

The constant periodic-replication condition (4.17) has an interesting "frequency domain" interpretation, that gives rise to the notion of a *dual lattice*.

Definition 4.2.3 (Dual lattice) *Lattices Λ and Λ^* in $\mathbb{R}^n$ are dual (or reciprocal) if the inner products of their points are integers, i.e., $\langle \lambda, \lambda^* \rangle \in \mathbb{Z}$ for all $\lambda \in \Lambda$, $\lambda^* \in \Lambda^*$. This is equivalent to the condition that the rows of the inverse G^{-1} of the generator matrix of Λ form a basis for Λ^*, i.e.,*

$$\Lambda^* = \Lambda(G^{-t}) \tag{4.18}$$

where $(\cdot)^{-t}$ denotes inverse-transpose.

Since the determinant of the inverse is the inverse of the determinant, $V(\Lambda^*) = 1/V(\Lambda)$.

Example 4.2.2 (Duality and self-duality) *The dual of the rectangular lattice $\{(2i, j) : i, j \in \mathbb{Z}\}$ in Figure 2.1 is the lattice $\{(1/2i, j) : i, j \in \mathbb{Z}\}$. The dual of the hexagonal lattice in Figure 2.2 is another hexagonal lattice, scaled by $1/\sqrt{12}$, and rotated 90 degrees (see Problem 4.4). The Gosset lattice E_8 is dual to itself, which is a property shared by all unimodular lattices (Section 2.5).*

It is convenient to think of a function $g : \mathbb{R}^n \to \mathbb{R}^1$ as a multi-dimensional *signal* (e.g., a picture is a two-dimensional signal). If g is not periodic, then its *Fourier transform* $G = \mathcal{F}\{g\}$ is defined as

$$G(\mathbf{f}) = \int_{\mathbb{R}^n} g(\mathbf{x}) \cdot e^{-j2\pi \langle \mathbf{f}, \mathbf{x} \rangle} d\mathbf{x}, \quad \text{for } \mathbf{f} \in \mathbb{R}^n, \tag{4.19}$$

where $j = \sqrt{-1}$. While if g is Λ-*periodic* (4.15), then its *Fourier series* $G = \mathcal{F}\{g\}$ is defined as

$$G(\lambda^*) = \frac{1}{V(\Lambda)} \cdot \int_{\mathcal{P}_0} g(\mathbf{x}) \cdot e^{-j2\pi \langle \lambda^*, \mathbf{x} \rangle} d\mathbf{x}, \quad \text{for } \lambda^* \in \Lambda^*, \tag{4.20}$$

where $\mathcal{P}_0 = \mathcal{P}_0(\Lambda)$ can be *any* fundamental cell of Λ, and Λ^* is the dual of Λ. We can think of the argument $\mathbf{f}$ of the Fourier transform (4.19) as spatial *frequency*, and of the vectors λ^* in (4.20) as the spatial *harmonies* of the Λ-periodic function g. By a standard argument in Fourier analysis, the Fourier series of a Λ-periodic replication of a function g is given by *sampling* the Fourier transform of g at the points of the dual lattice Λ^*, and normalizing by $1/V(\Lambda)$ [11].

The Fourier transform $\phi_{\mathbf{X}} = \mathcal{F}\{f_{\mathbf{X}}\}$, of a probability density function $f_{\mathbf{X}}$, is known as a *characteristic function*. Since $f_{\mathbf{X}}$ is non-negative and integrates to 1, we have $|\phi_{\mathbf{X}}(\mathbf{f})| \leq 1$ with equality at $\mathbf{f} = 0$. Due to the relation between modulo reduction and periodic replication (Lemma 4.2.2), it follows from the sampling property above that the characteristic functions of the folded and unfolded distributions coincide on the dual lattice:

$$\phi_{\mathbf{X} \bmod \Lambda}(\mathbf{f} = \lambda^*) = \phi_{\mathbf{X}}(\mathbf{f} = \lambda^*), \quad \text{for } \lambda^* \in \Lambda^*, \tag{4.21}$$

where $\phi_X = \mathcal{F}\{f_X\}$, and $\phi_{X \bmod \Lambda} = \mathcal{F}\{f_{X \bmod \Lambda}\}$.[6] See Problem 4.5. This relation implies a frequency-domain characterization for a generalized dither.

Lemma 4.2.4 (Generalized crypto lemma III) $U \sim f_U$ *is modulo-uniform if and only if its characteristic function* $\phi_U = \mathcal{F}\{f_U\}$ *is zero for all dual lattice points* $\lambda^* \in \Lambda^*$ *except at zero:*

$$\phi_U(\lambda^*) = \begin{cases} 1, & \lambda^* = 0 \\ 0, & \lambda^* \neq 0. \end{cases} \tag{4.22}$$

Proof See Problem 4.6. □

Nyquist pulses

The communication engineer may recognize the conditions in Lemmas 4.2.3 and 4.2.4 as the *Nyquist conditions* for zero inter-symbol interference in pulse-amplitude modulation (PAM). A "Nyquist pulse," sampled at the symbol rate, is zero for all non-zero sampling points; equivalently, the periodic replication of its Fourier transform is constant at all frequencies.

4.2.3   Imperfect dither

In various problems of interest, we may want to assess approximations for the modulo-uniformity condition of Lemmas 4.2.1, 4.2.3 and 4.2.4. The *high-resolution quantization* assumption (4.12), for example, implies that a source S is approximately modulo-uniform with respect to a sufficiently fine lattice Λ; hence, S can approximate a generalized dither for Λ.

Definition 4.2.4 (Flatness factor) *Given a lattice* Λ *and a random vector* $X \sim f_X(x)$, *the flatness factor is defined as the normalized* ℓ_∞ *distance between the folded distribution – i.e., the density of* $X \bmod \Lambda$ *with the modulo taken with respect to some fundamental cell* $\mathcal{P}_0$ *– and a uniform distribution:*

$$F(\Lambda, X) = \frac{\max_{x \in \mathcal{P}_0} |f_{X \bmod \Lambda}(x) - 1/V(\Lambda)|}{1/V(\Lambda)}. \tag{4.23}$$

The flatness factor $F(\Lambda, X)$ is *zero* if and only if X is modulo-uniform. Note that by Lemma 4.2.2, we can take the maximum in (4.23) over *all* x in $\mathbb{R}^n$, provided that we replace $f_{X \bmod \Lambda}$ by the periodic replication $f_{X \text{ rep } \Lambda}$. It follows that $F(\Lambda, X)$ is, in fact, *invariant* of the choice of the fundamental cell $\mathcal{P}_0$.

See Problems 4.8, 4.9 and 4.10 for more relations (frequency-domain analysis, Gaussian case and high-resolution quantization).

---

[6] By the replication-Fourier relation discussed above, the Fourier series of $f_{X \text{ rep } \Lambda}$ is given by normalizing (4.21) by $1/V(\Lambda)$.

4.3   White dither spectrum

Since a uniform dither plays the role of the equivalent "quantization noise" (Figure 4.5), it is useful to have better knowledge of its *second-order statistics*. We shall see that the dither which is uniform over the Voronoi region of the best lattice quantizer is white.

Definition 4.3.1 (Dither auto-correlation) *The auto-correlation matrix R_Q of a lattice quantizer $Q = \{\Lambda, \mathcal{P}\}$ is defined as*

$$R_Q \triangleq E\{\mathbf{U}\,\mathbf{U}^t\} = \frac{1}{V(\Lambda)} \int_{\mathcal{P}_0} \mathbf{u}\,\mathbf{u}^t d\mathbf{u}, \qquad (4.24)$$

where $\mathbf{U}$ is uniform over the fundamental cell $\mathcal{P}_0$, and $(\cdot)^t$ denotes transpose, i.e., R_Q is the $n \times n$ auto-correlation matrix of the dither. The second moment of the dither is given by the normalized trace of the auto-correlation matrix

$$\sigma_Q^2 = \frac{1}{n} E\{\|\mathbf{U}\|^2\} = \frac{1}{n}\text{trace}\{R_Q\}, \qquad (4.25)$$

where the trace of a square matrix is defined as the sum of its diagonal elements.

For example, under Voronoi partition, σ_Q^2 is the lattice second moment $\sigma^2(\Lambda)$; while under a centralized parallelepiped partition (where the center of $\mathcal{P}_0$ is at the origin), the dither $\mathbf{U}$ can be written as $\mathbf{U} = G \cdot \tilde{\mathbf{U}}$, where $\tilde{U}_1, \dots, \tilde{U}_n$ are i.i.d. samples uniform over $(-1/2, +1/2]$, so

$$R_Q = G \cdot E\{\tilde{\mathbf{U}}\,\tilde{\mathbf{U}}^t\}G^t = \frac{1}{12}GG^t.$$

Lemma 4.3.1 (NSM upper bound)

$$G(\Lambda) \leq \frac{\frac{1}{n}\text{trace}\{R_Q\}}{V^{2/n}(\Lambda)} \qquad (4.26)$$

with equality if and only if Q is the nearest-neighbor quantizer $Q_\Lambda^{(NN)} = \{\Lambda, \mathcal{V}\}$, in which case $\mathbf{U}$ is uniform over the fundamental Voronoi cell $\mathcal{V}_0$.

Proof The result follows since the fundamental Voronoi cell minimizes the second moment over all fundamental cells. □

Definition 4.3.2 (White dither) *A lattice quantizer Q is white if the samples of the dither are uncorrelated and have the same second moment, i.e.,*

$$R_Q = \sigma_Q^2 \cdot I, \qquad (4.27)$$

where σ_Q^2 is the second moment of the dither, and I is the identity matrix.

Note that uncorrelation does not imply independence; for example, a vector $\mathbf{U} = (U_1, U_2)$ which is uniform over a hexagon is white, i.e., $E\{U_1 \cdot U_2\} = 0$, although clearly U_1 and U_2 are *dependent*. Yet, in general, the dither is not white, even under nearest-neighbor quantization; take, for instance, a rectangular lattice, and rotate it less than 45 degrees.

Lemma 4.3.2 (NSM lower bound) *If* $Q_\Lambda^{(NN)}$ *is the nearest-neighbor quantizer of* Λ *(i.e., the dither* $\mathbf{U}$ *is uniform over* $\mathcal{V}_0$*), then*

$$G(\Lambda) \geq \frac{\det(R_Q)^{1/n}}{V^{2/n}(\Lambda)} \qquad (4.28)$$

with equality if and only if the dither is white.

Proof For the nearest-neighbor quantizer, (4.26) holds with equality. Furthermore, by the *arithmetic-geometric* means inequality for non-negative definite matrices (see, e.g., [53, Theorem 16.8.4]), the numerator of (4.26) is lower bounded by

$$\frac{1}{n}\text{trace}\{R_Q\} \geq \det(R_Q)^{1/n}$$

with equality if and only if R_Q is proportional to the identity matrix, and the lemma is proved. □

An *optimal* lattice quantizer in $\mathbb{R}^n$, denoted Q_n^{opt}, is a lattice quantizer with the minimal possible normalized second moment G_n over all lattices in $\mathbb{R}^n$. See Definition 3.2.2. Note that since $G(\Lambda)$ is invariant to scaling, Q_n^{opt} is determined up to scaling. The search for the optimum lattice quantizer can thus be done over a compact set of basis vectors, so the minimum is actually *achieved* by some lattice.

Clearly the quantizer Q_n^{opt} uses a Voronoi partition. Our main result states that it is also white.

Theorem 4.3.1 (The dither of an optimal lattice quantizer is white) *The optimal lattice quantizer* Q_n^{opt} *is white (Definition* 4.3.2*), and the auto-correlation matrix of its dither is given by*

$$R_Q = G_n \cdot V^{2/n}(\Lambda) \cdot I, \qquad (4.29)$$

where I is the identity matrix.

This theorem implies, for example, that the hexagonal lattice quantizer and the body-centered cubic (BCC) lattice quantizer, which are known to be optimal in dimensions $n = 2$ and 3, respectively, are *white*.

To prove Theorem 4.3.1, recall the definition of a linear transformation $\Lambda' = A \cdot \Lambda$ of a lattice Λ, and the property that $A \cdot \mathcal{P}_0(\Lambda)$ is a fundamental cell of Λ'; see Section 2.4. It is easy to see that if $\mathbf{U}$ is uniform over $\mathcal{P}_0(\Lambda)$, then $A\mathbf{U}$ is uniform over $\mathcal{P}_0(\Lambda') = A \cdot \mathcal{P}_0(\Lambda)$. This implies that the auto-correlation matrix of the transformed lattice quantizer is given by

$$R'_Q = E\{(A\mathbf{U})(A\mathbf{U})^t\} = AR_QA^t; \qquad (4.30)$$

i.e., the transformation A *shapes* the correlation structure of the quantization noise. For example, it can *whiten* a non-white lattice quantizer. Yet, a delicate point to note here is that even if $\mathcal{P}$ is the Voronoi partition of Λ, then the transformed partition $\mathcal{P}'$ is not necessarily the Voronoi partition of Λ' (see Section 2.4).

We can now proceed to the proof of the theorem.

Proof of Theorem 4.3.1 Let R_Q be the auto-correlation matrix of the dither of an optimal quantizer $Q_n^{\text{opt}} = \{\Lambda, \mathcal{P}\}$. We show below that if R_Q is not white, then we can get a better quantizer, i.e., a lower NSM. But we assume that Q_n^{opt} is optimal. Thus R_Q must be white.

Specifically, let $R_Q^{-1/2}$ be any inverse root of R_Q, i.e., a matrix satisfying $R_Q^{-1/2} R_Q (R_Q^{-1/2})^t = I$, and let us define the matrix

$$A = R_Q^{-1/2}. \tag{4.31}$$

As can be seen by substituting A into (4.30), transforming Q_n^{opt} by A results in a white lattice quantizer $Q' = \{\Lambda', \mathcal{P}'\}$, whose auto-correlation matrix is $R'_Q = I$, and whose volume is

$$V(\Lambda') = V(A\Lambda) = |\det(A)| \cdot V(\Lambda) = \frac{V(\Lambda)}{\sqrt{\det(R_Q)}}. \tag{4.32}$$

Thus,

$$
\begin{aligned}
G(\Lambda) &\overset{(a)}{\geq} \frac{\det^{1/n}(R_Q)}{V^{2/n}(\Lambda)} \\
&\overset{(b)}{=} \frac{1}{V^{2/n}(\Lambda')} \\
&\overset{(c)}{=} \frac{\frac{1}{n}\text{trace}(R_{Q'})}{V^{2/n}(\Lambda')} \\
&\overset{(d)}{\geq} G(\Lambda'),
\end{aligned} \tag{4.33}
$$

where (a) follows from Lemma 4.3.2 (with equality if and only if Q_n^{opt} is white); (b) follows from (4.32); (c) follows since $R_{Q'} = I$; and (d) follows from Lemma 4.3.1 (with equality if and only if Q' is a Voronoi partition). □

It follows from (4.33), that by iterating between *whitening* and *Voronoi partitioning*, the normalized second moment decreases (i.e., improves) monotonically. Thus, Theorem 4.3.1 may complement *Lloyd's necessary conditions* for an optimum quantizer; see the discussion in Section 5.7.

Any desired quantization noise spectrum, i.e., auto-correlation matrix R_Q (4.24), may be obtained by an appropriate linear transformation of a (non-degenerate) lattice quantizer, as seen in (4.30). This is useful, for example, if we want to modify an MSE-optimal quantizer to a *weighted square error* distortion measure of the form $(\hat{\mathbf{x}} - \mathbf{x})^t W (\hat{\mathbf{x}} - \mathbf{x})$.

4.4　　Wiener estimation

Before introducing the *Wiener estimator*, and its application to dithered quantization, let us discuss some intuitive background.

A common approach to reducing measurement noise is to average a number k of
independent measurements. Typically, the MSE of the average decreases like $1/k$.
But what if this MSE is comparable with (or even larger than) the variance of the
desired signal? For example, suppose we have a *single* measurement $Y = X + N$ of
a signal $X \sim N(a, 1)$, where the noise is $N \sim N(0, 10)$. Taking Y as the estimate of
X would give us an MSE ten times larger than the variance of the source! The *Wiener
filter*, or estimator, improves this undesirable situation by taking into account the
a priori information that $X = a$ with a unit variance. The estimator is given in this
case by $\hat{X} = a + \frac{1}{11}(Y - a)$, i.e., it effectively weights the a posteriori information,
the measurement Y, by how reliable it is with respect to the a priori information
about X. The resulting MSE $= E(\hat{X} - X)^2$ is given by

$$E\left(\frac{1}{11}N - \frac{10}{11}(X - a)\right)^2 = \frac{1}{11^2}10 + \frac{10^2}{11^2} = \frac{10}{11},$$

i.e., slightly *less* than the source variance.

The Wiener filter is a general tool for linear minimum mean-squared error estimation, when the joint second-order statistics of the desired signal and the measurements are known. If the desired signal and measurements are scalar variables,
then the Wiener filter becomes a simple scalar multiplier which depends on their
second moments (possibly with a bias, as in the example above). If these signals are
vectors or processes, then the Wiener filter takes the form of a matrix multiplication
or a linear time-invariant filter. However, we can constrain the Wiener filter to be a
scalar multiplier even in the *vector* and *process* cases, if we use the *average* second
moments of the signal and the measurements.

Let $\mathbf{X}$ and $\mathbf{Y}$ be two correlated n-dimensional random vectors, with aver-
age second moments $R_x = \frac{1}{n}E\|\mathbf{X}\|^2$ and $R_y = \frac{1}{n}E\|\mathbf{Y}\|^2$, average cross-correlation
$R_{x,y} = \frac{1}{n}E\{\mathbf{X}^t\mathbf{Y}\}$, and correlation coefficient

$$\rho = \frac{R_{x,y}}{\sqrt{R_x\,R_y}}.$$

To simplify the exposition, we shall ignore the means of $\mathbf{X}$ and $\mathbf{Y}$ (or assume
that they are zero), and thus restrict our attention to *pure multiplicative* (non-affine)
linear estimation. The scalar linear estimator of $\mathbf{X}$ from $\mathbf{Y}$ is defined as

$$\hat{\mathbf{X}} = \alpha\mathbf{Y}. \tag{4.34}$$

This results in an estimation error of $\hat{\mathbf{X}} - \mathbf{X} = \alpha \mathbf{Y} - \mathbf{X}$. The mean-squared error per dimension is given by

$$\text{MSE}(\alpha) = \frac{1}{n} E \|\alpha \mathbf{Y} - \mathbf{X}\|^2 = \alpha^2 R_y - 2\alpha R_{x,y} + R_x, \tag{4.35}$$

which is a quadratic expression in α, having a unique minimum.

Proposition 4.4.1 (Wiener estimator) *The scalar linear estimation MSE* (4.35) *is minimized by the* Wiener coefficient

$$\alpha^* = \frac{R_{x,y}}{R_y} = \rho \sqrt{\frac{R_x}{R_y}}, \tag{4.36}$$

and attains a minimum value of

$$\text{MSE}(\alpha^*) = R_x - \frac{R_{x,y}^2}{R_y} = (1 - \rho^2) R_x, \tag{4.37}$$

known as the linear minimum mean-squared error *(LMMSE) solution.*

An alternative way to obtain (4.36) is to use the *orthogonality principle*. This principle says that the error vector of an *optimal* estimator, whether it is linear or not, must be orthogonal to the measurements:

$$(\hat{\mathbf{X}} - \mathbf{X}) \perp \mathbf{Y}. \tag{4.38}$$

For *linear scalar* estimation (4.34), orthogonality amounts to $E\{(\alpha^* \mathbf{Y} - \mathbf{X})^t \mathbf{Y}\} = 0$, which implies (4.36). [7]

The restriction of the estimator (4.34) to be linear, however, does not come for free. Its MSE (4.37) is, in general, larger than the *minimum mean-squared error* (MMSE), or the *conditional variance*,

$$\text{Var}(\mathbf{X}|\mathbf{Y}) = \min_{g(\cdot)} \frac{1}{n} E \|g(\mathbf{Y}) - \mathbf{X}\|^2. \tag{4.39}$$

As is well known, this optimum estimation error is attained by the conditional mean

$$g^*(\mathbf{y}) = E\{\mathbf{X}|\mathbf{Y} = \mathbf{y}\} \tag{4.40}$$

which is, in general, a *non-linear* function of $\mathbf{y}$. Nevertheless, in the *jointly Gaussian* case the orthogonality principle (4.38) is also a sufficient condition for optimality, hence the Wiener (linear) estimator is optimal. In particular, in the memoryless case we have the following.

Proposition 4.4.2 (Gaussian MMSE) *If* $\mathbf{X}$ *and* $\mathbf{Y}$ *are memoryless jointly Gaussian vectors (i.e.,* $(X_1, Y_1), \ldots, (X_n, Y_n)$ *are i.i.d. pairs) with zero mean, then the*

---

[7] For general linear estimation $\hat{\mathbf{X}} = A\mathbf{Y}$ (where A is an $n \times n$ matrix), the orthogonality principle (4.38) amounts to a *matrix* equation $E\{(A^*\mathbf{Y} - \mathbf{X})\mathbf{Y}^t\} = 0$, that determines the matrix A^*.

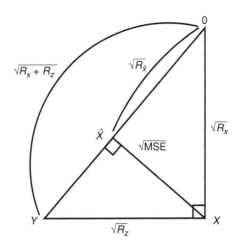

Figure 4.8 Pythagorean relations implied by the orthogonality principle in an additive-noise measurement channel: $R_y = R_x + R_z$ and $R_x = R_{\hat{x}} + \text{MSE}(\alpha^*)$.

conditional mean (4.40) is equal to the (scalar) Wiener estimator:

$$E\{\mathbf{X}|\mathbf{Y} = \mathbf{y}\} = \alpha^*\mathbf{y}. \tag{4.41}$$

In this case, the LMMSE (4.37) is also the MMSE (4.39).

Proof See Problem 4.12. □

In the *additive*-noise channel case, where $\mathbf{Y} = \mathbf{X} + \mathbf{Z}$, the noise $\mathbf{Z}$ is orthogonal to the input $\mathbf{X}$ (i.e., $E\{\mathbf{X}'\mathbf{Z}\} = 0$), so we have $R_y = R_x + R_z$, $R_{x,y} = R_x$ and

$$\rho = \sqrt{\frac{R_x}{R_x + R_z}}. \tag{4.42}$$

The scalar linear estimation error is given by

$$\hat{\mathbf{X}} - \mathbf{X} = \alpha\mathbf{Z} + (\alpha - 1)\mathbf{X}, \tag{4.43}$$

i.e., it is a *mixture* of the noise $\mathbf{Z}$ and the desired signal $\mathbf{X}$. The Wiener coefficient (4.36) minimizes the second moment of this mixture, and makes it scalarly orthogonal to $\mathbf{Y}$, as illustrated in Figure 4.8.

Proposition 4.4.3 (Additive-noise channel) *If the measurement is given by* $\mathbf{Y} = \mathbf{X} + \mathbf{Z}$, *where* $\mathbf{Z}$ *is orthogonal to* $\mathbf{X}$, *then the Wiener coefficient (4.36) is given by*

$$\alpha^* = \frac{R_x}{R_x + R_z} = \frac{\text{SNR}}{1 + \text{SNR}}, \tag{4.44}$$

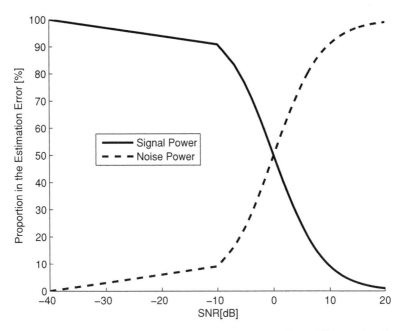

Figure 4.9 The proportion of signal and noise in the LMMSE of an additive-noise channel (i.e., (4.43) with the Wiener coefficient α^*), as a function of the SNR.

resulting in an MSE of

$$\text{MSE}(\alpha^*) = \frac{R_x \, R_z}{R_x + R_z} = \frac{R_z}{1 + 1/\text{SNR}}, \tag{4.45}$$

where

$$\text{SNR} \triangleq \frac{R_x}{R_z} \tag{4.46}$$

is the signal to noise ratio.

Note that α^* in (4.44) is between zero and one; it is zero when the SNR is equal to zero, and it approaches one when the SNR is much larger than one. Thus the error (4.43) is always a *convex mixture* of the noise and the desired signal. Note also that the *harmonic average* in (4.45) is strictly smaller than each of its arguments; thus the scaling by α^* reduces the mean-squared error $\text{MSE}(\alpha)$ below both the second moment of the signal (R_x) and the second moment of the noise (R_z). Figure 4.9 shows the relative contributions of noise and signal to $\text{MSE}(\alpha^*)$.

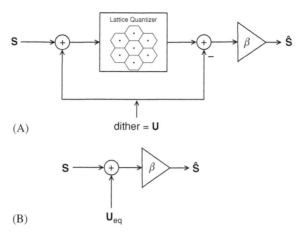

(A)

(B)

dither = **U**

U$_{eq}$

Figure 4.10 (A) Post-scaled dithered quantization, and (B) its equivalent channel (4.7).

4.5   Filtered dithered quantization

The equivalent additive-noise channel of the dithered lattice quantizer in Figure 4.5 calls for putting an estimator at the output of the channel, to reduce the overall MSE in reconstructing the source **S**. Consider the filtered reconstruction in Figure 4.10(A), where the quantizer output is multiplied by a linear coefficient β:

$$\hat{\mathbf{S}} = \beta[Q_\Lambda(\mathbf{S} + \mathbf{U}) - \mathbf{U}] \tag{4.47}$$

where **U** is a uniform (or a generalized, i.e., modulo-uniform) dither. By the equivalent channel model (Figure 4.10(B)), the joint distribution of $\hat{\mathbf{S}}$ and **S** is described by the relation (see (4.7)):

$$\hat{\mathbf{S}} \stackrel{\text{dist}}{=} \beta[\mathbf{S} + \mathbf{U}_{eq}], \tag{4.48}$$

where the equivalent dither $\mathbf{U}_{eq}$ is uniform over the reflected fundamental cell $(-\mathcal{P}_0)$ of the lattice quantizer and (by Theorem 4.1.1) statistically independent of the source.

Assume for simplicity that the source **S** and the equivalent dither $\mathbf{U}_{eq}$ have a zero mean, so their independence implies *orthogonality*, i.e., $E\{\mathbf{S}^t \mathbf{U}_{eq}\} = 0$.[8] The multiplication by β in (4.48) thus amounts to linear estimation of the source after an additive-noise channel, where the noise $\mathbf{U}_{eq}$ is orthogonal to the source. Applying Proposition 4.4.3, we have the following.

---

[8] If the mean of the source is not zero, then it is better to remove the mean before the quantization and add it back after the scaling by β. Note that if the lattice quantizer cells are *Voronoi* cells, then the equivalent dither has a zero mean.

Proposition 4.5.1 (Filtered dithered quantizer) *The optimum linear (Wiener) coefficient in the reconstruction* (4.47) *for a zero-mean source* **S** *is*

$$\beta^* = \frac{\sigma_s^2}{\sigma_Q^2 + \sigma_s^2},\tag{4.49}$$

and the resulting minimum distortion is

$$D^* = \frac{\sigma_Q^2 \cdot \sigma_s^2}{\sigma_Q^2 + \sigma_s^2},\tag{4.50}$$

where σ_Q^2 and σ_s^2 are the second moments of the dither (4.25) *and the source, respectively.*

For a *Voronoi* lattice quantizer, σ_Q^2 is equal to the lattice second moment $\sigma^2(\Lambda)$. The scaling by β^* thus reduces the distortion D^* below the second moment of the lattice (because the harmonic average in (4.50) is strictly smaller than each of its arguments).

Example 4.5.1 (Wiener coefficient and distortion) *Let the source variance and the lattice second moment be $\sigma_s^2 = 1$ and $\sigma^2(\Lambda) = \frac{1}{2}$, respectively. Then $\beta^* = \frac{1}{1+1/2} = \frac{2}{3}$, and*

$$D^* = \frac{1}{n} E \left\| \frac{2[Q_\Lambda(\mathbf{S} + \mathbf{U}) - \mathbf{U}]}{3} - \mathbf{S} \right\|^2 = \frac{1 \times \frac{1}{2}}{1 + \frac{1}{2}} = \frac{1}{3},$$

instead of $D = \sigma^2(\Lambda) = \frac{1}{2}$ which is achieved without post scaling (i.e., with $\beta = 1$).

Is random dither really necessary? It is too early to address this question fully, as we shall do later, in Section 5.8, but let us make a few insightful remarks regarding the dither.

1. **Memory** If the dither is *not white* (Definition 4.3.2), or the source has *memory*, then a *matrix* form of the Wiener filter can obtain a smaller mean-squared distortion.
2. **Non-Gaussianity** Since the dither is *not* Gaussian, an even better mean-squared distortion can be attained by the (non-linear) conditional mean estimator (4.40):

$$\hat{\mathbf{S}} = g^*(Q_\Lambda(\mathbf{S} + \mathbf{U}) - \mathbf{U})$$

where $g^*(\cdot)$ is calculated with respect to the equivalent additive-noise channel, i.e., $g^*(\mathbf{y}) = E\{\mathbf{S}|\mathbf{S} + \mathbf{U}_{\text{eq}} = \mathbf{y}\}$.
3. **Mismatch** The attenuated reconstruction $\beta Q_\Lambda(\mathbf{s})$ of (4.47) (assuming zero dither and some factor $\beta < 1$) means that the reconstruction codebook is the *finer* lattice $\beta\Lambda$, rather than Λ. The minimum-distortion reconstruction, given this finer codebook, is $Q_{\beta\Lambda}(\mathbf{s})$. But by the scaling law (2.43),

$$Q_{\beta\Lambda}(\mathbf{s}) = \beta Q_\Lambda(\mathbf{s}/\beta) \neq \beta Q_\Lambda(\mathbf{s}),$$

meaning that the encoder is *mismatched* to the decoder.

The next chapter will introduce another critical aspect of quantization: the *coding rate*. We shall see how the equivalent additive-noise channel of dithered quantization (Theorem 4.1.1) provides an analytic tool to compute the coding rate. Furthermore, in spite of the possible deficiencies above, the filtered dithered quantizer (4.47) provides a good – and even asymptotically optimal – trade-off between the coding rate and distortion.

Summary of Chapter 4

Crypto lemma If $\mathbf{U}$ is uniform on $\mathcal{P}_0$, then $[\mathbf{x} + \mathbf{U}] \bmod_{\mathcal{P}_0} \Lambda$ is uniform on $\mathcal{P}_0$.

Generalized dither The crypto lemma holds for any $\mathbf{U}$ which is *modulo-uniform*, i.e., $\mathbf{U} \bmod_{\mathcal{P}_0} \Lambda$ is uniform on $\mathcal{P}_0$.

Equivalent conditions for modulo-uniformity

1. $\sum_{\lambda \in \Lambda} f_{\mathbf{U}}(\mathbf{x} + \lambda)$ is constant for all $\mathbf{x}$;
2. $\phi_{\mathbf{U}}(\lambda^*) = 0$ for all non-zero dual lattice points $\lambda^* \in \Lambda^*$.

Generalized dithers (cases of modulo-uniformity)

1. $\mathbf{U} \sim \text{Unif}(\mathcal{P}_0')$ for some other fundamental cell $\mathcal{P}_0'$;
2. $\mathbf{U} \sim \text{Unif}(\mathcal{P}_0(\Lambda_c))$, where $\Lambda_c \subset \Lambda$;
3. $\mathbf{U} = \tilde{\mathbf{U}} + \tilde{\tilde{\mathbf{U}}}$, where $\tilde{\mathbf{U}}$ is modulo-uniform, and $\tilde{\tilde{\mathbf{U}}}$ is arbitrary.

White dither If $G(\Lambda) = G_n^*$ (i.e., Λ minimizes the NSM over all lattices in $\mathbb{R}^n$), then a uniform distribution on the fundamental Voronoi cell has a white covariance matrix.

Equivalent additive-noise channel for dithered quantization

$$Q_\Lambda(\mathbf{X} + \mathbf{U}) - \mathbf{U} \overset{\text{dist}}{=} \mathbf{X} + \mathbf{U}_{\text{eq}},$$

where $\mathbf{U}_{\text{eq}} \sim \text{Unif}(-\mathcal{P}_0)$ is independent of $\mathbf{X}$.

Filtered dithered quantizer

$$E \| \beta [Q_\Lambda(\mathbf{X} + \mathbf{U}) - \mathbf{U}] - \mathbf{X} \|^2 \text{ is minimized by } \beta^* = \frac{\sigma_x^2}{\sigma_x^2 + \sigma_u^2}.$$

Problems

P.4.1 (Modulo-uniformity condition) Prove the condition for a generalized dither (Lemma 4.2.1), based on the crypto lemma (Lemma 4.1.1) for dither which is uniform over the fundamental cell.

P.4.2 (Uniform dither over a sublattice cell) Show that if Λ_c is a sublattice of Λ, then periodic replication of $\mathcal{P}_0(\Lambda_c)$ by Λ induces exactly $|\Lambda/\Lambda_c|$ overlaps; hence if $f_{\mathbf{U}} \sim \mathrm{Unif}(\mathcal{P}_0(\Lambda_c))$, then its periodic replication is constant.

P.4.3 (Generalized dither by spreading) (i) Show that the third case of generalized dither in Example 4.2.1 ($\mathbf{U} = \tilde{\mathbf{U}} + \mathbf{X}$) holds also for $\tilde{\mathbf{U}}$ which is modulo-uniform. (ii) Give an example of generalized dither that is *not* a spreading of a uniform dither.

P.4.4 (Dual lattice) Show that in two dimensions, taking the dual of a unit-determinant lattice amounts to a rotation by 90 degrees.

P.4.5 (Replication as sampling in the frequency domain) Prove (4.21) in the one-dimensional case using standard Fourier analysis.

P.4.6 (Nyquist condition for generalized dither) Prove Lemma 4.2.4 by transforming the constant periodic-replication condition (4.17) to the frequency domain, and using the replication/sampling relation (4.21).

P.4.7 (Spreading in the frequency domain) Prove the third case of generalized dither in Example 4.2.1 using the fact that *convolution* in time (or space) amounts to *multiplication* in frequency; thus the Fourier transform of $\mathbf{U} = \tilde{\mathbf{U}} + \mathbf{X}$ can be expressed as the product $\mathcal{F}\{f_{\mathbf{U}}\} = \mathcal{F}\{f_{\tilde{\mathbf{U}}}\} \cdot \mathcal{F}\{f_{\mathbf{X}}\}$. Hence, $\mathcal{F}\{f_{\mathbf{U}}\}$ is zero on the dual lattice points (for any $f_{\mathbf{X}}$) if $\mathcal{F}\{f_{\tilde{\mathbf{U}}}\}$ is.

P.4.8 (Flatness factor bound in frequency domain) Use the Fourier expansion formula (or inverse Fourier transform) to prove that

$$F(\Lambda, \mathbf{X}) \leq \sum_{\lambda^* \in \Lambda^*, \lambda^* \neq 0} \left| \phi_{\mathbf{X}}(\lambda^*) \right| \tag{4.51}$$

where $\phi_{\mathbf{X}} = \mathcal{F}\{f_{\mathbf{X}}\}$ is the characteristic function of $\mathbf{X}$. Furthermore, show that equality holds if $\phi_{\mathbf{X}}$ is real and positive, in which case the maximum in Definition 4.2.4 is achieved for a lattice point in $\mathcal{P}_0$ (e.g., for $\mathbf{x} = 0$, if $\mathcal{P}_0$ contains the origin). Guidance: show that

$$f_{\mathbf{X} \bmod \Lambda}(\mathbf{x}) = \frac{1}{V(\Lambda)} \cdot \left[1 + \sum_{\lambda^* \in \Lambda^*, \lambda^* \neq 0} \phi_{\mathbf{X}}(\lambda^*) e^{j 2\pi \langle \lambda^*, \mathbf{x} \rangle} \right], \tag{4.52}$$

where Λ^* is the dual lattice, plug into the definition of $F(\Lambda, \mathbf{X})$, and note that $|e^{j 2\pi <\lambda^*, \mathbf{x}>}| \leq 1$.

P.4.9 (Gaussian flatness factor) Find the flatness factor of a white-Gaussian vector $\mathbf{X} \sim f(\mathbf{x}) = 1/\sqrt{2\pi\sigma^2} e^{\mathbf{x}^2/2\sigma^2}$. Guidance: show that $\phi_{\mathbf{X}}$ satisfies the condition for equality in Problem 4.8, and therefore the maximum in (4.23) is achieved by $\mathbf{x} = 0$; i.e., $F(\mathbf{X}, \Lambda)$ is given by summing $f(\mathbf{x})$ over all $\mathbf{x} \in \Lambda$, multiplying by $V(\Lambda)$ and subtracting one. Verify that you get the same result by summing $\phi_{\mathbf{X}}(\mathbf{f})$ over all $\mathbf{f} \in \Lambda^*$ and subtracting one.

Remark This formula can be expressed in terms of the *distance spectrum* of the lattice (see Section 13.5) or its *theta series*; see [167, 187].

P.4.10 (High-resolution quantization) Show that if $\mathbf{X}$ has a p.d.f., then $\lim_{\alpha \to \infty} F(\Lambda, \alpha \mathbf{X}) = 0$. Hint: examine the bound (4.51) as $\alpha \to \infty$.

P.4.11 (Weighted-quadratic distortion measure) Show that dither auto-correlation shaping is useful if we want to modify an MSE-optimal quantizer to a *weighted square error* distortion measure of the form $(\hat{\mathbf{x}} - \mathbf{x})^t W (\hat{\mathbf{x}} - \mathbf{x})$. In this case, we use in the quantizer scaling law (2.41) a matrix A satisfying $A A^t = W^{-1}$. See more on non-quadratic distortion measures in Section 5.4.

P.4.12 (Proposition 4.4.2: Gaussian MMSE) Show that if $\mathbf{X}$ and $\mathbf{Y}$ are memoryless jointly Gaussian vectors with zero mean, then the conditional mean (4.40) is equal to the Wiener estimator. Guidance: show, via direct calculation, that the conditional density $f(x|y)$ of X given $Y = y$ is Gaussian, whose mean and variance coincide with the Wiener estimator (4.36) and its MSE (4.37), respectively:

$$f(x|y) = \frac{1}{\sqrt{2\pi(1 - \rho^2)R_x}} \cdot e^{-\frac{(x - \alpha^* y)^2}{2(1 - \rho^2)R_x}}.$$

Historical notes

The idea of dithered quantization was proposed in the 1960s by Roberts [228], Schuchman [234] and Limb [160] for enhancing the perceptual effect of picture coding, and applied later to speech by Jayant and Rabiner [131]. See the tutorial papers by Lipshitz *et al.* [170], by Gray and Stockham [112] and by Gray and Neuhoff [111], for a comprehensive treatment of subtractive as well as non-subtractive dither. Lemma 4.2.4 in the one-dimensional case is known as Schuchman's condition, which parallels the Nyquist condition for zero inter-symbol interference [112, 170, 234]. Conway and Sloane [48] show how to generate a uniform distribution over a Voronoi cell using closet point search.

The spectra of quantized signals was analyzed under the high-resolution assumption by Bennett in 1948 [17], and extended to dithered quantization by Gray [109]. Zamir and Feder [289] showed that the dither spectrum (and hence also the subtractive-dither quantization error) of an optimal n-dimensional lattice quantizer is white. Extension to dither signals under non-quadratic measures was given by Gariby and Erez [98]. Non-uniform dithered quantization was proposed by Akyol and Rose [7].

The flatness factor is based on the smoothing parameter of Micciancio and Regev [187] in lattice complexity theory and cryptography. In the lattice coding literature, Ling *et al.* [167] used the flatness factor to measure the information leakage in lattice wiretap coding schemes; see [16, 168].

The incorporation of filtering into source coding goes back to the water-pouring solution of the rate-distortion function of a colored-Gaussian source [18]. Classical literature on signal compression (e.g., the books by Jayant and Noll [130] and Gersho and Gray [103]) provides tools for the analysis of (non-linear) quantization schemes involving linear operations (for transformation, prediction and estimation). But this analysis becomes inaccurate in the low-resolution regime; see, for example, [35]. A simple analysis for randomized quantization incorporating Wiener estimation, which is accurate at all resolution levels, was given by Zamir and Feder [290]. This idea was extended to various source coding problems; in particular, over-sampling [288], non-uniform sampling (Mashiach and Zamir [182]), successive refinement (Lastras and Berger [152], Zhao *et al.* [297]), predictive coding (Zamir *et al.* [291]), noise-shaped quantization and multiple descriptions (Chen *et al.* [39], Østergaard and Zamir [211] and Palgy *et al.* [212]).

Wiener-estimated dithered quantization was also applied in shaping (Voronoi coding with lattice decoding) and in dirty-paper channel coding [68, 71]; see Chapters 9 and 10.

5 Entropy-coded quantization

The elements of dithering and estimation provide tools to control the distribution of the quantization error, and to compute the average distortion. Another important parameter of source coding is the *coding rate*.

In this chapter we focus on the *quantizer entropy* as a measure for the coding rate. Although the lattice is *unbounded*, entropy coding keeps its coding rate finite. We examine the entropy-distortion trade-off for a general source and lattice quantizer in Sections 5.2–5.4, and compare it to Shannon's *rate-distortion function* $R(D)$ – the ultimate compression rate of any system achieving a distortion level D – in Sections 5.5–5.6. As we shall see, the redundancy above $R(D)$ is small for all sources; even for a simple *scalar* lattice quantizer it is at most $\approx 3/4$ bit, and only $\approx 1/4$ bit at high-resolution quantization. Furthermore, if we combine a Wiener filter at the quantizer output (as we did in Section 4.5), then the redundancy of a scalar ECDQ for a Gaussian source is at most $\approx 1/4$ bit at *any* resolution.

5.1 The Shannon entropy

In *fixed-rate* lossless coding, all elements of the data are mapped into binary codewords of identical length. Let $\mathcal{A}$ denote the data alphabet. Since there are 2^l binary words of length l, the codeword length must be at least the base-2 logarithm of the size of $\mathcal{A}$, rounded up to the nearest integer.

In contrast, *variable-rate* coding takes advantage of the non-uniformity of the source probability distribution; highly probable elements are mapped to short codewords, while rare elements are mapped to long codewords, so the *average* code length is reduced. The Shannon–Fano code, for example, maps the ith letter in the source alphabet to a codeword of length

$$\left\lceil \log_2 \left(\frac{1}{p_i} \right) \right\rceil \quad \text{bits}, \quad i \in \mathcal{A}, \tag{5.1}$$

where p_i is the letter probability, and $\lceil \cdot \rceil$ denotes rounding up to the nearest integer. This length is on the one hand large enough to allow reversible mapping (a result known as Kraft's inequality [53]). On the other hand, if we ignore the rounding up

Table 5.1 Various lossless source codes for a uniform (scalar lattice) quantizer of a Gaussian source; the Huffman code almost achieves the entropy $H(Q(S)) \approx 2.09$ bit

Symbol	p_i	Fixed length code	Shannon–Fano code	Huffman code
-2Δ	0.0606	000	01010	1110
$-\Delta$	0.2417	001	110	10
0	0.3829	010	00	0
Δ	0.2417	011	011	110
2Δ	0.0606	100	10101	11110
Overload	0.0125	101	1111111	11111
Average length		3	2.7475	2.1993

(which adds at most 1 bit), then the expected code length is equal to the source *entropy*,

$$H \triangleq \sum_{i \in \mathcal{A}} p_i \log_2 \left(\frac{1}{p_i} \right). \tag{5.2}$$

Information theory shows that the entropy (5.2) is the minimum number of bits required to describe unambiguously a long outcome of a memoryless source. The entropy function satisfies $H \leq \log_2 |\mathcal{A}|$, corresponding to rate saving with respect to fixed-rate coding, with equality if and only if the source is uniformly distributed over the alphabet $\mathcal{A}$. The Shannon–Fano code, along with other compression algorithms such as Huffman or arithmetic coding, are commonly known under the name *lossless* (or "entropy") *coding*. [1]

5.2 Quantizer entropy

Quantization amounts to *lossy* source coding, i.e., encoding with distortion. Yet, since the quantizer output $Q(\mathbf{S})$ is not necessarily equiprobable, it can be further losslessly encoded by a variable-length code, in order to approach its entropy $H(Q(\mathbf{S}))$; see Figure 5.1. For lattice quantization, such entropy coding is unavoidable, since the alphabet size – composed of all the lattice points – is infinite. [2] As in the previous chapter, we shall enhance the robustness and analysis of lattice quantization using dither.

---

[1] See Appendix A.1 for an overview of information theoretic measures. For an extension of (5.2) to a source with memory, see [53].
[2] Unless the source has a finite compact support, so only a finite number of lattice points are actually used in the quantization.

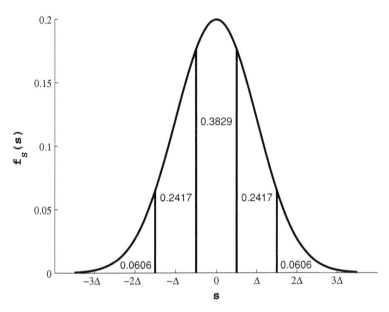

Figure 5.1 Uniform quantization of a Gaussian source implies highly non-uniform cell probabilities. Entropy coding reduces the rate of a 6-level quantizer (with thresholds at $\pm 5/2\Delta$, $\pm 3/2\Delta$ and $\pm\Delta/2$) from $\lceil \log_2 6 \rceil = 3$ bit to $H(Q(S)) \approx 2.09$ bit; see Table 5.1.

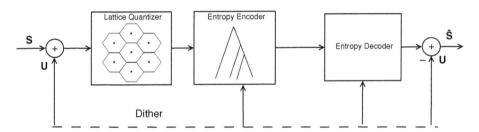

Figure 5.2 Entropy-coded dithered quantization.

Definition 5.2.1 (ECDQ) *An entropy-coded dithered quantizer is a dithered lattice quantizer* (4.3), *with a uniform (or modulo-uniform) dither, followed by lossless entropy coding of the quantizer output conditioned on the dither. See Figure* 5.2.

Let us explain the meaning of conditional entropy coding. Consider the dithered lattice quantizer output $Q_\Lambda(\mathbf{S} + \mathbf{U})$. For each dither value $\mathbf{U} = \mathbf{u}$, the variable-length code (5.1) is designed according to the probabilities

$$p_\lambda(\mathbf{u}) = \Pr\{\mathbf{S} + \mathbf{u} \in \mathcal{P}_\lambda\}, \quad \text{for } \lambda \in \Lambda, \tag{5.3}$$

i.e., the probabilities that the shifted source vector $\mathbf{S} + \mathbf{u}$ falls inside the lattice cells $\{\mathcal{P}_\lambda, \ \lambda \in \Lambda\}$. Thus, the expected code length – averaged with respect to both the source and the dither distributions – is equal (up to at most 1 bit) to the conditional entropy (A.3) of the quantizer output $Q_\Lambda(\mathbf{S} + \mathbf{U})$, given the dither $\mathbf{U}$:

$$H(Q_\Lambda(\mathbf{S} + \mathbf{U})|\mathbf{U}) = \int f_\mathbf{U}(\mathbf{u}) \cdot H(Q_\Lambda(\mathbf{S} + \mathbf{u}))d\mathbf{u}. \tag{5.4}$$

For a uniform (non-generalized) dither $\mathbf{U} \sim \mathrm{Unif}(\mathcal{P}_0)$, we can write this conditional entropy as (see (A.3))

$$H(Q_\Lambda(\mathbf{S} + \mathbf{U})|\mathbf{U}) = \frac{1}{V(\Lambda)} \int_{\mathcal{P}_0} H(Q_\Lambda(\mathbf{S} + \mathbf{u}))d\mathbf{u}$$

$$= -\frac{1}{V(\Lambda)} \int_{\mathcal{P}_0} \sum_\lambda p_\lambda(\mathbf{u}) \log p_\lambda(\mathbf{u})d\mathbf{u}. \tag{5.5}$$

Theorem 5.2.1 below implies that the same expression holds for a generalized (modulo-uniform) dither. The ECDQ *coding rate* in bits per sample is given by dividing the conditional entropy (5.5) by the lattice dimension n:

$$R_{\mathrm{ECDQ}} = \frac{1}{n} H(Q_\Lambda(\mathbf{S} + \mathbf{U})|\mathbf{U}). \tag{5.6}$$

Theorem 5.2.1 (ECDQ coding rate) *The ECDQ average code length (5.4) for a uniform (or modulo-uniform) dither is equal to the mutual information (A.6) in the equivalent additive-noise channel of Figure 4.5:*

$$H(Q_\Lambda(\mathbf{S} + \mathbf{U})|\mathbf{U}) = I(\mathbf{S}; \mathbf{S} + \mathbf{U}_{\mathrm{eq}}), \tag{5.7}$$

where $\mathbf{U}_{\mathrm{eq}} \sim \mathrm{Unif}(-\mathcal{P}_0)$ is the equivalent dither (4.8), and where $\mathcal{P}_0$ is the fundamental cell used by the lattice quantizer Q_Λ (4.1).

We see that the equivalent additive-noise channel of Figure 4.5 plays a dual role: it characterizes both the statistics of the quantization error (Theorem 4.1.1), and the average length of an ideal "entropy" code (Theorem 5.2.1).

Note that for a general n-dimensional lattice, the equivalent dither components are *dependent*. Thus, even if the source $\mathbf{S}$ is memoryless, the mutual information on the right-hand side cannot be broken into a sum of scalar mutual information terms.

We provide two proofs for Theorem 5.2.1: the first is based on a direct calculation, while the second is shorter but requires information theoretic tools. Also, the first proof assumes a uniform dither $\mathbf{U} \sim \mathrm{Unif}(\mathcal{P}_0)$, while the second proof covers the general (modulo-uniform) dither case.

We shall use the following lemma.

Lemma 5.2.1 *Given a source $\mathbf{S}$ and a region $\mathcal{A}$ in $\mathbb{R}^n$, let us define the vector $\mathbf{X} = \mathbf{S} - \mathbf{V}$, where $\mathbf{V}$ is uniform over $\mathcal{A}$ and independent of $\mathbf{S}$ (i.e., $f_\mathbf{V}(\mathbf{v}) = 1/\mathrm{Vol}(\mathcal{A})$ for $\mathbf{v} \in \mathcal{A}$, and zero elsewhere). Then, for any vector $\mathbf{t}$ in $\mathbb{R}^n$,*

$$\Pr\{\mathbf{S} \in \mathcal{A} + \mathbf{t}\} = \mathrm{Vol}(\mathcal{A}) \cdot f_\mathbf{X}(\mathbf{t}), \tag{5.8}$$

where $f_\mathbf{X}(\cdot)$ is the density of $\mathbf{X}$.

Proof See Problem 5.1.   □

Proof A of Theorem 5.2.1 Assume $\mathbf{U} \sim \text{Unif}(\mathcal{P}_0)$, and let $\mathbf{X} = \mathbf{S} + \mathbf{U}_{eq} = \mathbf{S} - \mathbf{U}$ denote the output of the equivalent additive-noise channel of Figure 4.5. Since $\mathbf{U}$ has a density, $\mathbf{X}$ must also have a density $f_{\mathbf{X}}(\cdot)$ (regardless of whether $\mathbf{S}$ has a density or not). By Lemma 5.2.1, the probability that the dithered source $\mathbf{S} + \mathbf{u}$ falls in some lattice quantizer cell $\mathcal{P}_\lambda = \mathcal{P}_0 + \lambda$ is given by

$$p_\lambda(\mathbf{u}) = \Pr\{\mathbf{S} \in \mathcal{P}_0 + \lambda - \mathbf{u}\} = V(\Lambda) \cdot f_{\mathbf{X}}(\lambda - \mathbf{u}). \tag{5.9}$$

Substituting (5.9) into (5.5), we can break the expression for the quantizer entropy into two terms:

$$-\frac{1}{V(\Lambda)} \int_{\mathcal{P}_0} \sum_{\lambda \in \Lambda} p_\lambda(\mathbf{u}) \cdot \log(V(\Lambda)) d\mathbf{u} \tag{5.10a}$$

and

$$-\int_{\mathcal{P}_0} \sum_{\lambda \in \Lambda} f_{\mathbf{X}}(\lambda - \mathbf{u}) \cdot \log(f_{\mathbf{X}}(\lambda - \mathbf{u})) d\mathbf{u}. \tag{5.10b}$$

The first term (5.10a) is equal to $-\log(V(\Lambda))$ (since $\sum_{\lambda \in \Lambda} p_\lambda(\mathbf{u}) = 1$ for all $\mathbf{u}$). As for the second term (5.10b), since $-\mathcal{P}_0$ is also a valid fundamental cell of Λ, the summation over $\lambda \in \Lambda$ and the integration over $-\mathcal{P}_0$ complement each other to integration over the entire space $\mathbb{R}^n$. The second term is thus equal to

$$-\int_{\mathbb{R}^n} f_{\mathbf{X}}(\mathbf{x}) \cdot \log(f_{\mathbf{X}}(\mathbf{x})) d\mathbf{x} = h(\mathbf{X}), \tag{5.11}$$

i.e., to the *differential entropy* of $\mathbf{X}$ (see (A.8) in the Appendix). Combining the two terms in (5.5), we obtain

$$H(Q_\Lambda(\mathbf{S} + \mathbf{U})|\mathbf{U}) = h(\mathbf{X}) - \log(V(\Lambda)) \tag{5.12}$$
$$= h(\mathbf{S} + \mathbf{U}_{eq}) - h(\mathbf{U}_{eq}) \tag{5.13}$$
$$= I(\mathbf{S}; \mathbf{S} + \mathbf{U}_{eq}), \tag{5.14}$$

where (5.13) follows since by (4.5) the dither is uniformly distributed over a region of volume $V(\Lambda)$ (the fundamental lattice cell $\mathcal{P}_0$), so its differential entropy is equal to the logarithm of this volume; and (5.14) follows from the decomposition of mutual information into a difference of entropies (see (A.11) in the Appendix).   □

Note again, that since $\mathbf{U}$ has a density, $\mathbf{X} = \mathbf{S} - \mathbf{U}$ also has a density, and hence a differential entropy, regardless of the nature of the distribution of the source $\mathbf{S}$.

An alternative proof, of a more "information theoretic flavor," is given below. It is shorter, though perhaps less intuitive.

Proof B of Theorem 5.2.1 Let us denote for short $Q_\Lambda \triangleq Q_\Lambda(\mathbf{S} + \mathbf{U})$. Using basic properties of entropy and mutual information (see Appendix A.1) we write,

$$
\begin{aligned}
H(Q_\Lambda | \mathbf{U}) &\overset{a}{=} H(Q_\Lambda | \mathbf{U}) - H(Q_\Lambda | \mathbf{U}, \mathbf{S}) \\
&\overset{b}{=} I(\mathbf{S}; Q_\Lambda | \mathbf{U}) \\
&\overset{c}{=} I(\mathbf{S}; Q_\Lambda - \mathbf{U} | \mathbf{U}) \\
&\overset{d}{=} I(\mathbf{S}; Q_\Lambda - \mathbf{U}) \\
&\overset{e}{=} I(\mathbf{S}; \mathbf{S} + \mathbf{U}_{eq}),
\end{aligned} \tag{5.15}
$$

where (a) follows since $Q_\Lambda(\mathbf{S} + \mathbf{U})$ is a deterministic function of $\mathbf{S}$ and $\mathbf{U}$, so its conditional entropy given $\mathbf{S}$ and $\mathbf{U}$ is zero; (b) follows by decomposing the mutual information into a difference of entropies (A.11); (c) follows since adding or subtracting the condition does not change the mutual information; and (e) follows from (4.7), the equivalent additive-noise channel of Figure 4.5. Step (d) is justified by the following three steps:

$$
\begin{aligned}
I(\mathbf{S}; Q_\Lambda - \mathbf{U} | \mathbf{U}) &\overset{d1}{=} I(\mathbf{S}; Q_\Lambda - \mathbf{U}) + I(\mathbf{S}; \mathbf{U} | Q_\Lambda - \mathbf{U}) - I(\mathbf{S}; \mathbf{U}) \\
&\overset{d2}{=} I(\mathbf{S}; Q_\Lambda - \mathbf{U}) + I(\mathbf{S}; \mathbf{U} | Q_\Lambda - \mathbf{U}) \\
&\overset{d3}{=} I(\mathbf{S}; Q_\Lambda - \mathbf{U})
\end{aligned} \tag{5.16}
$$

where $(d1)$ follows by using the chain rule (A.7) twice to decompose the joint mutual information $I(\mathbf{S}; Q_\Lambda - \mathbf{U}, \mathbf{U})$; $(d2)$ follows since the dither $\mathbf{U}$ is independent of the source $\mathbf{S}$, so $I(\mathbf{S}; \mathbf{U}) = 0$; finally, a uniform dither $\mathbf{U} \sim \mathrm{Unif}(\mathcal{P}_0)$ is a deterministic function of $Q_\Lambda - \mathbf{U}$ (because Q_Λ is a lattice point and $\mathbf{U} \in \mathcal{P}_0$, so $\mathbf{U} = -(Q_\Lambda - \mathbf{U}) \bmod \Lambda$), hence we also have $I(\mathbf{S}; \mathbf{U} | Q_\Lambda - \mathbf{U}) = 0$, implying $(d3)$.

In the case of a generalized (modulo-uniform) dither, $\mathbf{U}$ is *not* a deterministic function of $Q_\Lambda - \mathbf{U}$, so step $(d3)$ cannot be justified. This case is deferred to Problem 5.4. □

Remarks

1. The source $\mathbf{S}$ does not have to posses a probability density function.
2. The support of the source distribution can be infinite (in which case the entire infinite lattice must be mapped into codewords).

5.3   Joint and sequential entropy coding*

Theorem 5.2.1 can be extended to encoding of a sequence of vectors $\mathbf{S}_1, \mathbf{S}_2, \ldots$, where the dimension of each vector is the lattice dimension n. If the vectors are independent, then each vector can be encoded separately, so the coding rate per vector is given by Theorem 5.2.1. To take advantage of the possible dependence between the vectors, however, the entropy coding should be done *jointly*.

There are two options for drawing the dither along the sequence.

1. Constant dither, i.e., $\mathbf{U}_1 = \mathbf{U}_2 = \cdots = \mathbf{U}$, where $\mathbf{U}$ is uniform over $\mathcal{P}_0$.
2. Independent dithering, i.e., the dither vectors $\mathbf{U}_1, \mathbf{U}_2, \ldots$ are i.i.d., i.e., independent and each one is uniformly distributed over $\mathcal{P}_0$.

The second option turns out to be simpler to analyze.[3]

Let us assume quantizing k source vectors $\mathbf{s}_1, \ldots, \mathbf{s}_k$ of length n each, i.e., a total of $m = kn$ source samples. We denote the length-m "super" vector, generated by concatenating these n-length vectors, as $s^m = \mathbf{s}_1, \ldots, \mathbf{s}_k$. Observe that independent application of a lattice quantizer Q_Λ to the consecutive vectors $\mathbf{s}_1, \ldots, \mathbf{s}_k$ is equivalent to quantizing the vectors together using a single *super lattice quantizer* (a Cartesian product lattice quantizer) $Q_{\Lambda,m}$:

$$Q_\Lambda(\mathbf{s}_1), \ldots, Q_\Lambda(\mathbf{s}_k) = Q_{\Lambda,m}(s^m), \tag{5.17}$$

where the lattice and partition associated with $Q_{\Lambda,m}$ are Cartesian products of those of Q_Λ:

$$\Lambda_m = \Lambda \times \Lambda \times \cdots \times \Lambda \quad \text{and} \quad \mathcal{P}_{0,m} = \mathcal{P}_0 \times \mathcal{P}_0 \times \cdots \times \mathcal{P}_0. \tag{5.18}$$

Observe also that an m-dimensional vector U^m, which is uniform on the fundamental cell $\mathcal{P}_{0,m}$ of the super lattice quantizer $Q_{\Lambda,m}$, is equivalent to a set of independent dithers $\mathbf{U}_1, \ldots, \mathbf{U}_k$, each uniform over $\mathcal{P}_0$. Thus, luckily, we can use the ECDQ rate formula of Theorem 5.2.1 directly with respect to the joint ECDQ of the source super vector S^m:

$$H(Q_{\Lambda,1}, \ldots, Q_{\Lambda,k} | \mathbf{U}_1, \ldots, \mathbf{U}_k) = H(Q_{\Lambda,m}(S^m + U^m)|U^m) \tag{5.19}$$

$$= I(S^m; S^m + U^m_{\text{eq}}), \tag{5.20}$$

where U^m_{eq} is a concatenation of k equivalent dithers, and for short we denote $Q_\Lambda(\mathbf{S}_i + \mathbf{U}_i) = Q_{\Lambda,i}$, $i = 1, \ldots, n$.

5.3.1   Feedback coding

Dependence between the quantizer inputs can also be created due to feedback. In the system shown in Figure 5.3, past quantizer outputs are processed and combined with the current input. This situation is typical of *predictive coding*, where the goal is to encode only the *innovation* part of the source. Feedback is also used in *noise*

---

[3] Note that the concatenation of codewords must be *uniquely decodable* (UD), i.e., the decoder must be able to parse the code sequence into separate codewords. A delicate point in independent dithering is that the entropy code, which depends on the dither, must remain UD over time. A simple solution for that is to use an *instantaneous code*, also called a prefix code, where the decoder can identify the end of a codeword regardless of what comes next.

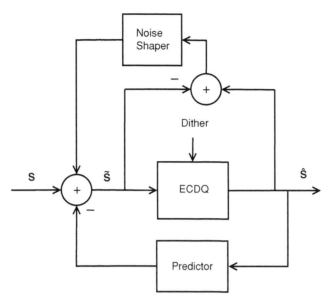

Figure 5.3 Typical feedback coding schemes.

shaping, where the goal is to shape the spectrum of the equivalent quantization noise.

In the presence of feedback, the current quantizer input depends on present and past values of the source and dither, and on past values of the reconstruction:

$$\tilde{S}_n = f(S_1, \ldots, S_n, U_1, \ldots, U_n, \hat{S}_1, \ldots, \hat{S}_{n-1}). \qquad (5.21)$$

One implication of this sequential structure is that it works only for a *scalar* quantizer, because $\hat{S}_{n-1}$ must be available before S_n is quantized. [4]

Another implication of (5.21) is that, although each new dither value is still independent of the previous quantizer inputs, the dither sequence $U_1, \ldots, U_m$ depends on the quantizer input sequence $\tilde{S}_1, \ldots, \tilde{S}_m$; hence formula (5.20) for the *joint* quantizer entropy is not valid (step (*d*2) in (5.16) cannot be justified). Nevertheless, it is easy to show that

$$H(Q_{\Lambda,1}, \ldots, Q_{\Lambda,k} \mid U_1, \ldots, U_k) = I(S^m; \tilde{S}^m + U_{\text{eq}}^m), \qquad (5.22)$$

where $Q_{\Lambda,i} = Q_\Lambda(\tilde{S}_i + U_i) =, i = 1, \ldots, k$, i.e. the equivalent channel input is the source *before* it is combined with the feedback, while the output is, as before, the sum of the quantizer input and the equivalent dither; see Problem 5.5. [5]

---

[4] See an extension to a *block-sequential* scheme in [291].

[5] An alternative form for (5.22) uses a sequential version of the mutual information, known as *directed mutual information* [183, 291].

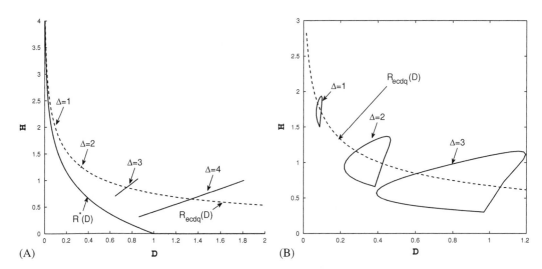

Figure 5.4 (A) Entropy-distortion curves for a Gaussian source $S \sim N(0, 1)$, and a uniform step-Δ quantizer with a *deterministic* dither u. Each curve corresponds to one step size value Δ (for $\Delta = 1, 2, 3$ and 4), and a full "cycle" of dither values in the range $-\Delta/2 < u < \Delta/2$. We see that in each curve, some dither values imply better entropy-distortion pairs than others. The dashed line shows the *average* performance over $U \sim \text{Unif}(-\Delta/2, \Delta/2)$, which is the function $R_{ECDQ}(D)$ of (5.6). Also shown for comparison is the Shannon rate-distortion function $R^*(D) = 0.5 \log(1/D)$. (B) The same curves for an exponential source distribution.

5.4   Entropy-distortion trade-off

For each value $\mathbf{u}$ of the random dither $\mathbf{U}$, the lattice gets an offset $\mathbf{u}$, and the ECDQ achieves a certain rate-distortion pair. Specifically, the rate per source vector is $H(Q_\Lambda(\mathbf{S} + \mathbf{u}))$. From (4.4), the quantization error is $Q_\Lambda(\mathbf{S} + \mathbf{u}) - \mathbf{u} - \mathbf{S}$, so the mean-squared distortion is $E\|Q_\Lambda(\mathbf{S} + \mathbf{u}) - \mathbf{u} - \mathbf{S}\|^2$, where the expectation is over the distribution of the source $\mathbf{S}$; see the example of scalar-lattice quantization of Gaussian and exponential sources in Figure 5.4. (Note that we do not use the Wiener estimator of Section 4.5 to reduce the distortion until Section 5.6.2.)

For a general lattice, the entropy and distortion for a specific value of the dither are, in general, hard to evaluate. Nevertheless, by combining Theorems 4.1.1 and 5.2.1 we can easily calculate the *expected* ECDQ rate-distortion performance.

Recall that the formula (5.12) for the quantizer conditional entropy can be written as (5.14):

$$I(\mathbf{S}; \mathbf{S} - \mathbf{U}) = h(\mathbf{S} - \mathbf{U}) - \log(V(\Lambda)). \tag{5.23}$$

Next, from (4.10), the mean-squared distortion per dimension *under Voronoi partition* is equal to the lattice second moment:

$$D = \sigma^2(\Lambda). \tag{5.24}$$

By (3.21), this can be written as $\sigma^2(\Lambda) = G(\Lambda) \cdot V^{2/n}(\Lambda)$, where $G(\Lambda)$ is the NSM of the lattice quantizer. [6] Thus, we can write the ECDQ coding rate (5.6) versus MSE trade-off as

$$R_{\text{ECDQ}}(D) = \frac{1}{n}h(\mathbf{S} - \mathbf{U}) - \frac{1}{2}\log\left(\frac{D}{G(\Lambda)}\right). \tag{5.25}$$

Example 5.4.1 *The NSM of the scalar uniform quantizer Q_Δ is $G(\Lambda) = 1/12$, and the dither U is uniform over the interval $[-\sqrt{3D}, +\sqrt{3D})$ (see (4.11)). Thus, if the source is a zero-mean Gaussian, $S \sim N(0, \sigma_s^2)$, the density of the independent sum $S - U$ is given by the convolution of the densities of its components*

$$f_{S-U}(x) = f_S(x) * f_U(x) = \frac{1}{2\sqrt{3D}} \int_{-\sqrt{3D}}^{\sqrt{3D}} \frac{1}{\sqrt{2\pi\sigma_s^2}} e^{t^2/2\sigma_s^2} dt, \tag{5.26}$$

and

$$H(Q_\Delta(S+U)|U) = -\int f_{S-U}(x)\log f_{S-U}(x)dx - \frac{1}{2}\log(12D). \tag{5.27}$$

5.4.1 ECDQ with high resolution

For a small distortion D, the convolution (5.26) does not change much the distribution of the source S. In this case, we have $h(S - U) \approx h(S)$, which simplifies (5.23) and the entropy-distortion trade-off of Example 5.4.1. This approximation holds, in fact, under very general conditions on the source $\mathbf{S}$, the lattice Λ, its associated partition $\mathcal{P}$, and the distortion measure $d(\mathbf{s}, \hat{\mathbf{s}})$.

Proposition 5.4.1 (Continuity of differential entropy [164]) *If $h(\mathbf{S})$ exists, and $\mathbf{U}$ has a finite second moment (e.g., $\mathbf{U}$ is a uniform dither over the fundamental lattice cell $\mathcal{P}_0$), then*

$$\lim_{\alpha \to 0} h(\mathbf{S} + \alpha\mathbf{U}) = h(\mathbf{S}).$$

It follows that for small distortion, we can approximate $h(\mathbf{S} - \mathbf{U})$ in (5.25) as $h(\mathbf{S})$.

Proposition 5.4.2 (Entropy for a small distortion) *For an NN lattice quantizer under mean-squared error:*

$$R_{\text{ECDQ}}(D) = \frac{1}{n}h(\mathbf{S}) - \frac{1}{2}\log\left(\frac{D}{G(\Lambda)}\right) + o(1), \tag{5.28}$$

where $o(1) \to 0$ as $D \to 0$.

---

[6] For a general partition $\mathcal{P}$ the distortion-volume trade-off will of course be worse.

The resulting rate-distortion trade-off is not only very simple to evaluate, but it also has a striking resemblance to Shannon's rate-distortion function, which we introduce in the next section.

Difference distortion measures

A similar analysis applies for more general *difference distortion measures*, i.e., distortion measures of the form

$$d(\mathbf{s}, \hat{\mathbf{s}}) = \rho(\mathbf{s} - \hat{\mathbf{s}}). \tag{5.29}$$

The average distortion, as for MSE (4.10), is only a function of the distribution of the dithered quantization error $Q_\Lambda(\mathbf{s} + \mathbf{U}) - \mathbf{U} - \mathbf{s}$ (4.4). So by Theorem 4.1.1 it is equal to $E\{\rho(\mathbf{U})\}$.

For example, the rth moment of a uniform distribution over the interval $(-\Delta/2, +\Delta/2)$ is given by $\Delta^r/\gamma(r)$, for $r > 0$, where $\gamma(r) = (r + 1)2^r$. We can thus modify (5.27), and characterize the entropy-distortion trade-off of the scalar ECDQ for a general rth power distortion measure, as (Problem 5.6)

$$H(Q_\Delta(S + U)|U) = -\int f_{S-U}(x) \log f_{S-U}(x) dx - \frac{1}{r} \log(\gamma(r)D). \tag{5.30}$$

At high-resolution conditions, the first term becomes $h(S)$, as in the squared-error case (5.28).

5.5   Redundancy over Shannon

Now that the two main performance figures of the dithered quantizer – distortion and coding rate – have been characterized, we turn to examine two important aspects: its distance from optimality and its sensitivity to the source parameters. Our reference for the first question will be the information theoretic bound on lossy compression: Shannon's *rate-distortion function*.

5.5.1   Shannon's rate-distortion function

The Shannon rate-distortion function of a source vector $\mathbf{S}$ under a distortion measure d, is defined as:

$$R_\mathbf{S}(D) = \frac{1}{n} \inf_{\hat{\mathbf{S}}:\, E\{d(\mathbf{S}, \hat{\mathbf{S}})\} \leq D} I(\mathbf{S}; \hat{\mathbf{S}}). \tag{5.31}$$

The minimization of the mutual information in (5.31) is taken over all conditional distributions $p(\hat{\mathbf{s}}|\mathbf{s})$ of $\hat{\mathbf{S}}$ given $\mathbf{S}$, called *"test channels,"* such that the distortion constraint $E\{d(\mathbf{S}, \hat{\mathbf{S}})\} \leq D$ is satisfied. As *information theory* shows, the rate-distortion function characterizes the minimum achievable rate, by any scheme, in coding a sequence of independent and identically distributed copies of the vector source $\mathbf{S}$ under the distortion measure d. It can be simplified to a *single-letter* quantity if the source is *memoryless* and the distortion measure is *per letter*, i.e.,

$$d(\mathbf{s}, \hat{\mathbf{s}}) = \frac{1}{n} \sum_{i=1}^{n} d(s_i, \hat{s}_i). \tag{5.32}$$

For example, the rate-distortion function of a zero-mean white-Gaussian source S with variance σ_s^2 under the squared-error (per letter) distortion measure $d(\mathbf{s}, \hat{\mathbf{s}}) = \frac{1}{n} \sum_{i=1}^{n} (s_i - \hat{s}_i)^2$, is given by

$$R^*(D) = \frac{1}{2} \log \left(\frac{\sigma_s^2}{D} \right), \quad \text{for } 0 < D \leq \sigma_s^2. \tag{5.33}$$

We can realize the quadratic-Gaussian rate-distortion function $R^*(D) = I(S; \hat{S})$ using a linear AWGN test channel $p(\hat{s}|s)$ of the form

$$\hat{S} = \beta(\alpha S + Z), \tag{5.34}$$

where $Z \sim N(0, \sigma_z^2)$, and where the linear coefficients α and β and the noise variance σ_z^2 are any triplet satisfying

$$\alpha\beta = 1 - \frac{D}{\sigma_s^2} \quad \text{and} \quad \sigma_z^2 = \frac{\alpha}{\beta} D. \tag{5.35}$$

Although, without loss of generality, we can set $\alpha = 1$ or $\beta = 1$ (and get a *two-parameter* representation), the "pre/post-scaling" form (5.34) proves to be useful in the sequel. [7]

5.5.2 The Shannon lower bound

For a general source, the quadratic rate-distortion function is greater than the following bound due to Shannon.

Proposition 5.5.1 (SLB) *For squared-error distortion measure, and a continuous source $\mathbf{S}$ having differential entropy $h(\mathbf{S})$,*

$$R_{\mathbf{S}}(D) \geq \frac{1}{n} h(\mathbf{S}) - \frac{1}{2} \log(2\pi e D). \tag{5.36}$$

The SLB follows from the *maximum entropy* property of the Gaussian distribution (A.13). The differential entropy of an i.i.d. Gaussian distribution with variance σ_s^2 is

$$h(\mathbf{S}) = \frac{n}{2} \log(2\pi e \sigma_s^2); \tag{5.37}$$

---

[7] Another simple equivalent representation for the joint distribution in (5.34) has a *backward channel* form: $S = \hat{S} + Z$, where $Z \sim N(0, D)$ is independent of $\hat{S}$.

thus, in view of (5.33), the SLB is tight for a white-Gaussian source. [8] Further-more, by a continuity argument, the SLB is *asymptotically tight* for small distortion levels.

Proposition 5.5.2 (Tightness of SLB) *If* $h(\mathbf{S})$ *exists, then*

$$R_\mathbf{S}(D) = \frac{1}{n}h(\mathbf{S}) - \frac{1}{2}\log(2\pi e D) + o(1), \qquad (5.38)$$

where $o(1) \to 0$ *as* $D \to 0$. *The optimum test channel which realizes (5.38) asymptotically is an AWGN channel:*

$$\hat{\mathbf{S}} = \mathbf{S} + \mathbf{Z}^*, \qquad (5.39)$$

where $\mathbf{Z}^*$ *is AWGN with variance D.*

Note that the existence of differential entropy implies that the source has a probability density function, i.e., no discrete values have a positive probability. [9]

Proof The channel $\mathbf{S} \to \hat{\mathbf{S}}$ of (5.39) satisfies the constraint of the minimization (5.31), and we have

$$
\begin{aligned}
I(\mathbf{S}; \hat{\mathbf{S}}) &= I(\mathbf{S}; \mathbf{S} + \mathbf{Z}^*) \\
&\overset{a}{=} h(\mathbf{S} + \mathbf{Z}^*) - h(\mathbf{Z}^*) \\
&\overset{b}{=} h(\mathbf{S} + \mathbf{Z}^*) - \frac{n}{2}\log(2\pi e D) \\
&\overset{c}{=} h(\mathbf{S}) - \frac{n}{2}\log(2\pi e D) + n \cdot o(1),
\end{aligned}
\qquad (5.40)
$$

where (*a*) follows as in (5.23); in (*b*) we substituted the Gaussian entropy; and (*c*) follows from Proposition 5.4.1. Thus, by definition (5.31) the rate-distortion function $R_\mathbf{S}(D)$ is bounded from above by the right-hand side of (5.38). In view of the SLB, this bound is, in fact, asymptotically tight.   □

5.5.3 Redundancy at high resolution

We see that for small distortion, the optimum test channel for the quadratic rate-distortion function takes the form of an AWGN channel (5.39). This is the same as the equivalent ECDQ channel of Figure 4.5, except that the dither is uniform and not Gaussian. Combining (5.28) and Proposition 5.5.2, we thus obtain the following.

---

[8] An alternative form for the SLB, which resembles the white-Gaussian rate-distortion function (5.33), is $R_\mathbf{S}(D) \geq \frac{1}{2}\log(P_E(\mathbf{S})/D)$, where $P_E(\mathbf{S}) = 2^{2h(\mathbf{S})/n}/2\pi e$ is the entropy power of $\mathbf{S}$ (A.15).

[9] For a discrete source, the differential entropy and hence the SLB are equal to $-\infty$, while the rate-distortion function is of course non-negative for all D.

Theorem 5.5.1 (ECDQ at high resolution) *For NN lattice quantization of a continuous source* **S**, *with differential entropy* $h(\mathbf{S})$, *under a squared-error distortion measure,*

$$\lim_{D \to 0} [R_{\mathrm{ECDQ}}(D) - R(D)] = \frac{1}{2} \log(2\pi e G(\Lambda)). \tag{5.41}$$

The beauty of Theorem 5.5.1 is that the redundancy term $\frac{1}{2} \log(2\pi e G(\Lambda))$ is *independent of the source* **S**, except for it being continuous.

Example 5.5.1 (Gap from Shannon) *The NSM of a scalar uniform quantizer is* $G(\Lambda) = G_1 = \frac{1}{12}$, *so its asymptotic redundancy (5.41) is*

$$\frac{1}{2} \log \left(\frac{2\pi e}{12} \right) \approx 0.254 \text{ bit.} \tag{5.42}$$

Indeed the gap between the curves $R_{\mathrm{ECDQ}}(D)$ *and* $R^*(D)$ *in Figure 5.4(A) is about a quarter of a bit (or 1.5 dB in the distortion axis* [10]*).*

Entropy-constrained quantization theory

The asymptotic redundancy result of Theorem 5.5.1 has the form of the redundancy of the optimum *entropy-constrained vector quantizer* (ECVQ) in high-resolution quantization theory [102, 103]. Indeed, if Gersho's conjecture is true, and the best space-filling polytope is a lattice cell (see Section 3.2), then the optimum n-dimensional ECVQ at the high-resolution regime takes the form of a lattice. The dither – as we may conclude from this coincidence – does not incur any loss or improvement in this regime.

Gaussian dither In Chapter 7 we shall relate the redundancy term $\frac{1}{2} \log(2\pi e G(\Lambda)$ with the distance of the dither from AWGN. We shall also observe that for a sequence of "good" lattices

$$G(\Lambda_n) \to \frac{1}{2\pi e}, \tag{5.43}$$

and therefore the dither, in the sense of this distance measure, becomes AWGN. The ECDQ thus amounts in this limit to an AWGN channel. Moreover, the asymptotic redundancy theorem above implies that for such good lattices, the ECDQ approaches Shannon's rate-distortion function in the double limit of high dimension and small distortion.

---

[10] Since the squared distortion scales like $D \propto 2^{-2R}$ (see the SLB (5.38)), each bit of information amounts to reduction of D by a factor of 4, corresponding to a $10 \log_{10}(4) \approx 6$ dB gap.

5.6   Optimum test-channel simulation

So far, we have concentrated on the asymptotic optimality of the ECDQ in the limit of high-resolution quantization. As we shall show in this section, a pre/post-filtered ECDQ can "simulate" the test channel (5.34) that realizes the quadratic-Gaussian rate-distortion function $R^*(D)$. This implies that for a Gaussian source, the ECDQ can, in fact, be optimal at *any* resolution.

5.6.1   Entropy bounds

We first develop some bounds on $R_{\text{ECDQ}}(D)$ for a *general resolution*, by bounding the entropy term $h(\mathbf{S} - \mathbf{U})$ in (5.25): an upper bound using the *maximum-entropy* property of the Gaussian distribution, and a lower bound via the *entropy power inequality* (EPI). See (A.13) and (A.16) in the Appendix.

Proposition 5.6.1 (Maximum-entropy bound) *Let* $\mathbf{S}$ *be a source with an average variance per dimension* $\sigma_s^2 = \frac{1}{n} E \|\mathbf{S} - E\{\mathbf{S}\}\|^2$. *Then, for NN lattice quantization,*

$$R_{\text{ECDQ}} \leq \frac{1}{2} \log \left(1 + \frac{\sigma_s^2}{\sigma_Q^2} \right) + \frac{1}{2} \log \left(2\pi e G(\Lambda) \right), \qquad (5.44)$$

where $\sigma_Q^2 = \sigma^2(\Lambda)$.

Proof Since the source $\mathbf{S}$ and the dither $\mathbf{U}$ are independent, the average variance per dimension of the vector $\mathbf{S} - \mathbf{U}$ is equal to the sum of their variances: $\sigma_s^2 + \sigma^2(\Lambda)$. By the *maximum-entropy* property of the Gaussian distribution (A.13), the entropy of this vector is bounded from above by the entropy of a white-Gaussian vector with the same variance, i.e.,

$$\frac{1}{n} h(\mathbf{S} - \mathbf{U}) \leq \frac{1}{2} \log(2\pi e (\sigma_s^2 + \sigma^2(\Lambda))). \qquad (5.45)$$

Substituting (5.45) in (5.25), we obtain the upper bound (5.44).   □

Proposition 5.6.2 (EPI-based bound) *If the source is white Gaussian, then*

$$R_{\text{ECDQ}} \geq \frac{1}{2} \log \left(1 + 2\pi e G(\Lambda) \frac{\sigma_s^2}{\sigma^2(\Lambda)} \right). \qquad (5.46)$$

Proof By the EPI (A.16),

$$2^{\frac{2}{n} h(\mathbf{S} - \mathbf{U})} \geq 2^{\frac{2}{n} h(\mathbf{S})} + 2^{\frac{2}{n} h(\mathbf{U})}.$$

Substituting $h(\mathbf{U}) = \log V(\Lambda) = \frac{n}{2} \log(\sigma^2(\Lambda)/G(\Lambda))$ and, for a white-Gaussian source, $h(\mathbf{S}) = \frac{n}{2} \log(2\pi e \sigma_s^2)$, the lower bound (5.46) follows.   □

Gaussian dither If $G(\Lambda_n) \to \frac{1}{2\pi e}$ (as already mentioned in (5.43), the case where the dither "becomes Gaussian"), then for a white-Gaussian source both bounds

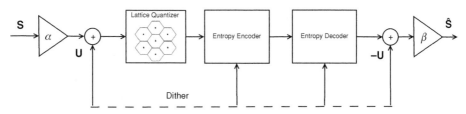

Figure 5.5 Pre/post-scaled ECDQ.

(5.44) and (5.46) coincide. Thus, for a white-Gaussian source and Gaussian dither, setting $\sigma^2(\Lambda) = D$, we obtain

$$R_{\text{ECDQ}}(D) = \frac{1}{2} \log \left(1 + \frac{\sigma_s^2}{D} \right). \tag{5.47}$$

Comparing (5.47) with the Gaussian rate-distortion function (5.33), we see that even in the ideal limit $G(\Lambda_n) \to \frac{1}{2\pi e}$ there is an "extra one" inside the logarithm, corresponding to some rate loss with respect to Shannon. We next show that this loss can be removed by appropriate *"pre/post-filtering"* of the ECDQ.

5.6.2 Pre/post-scaled ECDQ

Consider the post-filtered dithered quantizer of Section 4.5 (Figure 4.10). It is interesting to observe that the Wiener coefficient $\beta^* = \sigma_s^2/(\sigma_Q^2 + \sigma_s^2)$ (4.49), and distortion $D^* = \sigma_Q^2 \cdot \sigma_s^2/(\sigma_Q^2 + \sigma_s^2)$ (4.50), coincide with the parameters β and D of the linear-additive (*forward test-channel*) realization (5.34) of the quadratic-Gaussian rate-distortion function, for the case $\alpha = 1$. Let us now add a pre-scaling coefficient α and an entropy-coding stage, to obtain the pre/post-scaled ECDQ shown in Figure 5.5. By Theorems 4.1.1 and 5.2.1, this scheme is *equivalent* to the linear-additive test channel (5.34), except that the dither is uniform and not Gaussian. The equivalence is in terms of both rate and distortion, and it holds for all suitable pairs of pre/post-scaling coefficients α and β.

Lemma 5.6.1 (Pre/post-scaling invariance) *The rate*

$$R = \frac{1}{n} H(Q_\Lambda(\alpha \mathbf{S} + \mathbf{U})|\mathbf{U}) \tag{5.48}$$

and quadratic distortion

$$D = \frac{1}{n} E \| \beta \cdot (Q_\Lambda(\alpha \mathbf{S} + \mathbf{U}) - \mathbf{U}) - \mathbf{S} \|^2 \tag{5.49}$$

of a pre/post-scaled ECDQ are invariant to the choice of α and β, as long as they satisfy the conditions (5.35) of the linear-additive realization of $R^*(D)$ (with noise *power $\sigma_z^2 = \sigma^2(\Lambda)$):* [11]

$$\alpha\beta = 1 - \frac{D}{\sigma_s^2} \quad and \quad \sigma^2(\Lambda) = \frac{\alpha}{\beta}D. \tag{5.50}$$

Proof The result follows from the equivalent additive-noise channel of ECDQ, linearity, and the invariance of the mutual information to scaling.   □

Note that for $\alpha = 1$, we get the standard Wiener coefficient $\beta = \beta^*$ and distortion (4.49)–(4.50). Alternatively, we can use a symmetric scaling:

$$\alpha = \beta = \sqrt{1 - D/\sigma_s^2} \quad and \quad \sigma^2(\Lambda) = D, \tag{5.51}$$

i.e., *attenuate* the source (by $\alpha < 1$) prior to the quantization, and get the same rate and distortion.

To see the gain due to scaling more easily, let us assume the $\alpha = 1$ case. It follows from (5.50), that for a target distortion level D, we use a lattice with a second moment

$$\sigma^2(\Lambda) = \frac{1}{1/D - 1/\sigma_s^2}. \tag{5.52}$$

This lattice is *coarser* than the one used in the unscaled ECDQ, where $\sigma^2(\Lambda) = D$; see (5.24). Intuitively, a coarser lattice leads to a smaller coding rate; or in terms of the equivalent additive-noise channel of Figure 4.5, a stronger dither reduces the mutual information (5.7). The improvement is most dramatic in the low-resolution regime, as can be seen in Figure 5.6. We thus conclude that the reduction in distortion due to Wiener estimation translates into a rate saving for the *same* distortion.

The improvement in the entropy-distortion performance is also reflected in the upper bound (5.44). If we substitute the higher second moment (5.52) as σ_Q^2 in (5.44), we obtain

$$R_{\text{ECDQ}}^{\text{pre/post}}(D) \leq \frac{1}{2}\log\left(\frac{\sigma_s^2}{D}\right) + \frac{1}{2}\log\left(2\pi e G(\Lambda)\right). \tag{5.53}$$

That is, we have got rid of the extra "1" inside the log term in (5.47). Comparing with Shannon's rate-distortion function $R^*(D)$ of a white-Gaussian source (5.33), we conclude the following.

Theorem 5.6.1 (Redundancy of pre/post-scaled ECDQ) *For a white-Gaussian source and quadratic distortion, encoded with the pre/post-scaling parameters (5.50),*

$$R_{\text{ECDQ}}^{\text{pre/post}}(D) - R^*(D) \leq \frac{1}{2}\log\left(2\pi e G(\Lambda)\right), \tag{5.54}$$

at all *distortion levels D.*

---

[11] Note that in the sequel we assume an NN quantization rule (Voronoi partition), so $\sigma_Q^2 = \sigma^2(\Lambda)$.

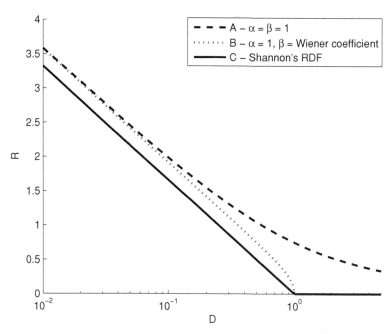

Figure 5.6 Scalar ECDQ of a Gaussian source $S \sim N(0, 1)$. By varying the quantizer step size Δ we obtain the entire entropy-distortion curve as shown in curve (A). Curve (B) shows the improvement in the entropy-distortion trade-off when the reconstruction is scaled by the Wiener coefficient. (In particular, the curve hits the zero-rate axis at $D = \sigma_s^2 = 1$, while in the unscaled ECDQ case the rate at this point is $R \approx 3/4$ bit.) For comparison, curve (C) shows the Shannon rate-distortion function $R^*(D) = 0.5 \log(1/D)$.

For example, for a scalar ECDQ the redundancy in (5.54) is at most $\frac{1}{2} \log \left(\frac{2\pi e}{12} \right) \approx$ 0.254 bit over the *entire* range of distortions, as can be seen by comparing curves (B) and (C) in Figure 5.6.

Gaussian dither If $G(\Lambda_n) \to \frac{1}{2\pi e}$ (5.43), then the redundancy in (5.54) goes to zero. In other words, in this case the pre/post-scaled ECDQ system approaches the white-Gaussian source rate-distortion function. Indeed, not only are the second-order parameters of this system (α, β and $\sigma^2(\Lambda)$) the same as those of the linear test-channel (5.35) realizing $R^*(D)$, but in the limit $G(\Lambda_n) \to \frac{1}{2\pi e}$, the system actually *simulates* this channel.

5.7   Comparison with Lloyd's conditions

It is interesting to compare the structure of the dithered lattice quantizer, which is independent of the source statistics, to that of the optimum vector quantizer matched to the source statistics. The comparison is based on Lloyd's (necessary) conditions,

which provide a simple characterization for the optimum quantizer. See [174] for
the scalar case and its extension in [161] for the vector quantizer case.

1. **Code vectors** The code vectors $\{c_i\}$ satisfy

$$i\text{th code vector} = \begin{cases} \text{"centroid" of }i\text{th cell,} & \text{for optimum quantization} \\ i\text{th lattice point,} & \text{for dithered quantization,} \end{cases}$$

where the "centroid" is the vector minimizing the conditional expected distortion
in the ith cell (for MSE distortion, it is equal to the conditional expectation).

2. **Implication for MSE** The quantization error $\hat{S} - S$ satisfies

$$\hat{S} - S \sim \begin{cases} \text{uncorrelated with }\hat{S}, & \text{for optimum quantization} \\ \text{uncorrelated with }S, & \text{for dithered quantization.} \end{cases}$$

3. Another implication for MSE:

$$E\|\hat{S}\|^2 = \begin{cases} \sigma_s^2 - D, & \text{for optimum quantization} \\ \sigma_s^2 + D, & \text{for dithered quantization.} \end{cases}$$

4. **Partition** For optimum fixed-rate quantization, the partition satisfies Lloyd's
   nearest-neighbor condition. For optimum variable-rate quantization, the partition
   is relative to a weighted distortion-codelength criterion. For the ECDQ, the
   partition much be a valid lattice partition.

5. **Distortion distribution** For optimum quantization, the distortion per cell varies
   between the cells. For dithered quantization, the distortion is the same for all
   cells.

6. **High-resolution conditions** In the limit as $D \to 0$, the structure of the optimum
   "entropy-constrained" quantizer converges to a lattice with Voronoi partition, and
   its entropy coincides with the ECDQ entropy. Note that in this limit, the entropy
   and distortion are invariant to the specific dither value $u \in \mathcal{V}_0$.

5.8   Is random dither really necessary?

Let us return to the question raised in the end of Chapter 4, now with the additional
perspective of the entropy-distortion trade-off. Observe from Figure 5.4 that any
specific dither value and lattice resolution corresponds to a different point in the
rate-distortion plane. The average of these points gives the performance curve
$R_{\text{ECDQ}}(D)$ of a (randomized) ECDQ. In contrast, the *lower convex envelope* of these
points gives us the best attainable rate-distortion performance with *non*-randomized
entropy-coded lattice quantization, which is strictly better than $R_{\text{ECDQ}}(D)$.

We should remember, though, that choosing a specific dither value means losing
the uniform (source-independent) distortion property of randomized quantization,
as promised by Theorem 4.1.1. Furthermore, the variation in distortion becomes

stronger if post-scaling (e.g., Wiener estimation) is used. To see that, observe that the total quantization error can be written as (see (4.43) and (5.49))

$$\hat{\mathbf{s}} - \mathbf{s} = \beta \cdot \mathbf{e}_q + (\beta - 1) \cdot \mathbf{s},$$

where $\mathbf{e}_q = -[\mathbf{s} + \mathbf{u}] \bmod \Lambda$, and β is the estimation coefficient. Note that $\mathbf{e}_q$ can take any value inside the lattice cell, by a suitable choice of the dither vector $\mathbf{u}$. Thus, for a given β and $\mathbf{s}$, some values of $\mathbf{u}$ will increase the absolute error $\|\hat{\mathbf{s}} - \mathbf{s}\|$, and some will decrease it.

Nevertheless, as can be seen in Figure 5.4, for small distortion the curves associated with different dither values shrink to a point. Thus, in the high-resolution regime all dither values (including zero) are equivalent. A similar phenomenon occurs for a sequence of "good" lattice quantizers, in the limit as $G(\Lambda_n) \to 1/2\pi e$. In this case, assuming a Gaussian source and Wiener estimation at the decoder, $R_{\text{ECDQ}}(D) \to R^*(D)$, as the lattice dimension n goes to infinity, i.e., the *average* performance of the randomized ECDQ is asymptotically optimal. Since no coding scheme can beat $R^*(D)$, it follows that *almost* all dither vectors must be asymptotically equivalent, and good. This, however, does not necessarily include a *zero* dither vector, so some non-zero translation of the lattice may be necessary for optimum performance.

5.9 Universal quantization*

Optimum compression of a source requires knowledge of its statistics. For quantization, this amounts to the joint design of the code vectors and the quantization cells. If entropy coding is allowed, then the source statistics is also used to design the mapping of quantizer outputs into binary codewords, for example, the Shannon–Fano code (5.1). In many real life scenarios, however, the source statistics is not known a priori.

The lattice quantizer is universal in the sense that its structure is independent of the source statistics. Theorem 4.1.1 showed us that dithered lattice quantization guarantees a target distortion level independent of the source statistics. Furthermore, Theorem 5.5.1 above and Theorem 5.9.1 below show that the rate redundancy of the ECDQ is bounded by universal constants. Can we achieve these results without knowing the source statistics in advance?

Indeed, there exist *universal lossless compression* schemes, for example, the Lempel–Ziv algorithm and others, which can sequentially approach the entropy of any stationary and ergodic source. Such schemes, if conditioned on the known (pseudo-) random dither sequence, can approach the conditional entropy of the quantizer (5.5). Thus, ECDQ with universal entropy coding is a universal lossy compression scheme, though with a slightly weaker notion of universality: it guarantees a fixed bound on the loss with respect to Shannon, universally for all sources.

5.9.1   Ziv's bound

The power of ECDQ, unlike high-resolution quantization theory, is that its behavior follows the same simple rules (Theorems 4.1.1 and 5.2.1) for all distortion levels D. A particularly interesting outcome of this continuity is the universal bound below on the redundancy of the ECDQ. This bound is slightly weaker than that of Theorem 5.5.1, nevertheless it holds for *any* source (not necessarily continuous) and for *all* distortion levels.

Definition 5.9.1 (Ziv's universal constant) *The* capacity C_{Ziv} *of the equivalent additive-noise channel* $\mathbf{Y} = \mathbf{X} + \mathbf{U}_{\mathrm{eq}}$ *of the ECDQ (4.7) at signal to dither ratio equal to 1 is defined as:*

$$C_{\mathrm{Ziv}}(\Lambda) = \frac{1}{n} \sup_{\mathbf{X}:\frac{1}{n}E\|\mathbf{X}\|^2 \leq \sigma^2(\Lambda)} I(\mathbf{X}; \mathbf{X} + \mathbf{U}_{\mathrm{eq}}). \tag{5.55}$$

Note that for NN quantization $\frac{1}{n}E\|\mathbf{U}\|^2 = \sigma^2(\Lambda)$ by Theorem 4.1.1, so the signal to noise ratio in the equivalent channel is indeed bounded by 1. Also, since the mutual information is invariant to scaling of its arguments, Ziv's constant is, in fact, independent of the scaling of the lattice Λ.

Theorem 5.9.1 (Universal bound on redundancy) *For any source* $\mathbf{S}$ *and squared distortion level D,*

$$R_{\mathrm{ECDQ}}(D) - R_{\mathbf{S}}(D) \leq C_{\mathrm{Ziv}}(\Lambda). \tag{5.56}$$

Furthermore, this bound is tight for some sources and distortion levels.

Before we prove Theorem 5.9.1 below, note that in the scalar case

$$C_{\mathrm{Ziv}}(\Lambda) \leq \frac{1}{2} + \frac{1}{2} \log(2\pi e/12) \approx 0.754 \text{ bit,}$$

while in the general lattice case

$$\frac{1}{2} \log(1 + 2\pi e G(\Lambda)) \leq C_{\mathrm{Ziv}}(\Lambda) \leq \frac{1}{2} + \frac{1}{2} \log(2\pi e G(\Lambda)), \tag{5.57}$$

where the bounds in (5.57) follow by substituting the special case $\sigma_s^2/\sigma^2(\Lambda) = 1$ (i.e., signal to dither ratio equal to 1) in the maximum-entropy and entropy-power bounds (Propositions 5.6.1 and 5.6.2) of Section 5.6.2. As can be seen in the numerical evaluation in Figure 5.7, the upper bound is quite tight. Furthermore, both upper and lower bounds converge to 1/2 bit if $G(\Lambda) \to 1/2\pi e$. Thus, Ziv's constant is roughly 1/2 bit larger than the asymptotic redundancy of Theorem 5.5.1.

Proof of Theorem 5.9.1 For any test channel $\mathbf{S} \to \hat{\mathbf{S}}$ in the definition of the rate-distortion function (5.31), and dither $\mathbf{U}$ such that $\frac{1}{n}E\|\mathbf{U}\|^2 = \sigma^2(\Lambda) = D$, we have

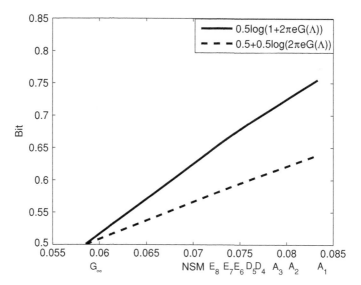

Figure 5.7 Bounds on Ziv's constant in the range $\frac{1}{2\pi e} \leq G(\Lambda) \leq \frac{1}{12}$.

the following chain of inequalities:

$$I(\mathbf{S}; \mathbf{S} - \mathbf{U}) - I(\mathbf{S}; \hat{\mathbf{S}}) \overset{a}{=} I(\mathbf{S}; \mathbf{S} - \mathbf{U}|\hat{\mathbf{S}}) - I(\mathbf{S}; \hat{\mathbf{S}}|\mathbf{S} - \mathbf{U})$$

$$\overset{b}{\leq} I(\mathbf{S}; \mathbf{S} - \mathbf{U}|\hat{\mathbf{S}})$$

$$\overset{c}{=} I(\mathbf{S} - \hat{\mathbf{S}}; \mathbf{S} - \hat{\mathbf{S}} - \mathbf{U}|\hat{\mathbf{S}})$$

$$\overset{d}{\leq} I(\mathbf{S} - \hat{\mathbf{S}}; \mathbf{S} - \hat{\mathbf{S}} - \mathbf{U})$$

$$\overset{e}{\leq} n \cdot C_{\text{Ziv}}(\Lambda), \tag{5.58}$$

where (a) follows using the chain rule to decompose the mutual information $I(\mathbf{S}; \mathbf{S} - \mathbf{U}, \hat{\mathbf{S}})$ in two ways; (b) follows from the non-negativity of the mutual information; (c) follows since subtracting the condition does not change the conditional mutual information; (d) follows since (because the dither $\mathbf{U}$ is independent of the source $\mathbf{S}$ and the output of the test channel $\hat{\mathbf{S}}$) the triple $\hat{\mathbf{S}} \to (\mathbf{S} - \hat{\mathbf{S}}) \to (\mathbf{S} - \hat{\mathbf{S}} - \mathbf{U})$ forms a Markov chain; finally (e) follows from the definition of C_{Ziv}, since $\frac{1}{n} E\|\mathbf{S} - \hat{\mathbf{S}}\|^2 \leq \sigma^2(\Lambda)$. The upper bound now follows from Theorem 5.2.1, and the definition of the rate-distortion function (5.31).

To see the condition for equality, consider a source $\mathbf{S}^*$ which achieves the maximization of Ziv's constant in (5.55). ECDQ encoding of this source at distortion level $D = \sigma^2(\Lambda)$ gives a rate of $R = R_{\text{ECDQ}}(D) = C_{\text{Ziv}}(\Lambda)$. On the other hand, since, by (5.55), D is greater than or equal to the source's second moment, the rate-distortion function of $\mathbf{S}^*$ at distortion level D is zero. (See the illustration in Figure 5.8 for the scalar-ECDQ case.) □

Like the SLB and the asymptotic redundancy of Theorem 5.5.1, Ziv's bound can be extended to general difference distortion measures.

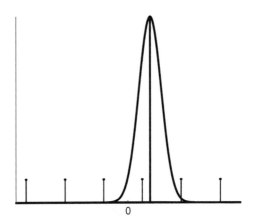

Figure 5.8 A case where ECDQ has maximum loss. The optimum quantizer contains a single code point at the mean (thus achieving the optimum pair $(R, D) = (0, \sigma_s^2)$), while the scalar lattice code points are randomly shifted by the dither.

Summary of Chapter 5

Rate $R_{\text{ECDQ}}(D) \triangleq \frac{1}{n} H(Q_\Lambda(\mathbf{X} + \mathbf{U}) | \mathbf{U})$ bit per source sample.

Equivalent channel (code-length formula)   If $\mathbf{U}$ is a uniform or modulo-uniform dither, then

$$H(Q_\Lambda(\mathbf{X} + \mathbf{U}) | \mathbf{U}) = I(\mathbf{X}; \mathbf{X} + \mathbf{U}_{\text{eq}}),$$

where $\mathbf{U}_{\text{eq}}$ is uniform over $-\mathcal{P}_0(\Lambda)$.

Redundancy at high resolution

$$R_{\text{ECDQ}}(D) - R(D) \to \frac{1}{2} \log(2\pi e G(\Lambda)) \text{ as } D \to 0.$$

Filtered ECDQ   For a Gaussian source and Wiener-estimated reconstruction,

$$R_{\text{ECDQ}}(D) - R(D) \leq \frac{1}{2} \log(2\pi e G(\Lambda)) \text{ for all } D.$$

Deterministic dither   If $G(\Lambda_n) \to 1/2\pi e$, then almost any dither value $\mathbf{U} = \mathbf{u}$ is rate-distortion optimal.

Problems

P.5.1 (Probability of a shifted region) Prove Lemma 5.2.1. Hint: assume that the source $\mathbf{S}$ has a density, and use the convolution rule for the sum of independent random variables $\mathbf{X} = \mathbf{S} - \mathbf{V}$ (where $\mathbf{V}$ is uniform over $\mathcal{A}$), to show that

$$f_{\mathbf{X}}(\mathbf{x}) = \frac{1}{\text{Vol}(\mathcal{A})} \cdot \int_{\mathcal{A}+\mathbf{x}} f_{\mathbf{S}}(\mathbf{s})d\mathbf{s}.$$

(If $\mathbf{S}$ does not have a density, then the derivation still holds replacing $f_{\mathbf{S}}(\mathbf{s})d\mathbf{s}$ by $d P_{\mathbf{S}}(\mathbf{s})$.)

P.5.2 (Sample-sum property) Show that the density $f_{\mathbf{X}}(\cdot)$ of (5.9) satisfies the property that its sum over any coset of Λ is equal to $1/V(\Lambda)$.

P.5.3 (Dithered quantization output statistics) We can use the argument in Proof A of Theorem 5.2.1 to calculate moments of the dithered quantizer *output* (as an alternative to using the equivalent additive-noise channel (4.7)). Show that the second moment is given by

$$E\|Q_\Lambda - \mathbf{U}\|^2 = \int_{\mathbb{R}^n} \|\mathbf{x}\|^2 f_{\mathbf{X}}(\mathbf{x})d\mathbf{x} = E\|\mathbf{S} - \mathbf{U}\|^2.$$

P.5.4 (Proof of Theorem 5.2.1 for a generalized dither) For a modulo-uniform dither $\mathbf{U}$, justify steps $(d0')$–(e') below, that replace steps $(d1)$–$(d3)$ and (e) in (5.16) and (5.15) in Proof B of Theorem 5.2.1:

$$I(\mathbf{S}; Q_\Lambda - \mathbf{U}|\mathbf{U}) \overset{d0'}{=} I(\mathbf{S}; Q_{\Lambda,e} - \mathbf{U}_e| \mathbf{U}_q, \mathbf{U}_e)$$
$$\overset{d1'}{=} I(\mathbf{S}; Q_{\Lambda,e} - \mathbf{U}_e) + I(\mathbf{S}; \mathbf{U}_q, \mathbf{U}_e| Q_{\Lambda,e} - \mathbf{U}_e)$$
$$-I(\mathbf{S}; \mathbf{U}_q, \mathbf{U}_e)$$
$$\overset{d2'}{=} I(\mathbf{S}; Q_{\Lambda,e} - \mathbf{U}_e) + I(\mathbf{S}; \mathbf{U}_q, \mathbf{U}_e| Q_{\Lambda,e} - \mathbf{U}_e)$$
$$\overset{d3'}{=} I(\mathbf{S}; Q_{\Lambda,e} - \mathbf{U}_e) + I(\mathbf{S}; \mathbf{U}_q| Q_{\Lambda,e}, \mathbf{U}_e)$$
$$\overset{d3''}{=} I(\mathbf{S}; Q_{\Lambda,e} - \mathbf{U}_e)$$
$$\overset{e'}{=} I(\mathbf{S}; \mathbf{S} + \mathbf{U}_{\text{eq}}) \tag{5.59}$$

where $\mathbf{U}_q = Q_\Lambda(\mathbf{U})$ and $\mathbf{U}_e = \mathbf{U} \bmod \Lambda$ (see (2.12)); and where $Q_\Lambda = Q_\Lambda[\mathbf{S} + \mathbf{U}]$ and $Q_{\Lambda,e} = Q_\Lambda[\mathbf{S} + \mathbf{U}_e]$.

Hint: to justify step $(d3'')$, show that if X is independent of (Y, Z), then $X \leftrightarrow \{Y, f(X, Y)\} \leftrightarrow Z$ form a Markov chain.

Alternative proof: show that $H(Q_\Lambda|\mathbf{U}) = H(Q_{\Lambda,e}|\mathbf{U}_e)$ for any $\mathbf{U}$.

P.5.5 (Quantizer entropy in the presence of feedback) Prove formula (5.22), by following the steps of Proof B of Theorem 5.2.1.

P.5.6 (ECDQ under difference distortion measures) Prove the entropy-distortion formula (5.30) for a scalar ECDQ and rth power distortion measure. Evaluate the small-distortion redundancy with respect to the SLB. (See the extension to a general lattice and norm-based distortion measure in [287].)

P.5.7 (Universal quantization) Extend Ziv's bound for universal quantization to this class of difference distortion measures.

P.5.8 (Colored-Gaussian source: pre/post-filtered ECDQ simulates the water-pouring solution) The quadratic rate-distortion function $R^*(D)$, of a discrete-time stationary Gaussian source with a power spectrum $S(e^{j2\pi f})$, $-1/2 < f \leq 1/2$, has a parametric form known as *reversed water-pouring* [18, 53]:

$$R^*(\theta) = \int_{f:S(e^{j2\pi f})>\theta} \frac{1}{2} \log\left(\frac{S(e^{j2\pi f})}{\theta}\right) df$$

$$D(\theta) = \int_{-1/2}^{1/2} \min\{\theta, \ S(e^{j2\pi f})\} df, \tag{5.60}$$

where the "water level" parameter θ varies in the range $0 < \theta \leq \max_f S(e^{j2\pi f})$. Show that if we replace the scalars α and β in the pre/post-scaled ECDQ (Figure 5.5) by complex-conjugate linear time-invariant filters $H_1(e^{j2\pi f}) = H_2^*(e^{j2\pi f})$, satisfying

$$|H_1(e^{j2\pi f})|^2 = |H_2(e^{j2\pi f})|^2 = \begin{cases} 1 - \frac{\theta}{S(e^{j2\pi f})}, & \text{if } S(e^{j2\pi f}) > \theta \\ 0, & \text{otherwise,} \end{cases} \tag{5.61}$$

and use a lattice with $\sigma^2(\Lambda) = D$, then this *pre/post-filtered ECDQ* (with joint entropy coding) achieves $R^*(D)$, up to a quantization loss of $\frac{1}{2} \log(2\pi e G(\Lambda))$ bit per sample.

Historical notes

Entropy-coded (variable-rate) quantization (sometimes called entropy-constrained quantization) was developed in parallel with fixed-rate quantization, for example, Pinkston [218]. Gish and Pierce [106] used Bennett's high-resolution analysis [17] to prove that the optimum entropy-constrained scalar quantizer asymptotically becomes a scalar lattice. The work of Zador [283] (previously an unpublished Bell Laboratories memo [282] which was corrected and generalized by Gray *et al.* [110]) considers high-resolution vector quantization. This analysis, together with Gersho's conjecture [102], imply that lattice quantizers are entropy optimal for each dimension. Lookabaugh and Gray [176] examined the vector-lattice quantizer advantage in the high-resolution regime.

Lloyd [174] gave necessary conditions for optimal fixed-rate scalar quantization (at general resolution), and proposed an iterative algorithm for quantizer design. Linde *et al.* [161] extended Lloyd's algorithm to vector quantization, while Chou *et al.* [40] proposed an efficient design for variable-rate ("entropy-constrained") vector quantization.

The idea to use ECDQ for universal variable-rate quantization was introduced by Ziv in 1985 [300]. This work was preceded by earlier work on universal fixed-rate quantization by Ziv himself [299], and by Neuhoff *et al.* [199]. Ziv found in [300] a universal bound on the excess rate of a randomized (one-time dithered) entropy-coded lattice quantizer above the optimal entropy-constrained quantizer. Several works followed this idea. Neuhoff [198] compared (universal) source coding strategies in terms of complexity, while Gutman [114] extended Ziv's bound to various distortion measures. The mutual-information formula for the rate of a randomized (multiple-dithered) entropy-coded lattice quantizer (Theorem 5.2.1) is due to Zamir and Feder [287]. They used this formulation of the ECDQ to examine sampling and quantization trade-offs [289], and pre/post-filtered quantization [290]. Linder and Zeger [165] examined ECDQ in the high-resolution limit. Other applications of ECDQ include multi-terminal source coding [285], noise shaping for non-uniform sampling [182], and multiple-description source coding [39, 95, 181, 211].

Rate-distortion theory is due to Shannon; see Berger's book [18]. The asymptotic tightness of the Shannon lower bound is due to Linkov [169], and Linder and Zamir [164].

6 Infinite constellation for modulation

Digital communication is the process of encoding an information source and sending it digitally over a noisy channel. Chapter 5 dealt with the source coding part: digitization of an analog signal using a lattice quantizer. In this chapter we shall consider the channel coding part: translation of digital information into an analog signal via lattice modulation, and detection of the information from a noisy version of the signal with a small probability of error.

In lattice modulation, the lattice points form the signal constellation, or codebook, where each lattice point carries a different message (Sections 1.1 and 3.3). Popular examples are pulse-amplitude modulation (PAM) and quadrature-amplitude modulation (QAM), which correspond to one-dimensional and two-dimensional lattice-like constellations. A general approach to constructing multi-dimensional lattice constellations is by *shaping*, i.e., by cutting a finite section from a lattice. The most popular shapes are a cube (uncoded constellation), a ball (spherical shaping) or the Voronoi cell of some coarse lattice (Voronoi shaping), as we shall see in Chapter 9.

In this chapter we want to keep the lattice *unbounded* – just as for the dithered lattice quantizer of Chapter 5. Thus we are faced with a problem: how do we define the transmission power and the coding rate of an infinite constellation?

We shall develop two alternative definitions: *rate per unit volume* in Sections 6.1–6.4, and *variable-rate modulation* in Sections 6.5–6.6. While the former amounts to avoiding shaping altogether, the latter amounts to a "probabilistic" form of shaping. Both definitions highlight the role of the NVNR (3.37) as a measure for the lattice-constellation goodness for an AWGN channel. Specifically, the capacity loss of a lattice constellation Λ, at a decoding error P_e, is shown to be $\frac{1}{2} \log[\mu(\Lambda, P_e)/2\pi e]$.

6.1 Rate per unit volume

The rate of an n-dimensional code $\mathcal{C}$ with M equally likely codewords $\mathbf{c}_1, \ldots, \mathbf{c}_M$ is defined as

$$R = \frac{1}{n} \log M \quad \text{bit per channel use.} \tag{6.1}$$

Table 6.1 Comparison of quantities associated with finite and infinite constellations

	Finite constellation	Infinite constellation
Dimension	n	n
Size/point density	M	γ
Rate	$R = \frac{1}{n}\log(M)$	$R_\infty = \frac{1}{n}\log(\gamma)$
Capacity (maximal rate of reliable communication)	C	C_∞

The average power of the code $\mathcal{C}$ is

$$P(\mathcal{C}) = \frac{1}{nM} \sum_{\mathbf{c}_i \in \mathcal{C}} \|\mathbf{c}\|^2. \tag{6.2}$$

Since the lattice is unbounded, the number of codewords M of the lattice code is infinite, so the coding rate $R = \infty$. Also, since the squared norm of the lattice points is unbounded, the power $P(\Lambda) = \infty$. How do we make these quantities finite?

There are several ways to define coding rate and capacity per unit volume of a general *infinite constellation* (IC), not necessarily a lattice. One simple way is to count the number of codewords per unit volume within a "large" cube and translate it into bits.

Definition 6.1.1 (Rate per unit volume) [1] *The coding rate per unit volume of an IC is defined as*

$$R_\infty(\mathbb{IC}) = \frac{1}{n} \limsup_{a \to \infty} \log\left(\frac{|\mathcal{C}_a|}{a^n}\right), \tag{6.3}$$

where

$$\mathcal{C}_a = \mathbb{IC} \cap \text{CUBE}(a) \tag{6.4}$$

is the intersection of the IC with the n-dimensional cube $\text{CUBE}(a) = [-a/2, a/2]^n.$ [2]

Note that as the logarithm of a *ratio*, $R_\infty(\mathbb{IC})$ does not have the significance of *information bits*, and may even be *negative* (if the number of codewords per unit volume is less than 1). Table 6.1 compares various information measures of infinite and finite constellations. In particular, while for a finite constellation $M = 2^{nR}$ is the *total* number of points (6.1), for an IC $2^{nR_\infty(\mathbb{IC})}$ is the point *density*.

---

[1] This quantity is sometimes called *normalized logarithmic density*.
[2] For "regular" ICs, the value of the limit (6.3) does not change if the cube in (6.4) is replaced by some other "simple body," say, a ball of radius a. The denominator in (6.3) is then the volume of this body.

Suppose now that the IC is a lattice Λ. Since the diameter d of the cells is finite, the number of codewords inside the cube is bounded between $(a - d)^n / V(\Lambda)$ and $(a + d)^n / V(\Lambda)$. Thus, $M(\Lambda_a) \approx a^n / V(\Lambda)$ for large a, so the rate per unit volume of a lattice code is

$$R_\infty(\Lambda) = \frac{1}{n} \log \left(\frac{1}{V(\Lambda)} \right) = \frac{1}{n} \log \gamma(\Lambda), \qquad (6.5)$$

where $\gamma(\Lambda)$ is the lattice point density (2.20).

6.2   ML decoding and error probability

Consider transmission over a general additive-noise channel

$$\mathbf{Y} = \mathbf{X} + \mathbf{Z} \qquad (6.6)$$

where the noise $\mathbf{Z}$ has some distribution $p_{\mathbf{Z}}(\cdot)$. The receiver may use either a *threshold* or a *comparative* rule to decode the transmitted codeword $\mathbf{c}_i$ from the channel output $\mathbf{Y}$. If all the codewords are *equally likely*, then the probability of error is minimized by *maximum likelihood* (ML) decoding, i.e., by the codeword $\mathbf{c}_i$ that maximizes the conditional probability of $\mathbf{Y}$ given $\mathbf{c}_i$. For an additive-noise channel (6.6), this conditional probability is equal to the probability that $\mathbf{Z} = \mathbf{Y} - \mathbf{c}_i$:

$$\hat{\mathbf{c}}_{\mathrm{ML}} = \arg \max_{\mathbf{c}_i \in \mathcal{C}} p_{\mathbf{Z}}(\mathbf{Y} - \mathbf{c}_i). \qquad (6.7)$$

The probability of error, given that the codeword $\mathbf{c}_i$ was transmitted, is then

$$P_e(\mathbf{c}_i) = \Pr\{\hat{\mathbf{c}}_{\mathrm{ML}} \neq \mathbf{c}_i \mid \mathbf{c}_i \text{ was transmitted}\}, \qquad (6.8)$$

where the probability is calculated with respect to the randomness of the channel (6.6).

Definition 6.2.1 (Average error probability) *The average decoding error probability of a* finite *constellation $\mathcal{C}$ with M codewords is defined as*

$$P_e(\mathcal{C}) = \frac{1}{M} \sum_{\mathbf{c}_i \in \mathcal{C}} P_e(\mathbf{c}_i). \qquad (6.9)$$

Definition 6.2.2 (Error probability of an IC) *The average decoding error probability of an IC is defined as the upper limit of (6.9) with respect to the truncated IC (6.4):*

$$P_e(\mathbb{IC}) = \limsup_{a \to \infty} P_e(\mathcal{C}_a). \qquad (6.10)$$

When the IC takes the form of a lattice Λ, the ML decoding rule (6.7) becomes

$$\hat{\lambda}_{\mathrm{ML}} = \arg \max_{\lambda \in \Lambda} p_{\mathbf{Z}}(\mathbf{Y} - \lambda). \qquad (6.11)$$

This amounts to a Voronoi (lattice) partition (Definition 2.2.1) of the space of received signals with respect to the lattice Λ, using $p_{\mathbf{Z}}(\cdot)$ as a "distance measure." [3]

Definition 6.2.3 (Noise-matched decoder) *A noise-matched decoder, for a lattice Λ and noise $\mathbf{Z}$, is a quantizer Q_Λ with partition $\{V_\lambda^{\mathbf{Z}} = \lambda + V_0^{\mathbf{Z}}\}$, where the fundamental (non-Euclidean) Voronoi cell $V_0^{\mathbf{Z}}$ is defined as*

$$V_0^{\mathbf{Z}}(\Lambda) = \left\{ \mathbf{y} : \ p_{\mathbf{Z}}(\mathbf{y}) \geq p_{\mathbf{Z}}(\mathbf{y} - \lambda) \ \text{for all } \lambda \in \Lambda, \ \lambda \neq \mathbf{0} \right\} \qquad (6.12)$$

and where ties are broken in a systematic manner (see Definition 2.2.1).

Due to the additivity of the channel (6.6), the probability of decoding error (6.8) is *independent* of the transmitted codeword $\lambda \in \Lambda$, and it is given for all $\lambda \in \Lambda$ by

$$P_e = P_e(\mathbf{Z}, \Lambda) = \Pr\{\mathbf{Z} \notin V_0^{\mathbf{Z}}\}. \qquad (6.13)$$

Thus, this is also the average error probability (6.10) of the lattice code.

In the case of AWGN, the noise distribution $p_{\mathbf{Z}}(\mathbf{z})$ is monotonic in the Euclidean norm $\|\mathbf{z}\|$. Hence, the ML rule (6.11) becomes a nearest-neighbor rule, i.e., $\hat{\lambda}_{\text{ML}}$ becomes $\hat{\lambda}_{NN}$ of (3.31); the fundamental cell $V_0^{\mathbf{Z}}$ of (6.12) becomes a Euclidean Voronoi cell (2.11); and the probability of error (6.13) (assuming AWGN with variance N) becomes $P_e(\Lambda, N)$ of (3.32).

While the partition $V^{\mathbf{Z}}$ is clearly unaffected by scaling of the noise $\mathbf{Z}$ by a constant, the error probability (6.13) does change. Specifically, it is always possible to scale the lattice Λ by a factor α, so that $P_e(\alpha\Lambda, N)$ is equal to some desired probability of error $0 < P_e < 1$. The resulting rate per unit volume is given by the following proposition.

Proposition 6.2.1 (Rate versus error) *If $P_e(\Lambda, N) = P_e$, then the rate per unit volume of Λ in the presence of AWGN of variance N is denoted $R_\infty(\Lambda, P_e)$, and it is given by*

$$R_\infty(\Lambda, P_e) = \frac{1}{2} \log \left(\frac{1}{N \cdot \mu(\Lambda, P_e)} \right), \qquad (6.14)$$

where $\mu(\Lambda, P_e)$ is the lattice NVNR (3.37).

Proof By the definition of the NVNR (which is a scale-invariant quantity of the lattice), the lattice volume after the scaling is $V(\Lambda) = [N\mu(\Lambda, P_e)]^{n/2}$, and the proposition follows by substituting in (6.5). □

How good is a lattice constellation in terms of its rate per unit volume for a general additive-noise channel? To answer that, we shall recall Shannon's notion of capacity.

---

[3] We assume that $p_{\mathbf{Z}}(\mathbf{z})$ satisfies some monotonicity condition, for example, with respect to some norm $\|\mathbf{z}\|$ of the noise $\mathbf{z}$, to guarantee that the boundary regions between cells have measure zero, so ties in the decoding rule (6.11) can be arbitrarily broken.

6.3   Gap to capacity

The Shannon capacity of an additive-noise channel is given by the maximum mutual information per dimension between the input $\mathbf{X}$ and the output $\mathbf{Y} = \mathbf{X} + \mathbf{Z}$,

$$C = \frac{1}{n} \max I(\mathbf{X}; \mathbf{X} + \mathbf{Z}), \qquad (6.15)$$

where the maximization is over all valid random channel inputs $\mathbf{X}$. In the case of a *power-constrained* channel, the maximization (6.15) is over all inputs satisfying

$$\frac{1}{n} E\{\|\mathbf{X}\|^2\} \leq P. \qquad (6.16)$$

As information theory shows, if the block length (the number of channel uses) n is large and the noise $\mathbf{Z}$ is a vector from a stationary and ergodic process, then C is the largest coding rate R (6.1) that allows *reliable communication* over the channel (6.6), i.e., there exists a sequence of rate-R codes, of a growing dimension n, such that the decoding error probability (6.9) goes to zero as n goes to infinity. Under the power constraint (6.16), C is an upper bound on the reliable rate of a code $\mathcal{C}$ satisfying $P(\mathcal{C}) \leq P$; see (6.2).

As for the rate-distortion function (5.31), the capacity formula (6.15) can be simplified in the case of a memoryless channel to a *single-letter* quantity – corresponding to a memoryless channel input.

For example, if the noise $\mathbf{Z}$ is AWGN with variance N, then (6.15) simplifies to

$$C_{\text{AWGN}} = \max_{E\{X^2\} \leq P} I(X; X + Z) = \frac{1}{2} \log(1 + \text{SNR}), \qquad (6.17)$$

where $\text{SNR} = P/N$ is the signal to noise ratio, and this capacity is achieved by a Gaussian input $X \sim N(0, P)$. Note that C_{AWGN} is finite for any value of the power constraint P, and it increases to infinity as P goes to infinity.

For a general additive-noise channel (6.6), the Shannon capacity satisfies the following simple bound.

Proposition 6.3.1 (The Shannon upper bound (SUB)) *For any additive noise* $\mathbf{Z}$ *with a differential entropy* $h(\mathbf{Z})$, *the power-constrained capacity of the channel* (3.28) *is bounded by*

$$C \leq C_{\text{SUB}} = \frac{1}{2} \log\left(\frac{P + \sigma_{\mathbf{Z}}^2}{P_E(\mathbf{Z})}\right), \qquad (6.18)$$

where

$$\sigma_{\mathbf{Z}}^2 = \text{Var}(\mathbf{Z}) = \frac{1}{n} E\{\|\mathbf{Z} - E\{\mathbf{Z}\}\|^2\} \qquad (6.19)$$

is the average variance of the noise $\mathbf{Z}$, *and*

$$P_E(\mathbf{Z}) = \frac{1}{2\pi e} 2^{\frac{2}{n}h(\mathbf{Z})} \tag{6.20}$$

is its entropy power (A.15).

Like the SLB (5.40) for the rate-distortion function, the SUB follows from a maximum entropy property. For an AWGN channel $P_E(\mathbf{Z}) = \sigma_{\mathbf{Z}}^2$, hence the SUB is tight and coincides with (6.17). For a general noise, we have $P_E(\mathbf{Z}) \leq \sigma_{\mathbf{Z}}^2$, with equality if and only if $\mathbf{Z}$ is AWGN; hence the SUB is always larger than or equal to the capacity of an AWGN channel with the same noise variance. Furthermore, the SUB is asymptotically tight at a large signal to noise ratio.

Proposition 6.3.2 (Tightness of the SUB) *If the differential entropy of the noise* $h(\mathbf{Z})$ *exists (equivalently,* $P_E(\mathbf{Z})$ *is positive and finite), then the SUB is asymptotically tight in the limit of large power:*

$$C = \frac{1}{2} \log\left(\frac{P}{P_E(\mathbf{Z})}\right) + o(1), \tag{6.21}$$

where $o(1) \to 0$ *as* $P \to \infty$. *Furthermore, C is asymptotically achieved by a white-Gaussian input of variance P:*

$$\mathbf{X} \sim N(0, P) \text{ with i.i.d. components.} \tag{6.22}$$

In particular, for an AWGN channel in the limit of high SNR,

$$C_{\text{AWGN}} = \frac{1}{2} \log(\text{SNR}) + o(1), \tag{6.23}$$

where $o(1) \to 0$ as $\text{SNR} = P/N \to \infty$.

The proof is similar to the proof of the tightness of the SLB (Proposition 5.5.2), and will be omitted.

6.3.1 Capacity per unit volume

So far we have assumed power-constrained channels, hence the capacity was finite. We now omit the power constraint (6.16), and turn to define the notion of *capacity per unit volume* of an unconstrained channel – known also as the *Poltyrev capacity*. The *operational* definition of this capacity is the largest rate per unit volume $R_\infty(\mathbb{IC})$ (6.3) of an IC that allows reliable communication (i.e., a vanishing error probability) over a large block of channel uses. In view of the error probability of an $\mathbb{IC}$ (6.10), the capacity per unit volume can be approximated by that of a large finite constellation.

The following theorem gives an information theoretic characterization for this capacity.

Theorem 6.3.1 (Capacity per unit volume) *The capacity per unit volume of a general (unconstrained) additive-noise channel is given by*

$$C_\infty = -\frac{1}{n}h(\mathbf{Z}) = \frac{1}{2}\log\left(\frac{1}{2\pi e P_E(\mathbf{Z})}\right), \tag{6.24}$$

where $h(\mathbf{Z})$ and $P_E(\mathbf{Z})$ are the differential entropy and the entropy power of the noise (6.20). In the special case of AWGN with variance N, the capacity per unit volume is

$$C_\infty = \frac{1}{2}\log\left(\frac{1}{2\pi e N}\right). \tag{6.25}$$

We see that like the high SNR capacity (6.21), the capacity per unit volume C_∞ is dictated by the noise entropy power.

Proof Let C_a denote the Shannon capacity when the channel input is restricted to CUBE(a). Similarly to the proof of the tightness of the SUB, we have $C_a = \log(a) - h(\mathbf{Z})/n + o(1)$, where $o(1) \to 0$ as $a \to \infty$. Now, by Shannon's theory, C_a is the highest reliable rate of the code $\mathcal{C}_a$ (see (6.4)). Hence, the maximum number of n-codewords $|\mathcal{C}_a|$ is 2^{nC_a}, which – by the asymptotic expression for C_a above – is equal to $a^n/2^{h(\mathbf{Z})-no(1)}$ for large a. Substituting that in the definition (6.3) of rate per unit volume $R_\infty(\mathbb{IC})$, the desired result follows. $\square$

6.3.2   Capacity loss and coding gain

We return to the question regarding the goodness of a lattice constellation Λ, and examine its gap to the capacity per unit volume C_∞. In view of Proposition 6.2.1 and Theorem 6.3.1, we have the following result.

Theorem 6.3.2 (Gap to capacity) *Suppose that P_e is the error probability (6.10) in ML decoding of a lattice constellation Λ in the presence of AWGN. Then, the gap to capacity is given by*

$$\Delta(\Lambda, C_\infty, P_e) \triangleq C_\infty - R_\infty(\Lambda, P_e) = \frac{1}{2}\log\left(\frac{\mu(\Lambda, P_e)}{2\pi e}\right), \tag{6.26}$$

where $\mu(\Lambda, P_e)$ is the lattice NVNR (3.37).

Note that for a small error probability, the NVNR is strictly greater than $2\pi e$ (see (3.43)). Thus, the gap to Shannon's capacity is positive, and it vanishes for "good" lattices, for which $\mu(\Lambda, P_e) = 2\pi e$ for all $P_e > 0$; see (3.44).

Example 6.3.1 (Cubic and hexagonal lattices) *In two dimensions, the NVNR of the $\mathbb{Z}^2$ lattice is roughly $15, 31$ and 48, for error probability $P_e = 10^{-1}, 10^{-2}$ and 10^{-3}, respectively. See Figure 3.9. Hence, the corresponding gap to capacity (6.26) is $-0.08, 0.43$ and 0.74 bit per channel use, respectively. Note that the first value is negative because the error probability is above the threshold; see (3.43). For* comparison, for the same error probabilities, the NVNR of the hexagonal lattice A_2, is roughly 17, 29 and 44 (i.e., worse at high P_e, but better for small P_e). See Figure 3.10. The corresponding gap to capacity is ≈ 0, 0.38 and 0.68 bit per channel use, respectively.

Note that the coding gain (3.38) of a lattice Λ, when expressed in bits (i.e., $\frac{1}{2}\log\Gamma_c(\Lambda, P_e)$), is the increase in rate per unit volume (or reduction of gap to capacity) when using Λ instead of a cubic lattice.

6.4   Non-AWGN and mismatch

It is possible to extend Theorem 6.3.2 above to a general additive noise $\mathbf{Z}$ by extending the definitions of the VNR (3.30) and NVNR (3.37). In view of the capacity per unit volume formula (6.24), it would be natural to let the noise entropy power $P_E(\mathbf{Z})$ (6.20) play the role of variance in these extensions. The generalized VNR – or *volume to noise entropy power ratio* (VNER) – of a lattice Λ in the presence of a noise $\mathbf{Z}$, is thus defined as:

$$\mu_E(\Lambda, \mathbf{Z}) = \frac{V^{2/n}(\Lambda)}{P_E(\mathbf{Z})}. \tag{6.27}$$

Assuming that the decoder is "matched" to the distribution of $\mathbf{Z}$ (i.e., uses a maximum likelihood decoding rule (6.11)), the error probability is given by $\Pr\{\mathbf{Z} \notin \mathcal{V}_0^{\mathbf{Z}}\}$; see (6.13). Based on that, we define the following.

Definition 6.4.1 (Noise-matched NVNR) *The noise-matched NVNR of a lattice constellation Λ, with respect to a general additive noise $\mathbf{Z}$ and probability of error $0 < P_e < 1$, is defined as*

$$\mu_{\text{matched}}(\Lambda, \mathbf{Z}, P_e) \overset{\Delta}{=} \mu_E(a \cdot \Lambda, \mathbf{Z}) = a^2 \cdot \frac{V^{2/n}(\Lambda)}{P_E(\mathbf{Z})}, \tag{6.28}$$

where $a = a(P_e)$ is chosen such that the scaled lattice $a\Lambda$ meets the target decoding error probability P_e with a noise-matched decoder (Definition 6.2.3).

The noise-matched NVNR (6.28) coincides with $\mu(\Lambda, P_e)$ when $\mathbf{Z}$ is AWGN. Like $\mu(\Lambda, P_e)$, it is a dimensionless number invariant to scaling of the lattice.[4] It is also invariant to transformation of both the lattice and the noise by the same non-singular matrix; see Problem 6.1.

Example 6.4.1 (Colored-Gaussian noise) *For $\mathbf{Z} \sim N(0, \Sigma)$, with a non-singular covariance matrix Σ, the noise-matched NVNR of the lattice constellation Λ (matched to the colored-Gaussian noise $\mathbf{Z}$) is equal to the NVNR with respect*

---

[4] Invariance to rotation holds only if the distribution of $\mathbf{Z}$ is circularly symmetric.

to AWGN of a transformed lattice constellation Λ':

$$\mu_{\text{matched}}(\Lambda, \mathbf{Z}, P_e) = \mu(\Lambda', P_e), \tag{6.29}$$

where $\Lambda' = \Sigma^{-1/2} \cdot \Lambda$; and $\Sigma^{1/2}$ is any square root of Σ, i.e., a matrix satisfying $\Sigma^{1/2}(\Sigma^{1/2})^t = \Sigma$.

The gap to capacity $C_\infty - R_\infty(\Lambda, P_e)$ for a general $\mathbf{Z}$ satisfies Theorem 6.3.2, only with the "Gaussian" NVNR (3.37) replaced by the noise-matched NVNR (6.28). Since the noise-matched NVNR is strictly greater than $2\pi e$ for small enough error probability, the gap to capacity of the lattice constellation Λ is positive. As we shall see in Chapter 7, for a noise sequence $\mathbf{Z}_n$ satisfying a "generalized AEP," there exists a sequence of lattices Λ_n such that $\mu_{\text{matched}}(\Lambda_n, \mathbf{Z}_n, P_e)$ goes to $2\pi e$, for all $P_e > 0$; hence these lattices approach the capacity per unit volume C_∞ of the channel $\mathbf{Y} = \mathbf{X} + \mathbf{Z}$.

6.4.1   Mismatched decoding

The decoder can use a general distance measure $d = d(\mathbf{x}, \mathbf{y})$, not necessarily matched to the noise distribution. For a lattice constellation Λ, the distance d induces a general (not necessarily Euclidean) Voronoi partition $\{\mathcal{V}_\lambda\}$ of $\mathbb{R}^n$ into decision cells (2.11), i.e., the "mismatched" decoder picks the lattice point that minimizes the distance d to the received vector:

$$\hat{\lambda}_d = \arg\min_{\lambda \in \Lambda} d(\mathbf{Y}, \lambda) \tag{6.30}$$

(with ties broken in a systematic manner). The "mismatched decoding" error probability is thus given by

$$P_e^{(d)} = \Pr\{d(\mathbf{Z}) > d(\mathbf{Z} - \lambda), \ \forall \lambda \in \Lambda, \lambda \neq \mathbf{0}\} \tag{6.31}$$

which is clearly greater than or equal to the "matched" error probability (6.13), by the definition of maximum likelihood lattice decoding (6.11).

From an information theoretic viewpoint, the *mismatched capacity* $C^{(d)}$ denotes the maximum achievable rate with a mismatched decoder d, and similarly $C_\infty^{(d)}$ denotes the corresponding capacity per unit volume.

A specific important case of mismatched decoding is the *nearest-neighbor decoder* (3.31), which uses the *Euclidean* norm as a distance measure, independent of the noise statistics. Hence, (6.31) becomes in this case

$$P_e^{(NN)} = \Pr\left\{\mathbf{Z} \notin \mathcal{V}_0^{NN}\right\}, \tag{6.32}$$

where $\mathcal{V}_0^{NN}$ is the usual (Euclidean) fundamental Voronoi cell. We can think of the NN decoder as *matched to AWGN* (even if the actual noise $\mathbf{Z}$ may have a different distribution). It is therefore natural (and proves useful) to normalize the corresponding VNR and NVNR by the *second moment* of the noise, rather than by its entropy power.

Definition 6.4.2 (Euclidean "mismatched" NVNR) *The NVNR of a lattice constellation Λ, for a general additive noise $\mathbf{Z}$ under a Euclidean nearest-neighbor decoding rule (3.31), is defined as*

$$\mu_{\text{euclid}}(\Lambda, \mathbf{Z}, P_e) = \mu(a \cdot \Lambda, \mathbf{Z}) = \frac{V^{2/n}(a \cdot \Lambda)}{\frac{1}{n}E\|\mathbf{Z}\|^2}, \tag{6.33}$$

where the scaling factor $a = a(P_e)$ is such that the NN (mismatched decoding) error probability (6.32) is equal to P_e.

We shall usually use the Euclidean mismatched NVNR in cases where the equivalent noise seen by the lattice decoder deviates only slightly from AWGN, so NN decoding is nearly optimal; see Chapter 9.

The mismatched capacity $C_\infty^{(NN)}$, associated with a stationary and ergodic noise $\mathbf{Z}$, is lower bounded by

$$C_\infty^{(NN)} \geq \frac{1}{2}\log\left(\frac{1}{2\pi e \text{Var}(\mathbf{Z})}\right) \triangleq \underline{C}_\infty^{\text{(euclid-th)}}, \tag{6.34}$$

where $\underline{C}_\infty^{\text{(euclid-th)}}$ is the lower *Euclidean-threshold* capacity, corresponding to a decoding rule of the form

$$\hat{\mathbf{c}} = \begin{cases} \mathbf{c}, & \text{if } \|\mathbf{Y} - \mathbf{c}\| < r_{\text{th}} \text{ for a unique } \mathbf{c} \in \mathcal{C}, \\ ?, & \text{otherwise.} \end{cases} \tag{6.35}$$

We shall discuss the asymptotic performance of such a decoder in Sections 7.7.2 and 7.8, and the Euclidean-threshold capacity (6.34) in Section 9.8.

6.5 Non-equiprobable signaling

An alternative approach to bound the power of an infinite lattice code is to assign *non-uniform probabilities* to the lattice points. Thus, although the power of the codewords is unbounded, the *expected* power of the codebook is finite. As we shall see, a desirable shape for the codeword distribution is *white Gaussian*.

Since each lattice point carries a message, the probability assignment amounts to transmitting messages which are *not* equally likely. It is possible to achieve that by letting each lattice point carry a *variable amount of information bits*; see Figure 6.1. We shall use the ECDQ of Chapter 5 to describe the mapping from a variable-length information bit string $\mathbf{b} = b_1, \ldots, b_k$ to a (dithered) lattice point $\lambda - \mathbf{U}$. However, in contrast to Chapter 5, we use the ECDQ in "reverse," i.e., switch the roles of the encoder and the decoder.

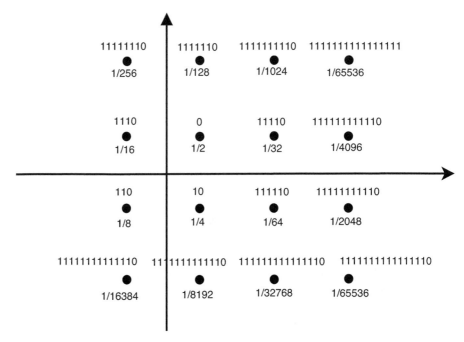

Figure 6.1 Probabilistic shaping of a lattice. The probability of a constellation point is equal to $(1/2)^\ell$, where ℓ is the length of the information string it represents.

Recall that the ECDQ encoder f adds the dither vector $\mathbf{u} \in \mathcal{P}_0$ to the source vector $\mathbf{s} \in \mathbb{R}^n$. It then quantizes the sum $\mathbf{s} + \mathbf{u}$ by the lattice quantizer Q_Λ (under the partition $\mathcal{P}$), and encodes the resulting lattice point λ by a variable-length code which depends on the value of $\mathbf{u}$. Thus, f is a mapping from a pair of source vector $\mathbf{s}$ and dither vector $\mathbf{u}$ into a variable-length bit string $\mathbf{b}^*$:

$$f : (\mathbb{R}^n, \mathcal{P}_0) \to \{0, 1\}^*, \tag{6.36}$$

where the superscript $*$ means a variable-length string. For a sequence of source vectors $\mathbf{s}_1, \mathbf{s}_2, \ldots$, the encoder concatenates the bit strings to form a sequence of code bits at a rate $R = \frac{1}{n}\overline{\{0, 1\}^*}$ bit per source sample (or dimension), where $\overline{\{0, 1\}^*}$ is the average length of the strings. If the mapping f is *uniquely decodable*, then the decoder f^{-1} can reproduce the original strings (conditional on the dither vectors), and map each pair of dither vector $\mathbf{u}$ and bit string $\mathbf{b}^*$ back into the lattice point $\lambda = f^{-1}(\mathbf{u}, \mathbf{b}^*)$.[5] The final source reconstruction $\hat{\mathbf{s}}$ is then given by $\lambda - \mathbf{u}$.

---

[5] This is possible if the variable-length code satisfies a prefix condition, for example, a Huffman code; see footnote 3 in Chapter 5.

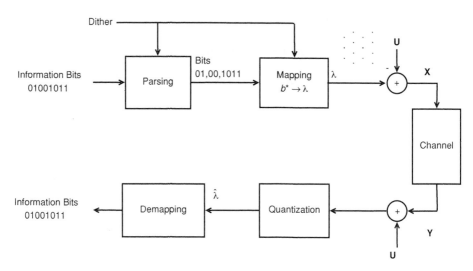

Figure 6.2 Variable-rate dithered modulation (VRDM) system (transmitter and receiver).

6.5.1 Variable-rate modulation

The ECDQ decoder f^{-1} can be used as a variable-length *channel encoder*. As shown in Figure 6.2, the information to be transmitted over the channel is represented by a sequence of bits $b_1, b_2, \ldots$. The function f^{-1} parses this information sequence into variable-length strings, and maps each string $\mathbf{b}^* \in \{0, 1\}^*$ into a lattice point λ. The average coding rate is thus given by $R = \frac{1}{n}\overline{\{0, 1\}^*}$ bit per channel use (or dimension). If we denote the length of the string mapped to λ (for a given dither $\mathbf{u}$) by $|\mathbf{b}^*| = \ell(\mathbf{u}, \lambda)$, then R is the average of $\ell(\mathbf{u}, \lambda)$ over both λ and $\mathbf{u}$. Finally, the encoder transmits the lattice point minus the dither vector,

$$\mathbf{x} = \lambda - \mathbf{u} = f^{-1}(\mathbf{u}, \mathbf{b}^*) - \mathbf{u}, \tag{6.37}$$

paralleling the dithered quantization reconstruction formula (4.3).

For each dither value $\mathbf{U} = \mathbf{u}$, the mapping f^{-1} induces a *non-uniform* distribution over the lattice points. If the information bit stream is completely random (i.e., Bernoulli(1/2)), then the probability to transmit a lattice point λ, is given by

$$p_\lambda(\mathbf{u}) = \left(\frac{1}{2}\right)^{\ell(\mathbf{u}, \lambda)}. \tag{6.38}$$

Hence, by varying the number of bits carried by each lattice point, we *probabilistically shape* the encoder output.

Assume that the codeword lengths $\{\ell(\mathbf{u}, \lambda)\}_{\lambda \in \Lambda}$ are ideally matched to the lattice point probabilities (5.3) after dithered quantization of a source $\mathbf{S}$, i.e., to the distribution of $Q_\Lambda(\mathbf{S} + \mathbf{u})$. Thus, by (5.9),

$$p_\lambda(\mathbf{u}) = \Pr\{\mathbf{S} + \mathbf{u} \in \mathcal{P}_\lambda\} = V(\Lambda) \cdot f_\mathbf{X}(\lambda - \mathbf{u}), \qquad (6.39)$$

where $f_\mathbf{X}(\cdot)$ is the density of $\mathbf{S} - \mathbf{U}$. We therefore call $\mathbf{S}$ the *shaping source*.

Example 6.5.1 *For $\Lambda = \mathbb{Z}^n$, and a memoryless shaping source S, the lattice point distribution (6.39) becomes a simple product $p_\lambda(\mathbf{u}) = \pi_{i=1}^n f_X(\lambda_i - u_i)$, where $f_X(x) = \int_{x-1/2}^{x+1/2} f_S(s)ds$.*

At high-resolution conditions the dither is negligible, hence $\mathbf{X} \approx \mathbf{S}$. For a white-Gaussian shaping source $\mathbf{S} \sim N(0, \sigma_s^2)$, the transmission probability (6.39) thus becomes

$$p_\lambda(\mathbf{u}) \approx p_\lambda(0) \approx V(\Lambda) \cdot f_\mathbf{S}(\lambda) \propto e^{-\|\lambda\|^2/2\sigma_s^2} \qquad (6.40)$$

which is known as a Gibbs distribution.

This *variable-rate dithered modulation* (VRDM) system has three basic parameters which we want to examine: power, coding rate, and probability of error under some decoding rule. Like the rate-distortion analysis of the ECDQ in Chapter 5, the analysis is made simple by averaging over the randomness of the dither $\mathbf{U}$.

6.5.2   Power, rate and error probability

We begin by considering the power of the VRDM. It follows from the equivalent channel of Theorem 4.1.1, that for a random (uniform or modulo-uniform) dither vector $\mathbf{U}$, the transmitter output $\mathbf{X}$ is distributed as the independent sum of $\mathbf{S}$ and $-\mathbf{U}$. In particular, for a Euclidean Voronoi partition $E\mathbf{U} = 0$ and $\frac{1}{n}E\|\mathbf{U}\|^2 = \sigma^2(\Lambda)$, so the transmission power is given by

$$P_{\text{VRDM}} = \frac{1}{n}E\|\mathbf{S} - \mathbf{U}\|^2 = \sigma_S^2 + \sigma^2(\Lambda), \qquad (6.41)$$

where the cross term vanishes due to the independence of $\mathbf{U}$ and $\mathbf{S}$. Thus choosing $\sigma_S^2 = P - \sigma^2(\Lambda)$ guarantees satisfying the power constraint P.

We next turn to the coding rate of the VRDM. As we saw in Chapter 5, if we choose an optimum variable-length code, then the average length of the strings $\overline{\{0, 1\}^*}$, parsed by f^{-1}, can approach the conditional entropy $H(Q_\Lambda(\mathbf{S} + \mathbf{U})|\mathbf{U})$ of the quantizer given the dither. By Theorem 5.2.1, this conditional entropy is equal to the mutual information between $\mathbf{S}$ and $\mathbf{S} + \mathbf{U}_{\text{eq}}$, where $\mathbf{U}_{\text{eq}} = -[\mathbf{U} \bmod \Lambda] \sim \text{Unif}(-\mathcal{P}_0)$. For a uniform (non-generalized) dither, we can set $\mathbf{U}_{\text{eq}} = -\mathbf{U}$; hence, the (ideal) coding rate of the VRDM is

$$R_{\text{VRDM}} = \frac{1}{n}\overline{\{0, 1\}^*} = \frac{1}{n}I(\mathbf{S}; \mathbf{S} - \mathbf{U}). \qquad (6.42)$$

By the analysis of the ECDQ (see (5.23)), this mutual information can be decomposed as

$$I(\mathbf{S}; \mathbf{S} - \mathbf{U}) = h(\mathbf{S} - \mathbf{U}) - \log V(\Lambda). \tag{6.43}$$

Thus, for a fixed lattice Λ, the rate R_{VRDM} is maximized by maximizing the entropy of the sum $\mathbf{S} - \mathbf{U}$ (over the distribution of the shaping source $\mathbf{S}$ and the partition $\mathcal{P}$), under the power constraint $P_{\text{VRDM}} = \sigma_S^2 + \sigma^2(\Lambda) \leq P$. For $\sigma^2(\Lambda)$ small compared to σ_S^2, the maximum is achieved by taking a white-Gaussian shaping source $\mathbf{S} \sim N(0, P)$.

Consider now the receiver, which knows the dither vector $\mathbf{U} = \mathbf{u}$, and receives the vector $\mathbf{Y} = \mathbf{x} + \mathbf{Z} = \lambda - \mathbf{u} + \mathbf{Z}$. A receiver with a "simple" structure ignores the non-uniform distribution of the lattice points; it adds the dither vector $\mathbf{u}$ and decodes to the most likely lattice point $\hat{\lambda}$ using the ML rule (6.7).[6] In the case where $\mathbf{Z}$ is AWGN, ML decoding becomes a nearest-neighbor (NN) rule (3.31):

$$\hat{\lambda} = \arg\min_{\lambda \in \Lambda} \|\mathbf{Y} + \mathbf{u} - \lambda\|, \tag{6.44}$$

with identical error probability $P_e(\Lambda, N)$ per transmitted message; see (3.32).[7]

6.5.3 The high SNR case

We are now in a position to examine the gap to capacity of the VRDM. We start with the simpler high SNR case.

Theorem 6.5.1 (Gap to capacity at high SNR) *Assume that the VRDM decoder uses the simple NN rule (6.44), and that $P_e(\Lambda, N)$ – the error probability in the presence of an AWGN of power N – is P_e. Then, in the limit of high signal to noise ratio SNR $= P/N$, the average coding rate R_{VRDM} in (6.42) is maximized by a white-Gaussian shaping source $\mathbf{S} \sim N(0, P)$, and the corresponding gap to the AWGN channel capacity is given by*

$$\lim_{P \to \infty} [C_{\text{AWGN}} - R_{\text{VRDM}}] = \frac{1}{2} \log \left(\frac{\mu(\Lambda, P_e)}{2\pi e} \right), \tag{6.45}$$

independent of the dither value $\mathbf{u}$, where $C_{\text{AWGN}} = \frac{1}{2} \log(1 + \text{SNR})$ is the AWGN channel capacity (6.17), and $\mu(\Lambda, P_e)$ is the NVNR (3.37).

Proof By (6.42)–(6.43), the average coding rate satisfies $n R_{\text{VRDM}} = h(\mathbf{S} - \mathbf{U}) - \log V(\Lambda)$. Now, by the definition of the NVNR, if the VRDM decoder satisfies the target error probability P_e, then the cell volume $V(\Lambda)$ must be (at least)

---

[6] Note that the receiver decision cells, i.e., partition $\mathcal{V}$, does not have to be the same as the lattice partition $\mathcal{P}$ used by the encoding/decoding function f (6.36): while the former matches the noise $\mathbf{Z}$, the latter aims at maximizing the entropy $h(\mathbf{S} - \mathbf{U})$ (see (6.43)).

[7] Although the error probability per *information* bit varies according to the length of the string.

$[N\mu(\Lambda, P_e)]^{n/2}$. Also, if $\mathbf{S}$ has a density, then by Proposition 5.4.1, $h(\mathbf{S} - \mathbf{U}) = h(\mathbf{S}) + o(1)$, where $o(1) \to 0$ as $P \to \infty$. (Otherwise the limit is $-\infty$.) Finally, by the maximum entropy property of a white-Gaussian distribution, (A.13),

$$h(\mathbf{S}) \le \frac{n}{2} \log(2\pi e \sigma_S^2) \tag{6.46}$$

$$= \frac{n}{2} \log(2\pi e(P - \sigma^2(\Lambda))) \tag{6.47}$$

$$= \frac{n}{2} \log(2\pi e P) + o(1), \tag{6.48}$$

where $o(1) \to 0$ as $P \to \infty$; equality in (6.46) holds if $\mathbf{S}$ is white Gaussian, and the second equality follows from the power constraint (6.41). Combining the above, we see that for a large signal to noise ratio, the average coding rate is maximized by a white-Gaussian shaping source $\mathbf{S}$, and it is given by

$$R_{\mathrm{VRDM}} = \frac{1}{2} \log(\mathrm{SNR}) - \frac{1}{2} \log\left(\frac{\mu(\Lambda, P_e)}{2\pi e}\right) + o(1) \tag{6.49}$$

where $o(1) \to 0$ as $\mathrm{SNR} \to \infty$. The gap to capacity now follows by comparing with the high SNR AWGN channel capacity (6.23). $\qquad\square$

This theorem can be easily extended to a general additive noise $\mathbf{Z}$ with noise-matched decoding, in which case the asymptotic (as $P \to \infty$) capacity loss is $\frac{1}{2} \log(\mu_{\mathrm{matched}}(\Lambda, \mathbf{Z}, P_e)/2\pi e)$, where $\mu_{\mathrm{matched}}(\Lambda, \mathbf{Z}, P_e)$ is the generalized NVNR of the lattice (6.28). In this extension, the entropy power of the noise (6.20) will replace the noise power N.

6.5.4   Separation of shaping and coding

To reduce the gap to capacity (6.45) of the VRDM (i.e., reduce the NVNR, or equivalently increase the coding gain, at some target error probability), the dimension of the lattice Λ must be large. In contrast, even a *scalar* lattice is sufficient for the task of probabilistic shaping.

A simple way to enjoy the reduced complexity of scalar lattice shaping is to separate the two aspects of the VRDM: shaping and coding. We can achieve this separation using a "good" (low NVNR) modulo-q lattice, where we shape only the "most significant bits," i.e., the bits defining a point in the scalar lattice $q\mathbb{Z}$.

To be more precise, let Λ_q be a modulo-q lattice, generated by some q-ary code $\mathcal{C} \subset (\mathbb{Z}/q\mathbb{Z})^n$ (2.56). We can write every point λ in Λ_q as a sum of a coarse point and a fine point:

$$\lambda = \lambda_c + \lambda_f, \tag{6.50}$$

where $\lambda_c \in q\mathbb{Z}^n$ and $\lambda_f \in \mathcal{C}$. As shown in Figure 6.3, the information bit stream is split into two sub-streams: *coded bits* (C.R. branch) – which define the λ_f component, and *uncoded bits* (V.R. branch) – which define the λ_c component. (See a

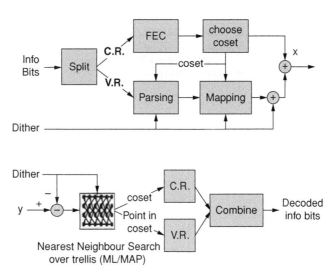

Figure 6.3 Splitting of shaping and coding in VRDM. The coding of the constant rate (C.R.) bit stream is done by the FEC (forward error correction) block.

multi-level code interpretation in Problem 2.13.) The mapping of coded bits to λ_f is done at a *constant rate*, while the mapping of uncoded bits to λ_c is done at a *variable rate*.

We see that shaping is done at a coarse scale, over the *scalar* lattice $q\mathbb{Z}$. Coding gain, on the other hand, is achieved at the finer scale of the n-dimensional code $\mathcal{C}$. This works well at high SNR conditions (and accordingly a large coding rate), where the coarse scale of $q\mathbb{Z}$ is still fine enough with respect to the signal amplitude $\sim \sqrt{P}$, so it can approximate a Gaussian input distribution $X \sim N(0, P)$. Another benefit of the high SNR regime is that the effect of the dither is negligible, so it can be ignored altogether.

At non-high SNR conditions, when the rate is low, the dither becomes crucial for smoothing out the discreteness of the coarse lattice $q\mathbb{Z}$. In this regime, the dependence of the distribution of $Q_\Lambda(\mathbf{S} + \mathbf{u})$ on the dither $\mathbf{u}$ becomes more significant; thus the variable-length code (in particular, the string length $\ell(\mathbf{u}, \lambda)$) must be conditioned on $\mathbf{u}$. Figure 6.4 shows the performance gain due to shaping and coding at a medium range of SNR.

6.5.5 The general SNR case

The asymptotic analysis in Section 6.5.3 does not apply to a general SNR. A more careful examination reveals that the gap to capacity depends on how far the distribution of the randomized transmitter output $\mathbf{X}$ is from a white-Gaussian distribution.

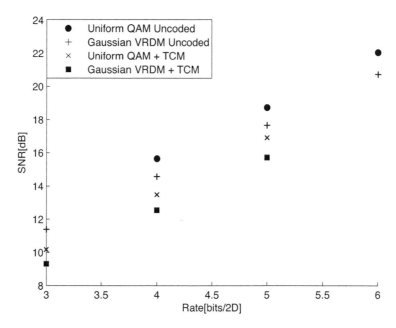

Figure 6.4 Performance of VRDM based on a trellis code at SER $=0.1$ [213].

Definition 6.5.1 (Divergence from white Gaussianity I)[8] *The divergence per sample of an n-dimensional random vector* $\mathbf{X}$ *from white Gaussianity, is defined as*

$$\bar{\mathcal{D}}(\mathbf{X}; \mathbf{X}^*) = \frac{1}{2} \log \left(\frac{\frac{1}{n} E \|\mathbf{X}\|^2}{P_E(\mathbf{X})} \right), \tag{6.51}$$

where $P_E(\mathbf{X}) = 2^{\frac{2}{n} h(\mathbf{X})} / 2\pi e$ *is its entropy power* (A.15).

Note that this distance measure is non-negative because (by the maximum entropy property (A.13)) $P_E(\mathbf{X}) \leq \frac{1}{n} E \|\mathbf{X}\|^2$, with equality if and only if $\mathbf{X}$ is (zero mean) white Gaussian.

Theorem 6.5.2 (Coding rate for a general SNR) *Suppose as in Theorem* 6.5.1, *that* $P_e(\Lambda, N) = P_e$, *and the VRDM decoder uses the simple NN rule* (6.44). *Let the shaping source* $\mathbf{S}$ *and the quantizer partition* $\mathcal{P}$ *be arbitrary (not necessarily Gaussian and Voronoi), so the average power for a uniform (or modulo-uniform) dither is given by* $P = \frac{1}{n} E \|\mathbf{S} - \mathbf{U}\|^2$, *where* $\mathbf{U} \sim \text{Unif}(\mathcal{P}_0)$. *Then, for any* $\text{SNR} = P/N$, *the coding rate* (6.42) *is given by*

$$R_{\text{VRDM}} = \frac{1}{2} \log(\text{SNR}) - \frac{1}{2} \log \left(\frac{\mu(\Lambda, P_e)}{2\pi e} \right) - \bar{\mathcal{D}} \left(\mathbf{S} - \mathbf{U}; (\mathbf{S} - \mathbf{U})^* \right). \tag{6.52}$$

---

[8] See another version in Definition 7.2.3.

Furthermore, for a white-Gaussian shaping source $\mathbf{S} \sim N(0, \sigma_s^2)$, *and* Q_Λ *with Voronoi partition, the coding rate is lower bounded by*

$$R_{\text{VRDM}} \geq \frac{1}{2} \log (\text{SNR}) - \frac{1}{2} \log \left(\mu(\Lambda, P_e) \cdot G(\Lambda) \right), \qquad (6.53)$$

where $G(\Lambda)$ *is the lattice NSM.*

Note that R_{VRDM} in (6.52) is *non-negative* for all SNR and P_e. See the scalar lattice case illustrated in Figure 6.5. To understand better the trade-off between these two parameters, recall that the rate loss term $\frac{1}{2} \log(\mu(\Lambda, P_e)/2\pi e)$ in (6.52) is positive, provided that P_e is below the threshold (3.43). If $P < N$ (low SNR conditions), then P_e is high above the threshold because, by (6.41), $\sigma^2(\Lambda) < P < N$. Hence, this rate "loss" term becomes negative, and it balances the log(SNR) term, to make the entire expression on the right-hand side of (6.52) non-negative.

Proof For the first part of the theorem, we follow the steps of the proof of Theorem 6.5.1. Only instead of approximating $h(\mathbf{S} - \mathbf{U})$ as $h(\mathbf{S})$ (which is only valid at high SNR), we use (6.51) to write it as

$$h(\mathbf{S} - \mathbf{U}) = \frac{n}{2} \log(2\pi e P_E(\mathbf{S} - \mathbf{U})) = \frac{n}{2} \log(2\pi e P) - n\bar{\mathcal{D}}(\mathbf{S} - \mathbf{U}; (\mathbf{S} - \mathbf{U})^*)$$

where we also used the fact that $P = \frac{1}{n} E \|\mathbf{S} - \mathbf{U}\|^2$. As for the second part of the theorem, it follows from the *divergence data-processing inequality* (A.19), that for a white-Gaussian shaping source $\mathbf{S}$, the divergence of $\mathbf{S} - \mathbf{U}$ from white Gaussianity is smaller than or equal to the divergence of $\mathbf{U}$ from Gaussianity:

$$\bar{\mathcal{D}}(\mathbf{S} - \mathbf{U}; (\mathbf{S} - \mathbf{U})^*) \leq \bar{\mathcal{D}}(\mathbf{U}; \mathbf{U}^*).$$

For a Voronoi partition, the dither $\mathbf{U}$ has zero mean and variance $\sigma^2(\Lambda)$. Combining with the fact that its entropy is $\log V(\Lambda)$, we conclude that $\bar{\mathcal{D}}(\mathbf{U}; \mathbf{U}^*) = \frac{1}{2} \log(2\pi e \sigma^2(\Lambda)) - \frac{1}{n} \log V(\Lambda) = \frac{1}{2} \log(2\pi e G(\Lambda))$, by the definition of the NSM. $\qquad \square$

Comparing (6.52) with the AWGN channel capacity $C_{\text{AWGN}} = \frac{1}{2} \log(1 + P/N)$ (6.17), we identify three capacity loss terms.

1. The missing "1" in (6.17), corresponding to a capacity loss of $\frac{1}{2} \log(1 + N/P)$, for $P \geq N$ (this term becomes negligible at high SNR conditions).
2. The term $\frac{1}{2} \log(\mu(\Lambda, P_e)/2\pi e)$, which also appeared in the high SNR case (Theorem 6.5.1).
3. The divergence of the equivalent transmitter output $\mathbf{X} = \mathbf{S} - \mathbf{U}$ from white Gaussianity. When the SNR is *not* high, the contribution of the dither $\mathbf{U}$ to this distribution is not negligible; thus, the transmitter output is not quite Gaussian, even for a Gaussian $\mathbf{S}$.

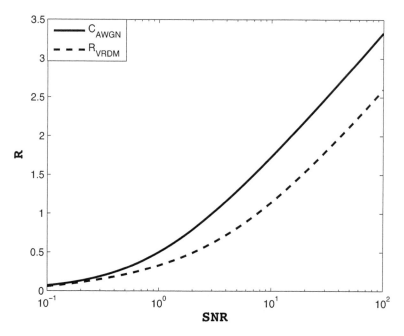

Figure 6.5 Performance of the scalar VRDM at $P_e = 0.001$ as a function of the SNR, compared to C_{AWGN}. Note that at high SNR the capacity loss is $\frac{1}{2} \log_2[\mu(\mathbb{Z}, 0.001)/2\pi e] \approx 0.7$ bit, in accordance with Theorem 6.5.1.

The second part of Theorem 6.5.2 becomes interesting when the SNR is near 1. It indicates that in this case, the lattice should be good both for modulation and for quantization.

If the two goodness measures – the NVNR and the NSM – go to their ideal values of $2\pi e$ and $1/2\pi e$, respectively (which as we shall see in Chapter 7 is indeed possible for good high-dimensional lattices), then we are only left with the first loss term above, i.e., the loss of "1" inside the log in the AWGN channel capacity. This residual capacity loss is at most half a bit (at SNR $= 1$). It is *not* an artifact of the lattice, but is due to the decoding mechanism. One way to gain back this extra half a bit is to modify the NN decoding rule (6.44), to take into account the non-uniform distribution of the lattice points. [9]

6.6 Maximum a posteriori decoding*

A decoding rule can give priority to one message over another. That is, the partition of the channel output space into decision cells can reduce $P_e(\mathbf{c})$, the probability of

---

[9] When the SNR is low, the Shannon capacity of the AWGN channel (6.17) enjoys the Gaussianity of the noise, whose power adds up on top of the transmitter power, and makes the effective codebook power $P + N$. However, an NN decoded unbounded lattice – even with probabilistic shaping – cannot enjoy this "natural shaping" gain. See Section 9.4.

error given that the codeword $\mathbf{c}$ was transmitted (6.8), at the cost of increasing it for some other codeword $\mathbf{c}'$ (or codewords). This is useful when the messages are *not* equally likely. [10]

When each message $\mathbf{c} \in C$ is transmitted with probability $p(\mathbf{c})$, the average error probability (6.9) is given by the weighted sum:

$$P_e(C) = \sum_{\mathbf{c} \in C} p(\mathbf{c}) \cdot P_e(\mathbf{c}), \tag{6.54}$$

where $P_e(\mathbf{c})$ is the probability of error given that the codeword $\mathbf{c}$ was transmitted (6.8). The optimum decoding rule that minimizes $P_e(C)$ in this case is the *maximum a posteriori* (MAP) rule:

$$\hat{\mathbf{c}}_{\text{MAP}} = \arg\max_{\mathbf{c} \in C} p_{\mathbf{X}|\mathbf{Y}}(\mathbf{c}|\mathbf{y}) \tag{6.55}$$

$$= \arg\max_{\mathbf{c} \in C} \left\{ p(\mathbf{c}) \cdot p_{\mathbf{Y}|\mathbf{X}}(\mathbf{y}|\mathbf{c}) \right\}. \tag{6.56}$$

The optimality of the MAP rule (6.55) is due to the fact that for each received $\mathbf{y}$, the conditional average error probability given $\mathbf{Y} = \mathbf{y}$ is minimized by picking the codeword $\mathbf{c}$ that maximizes the a posteriori probability $p(\mathbf{c}|\mathbf{y})$. And the second equality follows from the Bayes rule. For a continuous additive-noise channel $\mathbf{Y} = \mathbf{X} + \mathbf{Z}$, (6.56) becomes

$$\hat{\mathbf{c}}_{\text{MAP}} = \arg\max_{\mathbf{c} \in C} \{ p(\mathbf{c}) \cdot f_{\mathbf{Z}}(\mathbf{y} - \mathbf{c}) \} \tag{6.57}$$

where $f_{\mathbf{Z}}(\cdot)$ is the noise density. That is, the ML decoding rule (6.7) is modified by weighting the "forward" channel transition probability $p(\mathbf{y}|\mathbf{c})$ by the message probability $p(\mathbf{c})$.

For VRDM, the MAP rule (6.57), given a dither value $\mathbf{U} = \mathbf{u}$, becomes

$$\hat{\lambda}_{\text{MAP}} = \arg\max_{\lambda \in \Lambda} \{ p_\lambda(\mathbf{u}) \cdot f_{\mathbf{Z}}(\mathbf{y} + \mathbf{u} - \lambda) \} \tag{6.58a}$$

$$= \arg\max_{\lambda \in \Lambda} \{ f_{\mathbf{X}}(\lambda - \mathbf{u}) \cdot f_{\mathbf{Z}}(\mathbf{y} + \mathbf{u} - \lambda) \}, \tag{6.58b}$$

where $p_\lambda(\mathbf{u})$ is the probability of the lattice point λ given the dither (6.39); in the second line we used the formula $p_\lambda(\mathbf{u}) = V(\Lambda) \cdot f_{\mathbf{X}}(\lambda - \mathbf{u})$, (5.9), where $f_{\mathbf{X}}(\cdot)$ is the density of $\mathbf{X} = \mathbf{S} - \mathbf{U}$ (the output of the equivalent channel of Figure 4.5, which is the VRDM codebook distribution).

In the case of an AWGN channel, where $Z \sim N(0, N)$, and if also X is white Gaussian, i.e., $X \sim N(0, P)$ (which is possible asymptotically in the limit of high SNR, or when $G(\Lambda) \rightarrow \frac{1}{2\pi e}$), then (6.58b) can be simplified to the "weighted"

---

[10] Note that in the VRDM, the probabilities of the messages are determined by the system (the "f^{-1} mapping"), while the information bits are assumed to be completely random (i.e., Bernoulli(1/2)).

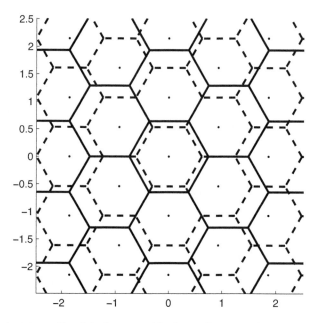

Figure 6.6 The decision cells of the hexagonal lattice under MAP decoding, for an AWGN channel at SNR = 7 dB, i.e., a Wiener coefficient $\alpha^* = 5/6$, based on the approximated rule of (6.59). The dashed line shows the ML partition for comparison.

nearest-neighbor rule (compare with (6.44)):

$$\hat{\lambda}_{\text{MAP}} \cong \arg\min_{\lambda \in \Lambda} \left\{ \frac{\|\mathbf{y} + \mathbf{u} - \lambda\|^2}{N} + \frac{\|\lambda - \mathbf{u}\|^2}{P} \right\}, \tag{6.59a}$$

$$= \arg\min_{\lambda \in \Lambda} \left\{ \|\alpha^* \mathbf{y} + \mathbf{u} - \lambda\|^2 \right\}, \tag{6.59b}$$

where

$$\alpha^* = \frac{\text{SNR}}{1 + \text{SNR}}$$

is the Wiener coefficient (4.44). Here, (6.59b) follows either by completing (6.59a) to a single quadratic term, or directly from (6.55) and the Gaussian a posteriori probability distribution $p_{\mathbf{X}|\mathbf{Y}}$; see Problem 6.2. Figure 6.6 shows the induced partition for a hexagonal lattice constellation at SNR = 7 dB.

Note that (6.59a) amounts to adding a new term, $\|\lambda - \mathbf{u}\|^2 / \text{SNR}$, to the NN rule (6.44), which "penalizes" the more remote lattice points. When the signal to noise ratio is high, this penalty term is negligible at the vicinity of the origin, and we are back with the NN rule (6.44).

How much do we gain in error probability by the MAP rule? Recall that Wiener (or Bayesian) estimation takes into account the input variance P (a priori information), and thus reduces the estimation MSE from N (the noise variance) to $PN/(P + N)$; see (4.45). That is, Wiener estimation gains a factor of $P/(P + N)$ in MSE. It is believed that the same gain occurs in MAP decoding, i.e., the probability of error in MAP decoding is the same as in ML decoding, only with a noise power smaller by a factor of $P/(P + N)$. If this conjecture is correct then, by (6.53), the gap to capacity of VRDM with MAP decoding is bounded by

$$C_{\text{AWGN}} - R_{\text{VRDM}} \leq \frac{1}{2} \log(\mu(\Lambda, P_e) \cdot G(\Lambda)) \qquad (6.60)$$

at *any* SNR; thus, MAP decoding gains back the missing "1" inside the logarithm in Theorem 6.5.2.

Summary of Chapter 6

Noise-matched (ML) lattice decoding For an additive-noise channel $\mathbf{Y} = \mathbf{x} + \mathbf{Z}$ with noise distribution $\mathbf{Z} \sim P_{\mathbf{Z}}(\cdot)$, if $\mathbf{x} = \lambda \in \Lambda$, then

$$\hat{\lambda}_{\text{ML}} = \arg\max_{\lambda \in \Lambda} p_{\mathbf{Z}}(\mathbf{Y} - \lambda).$$

Error probability

$$P_e = \Pr\{\mathbf{Z} \notin \mathcal{V}_0^{\mathbf{Z}}(\Lambda)\},$$

where $\mathcal{V}_0^{\mathbf{Z}}(\Lambda)$ is the (non-Euclidean) Voronoi region matched to $\mathbf{Z}$.

AWGN channel If Z is AWGN, then

$$\hat{\lambda}_{\text{ML}} = Q_\Lambda^{(NN)}(\mathbf{Y}) = \arg\min_{\lambda \in \Lambda} \|\mathbf{Y} - \lambda\|,$$

and $P_e = \Pr\{\mathbf{Z} \notin \mathcal{V}_0(\Lambda)\}$.

Rate per unit volume For AWGN $\sim N(0, N)$,

$$R_\infty(\Lambda, P_e) = \frac{1}{2} \log\left(\frac{1}{N \cdot \mu(\Lambda, P_e)}\right).$$

Gap to the infinite constellation capacity

$$C_\infty - R_\infty(\Lambda, P_e) = \frac{1}{2} \log\left(\frac{\mu(\Lambda, P_e)}{2\pi e}\right).$$

If $\mathbf{Z}$ is a general additive noise, then $\mu(\Lambda, P_e)$ is replaced by $\mu_{\text{matched}}(\Lambda, \mathbf{Z}, P_e)$ for noise-matched decoding, and by $\mu_{\text{euclid}}(\Lambda, \mathbf{Z}, P_e)$ for (mismatched) Euclidean decoding.

Variable-rate dithered modulation (VRDM)

$$\mathbf{x}(\mathbf{u}, b^*) = \lambda - \mathbf{u} = f^{-1}(\mathbf{u}, b^*) - \mathbf{u},$$

where $b^* = f(\lambda, \mathbf{u})$ is a variable-length code for $\lambda \in \Lambda$, conditioned on the dither $\mathbf{u}$.

Probabilistic shaping If f achieves the entropy $H(Q_\Lambda[\mathbf{S} + \mathbf{U}]|\mathbf{U})$ for a uniform dither $\mathbf{U}$, then

1. the transmitter output $\mathbf{X}$ is distributed as $\mathbf{S} - \mathbf{U}$;
2. the coding rate is $R_{\text{VRDM}} = I(\mathbf{S}; \mathbf{S} - \mathbf{U})/n$.

Rate at high SNR As $\text{SNR} = E\|\mathbf{X}\|^2/nN$ goes to infinity,

$$R_{\text{VRDM}} = \frac{1}{2}\log(\text{SNR}) - \frac{1}{2}\log\left(\frac{\mu(\Lambda, P_e)}{2\pi e}\right) + o(1),$$

where P_e is the probability in NN decoding.

MAP decoding at general SNR If MAP decoding amounts to Wiener estimation, then for all SNR

$$R_{\text{VRDM}} \geq \frac{1}{2}\log(1 + \text{SNR}) - \frac{1}{2}\log\left(\mu(\Lambda, P_e)G(\Lambda)\right).$$

Problems

P.6.1 (Noise-matched decoding for a colored noise) Prove that

$$\mathcal{V}_0^{T\mathbf{Z}}(T \cdot \Lambda) = T \cdot \mathcal{V}_0^{\mathbf{Z}}(\Lambda), \tag{6.61}$$

where $\mathcal{V}_0^{\mathbf{Z}}(\Lambda)$ is the (non-Euclidean) fundamental Voronoi cell associated with the noise $\mathbf{Z}$ (6.12), and T is a non-singular matrix. Use that to prove that

$$\mu_{\text{matched}}(T\Lambda, T\mathbf{Z}, P_e) = \mu_{\text{matched}}(\Lambda, \mathbf{Z}, P_e). \tag{6.62}$$

(Note that $P_E(T\mathbf{Z}) = |\det(T)|^{2/n}P_E(\mathbf{Z})$.) Find the decoding regions matched to a colored-Gaussian noise $\sim N(0, \Sigma)$ of a lattice Λ, based on the Euclidean fundamental Voronoi cell of a transformed lattice.

P.6.2 (MAP via Wiener estimation) Verify the transition from (6.59a) to (6.59b), in two ways: (i) by completion into a square; and (ii) using (6.55) and the a posteriori probability $p_{\mathbf{X}|\mathbf{Y}}$ in the Gaussian case (with $\mathbf{x} = \lambda - \mathbf{u}$); see Problem 4.12.

P.6.3 (Rate-distortion function per unit volume) Imagine a memoryless source which is uniformly distributed over a large interval. The *rate-distortion function per unit volume*, i.e., the minimum possible rate per unit volume of a large constellation that quantizes this source with a mean-squared distortion D, is given by $R_\infty(D) = -\frac{1}{2}\log(2\pi e D)$. Show that the rate redundancy of a lattice quantizer Q_Λ, satisfying the same distortion D, is given by $R_\infty(\Lambda) - R_\infty(D) = \frac{1}{2}\log(2\pi e G(\Lambda))$.

Historical notes

Poltyrev [221] proposed the setup of unconstrained channels. He defined the quantities of rate and capacity per unit volume (normalized logarithmic density), and derived the error exponent for random codes – as well as lattices – for the unconstrained AWGN channel case. His work is based on the Shannon capacity [240] and error probability analysis [244], on the Gallager error exponent [96, 97] and on the Siegel version of the Minkowski–Hlawka theorem [113].

The bounds on the capacity of general additive-noise channels are due to Shannon [240]; see the book by Cover and Thomas [53]. A comprehensive treatment of mismatched decoding and capacity can be found in the articles by Lapidoth *et al.* [148], and by Csiszár and Narayan [55].

Non-equiprobable signaling for the power-constrained AWGN channel by means of variable-rate modulation of a low-dimensional constellation was described by Forney *et al.* [91]. Kschischang and Pasupathy [145] combined Gaussian-shaped signaling with coding, while Ungerboeck [260] examined specifically Huffman shaping; see also Abrahams [1]. Palgy and Zamir [213] extended these "probabilistic" shaping schemes to the non-high SNR regime, by using reverse ECDQ (i.e., variable-rate dithered lattice modulation) and replacing the ML decoder by a MAP decoder. Ling and Belfiore [166] established asymptotic optimality, above some threshold SNR, for non-randomized high-dimensional lattice constellations with probabilistic shaping and MMSE decoding.

Other forms of shaping, based on fixed-rate high-dimensional modulation (shell mapping, trellis shaping and Voronoi modulation [30, 33, 133, 150]), will be addressed in Chapter 9.

7 Asymptotic goodness

As the dimension increases, lattices can form richer arrangements of points in space. Richness, though, comes at the cost of a higher complexity. Are high-dimensional lattices *better*?

The advantage of going to higher dimensions is questionable if we are only interested in sphere packing and covering. The *one*-dimensional lattice is already *perfect* for both problems, whereas higher-dimensional lattices are not. To be more specific, the *best* lattice packing efficiency ρ_{pack} gets worse (roughly monotonically) as the dimension increases, and decreases from 1 to the *Minkowski bound* of one-half. The *best* lattice-covering efficiency ρ_{cov} exhibits anomalous behavior. First it increases (i.e., deteriorates) but then, by the *Rogers bound*, it becomes asymptotically perfect again and approaches 1 as the dimension goes to infinity. See more on that later in this chapter.

While mathematicians paid attention to these two hard problems, Shannon and his followers were more interested in the "softer" questions of quantization and modulation. Shannon's theory demonstrates the advantage of high-dimensional source and channel codes. The underlying principle is the *law of large numbers*. Does this principle apply also to lattice codes?

Lattices indeed improve when it comes to the quantization and the modulation problems. As we saw in Chapter 5, the rate redundancy of the ECDQ over Shannon's rate-distortion function under a squared-error distortion measure is

$$\frac{1}{2} \log_2 (2\pi e G(\Lambda)) \quad \text{bit per source sample,} \tag{7.1}$$

where $G(\Lambda)$ is the NSM defined in (3.21). Because at high-resolution quantization MSE scales like 2^{-2R}, this is equivalent to a multiplicative MSE loss of $2\pi e G(\Lambda)$ with respect to the information theoretic optimum, or an additive MSE loss of $10 \log_{10} (2\pi e G(\Lambda))$ dB (which is $20 \log_{10} 2 \approx 6.02$ times the value in (7.1)). Indeed for the first three dimensions, $G(\mathbb{Z}) \approx 0.083$, $G(A_2) \approx 0.080$ (hexagonal lattice), and $G(A_3^*) = 0.078$ (BCC lattice); so the corresponding ECDQ redundancy -0.255, 0.227 and 0.212 bit (or MSE loss of 1.53, 1.36 and 1.27 dB) – decreases. Table 7.1 shows the quantization gain, corresponding to rate redundancy and MSE loss,

Table 7.1 The quantization "granular" gain with respect to the cubic lattice (3.25), $\Gamma_q(\Lambda) = 10 \log_{10} \left(\frac{1/12}{G(\Lambda)} \right)$ [dB], of some known lattices in dimensions 1 to 24 (Λ^* denotes the dual of Λ; see Definition 4.2.3)

The sphere upper bound (right column) is calculated with respect to the NSM of an n-dimensional sphere (G_n^*). Note that the gain is upper bounded by that of an infinite-dimensional sphere, $\Gamma_q^*(n = \infty) = 10 \log_{10} \left(\frac{2\pi e}{12} \right) \approx 1.53$ dB.

Dimension	Lattice		Γ_q [dB]	Sphere bound
1	$\mathbb{Z}$	integer	0	0
2	A_2	hexagonal	0.17	0.20
3	A_3	FCC	0.24	0.34
	A_3^*	BCC	0.26	0.34
4	D_4	(Example 2.4.2)	0.36	0.45
5	D_5^*		0.42	0.54
6	E_6^*		0.50	0.61
7	E_7^*		0.57	0.67
8	E_8	Gosset	0.65	0.72
12	K_{12}		0.75	0.87
16	BW_{16}	Barnes–Wall	0.86	0.97
24	Λ_{24}^*	Leech*	1.03	1.10
∞	?	?	1.53	1.53

for higher-dimensional lattices. The rate redundancy and MSE loss would vanish completely if the NSM could reach the value of $1/2\pi e$. Can it get there?

For modulation, we saw in Chapter 6 that the rate loss in lattice modulation with respect to the Shannon–Poltyrev capacity is

$$\frac{1}{2} \log_2 \left(\frac{\mu(\Lambda, P_e)}{2\pi e} \right) \text{ bit per channel use,} \tag{7.2}$$

where $\mu(\Lambda, P_e)$ is the NVNR at probability of error P_e defined in (3.37). Because power scales at high SNR as 2^{2R} (6.23), this is equivalent to a multiplicative power loss of $\mu(\Lambda, P_e)/2\pi e$ with respect to the information theoretic optimum, or an additive power loss of $10 \log_{10}(\mu(\Lambda, P_e)/2\pi e)$ dB. Like the NSM, the NVNR tends to decrease if we look at some good lattices – scalar, hexagonal and FCC – in the first three dimensions: $\mu(\mathbb{Z}, 10^{-4}) \approx 60.5$, $\mu(A_2, 10^{-4}) \approx 59.7$, and $\mu(A_3, 10^{-4}) \approx 58.8$, corresponding to a capacity loss of approximately 0.91, 0.90 and 0.89 bit, or a power loss of 5.49, 5.43 and 5.37 dB, respectively. The decrease is even stronger if we fix the symbol error rate (3.41), as can be seen in Table 7.2. Can the NVNR reach the ideal value of $2\pi e$?

Interestingly, the figures in the table indicate that a lattice which is good in one sense need not necessarily be good in another (in three dimensions, the FCC is a better modulation constellation while the BCC is a better quantizer). Nevertheless,

Table 7.2 The coding gain with respect to the cubic lattice, $\Gamma_c(\Lambda) = 10 \log_{10}\left(\frac{\mu(\mathbb{Z}^n, P_e)}{\mu(\Lambda, P_e)}\right)$ [dB], of some known lattices in dimensions 2 to 24; see (3.38)

For a fair comparison of different dimensions, the table uses the symbol error probability (SER) rather than the block error probability P_e, which are related by $1 - P_e = (1 - \text{SER})^n$; see (3.42). The number in brackets is a sphere upper bound, calculated with respect to the NVNR of an n-dimensional sphere $(\mu_n^*(P_e))$. The last line shows the coding gain of an infinite-dimensional sphere, $\Gamma_c^*(n = \infty) =$ $10 \log_{10}\left(\frac{\mu(\mathbb{Z}, P_e)}{2\pi e}\right)$ [dB], which is equivalent to the power loss (7.2) (in dB) of a scalar lattice constellation. More coding gains of high-dimensional lattice constellations ($n = 100 \div 1000$) appear in [125].

Dimension	Lattice	SER				
		10^{-1}	10^{-2}	10^{-3}	10^{-4}	10^{-5}
1	$\mathbb{Z}^1$	0	0	0	0	0
2	A_2	0.14 (0.16)	0.27 (0.33)	0.33 (0.45)	0.42 (0.54)	0.46 (0.6)
3	A_3	0.20 (0.27)	0.42 (0.56)	0.55 (0.78)	0.65 (0.93)	0.72 (1.05)
	A_3^*	0.20 (0.27)	0.40 (0.56)	0.52 (0.78)	0.59 (0.93)	0.61 (1.05)
4	D_4	0.29 (0.36)	0.60 (0.75)	0.82 (1.03)	0.95 (1.24)	1.00 (1.40)
8	E_8	0.50 (0.56)	1.08 (1.2)	1.49 (1.68)	1.80 (2.04)	2.00ʹ (2.30)
16	BW_{16}	0.63 (0.75)	1.47 (1.63)	2.09 (2.32)	2.52 (2.83)	2.80 (3.22)
24	Λ_{24}	0.75 (0.84)	1.76 (1.85)	2.51 (2.65)	3.08 (3.25)	3.50 (3.71)
∞	?	−2.0	1.9	4.0	5.5	6.6

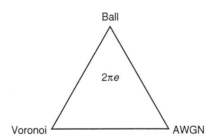

Figure 7.1 The asymptotic ball–AWGN–Voronoi connection.

there exists a sequence Λ_n^* of lattices of increasing dimension n which are asymptotically *simultaneously good* for quantization and modulation – i.e., $G(\Lambda_n^*) \to 1/2\pi e$ and $\mu(\Lambda_n^*, P_e) \to 2\pi e$ – as well as for covering and packing (in the sense of the Minkowski bound (3.15)). The existence of such good lattices is proved by *random coding* arguments. Therefore, unfortunately, we cannot construct them explicitly.

An NSM of $1/2\pi e$ amounts to the lattice Voronoi cell taking the shape of a high-dimensional ball, or equivalently, the dither (i.e., a uniform distribution over the cell) being a *white-Gaussian noise*. From a channel coding viewpoint, an NVNR of $2\pi e$ (with a small error probability) amounts to spherical Voronoi cells at high dimensions. Although such ideal spherical Voronoi cells cannot be achieved *exactly* by any lattice of dimension greater than one, good high-dimensional lattices can arbitrarily approach them.

This chapter establishes the triangular relation shown in Figure 7.1. Section 7.1 considers the ball–Voronoi edge of the triangle, and develops the sphere lower bounds on the NSM and NVNR used in Tables 7.1 and 7.2. Section 7.2 considers the ball–AWGN edge of the triangle (showing the asymptotic ball–AWGN equivalence), while Section 7.3 considers the Voronoi–ball and Voronoi–AWGN edges (showing the asymptotic Voronoi–ball/Voronoi–AWGN equivalence) for good lattice quantizers. To support that, Section 7.5 introduces the concept of a random ensemble of lattices, which is used in Sections 7.6–7.9 to prove that good lattices indeed exist.

7.1   Sphere bounds

Suppose a host is sitting a group of people in a concert hall, and wishes to make them happy about their location. A good strategy would be to fill in seats from best to least good, thus maximizing the average person satisfaction. A similar principle holds for answering the question: Among all bodies of a given volume in $\mathbb{R}^n$, which one minimizes the second moment? A spherical shell of a certain radius corresponds to a seat class in the concert hall – all points have the same square norm. Thus by "filling" shells from zero radius and up we minimize the average second moment for a given volume. This, and other unique properties of a sphere, are known as *iso-perimetric inequalities*.

In this section we use the iso-perimetric inequalities – and the properties of a sphere – to bound the performance of a lattice code Λ for quantization and modulation. As we shall see, $G(\Lambda)$ and $\mu(\Lambda, p_e)$ can be upper and lower bounded using the NSM and the NVNR of a sphere.

Recall the definition (3.1) of an n-dimensional ball of radius r centered at the origin:

$$\mathcal{B}_r = \{\mathbf{x} : \|\mathbf{x}\| \leq r\}$$

where $\|\cdot\|$ is the Euclidean norm. [1] Recall also the ball volume formula (3.2), $\mathrm{Vol}(\mathcal{B}_r) = V_n \cdot r^n$, where V_n is the *volume of the unit ball* (3.3).

Given a general body S in $\mathbb{R}^n$, let

$$V(S) = \int_S d\mathbf{x} \tag{7.3}$$

---

[1] Most results also apply to more general norms; see Problem 7.5.

denote its volume,

$$\sigma^2(S) = \frac{1}{nV(S)} \int_S \|\mathbf{x}\|^2 d\mathbf{x} \tag{7.4}$$

denote its *second moment* (i.e., the second moment per dimension of a uniform distribution over S),

$$\text{diameter}(S) = \max_{\mathbf{x}_1, \mathbf{x}_2 \in S} \|\mathbf{x}_2 - \mathbf{x}_1\| \tag{7.5}$$

denote its *diameter*,

$$r_{\text{pack}}(S) = \max\{r : \mathcal{B}_r \subset S\} \tag{7.6}$$

denote its *packing (or inner) radius* (with respect to a ball centered at the origin), and

$$V(\delta S) = \lim_{\epsilon \to 0} \frac{V(S + \mathcal{B}_\epsilon) - V(S)}{\epsilon} \tag{7.7}$$

denote its *surface area*, where $S + \mathcal{B}_\epsilon$ denotes the set-sum (or the Minkowski sum) of S and an epsilon ball. Define also the *Gaussian error probability* of S as the probability that a zero-mean white-Gaussian noise $\mathbf{Z}$ of variance σ^2 goes outside S:

$$P_e(S, \sigma^2) = \Pr\{\mathbf{Z} \notin S\} = 1 - \frac{1}{(2\pi\sigma^2)^{n/2}} \int_S e^{-\frac{\|\mathbf{z}\|^2}{2\sigma^2}} d\mathbf{z}. \tag{7.8}$$

Based on the second moment (7.4) and Gaussian error probability (7.8), define also two dimensionless quantities: the NSM (or "second-moment to volume ratio") of S,

$$G(S) = \frac{\sigma^2(S)}{V^{2/n}(S)} \tag{7.9}$$

and the NVNR of S,

$$\mu(S, P_e) = \frac{V^{2/n}(S)}{\sigma^2(P_e)}, \tag{7.10}$$

corresponding to earlier definitions for lattices in Sections 3.2 and 3.3, where $\sigma^2(P_e)$ is the value of σ^2 such that $P_e(S, \sigma^2) = P_e$; see (3.36).

For the n-dimensional ball $\mathcal{B}_r$, the first five quantities above are given explicitly by (see Problem 7.1):

$$\sigma^2(\mathcal{B}_r) = \frac{r^2}{n+2} \tag{7.11a}$$

$$\text{diameter}(\mathcal{B}_r) = 2r \tag{7.11b}$$

$$= 2r_{\text{pack}}(\mathcal{B}_r) \tag{7.11c}$$

$$\text{Vol}(\delta\mathcal{B}_r) = nV_n r^{n-1} \tag{7.11d}$$

$$P_e(\mathcal{B}_r, \sigma^2) = \Pr\{\|\mathbf{Z}\| > r\} \tag{7.11e}$$

$$= 1 - \chi^2_{\text{CDF}}\left(\frac{r^2}{\sigma^2}; n\right) \tag{7.11f}$$

where $\chi^2_{\text{CDF}}(x; n)$, for $x \geq 0$, is the *cumulative chi-square distribution* function with n degrees of freedom.

The ball characteristics above provide lower bounds for any body with the *same volume*, as stated by the following theorem.

Theorem 7.1.1 (Iso-perimetric inequalities) *A ball minimizes the second moment, diameter, surface area and error probability, and maximizes the packing radius, over all bodies of a given volume. That is, if $\text{Vol}(S) = \text{Vol}(\mathcal{B}_r)$, then*

$$\sigma^2(S) \geq \sigma^2(\mathcal{B}_r) \tag{7.12a}$$

$$\text{diameter}(S) \geq \text{diameter}(\mathcal{B}_r) \tag{7.12b}$$

$$\text{Vol}(\delta S) \geq \text{Vol}(\delta\mathcal{B}_r) \tag{7.12c}$$

$$P_e(S, \sigma^2) \geq P_e(\mathcal{B}_r, \sigma^2) \tag{7.12d}$$

$$r_{\text{pack}}(S) \leq r_{\text{pack}}(\mathcal{B}_r) \tag{7.12e}$$

with equality if and only if S is a ball. The characteristics of a ball in the right-hand side of (7.12) are given in (7.11a)–(7.11d).

Using the "shell filling" argument in the beginning of this section, but under diameter and packing radius constraints, we obtain the following.

Proposition 7.1.1 (Reverse iso-perimetric inequalities) *A ball maximizes the second moment over all convex zero-symmetric bodies with a given diameter, and maximizes the error probability over all convex bodies with a given packing radius. That is, if S is convex and symmetric about the origin, then $\sigma^2(S) \leq \sigma^2(\mathcal{B}_{\text{diameter}(S)/2})$, and $P_e(S, \sigma^2) \leq P_e(\mathcal{B}_{r_{\text{pack}}(S)}, \sigma^2)$, with equality if and only if S is a ball.*

In particular, letting S in (7.12a), (7.12d) and Proposition 7.1.1 be the fundamental Voronoi region $\mathcal{V}_0$ of the lattice Λ (which is convex and symmetric about the origin, but it is not a ball for dimension greater than 1), we obtain the following.

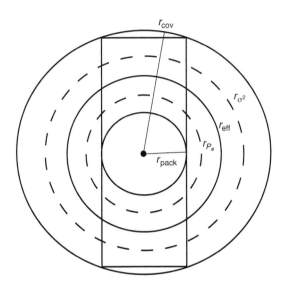

Figure 7.2 Equivalent radiuses of a rectangular cell. The figure shows r_{cov}, r_{pack} and r_{eff}, as well as r_{σ^2} (the radius of a ball with the same second moment) and r_{P_e} (the radius of a ball with the same error probability). We clearly have $r_{\mathrm{pack}} < r_{P_e} < r_{\mathrm{eff}} < r_{\sigma^2} < r_{\mathrm{cov}}$.

Corollary 7.1.1 *The lattice second moment* (3.20) *and error probability* (3.32) *are lower and upper bounded by*

$$\sigma^2(\mathcal{B}_{r_{\mathrm{cov}}(\Lambda)}) \geq \sigma^2(\Lambda) \geq \sigma^2(\mathcal{B}_{r_{\mathrm{eff}}(\Lambda)}) \tag{7.13}$$

and

$$P_e(\mathcal{B}_{r_{\mathrm{pack}}(\Lambda)}, \sigma^2) \geq P_e(\Lambda, \sigma^2) \geq P_e(\mathcal{B}_{r_{\mathrm{eff}}(\Lambda)}, \sigma^2), \tag{7.14}$$

and equality holds (only) for a scalar lattice.

Figure 7.2 reflects these bounds in terms of "equivalent radiuses."

Corollary 7.1.1 implies that the NSM and NVNR of a lattice in $\mathbb{R}^n$ are bounded from below by the NSM and NVNR of an n-dimensional ball. Using (7.9), we denote the NSM of an n-dimensional ball as

$$G_n^* \triangleq G(\mathcal{B}_r) = \frac{1}{(n+2)V_n^{2/n}}, \tag{7.15}$$

where $V_n = \pi^{n/2}/(n/2)!$ is the unit-ball volume (3.3), and the second equality follows from (7.11a) and (3.2). See Figure 7.3. In particular, $G_1^* = 1/12 \approx 0.0833$, $G_2^* = 1/(4\pi) \approx 0.0802$, and $G_3^* = 0.0779$. Also, by a lower bound on $n!$ we have that

$$G_n^* > \frac{1}{2\pi e} \quad \text{for all } n. \tag{7.16}$$

See Table 7.1 and Problem 7.2.

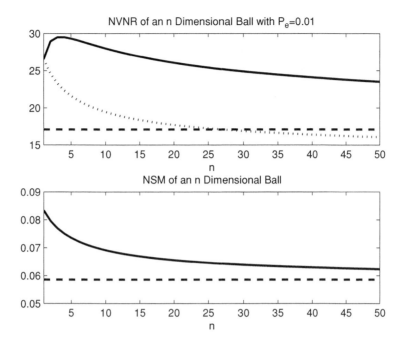

Figure 7.3 The NSM G_n^* and the NVNR $\mu_n^*(P_e)$ (at $P_e = 0.01$) of an n-dimensional ball, as a function of the dimension n. The dotted line shows μ_n^* for a fixed symbol error rate SER $= 0.01$.

The NVNR of an n-dimensional ball, denoted $\mu_n^*(P_e)$, unfortunately does not have a simple characterization. Using (7.10), we have

$$\mu_n^*(P_e) \triangleq \mu(\mathcal{B}_r, P_e) = V_n^{2/n} \cdot \chi_{\text{ICDF}}^2 (1 - P_e; n), \tag{7.17}$$

where the second equality follows from (7.11) and (3.2), and where $\chi_{\text{ICDF}}^2 (p; n)$, for $0 \leq p \leq 1$, is the *inverse* cumulative chi-square distribution with n degrees of freedom, i.e.,

$$\chi_{\text{ICDF}}^2 (p; n) = x \quad \text{if} \quad \chi_{\text{CDF}}^2(x; n) = p. \tag{7.18}$$

In particular, in the one-dimensional case $\chi_{\text{CDF}}^2(x; 1) = 1 - 2Q(\sqrt{x}) \approx 1 - e^{-x/2}$, where $Q(\cdot)$ is the Q-function, and the approximation holds for $x \gg 1$ [62, section 1; 272, p. 83]. Thus $\mu_1^*(P_e) = 4[Q^{-1}(P_e/2)]^2 \approx 8 \ln(1/P_e)$. In two dimensions, $\chi_{\text{CDF}}^2(x; 2) = 1 - e^{-x/2}$ for all x (in this case, it is the *Rayleigh distribution*), so $\mu_2^*(P_e) = 2\pi \ln(1/P_e)$. See Figure 7.3 and Problem 7.3. Furthermore, the ball error probability (7.11e) can be lower bounded, for *all* dimensions n, by

$$P_e(\mathcal{B}_r, \sigma^2) > Q \left(\frac{r^2 - n\sigma^2}{\sigma^2 \sqrt{2n}} \right) - \frac{2}{\sqrt{n}}; \tag{7.19}$$

see the application of the Berry–Esseen theorem (a refinement of the central limit theorem) in [125, Lemma 8] and [222]. This implies, among others, the uniform

$2\pi e$ lower bound (3.43) on the ball NVNR:

$$\mu_n^*(P_e) > 2\pi e \quad \text{for } P_e < 0.03, \text{ for all } n. \tag{7.20}$$

See Table 7.2 and Problem 7.3.

Theorem 7.1.2 (Sphere bounds on NSM and NVNR) *If Λ is an n-dimensional lattice, $n > 1$, then*

$$\rho_{\text{cov}}^2(\Lambda) \cdot G_n^* > \quad G(\Lambda) \quad > G_n^* \tag{7.21}$$

$$\frac{1}{\rho_{\text{pack}}^2(\Lambda)} \cdot \mu_n^*(P_e) > \mu(\Lambda, P_e) > \mu_n^*(P_e), \tag{7.22}$$

where $\rho_{\text{cov}}(\Lambda) = r_{\text{cov}}(\Lambda)/r_{\text{eff}}(\Lambda)$, and $\rho_{\text{pack}}(\Lambda) = r_{\text{pack}}(\Lambda)/r_{\text{eff}}(\Lambda)$ are the covering and packing efficiencies of Λ, defined in (3.17) and (3.11), respectively.

Proof The bounds follow directly from the iso-perimetric inequalities of Corollary 7.1.1. See Figure 3.3. □

Since the Voronoi cell of a multidimensional lattice is a polytope and not a ball, all four inequalities in (7.21) and (7.22) are strict. Yet, Theorem 7.1.2 implies that if a lattice is "good" for packing or covering, in the sense that its packing or covering efficiencies are close to 1, then it is also "good" for quantization or modulation, i.e., its NVNR and NSM are close to the sphere lower bounds.

7.2 Sphere-Gaussian equivalence

So far, out of the three main entities of this chapter – a Voronoi cell, a ball and a white-Gaussian vector (all living in the n-dimensional space) – we have only related the first pair. Our next objective is to show an asymptotic relationship between the second pair, a ball and a Gaussian vector, as the dimension goes to infinity.

The probability density function f_W of a zero-mean white-Gaussian vector $\mathbf{W} = (W_1, \ldots, W_n)$,

$$f_W(\mathbf{w}) = \frac{1}{(2\pi\sigma^2)^{n/2}} e^{-\frac{\|\mathbf{w}\|^2}{2\sigma^2}} \tag{7.23}$$

is *isotropic* in space. That is, it depends only on the Euclidean norm $\|\mathbf{w}\|$ of the vector, hence it is constant over spherical shells around the origin. On the other hand, the law of large numbers (LLN) implies that the normalized squared norm of the vector, $\frac{1}{n}\|\mathbf{W}\|^2$, goes in probability to its variance σ^2, as the dimension n goes to infinity. These two facts imply that a zero-mean white-Gaussian vector tends to be *uniformly distributed* over a *spherical shell* of radius $\sqrt{n\sigma^2}$. By (3.4), this is equivalent asymptotically to a uniform distribution over a *ball* of the same radius.

This statement actually goes both ways. The per component marginal of a uniform distribution over a ball converges to a Gaussian density. Moreover, the joint distribution of any fixed subset of the components converges to a white-Gaussian distribution.

7.2.1   Gaussian becomes a sphere

We follow classical definitions and results from information theory: the *typical set* and the *asymptotic equipartition property (AEP)*, see [53].

Definition 7.2.1 (Typical set) *The weakly ϵ-typical set $A_\epsilon^{(n)}$ of a memoryless random vector $\mathbf{W} \sim f_W$ is defined as the set of vectors $\mathbf{w}$ whose normalized log-density is close to the entropy:*

$$A_\epsilon^{(n)} = \left\{ \mathbf{w} : \ \left| -\frac{1}{n} \log(f_W(\mathbf{w})) - h(W) \right| < \epsilon \right\}, \tag{7.24}$$

where $h(W)$ is the differential entropy of each component of $\mathbf{W}$.

By substituting f_W from (7.23), and $h(W) = \frac{1}{2}\log(2\pi e\sigma^2)$, we obtain for a white-Gaussian vector

$$A_\epsilon^{(n)} = \left\{ \mathbf{w} : \ \|\mathbf{w}\| \approx \sqrt{n\sigma^2} \right\}, \tag{7.25}$$

where $\approx$ is in the sense that $(1 - \epsilon')\sqrt{n\sigma^2} < \|\mathbf{w}\| < (1 + \epsilon')\sqrt{n\sigma^2}$, and where $\epsilon' = 2\epsilon/\log(e) = O(\epsilon)$. Thus, the ϵ-*typical set of a white-Gaussian vector is an "ϵ' spherical shell"* of radius $r = \sqrt{n\sigma^2}$.

The significance of the typical set comes from the AEP, whose Gaussian version is stated below.

Theorem 7.2.1 (Gaussian AEP) *For any $\epsilon > 0$,*

(i) *All vectors $\mathbf{w}$ in the ϵ-typical set $A_\epsilon^{(n)}$ have roughly the same density $f_W(\mathbf{w}) \doteq 2^{-nh(W)} = (2\pi e\sigma^2)^{-n/2}$.*

(ii) *The probability that $\mathbf{W}$ falls in $A_\epsilon^{(n)}$ is close to 1 for sufficiently large n.*

(iii) *The volume of $A_\epsilon^{(n)}$ is $\doteq 2^{nh(W)} = (2\pi e\sigma^2)^{n/2}$.*

Here $\doteq$ denotes "equality to the first order in the exponent," i.e., $a_n \doteq 2^{nE}$ means that

$$\lim_{n \to \infty} \frac{1}{n} \log(a_n) = E. \tag{7.26}$$

Proof (i) follows directly from Definition 7.2.1; (ii) follows from the LLN, since, as $n \to \infty$,

$$\frac{1}{n} \sum_{i=1}^{n} W_i^2 \to E\{W_1^2\} = \sigma^2, \text{ in probability;}$$

finally (iii) follows by combining (i) and (ii).   □

We see that a white-Gaussian distribution is asymptotically concentrated on a spherical shell of radius $\approx \sqrt{n\sigma^2}$.

Definition 7.2.2 (Typical AWGN radius) *The typical radius of an n-dimensional white-Gaussian vector with variance σ^2, is defined as*

$$r_{\text{noise}} = \sqrt{n\sigma^2}. \tag{7.27}$$

Interestingly, although the typical set $A_\epsilon^{(n)}$ in (7.25) is a spherical *shell*, part (iii) of the AEP coincides with the asymptotic volume of a unit *ball*,

$$V_n = \frac{\pi^{n/2}}{(n/2)!} \sim \left(\frac{2\pi e}{n}\right)^{n/2} \tag{7.28}$$

which follows from the *Stirling approximation* formula,

$$n! \sim \sqrt{2\pi n} \left(\frac{n}{e}\right)^n. \tag{7.29}$$

In the above, $(n/2)!$ for odd n is defined through the gamma function continuation of the factorial, i.e., $(n/2)! = \Gamma(n/2 + 1)$ for all n. See (3.3). And the approximation $\sim$ is in the sense that the ratio goes to 1. Thus, for a ball of radius $r = \sqrt{n\sigma^2}$, we get by (3.2)

$$\text{Vol}(\mathcal{B}_{\sqrt{n\sigma^2}}) \approx (2\pi e\sigma^2)^{n/2}, \tag{7.30}$$

which is the same as the volume of $A_\epsilon^{(n)}$. This coincidence is not surprising because, as we saw earlier in (3.4), a ball and a spherical shell of the same radius are equivalent volume-wise for a large dimension n (provided the shell thickness is at a fixed proportion of the radius).

Substituting (7.28) in the ball NSM formula (7.15), we conclude that the NSM of a ball achieves its lower bound (7.16) as the dimension goes to infinity.

Corollary 7.2.1 (Asymptotic NSM of a ball)

$$G_n^* \to \frac{1}{2\pi e}, \quad \text{as } n \to \infty \tag{7.31}$$

at a rate of

$$\log(2\pi e G_n^*) = O\left(\frac{\log n}{n}\right). \tag{7.32}$$

The rate of convergence follows from a refined Stirling approximation (Problem 7.6).

Part (ii) of the AEP (Theorem 7.2.1) amounts to the concentration of a white-Gaussian vector of variance σ^2 on a ball of radius $\sqrt{n\sigma^2}$. This concentration implies the following.

Corollary 7.2.2 (Asymptotic error probability of a ball) *The Gaussian error probability* (7.11), *of an n-dimensional ball* $\mathcal{B}_r$ *of radius* $r = \sqrt{nt}$, *satisfies asymptotically, as* $n \to \infty$,

$$P_e(\mathcal{B}_{\sqrt{nt}}, \sigma^2) \to \begin{cases} 1, & t < \sigma^2 \\ 0, & t > \sigma^2. \end{cases} \tag{7.33}$$

In terms of the cumulative chi-square distribution, Corollary 7.2.2 is equivalent to a threshold effect for large n:

$$\chi^2_{\text{CDF}}(nt; n) \approx \begin{cases} 1, & t > 1 \\ 0, & t < 1, \end{cases} \tag{7.34}$$

or, conversely, the inverse cumulative chi-square distribution (7.18) satisfies $\chi^2_{\text{ICDF}}(p; n) \approx n$ for large n, for *all* $0 < p < 1$. Combining this fact with (7.17) and (7.28), we obtain the following.

Corollary 7.2.3 (Asymptotic NVNR of a ball) *The NVNR* (7.17) *of an n-dimensional ball satisfies asymptotically*

$$\mu_n^*(P_e) \to 2\pi e, \quad \text{as } n \to \infty, \quad \text{for all } 0 < P_e < 1. \tag{7.35}$$

7.2.2 Sphere becomes Gaussian

We now address the reverse direction: tendency to Gaussianity of a uniform distribution over a ball. Let $\mathbf{U} \sim \text{Unif}(\mathcal{B}_r)$, where $\mathcal{B}_r$ is an n-dimensional ball of radius r centered at the origin. That is, the density of $\mathbf{U}$ is

$$f_U^{(n)}(\mathbf{u}) = \begin{cases} (V_n r^n)^{-1}, & \text{if } \mathbf{u} \in \mathcal{B}_r \\ 0, & \text{if } \mathbf{u} \notin \mathcal{B}_r, \end{cases} \tag{7.36}$$

where V_n is the volume of the unit ball (3.3). We shall address two forms of convergence of $f_U^{(n)}$: the distribution of a subset of components, and the normalized divergence.

If $\mathbf{u}$ belongs to the ball $\mathcal{B}_r$, and we fix the first component u_1, then the remaining components belong to a ball cut, which is an $(n-1)$-dimensional ball of radius $r' = \sqrt{r^2 - u_1^2}$. Similarly, if we fix the first k components $(u_1, \ldots, u_k)$, then the remaining components belong to an $(n-k)$-dimensional ball of radius $r' = \sqrt{r^2 - u_1^2 - \ldots - u_k^2}$. See the illustration in Figure 7.4.

Theorem 7.2.2 (Ball projection) *Let* $f_U^{(n)}$ *be the uniform density* (7.36). *If the ball radius* r *is growing as the square root of the dimension, i.e.,* $r = \sqrt{tn}$, *then, as* $n \to \infty$, *the projection of* $f_U^{(n)}$ *on the first coordinate converges point-wise to a Gaussian distribution with variance* t. *And in general, the projection of* $f_U^{(n)}$ *on any*

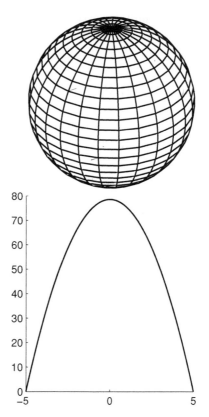

Figure 7.4 The marginal distribution of a uniform distribution over a (three-dimensional) ball.

fixed *number of coordinates k, converges to a white-Gaussian distribution:*

$$\lim_{n \to \infty} \int_{(u_{k+1},\ldots,u_n) \in \mathcal{B}_{r'}} f_U^{(n)}(u_1, \ldots, u_n) du_{k+1} \ldots du_n = f_W(u_1, \ldots, u_k)$$

where $\mathcal{B}_{r'}$ is the ball cut of radius $r' = \sqrt{r^2 - u_1^2 - \ldots - u_k^2}$ defined above, and f_W is the density of a white-Gaussian vector (7.23) with variance $\sigma^2 = t$.

Proof See Problem 7.8.   □

Another natural measure for the distance to a Gaussian distribution is the *information divergence*, also known as "relative entropy" or "Kullback–Leibler distance" (see Definition 6.5.1).

Definition 7.2.3 (Divergence from white Gaussianity II)

$$\mathcal{D}(\mathbf{U}; \mathbf{W}) = \mathcal{D}(f_{\mathbf{U}} \| f_{\mathbf{W}}) = \int_{\mathcal{R}^n} f_{\mathbf{U}}(\mathbf{x}) \log \frac{f_{\mathbf{U}}(\mathbf{x})}{f_{\mathbf{W}}(\mathbf{x})} d\mathbf{x}, \qquad (7.37)$$

where $\mathbf{W}$ *is a zero-mean white-Gaussian vector with the same average power as* $\mathbf{U}$, *i.e.,* $\sigma_W^2 = \overline{U^2} \triangleq \frac{1}{n} \sum_{i=1}^{n} E\{U_i^2\}$.

Proposition 7.2.1 (Divergence-entropy relation) *The divergence from white Gaussianity can be written as a difference between differential entropies:*

$$\mathcal{D}(\mathbf{U}; \mathbf{W}) = h(\mathbf{W}) - h(\mathbf{U}) = \frac{n}{2} \log(2\pi e \overline{U^2}) - h(\mathbf{U}). \qquad (7.38)$$

Proof The proof follows from the definition (7.37), and since the entropy of $\mathbf{W}$ can written as $h(\mathbf{W}) \triangleq - \int f_{\mathbf{W}} \log f_{\mathbf{W}} = - \int f_{\mathbf{U}} \log f_{\mathbf{W}}$. This is due to the special form of the white-Gaussian density (7.23), i.e., $\log f_{\mathbf{W}}$ is an affine function of $\|\mathbf{w}\|^2$, and since $\mathbf{U}$ and $\mathbf{W}$ have the same average power. $\qquad \square$

Lemma 7.2.1 (Divergence of uniform from white Gaussianity) *If* $\mathbf{U}$ *is uniform on a body S, then the normalized (per dimension) divergence of* $\mathbf{U}$ *from white Gaussianity is given by*

$$\frac{1}{n}\mathcal{D}(\mathbf{U}; \mathbf{W}) = \frac{1}{2} \log(2\pi e G(S)),$$

where $G(S)$ *is the NSM (7.9) of the body S.*

Proof The result follows by substituting $h(\mathbf{U}) = \log(V)$ in (7.38), and using the definition of $G(S)$. $\qquad \square$

Combining with Corollary 7.2.1, we thus conclude that a high-dimensional ball tends to Gaussianity also in the normalized divergence sense.

Theorem 7.2.3 (Divergence of a ball from white Gaussianity) *The divergence from white Gaussianity of a vector* $\mathbf{U}_n$ *which is uniform over an n-dimensional ball vanishes asymptotically:*

$$\frac{1}{n}\mathcal{D}(\mathbf{U}_n; \mathbf{W}) = \frac{1}{2} \log(2\pi e G_n^*) \to 0, \qquad (7.39)$$

as n goes to infinity, where G_n^* *is the ball NSM given in (7.15), and the rate of convergence is the same as in Corollary 7.2.1.* [2]

7.3   Good covering and quantization

So far we have revealed two out of the three links of the "$2\pi e$ connection" in Figure 7.1, between a ball, AWGN and a lattice Voronoi cell. We developed relations between sphere packing/covering and modulation/quantization (Section 7.1), and showed the tendency of large-dimensional spheres to a white-Gaussian

---

[2] Since divergence can upper bound the ℓ_1-distance between two distributions, Theorem 7.2.3 implies an average form of the ball projection theorem (Theorem 7.2.2).

distribution (Section 7.2). We now wish to close the triangle, and show that the cells of good lattices become ball-like, and hence their NSM and NVNR converge to the ideal values of $1/2\pi e$ and $2\pi e$. Somewhat surprisingly, however, the duality between the pairs packing-modulation and covering-quantization breaks. To show the desired limits we shall have to take different paths.

We start by stating a fundamental theorem, due to Rogers [231], regarding the existence of a sequence of lattices with sphere covering efficiency going to 1; this same sequence of lattices has an asymptotic NSM of $1/2\pi e$. Recall from (3.17) that the covering efficiency of a lattice is defined as $\rho_{\text{cov}}(\Lambda) = r_\Lambda^{\text{cov}}/r_{\text{eff}}(\Lambda)$.

Theorem 7.3.1 (Good covering) *There exists a sequence of lattices Λ_n of increasing dimension n such that*

$$\rho_{\text{cov}}(\Lambda_n) \to 1. \tag{7.40}$$

The convergence rate is such that $\log \rho_{\text{cov}}(\Lambda_n) = O(\log(n)/n)$.

The proof is outside the scope of this book, and can be found in [66, 231]. One approach is to use the random ensemble of construction A lattices of Section 7.9.

Definition 7.3.1 (Rogers-good lattices) *A sequence Λ_n of lattices satisfying (7.40) is called "Rogers good," or good for covering.*

Theorem 7.3.1 implies that in a *minimal sphere covering* by a *Rogers-good lattice*, the volume of the intersection between the spheres is negligible with respect to the volume of the spheres themselves.[3] We may visualize this as if the Voronoi cells of a Rogers-good lattice do not have *sharp corners*.

In view of the sphere bound of Theorem 7.1.2, these lattices also approach the ideal value of the NSM.

Corollary 7.3.1 (Covering implies quantization) *A sequence of Rogers-good lattices Λ_n achieves the minimum ball NSM (7.16)*

$$G(\Lambda_n) \to \frac{1}{2\pi e}, \tag{7.41}$$

as $n \to \infty$. It follows that the sequence G_n of minimum possible NSMs (Definition 3.2.2) converges to $1/2\pi e$ as well. The convergence rate is as in Corollary 7.2.1.

Definition 7.3.2 (Good lattice quantizers) *A sequence Λ_n of lattices satisfying (7.41) is said to be* good for quantization *under the mean square distortion measure.*

---

[3] For example, if the radius increases with the dimension like $\sim\sqrt{n}$, then the volume of the sphere increases exponentially with n, while the volume of its intersection with other spheres increases *sub-exponentially* with n.

Recall that $\frac{1}{2}\log(2\pi e G(\Lambda))$ is the redundancy of ECDQ above Shannon's rate-distortion function at high-resolution conditions (Theorem 5.5.1), as well as at *any* resolution for a white-Gaussian source with suitable post-scaling (Theorem 5.6.1).

Corollary 7.3.2 *ECDQ using the sequence Λ_n above approaches the rate-distortion function as the dimension n goes to infinity, with redundancy on the order of $\log(n)/n$.*

7.3.1 Tendency of dither to a ball and AWGN

As the dimension increases, the cell of a lattice which is good for quantization becomes "ball-like," in the sense that its second moment does not go much beyond that of a ball with the same *volume*. Does this statement extend to other forms of similarity to a ball?

Since the dither is uniform over the lattice cell, its closeness to a ball can also be measured by the chance that it leaves a ball of about the same radius as the lattice (effective) radius.

Theorem 7.3.2 (Chance of exceeding a ball) *If* $U_n \sim \mathrm{Unif}(\mathcal{V}_0(\Lambda_n))$, *and* $G(\Lambda_n) \to 1/2\pi e$ *as* $n \to \infty$, *then*

$$\Pr\{U_n \notin \mathcal{B}(0, (1+\epsilon)r_{\mathrm{eff}}(\Lambda_n))\} \to 0, \quad \text{for all } \epsilon > 0. \tag{7.42}$$

The proof below shows that, up to a rare event, the dither of a good lattice quantizer is uniform over a region that is contained in a ball with a slightly larger effective radius.

Proof For any $r > 0$, let $p_r = \Pr\{U \notin \mathcal{B}(0, r)\}$ denote the probability that the dither of a lattice Λ leaves a ball of radius r about the origin. We break the second moment of the dither into two terms, corresponding to the event $U \notin \mathcal{B}(0, r)$ (denoted "out") and its complement (denoted "$\overline{\mathrm{out}}$"):

$$\frac{1}{n}E\|U\|^2 = (1 - p_r) \cdot \frac{1}{n}E\left\{\|U\|^2 \mid \overline{\mathrm{out}}\right\} + p_r \cdot \frac{1}{n}E\left\{\|U\|^2 \mid \mathrm{out}\right\} \tag{7.43}$$

$$\geq (1 - p_r) \cdot \frac{[(1 - p_r)V(\Lambda)/V_n]^{2/n}}{n+2} + p_r \cdot \frac{r^2}{n}, \tag{7.44}$$

where the lower bound for the first term follows from the iso-perimetric inequality (7.12a), noting that the conditional distribution of U given that it falls inside $\mathcal{B}(0, r)$ is uniform over a region with volume $(1 - p_r)V(\Lambda)$. By the definition of the NSM (3.21) and the effective radius (3.10), we thus have

$$G(\Lambda) \geq \frac{(1 - p_r)^{1+2/n}}{(n+2)V_n^{2/n}} + p_r \frac{r^2/r_{\mathrm{eff}}^2(\Lambda)}{n V_n^{2/n}} \tag{7.45}$$

for all $r > 0$. Setting $r = r(\epsilon) = (1 + \epsilon)r_{\text{eff}}(\Lambda)$, and noting that $nV_n^{2/n} \to 2\pi e$ as $n \to \infty$, the lower bound above becomes

$$G(\Lambda) \geq \frac{1 + ((1 + \epsilon)^2 - 1)p_{r(\epsilon)}}{2\pi e} - o(1) \tag{7.46}$$

(where $o(1) \to 0$ as $n \to \infty$), which is strictly greater than $1/2\pi e$ for all $\epsilon > 0$ for a sufficiently large n, unless $p_{r(\epsilon)} = 0$. Thus, $G(\Lambda_n) \to 1/2\pi e$ implies that $p_{r(\epsilon)}(\Lambda_n)$ goes to zero for all $\epsilon > 0$. $\qquad\square$

Another characterization for the goodness of a lattice quantizer is the divergence of the dither from white Gaussianity. Lemma 7.2.1 showed earlier that the NSM of a body determines its divergence from white Gaussianity. Theorem 7.2.3 then showed that the divergence from white Gaussianity of a ball vanishes asymptotically with the dimension. Here we state a parallel result with respect to the Voronoi cells of a good lattice quantizer.

Theorem 7.3.3 (Divergence from white Gaussianity) *If the dither* $\mathbf{U} \sim$ Unif$(\mathcal{V}_0(\Lambda))$ *is uniform over the fundamental Voronoi cell of a lattice* Λ, *then its divergence from white Gaussianity (Definition 7.2.3) is given by*

$$\frac{1}{n}\mathcal{D}(\mathbf{U}; \mathbf{W}) = \frac{1}{2}\log(2\pi e G(\Lambda)), \tag{7.47}$$

where $\mathbf{W}$ *is the corresponding white-Gaussian noise. If* Λ_n *is a sequence of good lattice quantizers (Definition 7.3.2), then the divergence* $\frac{1}{n}\mathcal{D}(\mathbf{U}_n; \mathbf{W})$ *vanishes asymptotically as n goes to infinity. The rate of convergence is the same as in Corollary 7.2.1.*

In Section 4.3 we saw that in each dimension, the dither of a good lattice quantizer must be white (though not Gaussian). We can thus interpret Theorems 4.3.1, 7.3.2 and 7.3.3 that asymptotically for a large dimension:

> *the dither of good lattice quantizers converges to AWGN.*

One interesting implication of this result is that the divergence from white Gaussianity of the equivalent ECDQ noise (Theorems 4.1.1 and 5.2.1) is equal to $\frac{1}{2}\log(2\pi e G(\Lambda))$; thus, for good lattice quantizers the ECDQ becomes equivalent to an AWGN channel.

7.4   Does packing imply modulation?

One may expect that a parallel result will hold for packing and modulation: that for some "good" sequence of lattices Λ_n the packing efficiency (3.11) goes to 1, thus implying by (7.22) that the NVNR goes to its ideal value of $2\pi e$. This wishful parallelism, however, does not hold; packing turns out to be harder than covering.

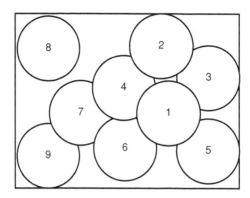

Figure 7.5 Modulation code design by maximal soft packing with overlap of $\epsilon \approx 10\%$.

Specifically, the best asymptotic value of (3.11) is known to be strictly worse than 1, and is conjectured to be equal to *one-half*. It follows that the volume of the packed balls occupies only $(1/2)^n$ of the space. In view of the sphere bound of Theorem 7.1.2, this also implies an asymptotic NVNR of $4\pi e$, i.e., *twice* the ideal value. Can we expect to close this gap?

Luckily, it turns out that modulation is easier than packing as it amounts to "soft packing." A small overlap between the noise balls (corresponding to a small ambiguity at the decoder) allows good lattices to achieve an asymptotic packing efficiency equal to 1, hence an NVNR of $2\pi e$.

7.4.1   Soft packing

Consider filling a large cube with balls, allowing up to a fraction ϵ of the volume of each ball to overlap previous balls, as illustrated in Figure 7.5. The filling process continues until it is impossible to add a new ball. It can be shown by a random translation argument that as long as the total volume occupied by balls is less than a fraction ϵ of the cube, we can find a location for a new ball with at most ϵ overlap. Thus, when the process stops, at least ϵ of the volume is packed with balls. Although this fraction can be small, the packing efficiency (3.12) is at least $\epsilon^{1/n}$, which approaches *one*, as n goes to infinity, for all $\epsilon > 0$.[4]

By the AEP (Theorem 7.2.1), an AWGN vector of large dimension concentrates on a ball. If the probability to exceed this ball can be tolerated by the communication system, then the design of an unbounded constellation for the AWGN channel amounts to "soft packing," i.e., packing of "noise balls" with a small overlap, and using the ball centers as codewords.

---

[4] In logarithmic terms, the "gap to capacity" is $\log_2(1/\epsilon)$ bit, which is just $\frac{1}{n}\log_2(1/\epsilon)$ bit per dimension, i.e., negligible for large n.

The *maximal code* construction above underlines Feinstein's proof of the channel coding theorem [77]. It applies, in fact, to any shape, not necessarily a ball. From an information theoretic viewpoint, this idea thus proves the achievability of capacity per unit volume of general additive-noise channels.

The drawback of this construction, however, is that it does not impose any structure on the positions of the centers of the packed shapes. Can a soft packing achieve the same efficiency if the centers must form a *lattice*? To show that indeed it can, we introduce below the notion of a random ensemble of lattices.

7.5   The Minkowski–Hlawka theorem

It is insightful to think of a lattice Λ as a set with a "uniform" point density in space, that is, a set with

$$\gamma = \frac{1}{V(\Lambda)}$$

points per unit volume (see (2.20)). This cannot be quite so, however, for a *specific* lattice. One reason for this is that the lattice density is not necessarily the same in all directions; for example, in the rectangular grid of Figure 2.1 the vertical point density is twice that of the horizontal. Also, a uniform point density would imply that the number of points inside a body S is exactly γ times the volume of S; but this can only hold as an approximation for bodies which are large compared to the scale of the lattice cells (see, e.g., the proof of the cell volume formula in Proposition 2.2.1).

In statistical terms, it is possible to define a *random set* of points which is uniform over a large portion of space. For example, draw γa^n independent points uniformly over the n-cube $[-a/2, a/2]^n$, and take a to infinity. We thus obtain in the limit a random set whose *expected* number of points in any bounded body S is equal *exactly* to $\gamma \mathrm{Vol}(S)$. [5] See Figure 7.6.

A more general property of such a random set is that it allows sampling to be replaced by integration. That is, for any integrable function $f(\mathbf{x})$, the expected sum of the samples of f over the set is equal to $\gamma \int f(\mathbf{x})d\mathbf{x}$. [6] If f is the indicator function

---

[5] The limiting random set can be modeled as a *Poisson point process* [9].
[6] For a random point $\mathbf{X}$ uniform on CUBE $= [-a/2, a/2]^n$,

$$E\{f(\mathbf{X})\} = \frac{1}{a^n} \int_{\mathrm{CUBE}} f(\mathbf{x})d\mathbf{x}.$$

Thus for M random points

$$E\left\{\sum_{i=1}^{M} f(\mathbf{X}_i)\right\} = \frac{M}{a^n} \int f(\mathbf{x})d\mathbf{x} = \gamma \int f(\mathbf{x})d\mathbf{x}. \tag{7.48}$$

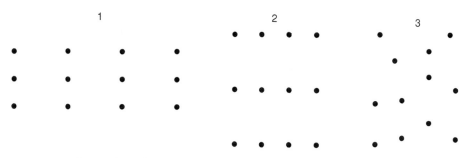

Figure 7.6 Lattices and a uniform random set of the same point density.

of the body S (i.e., $f(\mathbf{x})$ is 1 inside S and 0 outside), then the sum of the samples reduces to the number of points in S discussed earlier.

A remarkable result due to Minkowski, Hlawka and Siegel [36, 113] shows that a set satisfying these properties does not have to be completely random; in fact, it can be a *random lattice*. There are, however, two exceptions to the rule: the dimension must be greater than one, and the zero lattice point $\lambda_0 = \mathbf{0}$ must be excluded. Indeed, the zero lattice point is fixed and cannot be randomized, while a one-dimensional lattice is uniquely determined by its step size and cannot be random.

To simplify the notation, we use $N_S(\Lambda)$ for the number of lattice points other than zero that fall in a set S:

$$N_S(\Lambda) = |S \cap (\Lambda \setminus \lambda_0)| \tag{7.49}$$

$$= \sum_{\lambda \in \Lambda, \lambda \neq 0} 1_{\{\lambda \in S\}} \tag{7.50}$$

where $1_{\{A\}}$ denotes the indicator of the event A (i.e., $1_{\{A\}}$ is equal to 1 if A occurs and 0 otherwise).

Theorem 7.5.1 (Minkowski–Hlawka–Siegel [113]) *For each dimension $n > 1$, there exists a random ensemble $\mathbb{L} = \{\Lambda\}$ of lattices of unit determinant (i.e., $V(\Lambda) = 1$ for all $\Lambda \in \mathbb{L}$), such that for any bounded measurable set S, the expected number of non-zero lattice points in S is equal to the volume of S:*

$$E_{\mathbb{L}}\{N_S(\Lambda)\} = \mathrm{Vol}(S), \tag{7.51}$$

where $E_{\mathbb{L}}$ denotes expectation with respect to the random ensemble. [7]

---

[7] A modification of the theorem deals with the intersection of a body with only *primitive lattice points* (that is, counting a single lattice point in each direction) [36]. This allows us to strengthen slightly some of the results in the sequel.

By scaling the lattice ensemble by $1/\sqrt[n]{\gamma}$, we obtain an ensemble $\mathbb{L}$ where each of its members has a point density γ and cell volume $V(\Lambda) = 1/\gamma$, so that

$$E_{\mathbb{L}}\{N_S(\Lambda)\} = \gamma \cdot \text{Vol}(S). \tag{7.52}$$

Example 7.5.1 (Intersection with a ball) *If S in (7.52) is an n-dimensional ball of radius r, then,*

$$E_{\mathbb{L}}\{N_{\mathcal{B}_r}(\Lambda)\} = \gamma V_n r^n = \left(\frac{r}{r_{\text{eff}}}\right)^n, \tag{7.53}$$

where $r_{\text{eff}} = r_{\text{eff}}(\mathbb{L}) = \sqrt[n]{1/\gamma V_n}$ is the effective radius (3.10) of the ensemble $\mathbb{L}$.

A more general form of the theorem states that for any Riemann integrable function $f : \mathbb{R}^n \to \mathbb{R}$ with a bounded support,

$$E_{\mathbb{L}}\left\{ \sum_{\lambda \in \Lambda, \lambda \neq 0} f(\lambda) \right\} = \gamma \int_{\mathbb{R}^n} f(\mathbf{x}) d\mathbf{x}, \tag{7.54}$$

paralleling (7.48). As discussed earlier, this form specializes to (7.52) if f is the indicator function of the body S. And in fact, the two forms are *equivalent*; see Problem 7.9.

Definition 7.5.1 (MHS ensemble) *An ensemble of lattices $\mathbb{L}$ which satisfies Theorem 7.5.1, is called an* MHS ensemble.

There are several ways to construct an MHS ensemble and prove the Minkowski–Hlawka–Siegel theorem. In Section 7.9 we shall describe one such ensemble – a randomized version of the (generalized) construction A of Section 2.5 – which provides some "engineering intuition."

The MHS ensemble can be viewed as a random collection of points, which is *uniform* over the entire space. As we shall see, this uniformity implies asymptotically good sphere packing and modulation.

7.6   Good packing

Like in many problems in information theory, the existence of good lattice codes is most easily established using random coding arguments: we build a random ensemble of codes whose *average* performance is good. This implies that at least one member in the ensemble is good (and usually, most members are good). Although this technique is non-constructive (we cannot say which codes in the ensemble are the good ones), it is very simple and intuitive.

We start with the minimum distance (3.8) and goodness for packing.

Theorem 7.6.1 (Minkowski) *For dimension $n > 1$, and any measurable bounded set S in $\mathbb{R}^n$ with volume less than V, there exists a lattice Λ, with volume $V(\Lambda) = V$,* *that does not intersect S except possibly at the origin, i.e.,*

$$N_S(\Lambda) = 0. \tag{7.55}$$

If S is symmetric about the origin, then (7.55) holds for some twice denser lattice, i.e., with $V(\Lambda) = V/2$. [8]

Proof By (7.52), for any body S with volume less than $1/\gamma$, the expected number of non-zero points of a lattice in the MHS ensemble in S is less than 1:

$$E_{\mathbb{L}}\{N_S(\Lambda)\} = \gamma \text{Vol}(S) < 1.$$

Thus by the definition of expectation, there must be at least one member Λ in $\mathbb{L}$ with $N_S(\Lambda)$ strictly smaller than 1. Since $N_S(\Lambda)$ is an integer, it must be equal to zero for that Λ.

If the body is symmetric about the origin, then $N_S(\Lambda)$ must be *even* (if $\lambda \in S$ then also $-\lambda \in S$); thus it is enough that $\text{Vol}(S) < 2/\gamma$ (implying $E_{\mathbb{L}}\{N_S(\Lambda)\} < 2$) to conclude that $N_S(\Lambda) = 0$ for some $\Lambda \in \mathbb{L}$. □

By specializing Theorem 7.6.1 to a ball, we conclude that in each dimension there exists a unit-determinant lattice, whose all non-zero points are outside a unit-volume ball centered at the origin. [9] This means that for this "good" lattice, the shortest vector must be at least as long as the effective radius (3.10), i.e., $d_{\min}(\Lambda) \geq r_{\text{eff}}(\Lambda)$. By (3.9), this is equivalent to a packing radius $r_{\text{pack}}(\Lambda)$ larger than or equal to half the effective radius.

Recall that the packing efficiency of a lattice Λ (Definition 3.1.2) is defined as $\rho_{\text{pack}}(\Lambda) = r_{\text{pack}}(\Lambda)/r_{\text{eff}}(\Lambda)$.

Corollary 7.6.1 (Packing efficiency of at least one-half) *In each dimension $n \geq 1$ there exists a lattice whose minimum distance $d_{\min}(\Lambda)$ is greater than or equal to its effective radius $r_{\text{eff}}(\Lambda)$; or equivalently, its packing efficiency $\rho_{\text{pack}}(\Lambda)$ is at least $1/2$.*

The lower bound of Corollary 7.6.1 holds in particular in the limit of large n. Since the packing efficiency is the nth root of the volume ratio (i.e., of the packing density $\Delta(\Lambda)$ (3.12)), the stronger version in (7.56) does not improve the bound in the limit as $n \to \infty$. In fact, it is believed that this bound is asymptotically tight,

---

[8] The stronger version of the Minkowski–Hlawka–Siegel theorem, discussed in footnote 7, deals with *star-shaped bodies* which are symmetric about the origin (e.g., balls). For such bodies, the lattice volume needs to satisfy the weaker condition

$$V(\Lambda) > \frac{\text{Vol}(S)}{2\zeta(n)}, \tag{7.56}$$

where $\zeta(n) = \sum_{k=1}^{\infty} \frac{1}{k^n}$ is the *Reimann zeta function*. Note that $\zeta(n) \to 1$ as $n \to \infty$.
[9] This holds trivially also in dimension $n = 1$.

i.e., that in large dimensions it is impossible to get packing efficiencies better than one-half. [10]

A packing efficiency of one-half means that parallel sides of a Voronoi cell are not too close; if the lattice is also "Rogers good" (Definition 7.3.1), then they are at least half the diameter apart.

But a packing efficiency of one-half also means that only a fraction $(1/2)^n$ of space is occupied by balls. To achieve the Shannon capacity of the AWGN channel, a denser packing of space with "noise balls" is needed. [11]

7.7   Good modulation

In the context of digital modulation in the presence of AWGN, a good constellation in high dimensions is equivalent to a good "soft" packing of spheres (Section 7.4). We shall see that, as for "hard" packing, the MHS ensemble can demonstrate the existence of a sequence of lattices with a good soft-packing efficiency. The NVNR (3.37) of these lattices goes to the ideal value of $2\pi e$, the NVNR of a high-dimensional ball, for any $0 < P_e < 1$.

Theorem 7.7.1 (Asymptotically optimal NVNR) *There exists a sequence of lattices Λ_n of increasing dimension n, whose NVNR approaches $2\pi e$ for any non-zero error probability:*

$$\lim_{n \to \infty} \mu(\Lambda_n, P_e) = 2\pi e \quad \text{for all } 1 > P_e > 0 \tag{7.57}$$

with a rate of convergence given by

$$\log\left(\frac{\mu(\Lambda_n, P_e)}{2\pi e}\right) = O\left(\frac{1}{\sqrt{n}}\right). \tag{7.58}$$

The converse statement (i.e., that for any lattice sequence Λ_n of increasing dimension n, $\liminf_{n \to \infty} \mu(\Lambda_n, P_e) \geq 2\pi e$) is an obvious consequence of the sphere lower bound (7.22) and the asymptotic NVNR of a ball (7.35). Thus, Theorem 7.7.1 implies in particular that the sequence $\mu_n(P_e)$ of optimal NVNRs (Section 3.3) converges to $2\pi e$.

---

[10] The best known asymptotic upper bound on the packing efficiency of *any* arrangement of balls (not necessarily on a lattice) was found by Kabatiansky and Levenshtein [132] (see [49]): for any sequence of lattices Λ_n of increasing dimension

$$\limsup_{n \to \infty} \rho_{\text{pack}}(\Lambda_n) \leq 0.6603.$$

[11] A similar gap between hard and soft packing exists in the binary space; the *Gilbert–Varshamov bound* on the maximum rate of a binary code which can correct ϵn errors in a block of n bits is $1 - H_B(2\epsilon)$, while the *Hamming bound* on this rate (which meets the Shannon capacity of a binary-symmetric channel with cross-over probability ϵ) is $1 - H_B(\epsilon)$, where $H_B(\epsilon) = -\epsilon \log \epsilon - (1 - \epsilon) \log(1 - \epsilon)$ is the binary entropy.

The proof of the direct part of Theorem 7.7.1, which is the main result of this section, is given after discussing some implications and connections.

Definition 7.7.1 (Good lattice constellation) *A sequence Λ_n of lattices satisfying (7.57) is said to be "good for coding" over the AWGN channel.*

Recall that the quantity $\frac{1}{2}\log(\mu(\Lambda, P_e)/2\pi e)$ appeared in Chapter 6 as the gap to capacity of an infinite constellation, in two different setups.

Corollary 7.7.1 (Closing the gap to capacity) *A sequence Λ_n which is good for AWGN channel coding (Definition 7.7.1) approaches the AWGN channel capacity per unit volume (Theorem 6.3.2) as the dimension n goes to infinity. Furthermore, a VRDM scheme based on Λ_n approaches the power-constrained capacity of the AWGN channel in the limit of high SNR (Theorem 6.5.1). If Λ_n also satisfies (7.58), then the gap to capacity in both cases is $O(1/\sqrt{n})$.*

7.7.1   Typicality decoding of a random code

The proof of Theorem 7.7.1 is in the spirit of the proof of the channel coding theorem in information theory. The idea is to combine "random coding" with "typicality decoding." Although this scheme is *sub*optimal (compared to optimal coding with ML decoding), it asymptotically achieves the channel capacity, as the code dimension goes to infinity. [12]

Recall the information theoretic view of channel coding, discussed in Section 1.2. For some input distribution $p(x)$, generate $M = 2^{nR}$ independent codewords $\mathbf{X}(1), \ldots, \mathbf{X}(M)$ at random, with probability $p(x)$, and reveal them to the encoder and the decoder. Message $1 \leq i \leq M$ is encoded as $\mathbf{X}(i)$, and transmitted over the channel $p(y|x)$. The decoder receives the resulting channel output $\mathbf{Y}$, and searches for a *unique* codeword $\mathbf{X}(\cdot)$, which is "jointly typical" with $\mathbf{Y}$. For an additive-noise channel, and a uniform input distribution $p(x)$, typicality amounts to the difference $\mathbf{Y} - \mathbf{X}(i)$ being in the typical set of the noise, see Definition 7.2.1. It follows that the decoder declares (or unknowingly makes) an error if either (i) the noise is atypical, or (ii) some competing codeword "pretends" to be jointly typical with $\mathbf{Y}$. By the union bound, the average decoding error probability $\bar{P}_e$ (averaged over the M messages, the possible codes, and the channel statistics), is thus bounded by

$$\bar{P}_e \leq \Pr(\text{atypical noise}) + \underbrace{(2^{nR} - 1) \cdot 2^{-n(I(X;Y)-\epsilon)}}_{\text{expected number of competing codewords}} \tag{7.59}$$

---

[12] The advantage of ML decoding over typicality decoding will be evident from the finer analysis in Chapter 13 of the decay of the error probability as a function of the lattice dimension.

where $\approx 2^{-nI(X;Y)}$ is the probability that an (independent) competing codeword $\mathbf{X}(j)$, for some $j \neq i$, pretends to be jointly typical with $\mathbf{Y}$. [13] Since the first term vanishes by the AEP (Theorem 7.2.1), it follows that $\bar{P}_e$ vanishes for large n, provided that R is smaller than the mutual information $I(X; Y)$. [14] And for $p^*(x)$ (the maximizing input distribution), $\bar{P}_e$ vanishes for all rates R smaller than the capacity C. The punch line of the random coding argument is that there must be *at least one* code in the ensemble whose error probability is not larger than the ensemble average $\bar{P}_e$.

7.7.2   Threshold decoding of a random lattice

The idea of typicality decoding of a random code translates, in our framework, to *sphere* (or *threshold*) decoding of a lattice from the MHS ensemble of Section 7.5. This decoder searches for a *unique* codeword within a radius r_{th} around the received vector $\mathbf{Y} = \lambda_0 + \mathbf{Z}$:

$$\hat{\lambda} = \begin{cases} \lambda, & \text{if } \|\mathbf{Y} - \lambda\| < r_{th} \text{ for a unique } \lambda \in \Lambda, \\ ?, & \text{otherwise.} \end{cases} \tag{7.60}$$

As shown in Figure 7.7, the decoding region Ω_λ associated with λ consists of all $\mathbf{y}$ vectors that belong to the first case of (7.60). And an error $\hat{\lambda} \neq \lambda_0$ occurs either if the noise exceeds the search range, or if there is another (a competing) codeword inside this range.

We first consider the error probability for a *specific* (non-random) lattice.

Lemma 7.7.1 (Threshold decoding of a specific lattice) *The error probability of the sphere decoder (7.60) of a lattice code Λ, in the presence of noise $\mathbf{Z}$, is bounded, for any threshold $r_{th} > 0$, by*

$$\Pr\{\hat{\lambda} \neq \lambda_0\} \leq \Pr\{\|\mathbf{Z}\| > r_{th}\} + E_{\mathbf{Z}}\{N_{\mathcal{B}(\mathbf{Z}, r_{th})}(\Lambda)\}, \tag{7.61}$$

where $N_S(\Lambda)$ is defined in (7.49), $\mathcal{B}(\mathbf{Z}, r_{th})$ is a ball of radius r_{th} centered at $\mathbf{Z}$, and the expectation is over the noise $\mathbf{Z}$.

Proof By the union bound, the decoding error probability $\Pr\{\hat{\lambda} \neq \lambda_0\}$ in (7.61) is upper bounded by

$$\Pr\{\|\mathbf{Z}\| > r_{th}\} + \Pr\{\mathcal{B}(\mathbf{Y}, r_{th}) \text{ contains lattice points other than } \lambda_0\}. \tag{7.62}$$

Assuming (without loss of generality, due to the lattice symmetry) that the zero codeword was transmitted (i.e., $\mathbf{X} = \mathbf{0}$ and $\mathbf{Y} = \mathbf{Z}$), we thus have:

$$\Pr\{\hat{\lambda} \neq \lambda_0\} \leq \Pr\{\|\mathbf{Z}\| > r_{th}\} + \Pr\{N_{\mathcal{B}(\mathbf{Z}, r_{th})}(\Lambda) \geq 1\}. \tag{7.63}$$

---

[13] See more on joint typicality in Section 9.4.3.

[14] See the success-threshold exponent in Appendix A.2, for the asymptotic "0/1-behavior" of the error around the point $R = I(X; Y)$.

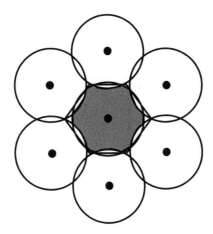

Figure 7.7 The gray area shows the decoding region Ω_0 of the threshold decoder for a hexagonal lattice, where the threshold r_{th} is in the range $r_{\text{pack}}(\Lambda) < r_{\text{th}} < r_{\text{cov}}(\Lambda)$. If the noise $\mathbf{Z}$ falls inside $\mathcal{V}_0$, then selecting $r_{\text{th}} = r_{\text{cov}}(\Lambda)$ guarantees that the true codeword λ_0 is in the search range, but there may be an ambiguity; selecting $r_{\text{th}} = r_{\text{pack}}(\Lambda)$ guarantees that there is no ambiguity, but the search range may be empty. If the noise $\mathbf{Z}$ falls outside $\mathcal{V}_0$, then there is an error for any choice of r_{th}.

Finally, since for any non-negative integer-valued random variable N

$$\Pr\{N \geq 1\} \leq \sum_{n \geq 1} n \cdot \Pr\{N = n\} = E\{N\}, \tag{7.64}$$

the second term in (7.63) is upper bounded by the expected value of $N_{\mathcal{B}(\mathbf{Z}, r_{\text{th}})}(\Lambda)$.

$\square$

The decision region of the sphere decoder (7.60) is a subset of the fundamental Voronoi region $\mathcal{V}_0$.[15] Thus, for *any* noise distribution, the NN decoder (3.31) is at least as good as the threshold decoder, hence its error probability is also upper bounded by Lemma 7.7.1. In the special case of a white-Gaussian noise, the NN decoding rule becomes ML, while the spherical search of (7.60) corresponds, by the AEP (7.25), to *typicality decoding* (provided the search range is roughly equal to the typical radius (7.27): $r_{\text{th}} \approx r_{\text{noise}} = \sqrt{n\sigma^2}$).

We do not try to optimize the threshold r_{th} for a *specific* lattice. Rather, we turn to examine a *random* lattice in the MHS ensemble $\mathbb{L}$. The ensemble average of the error probability, under NN decoding, is defined as

$$P_e(\mathbb{L}, \mathbf{Z}) \overset{\Delta}{=} E_{\mathbb{L}}\{\Pr(\mathbf{Z} \notin \mathcal{V}_0(\Lambda))\}, \tag{7.65}$$

where $E_{\mathbb{L}}\{\cdot\}$ denotes expectation with respect to the ensemble.

---

[15] If $\|\mathbf{Y} - \lambda\| < r_{\text{th}}$ for a *unique* $\lambda \in \Lambda$, then $\mathbf{Y}$ must be closer to λ than to any other lattice point (but there is no r_{th} for which the converse is true for all $\mathbf{y}$). See Problem 7.11.

Lemma 7.7.2 (Threshold decoding of the MHS ensemble) *The average error probability (7.65) of a lattice in the MHS ensemble* $\mathbb{L}$ *with a point density* γ, *is bounded by*

$$P_e(\mathbb{L}, \mathbf{Z}) \leq \underbrace{\Pr\{\|\mathbf{Z}\| > r_{\text{th}}\}}_{\text{atypical noise}} + \underbrace{\left(\frac{r_{\text{th}}}{r_{\text{eff}}}\right)^n}_{\text{competing codewords}} \tag{7.66}$$

for any $r_{\text{th}} > 0$, *where* $r_{\text{eff}} = \sqrt[n]{1/\gamma V_n}$ *is the ensemble effective radius (7.53).*

Observe that if $\mathbf{Z}$ is AWGN, and r_{th} is (about) its typical radius (7.27), then the two terms in the upper bound (7.66) correspond to the probability of an atypical noise behavior, and the average number of competing codewords, respectively. Thus, (7.66) resembles the bound (7.59) on the average error probability of a random code with typicality decoding.

Proof Since the NN decoder is at least as good as the threshold decoder (7.60), Lemma 7.7.1 implies that the average error probability (7.65) is bounded by (letting $r_{\text{th}} = r$ for convenience of notation)

$$P_e(\mathbb{L}, \mathbf{Z}) \leq \Pr\{\|\mathbf{Z}\| > r\} + E_{\mathbb{L}} E_{\mathbf{Z}}\{N_{\mathcal{B}(\mathbf{Z},r)}(\Lambda)\} \tag{7.67}$$

$$= \Pr\{\|\mathbf{Z}\| > r\} + E_{\mathbf{Z}} E_{\mathbb{L}}\{N_{\mathcal{B}(\mathbf{Z},r)}(\Lambda)\} \tag{7.68}$$

$$= \Pr\{\|\mathbf{Z}\| > r\} + E_{\mathbb{L}}\{N_{\mathcal{B}(0,r)}(\Lambda)\}, \tag{7.69}$$

where (7.69) follows since for the MHS ensemble $\mathbb{L}$, the expected number of non-zero lattice points in a ball depends only on the *volume* of the ball (see Theorem 7.5.1), so the dependence on the ball center $\mathbf{Z}$ disappears. The lemma now follows from (7.53). □

7.7.3   Discussion and proof of the main result

We still have a free parameter – the decoding threshold radius r_{th} – that we can optimize. As Figure 7.8 shows, for a small r_{th} the probability of the noise exceeding the threshold (the first term in (7.66)) is dominant, while for a large r_{th} the average number of competing codewords (the second term) is dominant.

For AWGN of variance σ^2, we observe an asymptotic "0/1-behavior" as the dimension n gets large, similar to that in random coding (7.59), around the point where the ensemble effective radius r_{eff} is equal to the typical noise radius $r_{\text{noise}} = \sqrt{n\sigma^2}$ (7.27). Specifically, if we let r_{th} grow as $\sqrt{n}$, then by the Gaussian AEP, the first term of (7.66) goes to

$$\Pr\{\|\mathbf{Z}\| > r_{\text{th}}\} \rightarrow \begin{cases} 1, & \text{if } r_{\text{th}} < r_{\text{noise}} \\ 0, & \text{if } r_{\text{th}} > r_{\text{noise}}, \end{cases} \tag{7.70}$$

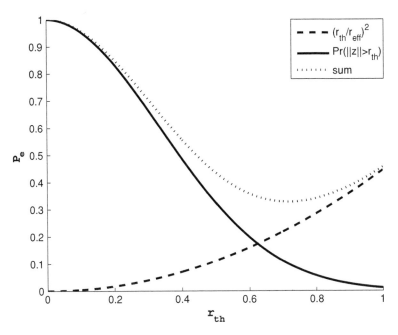

Figure 7.8 The behavior of the two terms in (7.66) and of their sum, for a two-dimensional MHS ensemble, as a function of the threshold radius r_{th}.

whereas the second term of (7.66) goes to

$$(r_{th}/r_{eff})^n \rightarrow \begin{cases} 0, & \text{if } r_{th} < r_{eff} \\ \infty, & \text{if } r_{th} > r_{eff}. \end{cases} \tag{7.71}$$

It follows that for an n-dimensional MHS ensemble $\mathbb{L}_n$, if we keep the effective radius $r_{eff} > r_{noise}$, then for any threshold value in the range $r_{noise} < r_{th} < r_{eff}$, both terms in (7.66) vanish, as $n \rightarrow \infty$. Hence, if $r_{eff\,n} = \sqrt{n\sigma^2}(1+\delta)$, then the ensemble average error probability (7.65) vanishes: [16]

$$P_e(\mathbb{L}_n, \mathbf{Z}) \rightarrow 0, \quad \text{as } n \rightarrow \infty \tag{7.72}$$

for any $\delta > 0$.

Theorem 7.7.1 is a simple consequence of this fact.

Proof of Theorem 7.7.1 It follows from (7.72) that, for any target error probability $P_e > 0$, and for $r_{eff\,n} = (1+\delta)r_{noise}$ for some $\delta > 0$, we have $P_e(\mathbb{L}_n, \mathbf{Z}) < P_e$ for sufficiently large n. Let Λ_n be a lattice in $\mathbb{L}_n$ satisfying this error probability (there must be at least one such lattice, since $P_e(\mathbb{L}_n, \mathbf{Z})$ is the ensemble average). The

---

[16] Note that $r_{eff\,n} = \sqrt{n\sigma^2}$ implies that the volume of the lattice cells grows exponentially with the dimension; see (7.30).

NVNR of this lattice is better than

$$\mu(\Lambda_n, P_e) \leq \frac{[V_n r_{\text{eff}}^n(\Lambda_n)]^{2/n}}{\sigma^2}$$
$$= V_n^{2/n} n(1 + \delta)^2$$
$$\rightarrow 2\pi e(1 + \delta)^2. \tag{7.73}$$

The first part of the theorem now follows because δ can be arbitrarily small.

The proof of the second part of the theorem (7.58) requires a finer characterization of the probability that $\|\mathbf{Z}\| > r$ (the first term in (7.66)), and shall be omitted; see Ingber *et al.* [125]. □

7.8   Non-AWGN

For future analysis of dithered lattice coding schemes, we extend Theorem 7.7.1 to noise-matched decoding in the presence of a general additive noise $\mathbf{Z}$. Under some conditions on the noise distribution, we shall see that the *generalized NVNR* $\mu_{\text{matched}}(\Lambda_n, \mathbf{Z}, P_e)$ goes to $2\pi e$ for some lattice sequence Λ_n of increasing dimension.

The extension requires modifying the spherical decoding regions in Lemmas 7.7.1 and 7.7.2. The spherical region fits well a circularly symmetric joint p.d.f. as in the case of AWGN, but is inappropriate for a general additive noise.

We consider (again a suboptimal) decoder, which uses lattice shifts of some set $\mathcal{S}$ as decision cells. Note that unlike a lattice partition, these shifts may overlap. Per reception of the vector $\mathbf{y}$, the decoder output is given by

$$\hat{\lambda}_{\mathcal{S}} = \begin{cases} \lambda, & \text{if } \mathbf{y} \text{ belongs to } \lambda + \mathcal{S}, \text{ for a unique } \lambda \in \Lambda \\ ?, & \text{otherwise.} \end{cases} \tag{7.74}$$

The decoding rule (7.74) coincides with ML decoding if $\mathcal{S}$ is a noise-matched Voronoi cell (6.12), and with *threshold decoding* (7.60) if $\mathcal{S}$ is a ball.

Per transmission of the zero lattice point $\lambda = \mathbf{0}$, decoding will be correct if and only if the noise vector $\mathbf{Z}$ falls inside the set $\mathcal{S}$ and outside any other (non-zero) lattice shift of the set $\mathcal{S}$. Thus, an error occurs if either $\mathbf{Z} \notin \mathcal{S}$, or $\mathbf{Z} \in \lambda + \mathcal{S}$ for some non-zero $\lambda \in \Lambda$. The second case amounts to $\lambda \in \mathbf{Z} - \mathcal{S}$, which can be written as $N_{\mathbf{z}-\mathcal{S}}(\Lambda) \geq 1$. By the symmetry of the lattice and the decoding rule (7.74), the error probability is invariant of the actual transmitted lattice point. Thus, a straightforward extension of Lemmas 7.7.1 and 7.7.2 implies the following bound on the error probability of the $\mathcal{S}$ decoder above.

Lemma 7.8.1 ($\mathcal{S}$-bound on error probability) *For any measurable set $\mathcal{S}$, the ML decoding error probability of a lattice constellation Λ in the presence of additive* noise $\mathbf{Z}$, is bounded from above by

$$\Pr\{\hat{\lambda}_S \neq \lambda\} \leq \Pr\{\mathbf{Z} \notin \mathcal{S}\} + E_{\mathbf{Z}}\{N_{\mathbf{Z}-S}(\Lambda)\}. \tag{7.75}$$

Furthermore, for an MHS lattice ensemble $\mathbb{L}$ with a point density γ,

$$P_e^{\mathrm{matched}}(\mathbb{L}, \mathbf{Z}) \leq \Pr\{\mathbf{Z} \notin \mathcal{S}\} + \gamma \cdot \mathrm{Vol}(\mathcal{S}) \tag{7.76}$$

where $P_e^{\mathrm{matched}}(\mathbb{L}, \mathbf{Z})$ denotes the ensemble average of the ML decoding error probability. (As opposed to $P_e(\mathbb{L}, \mathbf{Z})$ in (7.65), which corresponds to NN decoding.)

This lemma can be used to characterize the asymptotic generalized NVNR (6.28) for a class of noise distributions.

Definition 7.8.1 (Generalized AEP) *We say that a sequence of random vectors $\mathbf{Z}_n$ of growing dimension n (not necessarily coming from a fixed process) satisfies a generalized AEP, if*

(i) as $n \to \infty$,

$$\frac{1}{n} h(\mathbf{Z}_n) \to \bar{h}; \tag{7.77}$$

(ii) for every $\epsilon > 0$, there exists a sequence of bounded sets $S_\epsilon^{(n)} \subset \mathbb{R}^n$, such that for sufficiently large n,

$$\Pr\{\mathbf{Z}_n \in S_\epsilon^{(n)}\} > 1 - \epsilon \tag{7.78}$$

and

$$\left| \frac{1}{n} \log\left(\mathrm{Vol}(S_\epsilon^{(n)})\right) - \bar{h} \right| < \epsilon. \tag{7.79}$$

Definition 7.8.2 (Semi-spherical noise) *If the sets in Definition 7.8.1 are balls*

$$S_\epsilon^{(n)} = \mathcal{B}\left(0, (1+\epsilon)\sqrt{n\sigma^2}\right) \tag{7.80}$$

(in which case $\bar{h} = \frac{1}{2}\log(2\pi e\sigma^2)$), then we say that the sequence $\mathbf{Z}_n$ is semi-spherical.

Intuitively, for a large dimension n, the vector $\mathbf{Z}_n$ tends to be uniform over the set $S_{0^+}^{(n)}$. For example, i.i.d. noise vectors satisfy the generalized AEP, in which case $S_\epsilon^{(n)}$ is the weakly ϵ-typical set (7.24). Another example is a sequence of sets $\mathcal{S}_n$ whose normalized log-volume $\frac{1}{n}\log(\mathrm{Vol}(S_n))$ converges to a limit, and where each noise vector $\mathbf{Z}_n$ is exactly *uniform* over $\mathcal{S}_n$.

An example of semi-spherical noise is the (uniform) dither of a good lattice quantizer Λ_n (Definition 7.3.2). The ball (7.80) in this case is $\mathcal{B}(0, (1+\epsilon)r_{\mathrm{eff}}(\Lambda_n))$, i.e., its normalized log-volume is slightly larger than that of the lattice; see Theorem 7.3.2. Another example is the sum of the dither of a good lattice quantizer and AWGN; see Problem 7.12. Note that, like the dither of good lattice quantizers, semi-spherical noise tends to AWGN in the divergence rate sense (Theorem 7.3.3).

Theorem 7.8.1 (Good noise-matched NVNR) *For any sequence of noise vectors of increasing dimension* $\mathbf{Z}_1, \mathbf{Z}_2, \ldots$ *satisfying the generalized AEP of Definition 7.8.1, there exists a sequence of lattices* Λ_n *such that*

$$\lim_{n \to \infty} \mu_{\text{matched}}(\Lambda_n, \mathbf{Z}_n, P_e) = 2\pi e \;\; \text{for all } P_e > 0, \tag{7.81}$$

where $\mu_{\text{matched}}(\Lambda, \mathbf{Z}, P_e)$ *is the* noise-matched NVNR (6.28) *of the lattice* Λ *with respect to the noise* $\mathbf{Z}$.

The lattice sequence Λ_n in Theorem 7.8.1 depends, in general, on the noise distribution. Nevertheless, for the class of semi-spherical noises, there exists a *universal* lattice sequence Λ_n^* which is good (under NN decoding) for the entire class.

Theorem 7.8.2 (Euclidean NVNR for semi-spherical noise) *There exists a sequence of lattices* Λ_n^*, *such that if the noise sequence* $\mathbf{Z}_n$ *is* semi-spherical, *then*

$$\lim_{n \to \infty} \mu_{\text{euclid}}(\Lambda_n^*, \mathbf{Z}_n, P_e) = 2\pi e \;\; \text{for all } P_e > 0, \tag{7.82}$$

where $\mu_{\text{euclid}}(\Lambda, \mathbf{Z}, P_e)$ *is the Euclidean mismatched NVNR (6.33).*

The proofs of Theorems 7.8.1 and 7.8.2 are simple extensions of the proof of Theorem 7.7.1, based on Lemma 7.8.1; see Problems 7.13 and 7.14.

Definition 7.8.3 (Good lattice constellation: general noise) *A sequence of lattices satisfying Theorem 7.8.1 or Theorem 7.8.2 is said to be "good for coding" over an additive-noise channel with general noise (under noise-matched decoding), or semi-spherical noise (under NN decoding).*

7.9   Simultaneous goodness

The Rogers construction shows the existence of lattices which are good for both covering and quantization (Section 7.3). As we saw in Sections 7.6–7.7, the MHS ensemble contains lattices which are good for both packing and modulation (in the presence of AWGN or more general additive noises). Can the *same* lattice be good under all four criteria?

The answer is yes, as stated by the next theorem.

Theorem 7.9.1 (Simultaneous goodness) *There exists a sequence of lattices of increasing dimension* Λ_n, *which satisfies* $\rho_{\text{cov}}(\Lambda_n) \to 1$, $G(\Lambda_n) \to$

$1/2\pi e$, $\mu(\Lambda_n, P_e) \to 2\pi e$ for all $P_e > 0$, and $\liminf \rho_{\text{pack}}(\Lambda_n) \geq 1/2$, as $n \to \infty$. Moreover, the same lattice sequence satisfies $\mu_{\text{euclid}}(\Lambda_n, \mathbf{Z}_n, P_e) \to 2\pi e$, for any semi-spherical noise sequence $\mathbf{Z}_n$.

Proof The proof is based on the properties of a random ensemble of construction A lattices (Section 2.5). We show below how such an ensemble mimics the uniformity property of the MHS ensemble (Theorem 7.5.1), implying goodness in expectation for packing and modulation (by Theorems 7.6.1 and 7.7.1). Moreover, this ensemble allows for a finer analysis of the point density distribution, that proves its goodness in expectation also for covering and quantization. (The proof for covering is not included; see [66].) A random lattice in the ensemble is good with high probability under each of these criteria, and hence, by the union bound, under all four criteria. □

7.9.1 Why random construction A?

As we saw in the previous sections, a random lattice ensemble is a useful analytical tool. Its main characteristic is that the lattice points are uniformly distributed in space. This is a key property for showing that the lattice ensemble will achieve, on the average, as the dimension goes to infinity, all the desired bounds on packing, covering, quantization (NSM) and modulation (NVNR) efficiencies.

Our derivation so far has been based on the existence of the MHS ensemble, and its simple uniformity property (Theorem 7.5.1). Making a precise definition and analysis of this ensemble, however, requires measure theoretic tools, which are beyond the scope of this book; see [113].

In this section we present an alternative ensemble, which is closer to the way common lattices are constructed in practice. The ensemble is based on a randomized *construction A* (Section 2.5). A lattice is constructed by lifting a linear q-ary code into the Euclidean space, and replicating it modulo q. By randomizing the linear code, the lifted points become uniform over a finite grid in space. Then, by an appropriate scaling, this grid becomes finer and wider as the alphabet size q increases. The resulting lattice thus tends to be uniform over the whole space.

The random construction A ensemble also allows an assessment of the sensitivity of the results to the alphabet being discrete and finite (of size q). In Section 7.9.5 we will see that a *binary* alphabet is nearly optimal for lattice quantization. As for modulation, if one is interested in a *finite cubic-shaped* lattice constellation – as in many digital modulations at low SNR – then a small alphabet (e.g., $q = 2$) is sufficient as well. However, an infinite lattice constellation, or a *spherically shaped* lattice constellation (for a general SNR), requires a large alphabet. More on that in Chapters 8 and 9.

7.9.2   Random linear code

Consider a q-ary linear code $\mathcal{C} = \{G\mathbf{w} : \mathbf{w} \in \mathbb{Z}_q^k\}$ as defined in (2.54), where G is an $n \times k$ matrix, $1 \leq k < n$, with elements in the modulo-q group $\mathbb{Z}_q$, and where the product is defined modulo q. Using the generator matrix representation of $\mathcal{C}$, we obtain a random ensemble of codes by drawing each element of the matrix G at *random, independently* of the other elements and *uniformly* over $\mathbb{Z}_q$. The resulting ensemble contains q^{nk} equally likely members (not all of which are distinct).

Unlike a truly random (non-linear) code, the codewords of a random *linear* code are *not* statistically independent. In fact, the entire code is completely determined by a set of k codewords. [17] In spite of that, if we limit the discussion to a *prime* $q = p$, then the random linear code $\mathcal{C}$ has a few useful properties.

Lemma 7.9.1 (Properties for a prime alphabet) *If the elements of G are i.i.d. and uniform over $\mathbb{Z}_p$, then we have the following.*

(i) *Every codeword $\mathbf{c} = G\mathbf{w}$, for $\mathbf{w} \neq \mathbf{0}$, is uniformly distributed over $\mathbb{Z}_p^n$.* [18]
(ii) *The chance of a non-full-rank generator matrix G is smaller than $1/p^{n-k}$. Thus, with probability of at least $1 - 1/p^{n-k}$ the code $\mathcal{C}$ has $|\mathcal{C}| = p^k$ distinct codewords.*
(iii) *Let S_p be a set in $\mathbb{Z}_p^n$, and let $N_{S_p}(\mathcal{C})$ denote the number of non-zero vectors $\mathbf{w}$, whose codeword $\mathbf{c} = G\mathbf{w}$ is in S_p.* [19] *Then,*

$$E\{N_{S_p}(\mathcal{C})\} = \frac{p^k - 1}{p^n}|S_p|, \tag{7.83}$$

where $E\{\cdot\}$ denotes expectation with respect to the random code, and $|S_p|$ is the size of the set S_p.

Proof By (2.54), any codeword $\mathbf{c}$ is a linear combination, modulo q, of the column vectors of $G = [\mathbf{g}_1| \ldots |\mathbf{g}_k]$ by the elements of the vector $\mathbf{w} = (w_1, \ldots, w_k)$:

$$\mathbf{c} = w_1\mathbf{g}_1 \oplus \ldots \oplus w_k\mathbf{g}_k. \tag{7.84}$$

Since each element of the matrix G is drawn uniformly over $\mathbb{Z}_p$, each column $\mathbf{g}_i$ is a random vector which is uniform over $\mathbb{Z}_p^n$. This makes the linear combination above uniform over $\mathbb{Z}_p^n$ as well, provided at least one w_i is non-zero. Property (ii) follows since G is non-full-rank if and only if there exists a *non-zero* vector $\mathbf{w}$ such that $G\mathbf{w}$ is equal to zero; due to property (i) the probability of such a G for a *specific* $\mathbf{w}$ is

---

[17] For $q = 2$, the codewords are *pairwise independent*, which is a weak notion of independence. See Problem 7.15. For a prime $q > 2$, however, if $\mathbf{c} \in \mathcal{C}$ is a non-zero codeword, then $2\mathbf{c}, 3\mathbf{c}, \ldots, (q-1)\mathbf{c}$ are also non-zero codewords that are a function of $\mathbf{c}$; thus they are not independent in pairs.
[18] Note that the zero codeword $\mathbf{c} = \mathbf{0}$ can result from a non-zero $\mathbf{w}$ only if G is a non-full-rank matrix.
[19] We count every codeword according to how many distinct $\mathbf{w}$ vectors generate it. If G has full rank, then each codeword is generated by a *unique* $\mathbf{w}$ (the relation between $\mathbf{w}$ and $\mathbf{c} = G\mathbf{w}$ is one-to-one); hence $N_{S_p}(\mathcal{C})$ is the number of distinct non-zero *codewords* in S_p.

p^{-n}, hence by the union bound the probability for *any* non-zero $\mathbf{w}$ in $\mathbb{Z}_p^k$ is at most $(p^k - 1)p^{-n}$. Finally, property (iii) follows since by the codeword uniformity (the first property), the probability (with respect to the randomness of G) that a non-zero codeword $\mathbf{c} = G\mathbf{w}$ falls inside the set S_p is given by

$$\Pr\{\mathbf{c} \in S_p\} = \frac{|S_p|}{p^n}. \tag{7.85}$$

Writing

$$N_{S_p}(\mathcal{C}) = \sum_{\mathbf{w} \neq 0} 1_{\{G\mathbf{w} \in S_p\}} \tag{7.86}$$

and noting that the expectation of the event indicator $1_{\{G\mathbf{w} \in S_p\}}$ is equal to the probability (7.85), and that the number of non-zero $\mathbf{w}$ vectors is $p^k - 1$, proves (7.83). $\qquad\qquad\qquad\qquad\qquad\qquad\qquad\qquad\qquad\qquad\qquad\qquad\qquad\square$

Note that (7.85) and (7.83) depend on the set S_p only through its cardinality, similarly to the property (Theorem 7.5.1) of the MHS ensemble.

7.9.3 Random ensemble of lattices

We turn to define an ensemble of construction A lattices, based on the random linear code $\mathcal{C}$ above. Before that, we make the following definition associated with construction A based lattices.

Definition 7.9.1 (Period and resolution) *The period of a lattice Λ is the smallest number L such that $(\lambda \bmod L) \in \Lambda$, for all $\lambda \in \Lambda$. (If $L = \infty$ then the lattice is aperiodic.) The lattice resolution is the largest number δ such that Λ is contained in the scaled cubic lattice $\delta \mathbb{Z}^n$.*

Clearly, a modulo-q integer lattice $\Lambda_{\mathcal{C}}$ (2.56) has a period (at most) q, and a resolution (at least) 1. If $\Lambda_{\mathcal{C}}$ is scaled by a scalar a, then its period and resolution are both scaled by a.

Definition 7.9.2 (The $\mathbb{L}_{n,k,p,\gamma}$ (Loeliger) ensemble [175]) *For a prime p, let $\Lambda_{\mathcal{C}}$ be a modulo-p lattice (Definition 2.5.2) induced by the random (n, k, p) linear code $\mathcal{C}$ described above. (That is, $\mathcal{C}$ is generated by an $n \times k$ matrix G, where $n \geq 2$ and $1 \leq k < n$, whose elements are statistically independent and uniform over $\mathbb{Z}_p$.) Then, the $\mathbb{L}_{n,k,p,\gamma}$ ensemble is a scaled version of the random lattice $\Lambda_{\mathcal{C}}$:*

$$\mathbb{L}_{n,k,p,\gamma} = \frac{1}{\gamma^{1/n}} \cdot \frac{1}{p^{1-k/n}} \cdot \Lambda_{\mathcal{C}}. \tag{7.87}$$

The scaling in (7.87) guarantees that the point density of (almost) all lattices in the ensemble is γ, independent of p, k and n. Specifically, every lattice Λ in $\mathbb{L}_{n,k,p,\gamma}$

satisfies the following properties (see Problem 7.16):

$$L = \text{period of } \Lambda = \frac{1}{\gamma^{1/n}} \cdot p^{k/n}. \tag{7.88}$$

If $G \neq \mathbf{0}$ (which occurs with probability $1 - 1/p^{nk}$), then

$$\delta = \text{resolution of } \Lambda = \frac{1}{\gamma^{1/n}} \cdot \frac{1}{p^{1-k/n}}. \tag{7.89}$$

And if G has a full rank (an event that occurs with probability of at least $1 - 1/p^{n-k}$), then the cell volume is

$$V(\Lambda) = \frac{\text{period}^n}{|\mathcal{C}|} = \frac{1}{\gamma}. \tag{7.90}$$

Ensemble parameters

---

n lattice dimension
p alphabet size (prime)
k information dimension of underlying code $\mathcal{C}$
γ point density, $1/\gamma$ cell volume
L period
δ resolution

In the limit of a large alphabet p (for fixed k, n and γ), the ensemble period goes to infinity, while its resolution goes to zero. These properties lead us to a uniform property of the $\mathbb{L}_{n,k,p,\gamma}$ ensemble, similar to that of the MHS ensemble. To state this property, we make the following definition.

Definition 7.9.3 (Simple body) *A bounded set S in $\mathbb{R}^n$ is a* simple body, *if its volume can be arbitrarily approximated by the size of its intersection with a fine grid*

$$\text{Vol}(S) = \lim_{\delta \to 0} \delta \cdot |S \cap \delta \mathbb{Z}^n|. \tag{7.91}$$

Note that any Jordan measurable set is simple [61]. In particular, any bounded convex set is simple.

Lemma 7.9.2 (Uniformity of the $\mathbb{L}_{n,k,p,\gamma}$ ensemble) *For every $n \geq 2$, and $1 \leq k < n$, the expected number (over the ensemble $\mathbb{L}_{n,k,p,\gamma}$) of non-zero lattice points in a simple body $S \in \mathbb{R}^n$ satisfies:*

$$E_{\mathbb{L}_{n,k,p,\gamma}}\{N_S(\Lambda)\} \to \gamma \cdot \text{Vol}(S), \quad \text{as } p \to \infty. \tag{7.92}$$

Specifically, if S is an n-dimensional ball of radius $r = t\sqrt{n}$, then for every p

$$\sqrt[n]{\frac{E_{\mathbb{L}_{n,k,p,\gamma}}\{N_S(\Lambda)\}}{\gamma \cdot \mathrm{Vol}(S)}} = 1 \pm \delta/t, \tag{7.93}$$

where $\delta = 1/p^{(1-k/n)}$ is the resolution (7.89).

Proof The first part of the lemma is a simple consequence of the properties of a random linear code (Lemma 7.9.1), and the definition (7.91) of a simple body. The second part hinges on more specific properties of a ball. See Problem 7.17. □

We see that the ensemble $\mathbb{L}_{n,k,p,\gamma}$ approaches the basic uniformity property of the MHS ensemble (Theorem 7.5.1), in the limit as the alphabet size p goes to infinity. Thus, as $p \to \infty$, it also satisfies the implied MHS properties of good packing (for every n) and modulation (as n also goes to infinity); see Sections 7.6 and 7.7. The effect of limiting the size of the alphabet p is discussed at the end of the next section.

7.9.4 Goodness for quantization

Proving goodness for quantization (Definition 7.3.2) requires a more refined analysis. Let $\mathbf{U}$ be uniform over a large region in space, and let $N_{B(\mathbf{U},r)}(\Lambda)$ denote the number of non-zero lattice points within a distance r from $\mathbf{U}$. The condition $G(\Lambda_n) \approx 1/2\pi e$ (for a large dimension n) means that (i) $N_{B(\mathbf{U},r)}(\Lambda_n) \geq 1$ with a high probability for r close to $r_{\mathrm{eff}}(\Lambda_n)$; and (ii) the contribution of the (rare) event $N_{B(\mathbf{U},r)}(\Lambda_n) = 0$ to the mean-squared quantization error is negligible. The ensemble uniformity property implies that (see (7.53))

$$E\{N_{B(\mathbf{U},r)}(\Lambda)\} = \left(\frac{r}{r_{\mathrm{eff}}}\right)^n \to \begin{cases} 0, & \text{if } r < r_{\mathrm{eff}} \\ \infty, & \text{if } r > r_{\mathrm{eff}} \end{cases} \tag{7.94}$$

as $n \to \infty$, where the expectation is over both $\mathbf{U}$ and the ensemble, and where r_{eff} is the ensemble effective radius. But this property alone does not imply the two conditions for quantization goodness; in particular, even if the expected number of points for $r > r_{\mathrm{eff}}$ is large, we cannot conclude that the probability that this number is equal to zero is small.

To prove that the $\mathbb{L}_{n,k,p,\gamma}$ ensemble is good for quantization, we shall need stronger conditions than those in Lemma 7.9.2 on the growth of p and k as a function of the lattice dimension n. As we shall see, the derivation is more sensitive to the fact that, even for a random generating matrix G, the lattice points are *not* statistically independent.

Let

$$G(\mathbb{L}_{n,k,p,\gamma}) \triangleq \gamma^{2/n} E_{\mathbb{L}_{n,k,p,\gamma}}\{\sigma^2(\Lambda)\} \tag{7.95}$$

denote the expected NSM of the $\mathbb{L}_{n,k,p,\gamma}$ ensemble (7.87).

Theorem 7.9.2 (Ensemble of good lattice quantizers) *If as $n \to \infty$*

$$\frac{n}{2k}[\log(n) + O(1)] < \log(p) < o(n), \tag{7.96}$$

then the $\mathbb{L}_{n,k,p,\gamma}$ ensemble (7.87) is good for quantization in expectation, i.e.,

$$G(\mathbb{L}_{n,k,p,\gamma}) \to \frac{1}{2\pi e}. \tag{7.97}$$

Although the ensemble point density γ is just a scaling parameter that does not affect the NSM, it is convenient to keep the lattice second moment fixed, say, $\sigma^2(\Lambda) = D$, so $\gamma \approx [2\pi e D]^{-n/2}$ is exponential with the dimension n (like for channel coding goodness; see footnote 16).

Condition (7.96) requires that p grows sub-exponentially with n, and that k grows faster than $\log(n)$. For example, $k = \sqrt{n}$, implying $p \sim n^{\sqrt{n}}$; or k can be linear, $k = rn$, in which case p is polynomial $p = n^{1/2r}$. (Since p must be a prime number, n grows along a suitable sub-sequence of the integer numbers.) Note that these conditions are more restrictive than those for modulation goodness (Lemma 7.9.2); for example, we cannot take p to infinity for a fixed dimension n, and we cannot take $k = 1$.

This theorem follows from a sequence of lemmas, whose proof is deferred to Problem 7.18. To guarantee a worst-case mean-squared quantization error D, the covering radius $r_{\text{cov}}(\Lambda)$ should be $\sqrt{nD}$. But as we discussed earlier, a "softer" probabilistic requirement is sufficient for our purpose: the number of lattice points N, inside a ball of radius $\sqrt{nD}$ located anywhere in space, should be with high probability greater (but not much greater) than one. For that, the expected number $E\{N\}$ should be slightly larger than one, while the standard deviation $\sqrt{\text{Var}\{N\}}$ should be small compared to $E\{N\}$ (to make N "nearly deterministic").

The relations between the probability and standard deviation of N are made clear by the following lemma.

Lemma 7.9.3 (Chebishev) *For a non-negative variable N,*

$$\Pr\{N = 0\} \leq \frac{\text{Var}\{N\}}{E^2\{N\}}. \tag{7.98}$$

Secondly, the condition for a small variance to mean-square ratio will follow from the following lemma.

Lemma 7.9.4 (Variance of intersection) *If the body S satisfies $V(S \bmod L) = V(S)$, i.e., S is packed (without overlap) by the lattice $L\mathbb{Z}^n$ (in particular, if S is contained in $[0, L)^n$), then the expectation and variance (with respect to the* *ensemble $\mathbb{L}_{n,k,p,\gamma}$ (7.87)) of the number of non-zero lattice points in S, satisfies* [20]

$$\text{Var}\,\{N_S(\Lambda)\} \leq p \cdot E\{N_S(\Lambda)\}. \tag{7.99}$$

Thirdly, to ensure that the packing condition of Lemma 7.9.4 holds for a body S which is equal to the "distortion ball" $\mathcal{B}(\mathbf{x}, \sqrt{nD})$, we require the following condition for each p, k and n.

Lemma 7.9.5 (Ball in lattice period) *If the lattice period $L = p^{k/n}/\gamma^{1/n}$ is greater than $2\sqrt{nD}$, then the ball $B_D = \mathcal{B}(\mathbf{x}, \sqrt{nD})$ is packed without overlap by the lattice $L\mathbb{Z}^n$, i.e., $V(B_D \bmod L) = V(B_D)$.*

Finally, the following property ensures that the cost in distortion does not explode even in the rare event that the ball B_D is empty.

Lemma 7.9.6 (Bounded quantization error) *The covering radius of a period-L construction A lattice Λ is bounded by $r_{\text{cov}}(\Lambda) \leq \sqrt{n}L/2$. Thus for any point $\mathbf{x}$, the quantization squared error is bounded by $\frac{1}{n}\|\mathbf{x} - Q_\Lambda(\mathbf{x})\|^2 \leq (L/2)^2$.*

The combination of the four lemmas above proves Theorem 7.9.2; see Problem 7.18.

A slightly more refined analysis shows that with an appropriate growth of p and k as a function of n, the ensemble $\mathbb{L}_{n,k,p,\gamma}$ also proves the existence of lattices good for *covering* (Section 7.3) [66]. Since covering implies quantization (Section 7.3), it is not surprising that the parameters (p_n, k_n) for covering goodness are a subset of those for quantization goodness (Theorem 7.9.2); and luckily, both are subsets of those for packing and modulation (Lemma 7.9.2). Hence, for this set of parameters, the ensemble $\mathbb{L}_{n,k,p,\gamma}$ (7.87) must contain lattices which are simultaneously good under all four goodness criteria, thus proving Theorem 7.9.1.

7.9.5 Modulo-2 lattices

The requirement that the alphabet size p must go to infinity complicates the practical implementation of the lattice construction above. On the other hand, keeping the alphabet size fixed means that the Voronoi cells become "skewed" as the lattice dimension n grows. This skewness is due to the cubic-periodic structure of construction A, where the distance to the nearest neighbor in the *Cartesian directions* remains constant (at most p), while in the non-Cartesian directions it typically grows as $\sqrt{n}$.

For modulation in the presence of AWGN $\sim N(0, \sigma^2)$, a bounded minimum distance implies that a lattice constellation with a finite alphabet suffers from a "symbol error rate floor" of $\text{SER}_{\text{min}} \approx 2Q((p/2)/\sigma)$, implying that the block error probability goes to 1 as the dimension goes to infinity. Hence, it cannot approach

---

[20] As usual, we do not count the zero lattice point and its associated coset $L\mathbb{Z}^n$, whose location is not randomized.

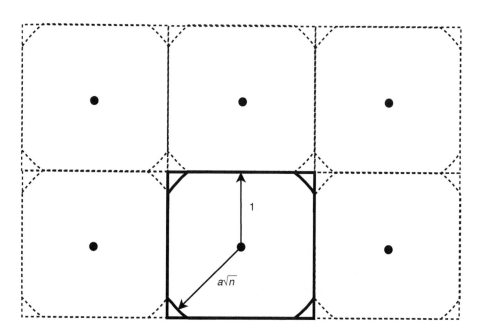

Figure 7.9 Truncated balls in the cubic lattice $2\mathbb{Z}^n$. The ball radius increases as $\sqrt{n}$, while the cube edge stays constant equal to 2. The Voronoi cells of a modulo-2 lattice have a similar ("skewed") behavior. Therefore, they can be good for quantization but not for modulation in the presence of AWGN.

capacity with an arbitrarily small error probability, as desired. See Figure 7.9. We shall return to the channel coding goodness of a finite lattice constellation in Chapter 8.

Yet from a quantization goodness viewpoint, too close neighbors are harmless. In fact, near-optimum quantization performance can be achieved with a finite, and even binary, alphabet, i.e., with a modulo-2 lattice. Specifically, suppose an underlying (n, k) linear binary code $\mathcal{C}$, with a rate $R = k/n$ bit per dimension, and average mean-squared distortion

$$D = \frac{1}{n} \sum_{i=1}^{n} E(\hat{X}_i - X_i \bmod 2)^2, \qquad (7.100)$$

where the reconstruction letters $\hat{X}_i$ are in $\{0, 1\}$, and where the mod 2 corresponds to the fact that all shifts $\hat{\mathbf{x}} + 2\mathbb{Z}^n$ of a codeword $\hat{\mathbf{x}}$ are valid codewords. Note that if the X_i are i.i.d. and uniform over $(0, 2)$, then D is equal to the lattice second moment $\sigma^2(\Lambda_C)$; and since the lattice cell volume is $V(\Lambda_C) = 2^n/2^k = 2^{n(1-R)}$, the NSM is given by (3.21),

$$G(\Lambda_C) = \frac{1}{4} D \cdot 2^{2R}. \qquad (7.101)$$

The optimum (R, D) performance is given by the rate-distortion function (5.31) of a source $X \sim \text{Unif}(0, 2)$, with a two-letter reconstruction alphabet $\hat{\mathcal{X}} = \{0, 1\}$, under the squared-modulo distortion measure (7.100); see, for example, [18, 79, 232] for the computation of this rate-distortion function. Thus, the NSM of a *random* binary code based quantizer, $G(p = 2, R = k/n)$, is optimized by the (R, D) pair on the rate-distortion curve with the minimal product $D \cdot 2^{2R}$. Kudryashov and Yurkov [146] showed that a *linear* binary code based quantizer (i.e., a modulo-2 lattice quantizer) can achieve the same performance. Furthermore, the optimum operation point is at $R \approx 0.41$, and the resulting NSM (7.101) is only slightly (≈ 0.1 dB) above the ideal value of $1/2\pi e$.

Summary of Chapter 7

NSM and NVNR of an n-dimensional sphere, as $n \to \infty$

$$G_n^* \to \frac{1}{2\pi e}; \quad \mu_n^*(P_e) \to 2\pi e, \quad \text{for } 0 < P_e < 1.$$

Good n-dimensional lattice quantizers, as $n \to \infty$

$$G(\Lambda_n) \to \frac{1}{2\pi e}.$$

Divergence from white Gaussianity If $U_n \sim \text{Unif}(\mathcal{V}_0(\Lambda_n))$ for a good lattice quantizer Λ_n, then

$$\frac{1}{n} D(U_n \ ; \ N(0, \sigma^2(\Lambda_n))) \to 0.$$

Tendency to a ball For the U_n above,

$$\Pr(U_n \notin \mathcal{B}(0, r_n)) \to 0, \quad \text{where } r_n = (1 + \epsilon) r_{\text{eff}}(\Lambda_n).$$

Good lattice constellations (for channel coding)

$$\mu(\Lambda_n, P_e) \to 2\pi e, \quad \text{for } 0 < P_e < 1;$$

$$\mu_{\text{euclid}}(\Lambda_n, Z_n, P_e) \to 2\pi e, \quad \text{for semi-spherical noise } Z_n;$$

$$\mu_{\text{matched}}(\Lambda_n, Z_n, P_e) \to 2\pi e, \quad \text{for noise satisfying the generalized AEP.}$$

Existence of (simultaneously) good lattices

(i) $G(\Lambda_n) \to 1/2\pi e$ for Λ_n in the Rogers ensemble;
(ii) $\mu(\Lambda_n, P_e) \to 2\pi e$ for Λ_n in the Minkowski–Hlawka–Siegel ensemble;
(iii) both (i) and (ii) for a typical Λ_n in the $\mathbb{L}_{n,k,p,\gamma}$ ensemble (random construction A).

Problems

P.7.1 (Ball characterization) Prove the formulas for the characteristics of the ball (7.11).

P.7.2 (Ball NSM lower bound) Use the Stirling lower bound for $n!$ to show (7.16).

P.7.3 (Ball NVNR lower bound) Prove that $\mu_1^*(P_e) \approx 8 \ln(1/P_e)$ and $\mu_2^*(P_e) \approx \pi \ln(1/P_e)$, for small P_e, using approximations for the chi-square distribution. Use the uniform lower bound (7.19) to show (7.20) numerically, i.e., $\mu_n^*(P_e) \geq 2\pi e$ for all n, for P_e smaller than some fixed threshold.

P.7.4 (Sphere bounds on NSM and NVNR) Elaborate the proof of Theorem 7.1.2.

P.7.5 (Volume of rth norm ball) Let $V_{n,r}$ denote the volume of a unit ball with respect to the rth-power norm (i.e., the Euclidean ball corresponds to the case $r = 2$). Find an asymptotic approximation for $V_{n,r}$ for large n (paralleling $V_n \approx (2\pi e/n)^{n/2}$). Extend the volume-wise equivalence of a sphere and a Gaussian to rth power: use the AEP to show that an i.i.d. distribution of the form $Ke^{\beta x^r}$ belongs asymptotically to a norm-r ball, and that the AEP volume formula 2^{nh} (where h is the differential entropy of this distribution) is exponentially the same as the volume of this ball.

P.7.6 (Convergence of NSM) Use (7.29) to prove the convergence rate in Corollary 7.2.1.

P.7.7 (Divergence from Gaussianity) Show equivalence of Definitions 6.5.1 and 7.2.3. Show that the divergence from white Gaussianity is larger than the divergence from a Gaussian distribution having the same first and second order statistics.

P.7.8 (Projection of a sphere) Prove Theorem 7.2.2 using integration over a cut of a sphere.

P.7.9 (Minkowski–Hlawka–Siegel ensemble property) Show that the simple form (7.51) of Theorem 7.5.1 implies the stronger form (7.54). Guidance: use a "fine" partition of $\mathbb{R}^n$ into "small" cells Δ_i. Use this partition and the mean-value theorem to express the function f as $f(\mathbf{x}) = \sum_i f_i \cdot I_i(\mathbf{x})$, where $I_i(\mathbf{x})$ is the indicator of the cell Δ_i, and f_i is some "representative" of the function in this cell. Write the left-hand side of (7.54) in this form, apply (7.51), and use the Riemann integral approximation.

P.7.10 Explain the random translation argument discussed in Section 7.4.1.

P.7.11 (Threshold versus NN decoding) Prove that the threshold decoding bound (Lemma 7.7.1) applies also to NN decoding. That is, the decision region $\Omega_0 = \{\mathbf{y} : \hat{\lambda}(\mathbf{y}) = 0\}$ is a subset of the fundamental Voronoi region.

P.7.12 (Semi-spherical noise) Prove that the sum of a semi-spherical noise and AWGN is a semi-spherical noise.

P.7.13 (Goodness for non-AWGN with noise-matched decoding) Prove Theorem 7.8.1 using the decoding rule (7.74), where the decoding region S is equal to the generalized typical set $S_\epsilon^{(n)}$ of the noise $\mathbf{Z}_n$ (Definition 7.8.1).

P.7.14 (Goodness for non-AWGN with Euclidean decoding) Prove Theorem 7.8.2 by verifying that if a lattice has a good NVNR for AWGN (Theorem 7.7.1), then this lattice is a good constellation for any semi-spherical noise (i.e., the Euclidean mismatched NVNR is close to $2\pi e$).

P.7.15 (Random binary linear code) For $q = 2$, show that the codewords are *pairwise independent*. That is, if $\mathbf{w}' \neq \mathbf{w}$, then $G\mathbf{w}'$ is statistically independent of $G\mathbf{w}$, where G is a random generator matrix.

P.7.16 (Period, resolution and cell volume of the $\mathbb{L}_{n,k,p,\gamma}$ ensemble) Prove properties (7.88), (7.89) and (7.90) of the $\mathbb{L}_{n,k,p,\gamma}$ ensemble.

P.7.17 (Proof of Lemma 7.9.2: uniformity of the $\mathbb{L}_{n,k,p,\gamma}$ ensemble) Show that $E\{N_S(\Lambda_C)\} \to \mathrm{Vol}(S)$, as $p \to \infty$, where $E\{\cdot\}$ denotes average over the $\mathbb{L}_{n,k,p,\gamma}$ ensemble.

P.7.18 (Goodness for quantization of the $\mathbb{L}_{n,k,p,\gamma}$ ensemble) Part I: prove Lemmas 7.9.3, 7.9.4, 7.9.5, and 7.9.6.

Guidance for Lemma 7.9.4 (variance of intersection). (A) Let $S_p = 1/\delta[(S \cap \delta\mathbb{Z}^n) \bmod L]$ denote the intersection of the scaled and folded body S with the grid, where δ is the resolution (7.89). Show that $N_S(\Lambda) = N_{S_p}(C)$ defined in Lemma 7.9.1. (B) Use Lemma 7.9.1 to calculate the expectation of $N_S(\Lambda)$ over the ensemble. (C) For $\mathbf{w} \in \mathbb{Z}_p^k$, let $\xi_\mathbf{w} = 1_{\{G\mathbf{w}\in S_p\}}$, as in (7.86). Show that $\xi_\mathbf{w}$ and $\xi_{\mathbf{w}'}$ are independent, unless $\mathbf{w} = g \cdot \mathbf{w}'$ for some $g \in \mathbb{Z}_p$. (D) Conclude that for a non-zero $\mathbf{w}$, each indicator $\xi_\mathbf{w}$ depends on at most $p - 1$ indicators. (E) Now use (7.86), and write the variance as a double sum over pairwise covariances, to upper bound the variance in terms of $E\{N_S(\Lambda)\}$ and establish the lemma.

Part II: prove Theorem 7.9.2 (ensemble of good lattice quantizers).

Guidance: Restrict attention to a ball $B_D = \mathcal{B}(\mathbf{x}, \sqrt{nD})$ for some arbitrary $\mathbf{x}$. (A) Use Lemma 7.9.6 to show that

$$E\left\{\tfrac{1}{n}\|\mathbf{x} - Q_\Lambda(\mathbf{x})\|^2\right\} \leq D + (L/2)^2 \cdot \Pr\{N_{B_D}(\Lambda) = 0\}$$

for any point $\mathbf{x}$, where both the expectation and the probability are measured with respect to the ensemble $\mathbb{L}_{n,k,p,\gamma}$. (B) Prove Lemma 7.9.5 i.e., find a condition on L such that Lemma 7.9.4 can be applied with respect to B_D. (C) Assuming this condition holds, apply Lemmas 7.9.2, 7.9.3 and 7.9.4, to show that

$$\Pr\{N_{B_D}(\Lambda) = 0\} \leq p \cdot \frac{1}{(1 - \frac{\delta}{\sqrt{D}})^n \gamma V(B_D)}$$

where probability is measured with respect to the ensemble. (D) Show that the expected ensemble NSM $G(\mathbb{L}_{n,k,p,\gamma})$ in (7.95) is upper bounded by $\gamma^{2/n}[D + (L/2)^2 \cdot \Pr\{N_{B_D}(\Lambda) = 0\}]$. For that, let $\mathbf{X}$ be a generalized dither which is uniform over the lattice period $[0, L)^n$ (note that this is a sublattice of any lattice in the ensemble; see Example 4.2.1); write the lattice second moment $\sigma^2(\Lambda)$ in the definition of expected ensemble NSM (7.95) in terms of the expected quantization error of this generalized dither; switch the order of expectations; and apply the bounds above. (E) Let $\gamma = (1 + \epsilon)^n / V(B_D)$ for some small $\epsilon > 0$. Use the asymptotic behavior of the volume of an n-dimensional ball to show that the second term of (D) vanishes if p grows sub-exponentially with n. Then show that (B) is satisfied for this choice of γ under the condition in the left-hand side of (7.96). Conclude that $G(\mathbb{L}_{n,k,p,\gamma})$ must then go to $1/2\pi e$.

Historical notes

The Minkowski–Hlawka theorem [36, 49, 113] plays an analogous role to Shannon's random coding technique in the case of lattice codes: a non-constructive tool for proving the existence of "good" lattice codes. Its first role was to show that lattices can form dense sphere packings: the best packing density in each dimension n is $(1/2)^n$, i.e., a packing efficiency of at least one-half. As for covering, Rogers [229, 230, 231] used a variation of the Minkowski–Hlawka theorem to prove that there exist lattices for which the proportion of overlap between the spheres can grow polynomially with the dimension, implying that their covering efficiency approaches 1.

In the coding literature, averaging bounds were already used to show that linear and convolutional codes can approach the capacity of a BSC (and more general symmetric channels) through the 1950s and 1960s; see Elias [65], Gallager's book [97], and Forney et al. [92]. de Buda [57, 58] used the Minkowski–Hlawka theorem to show a parallel result for lattices in the Euclidean space, i.e., that spherically shaped lattice codes can approach the power-constrained AWGN channel capacity. (See Linder et al. [163] for a corrected proof of de Buda's theorem.) Poltyrev [221] used the Siegel version of the Minkowski–Hlawka theorem [36, 113] (called the "MHS ensemble" in this book) to show that high-dimensional lattice codes can be "good for AWGN channel coding," i.e., approach the capacity per unit volume (maximum logarithmic density), $C_\infty = \frac{1}{2} \log(1/2\pi e N)$, of an unconstrained AWGN channel. Loeliger [175] made an important step towards implementation, by showing that the same result holds for a random ensemble of construction A (modulo-p) lattices. While Poltyrev and Loeliger arrived at capacity through a more elaborate error exponent analysis, Forney *et al.* [92] gave a direct existence proof, based on Shannon's random coding and typicality arguments, for what they called "sphere-bound achieving codes." The proof in this book combines the power of the MHS ensemble with the simplicity of Shannon's typicality arguments. See also the dispersion analysis of Ingber *et al.* [125].

Spherical bounds for quantization appear in the seminal work of Zador [283]; see an extension to general norm-based distortion measures in Yamada *et al.* [277]. The existence of good lattice quantizers can be deduced from Zador's high-resolution random-quantization analysis, combined with Gersho's conjecture [102] (specialized to lattices). Poltyrev pointed out in [289] that good covering implies good quantization; thus, the NSM of Rogers-good lattices can approach $1/2\pi e$. Zamir and Feder [289] showed that the dither of good lattice quantizers converges to AWGN in the divergence rate sense and in the sliding window ℓ_1-distance sense, while Ordentlich and Erez [206] showed that such a dither lies with high probability inside a ball of a similar volume.

The effect of a finite reconstruction alphabet on the rate-distortion performance of continuous sources was analyzed by Finamore and Pearlman [79]; see also Rose [232]. For lattice-based and trellis-based quantizers, this effect was examined by Calderbanck, Fishburn and Rabinovich [31, 32], and by Eyuboglu and Forney [76]. Kudryashov and Yurkov [146] gave a direct existence proof for quantization goodness (based on Loeliger's construction A ensemble), and analyzed the NSM loss of modulo-q lattice quantizers as a function of the alphabet size q.

The duality between transmission and quantization is central to Conway and Sloane [45, 47]. Forney [88] considered this duality from the perspective of lattice goodness. Erez *et al.* [66] combined all four goodness criteria above into a single sequence of lattices that are "good for everything": packing, covering, channel coding and quantization. The proof of quantization goodness in Section 7.9 follows Kudryashov and Yurkov [146] and Ordentlich and Erez [206].

8 Nested lattices

The magic of lattices in modern communication and information theory is their potential for coordinated operation. Many of the coding techniques which we shall see from this point on, *shaping, binning and alignment*, are based on the interplay between two (or more) lattices. These lattices generate a code for a *single* communication link, or a system of codes for a *distributed* communication setup.

This short chapter provides the necessary background on nested lattices. It will be used in Chapter 9 for lattice-shaped modulation over the AWGN channel, and extended in Chapters 10–12 to multi-terminal setups and joint source-channel coding.

Imagine a ruler with a coarse scale and a finer scale, numbered on a binary basis: the coarse scale is numbered by the most significant bits (MSBs), while the least significant bits (LSBs) identify the fine markings in between; see Figure 8.1. A set of fine markings between two coarse markings – which share the same MSBs – can be thought of as a "finite codebook." The union of all such finite codebooks gives the whole ruler. An alternative partition of the ruler is into sets of markings with the *same LSBs*. Each such *coset* is a shift of the coarse scale by a point in the fine scale, and the union over all such cosets gives again the whole ruler.

A similar picture – only in multiple dimensions – is created by a pair of *nested lattices*. Construction A (modulo-q lattice), for example, is a special case of nested lattices, where the coarse lattice is cubic ($q\mathbb{Z}^n$) (see Section 2.5). The coarse cell of a general nested lattice pair can *shape* a codebook consisting of points from the fine lattice. Or, as we shall see in Chapter 10, the cosets of the coarse lattice (shifted by fine lattice points) generate a structured binning ("coloring") scheme for coding with side information.

The nested lattice structure plays also an important role in coding for a multi-user system. In a distributed scenario, *aligned* lattice codes can coordinate signaling and quantization at remote locations. Such a coordination allows for interference mitigation, distributed compression, parallel relaying, and more.

This chapter brings together the basic definitions and properties of nested lattices, which will be used in the chapters to come.

...00   ...01   ...10   ...11

Figure 8.1 A ruler with coarse and fine scales.

8.1 Definition and properties

A pair of n-dimensional lattices (Λ_1, Λ_2) is called *nested* if

$$\Lambda_2 \subset \Lambda_1 \tag{8.1}$$

i.e., Λ_2 is a *sublattice* of Λ_1. It follows that each basis vector $\boldsymbol{g}_{2,i}$ of Λ_2 is an integer combination

$$\boldsymbol{g}_{2,i} = \sum_{k=1}^{n} j_{k,i} \boldsymbol{g}_{1,k}$$

of the basis vectors $\boldsymbol{g}_{1,1}, \ldots, \boldsymbol{g}_{1,n}$ of Λ_1; hence the corresponding generator matrices G_1 and G_2 satisfy

$$G_2 = G_1 \cdot J, \tag{8.2}$$

where the *nesting matrix* $J = \{j_{k,i}\}$ is an $n \times n$ integer matrix whose determinant is greater than or equal to 1, with equality if and only if the two lattices are identical (Proposition 2.1.1).[1] We call Λ_1 the *fine* lattice, and Λ_2 the *coarse* lattice. See two two-dimensional examples in Figure 8.2.

As in Chapter 2, we assume that both lattices are non-degenerate, so the matrices G_1, G_2 and J are non-singular. By the cell volume formula (Proposition 2.2.1), the volumes of the cells of Λ_1 and Λ_2 satisfy

$$V(\Lambda_2) = \det(J) \cdot V(\Lambda_1). \tag{8.3}$$

Hence, the determinant of the nesting matrix J is *invariant* of the choice of generating matrices for Λ_1 and Λ_2.

Definition 8.1.1 (Nesting ratio) *The nesting ratio Γ of $\Lambda_2 \subset \Lambda_1$ is defined as*

$$\Gamma(\Lambda_1, \Lambda_2) \triangleq \sqrt[n]{\frac{V(\Lambda_2)}{V(\Lambda_1)}} = \sqrt[n]{\det(J)}. \tag{8.4}$$

The nesting ratio is invariant to scaling of both lattices by the *same* factor, $\Gamma(\alpha\Lambda_1, \alpha\Lambda_2) = \Gamma(\Lambda_1, \Lambda_2)$ or, more generally, to a linear transformation by the same (non-singular) matrix: $\Gamma(T\Lambda_1, T\Lambda_2) = \Gamma(\Lambda_1, \Lambda_2)$.

---

[1] Due to the symmetry of the lattices, we can always negate a basis vector to make the determinant non-negative.

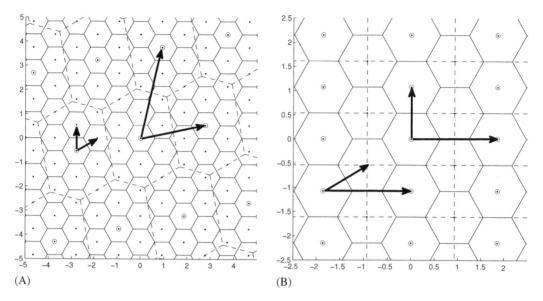

Figure 8.2 Nested two-dimensional lattices. The nesting ratio is 3 in (A) and $\sqrt{2}$ in (B).

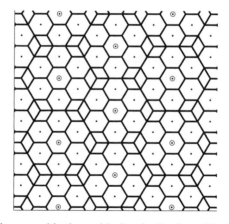

Figure 8.3 Self-similar hexagonal lattices with $J = 3 \cdot I$, where I is the 2×2 identity matrix, i.e., nesting ratio $\Gamma = 3$.

Many of the relations between nested lattices are due to their algebraic definition (rather than to their geometry), and hence are similar to properties of general sub-groups. This similarity is enhanced by the fact that the quotient group Λ_1/Λ_2 (i.e., the set of shifts of the coarse lattice by points from the fine lattice) is finite.

Figure 8.3 shows the special case of *self-similar* nested lattices, $G_2 = jG_1$, with j being an integer (i.e., a nesting matrix $J = jI$, with I being the identity matrix), in which case [2]

$$\Lambda_2 = j \cdot \Lambda_1. \tag{8.5}$$

For general nested lattices, we can always find generator matrices such that the nesting matrix is *diagonal*. This will greatly simplify codeword and coset enumeration, as we shall see in the next section.

Proposition 8.1.1 (Diagonal nesting) *Let $\Lambda_2 \subset \Lambda_1$ be a nested lattice pair. There exist generator matrices G_1 and G_2 of Λ_1 and Λ_2, respectively, such that*

$$G_2 = G_1 \cdot \mathrm{diag}(j_1, \ldots, j_n), \tag{8.6}$$

where the j_i are positive integers. That is, each basis vector of Λ_2 is an integer multiple of a single *basis vector of Λ_1.*

Proof Suppose that G_1 and $G_2 = G_1 \cdot J$ are arbitrary generator matrices of the lattices Λ_1 and Λ_2, respectively, where J is a general integer matrix. By the *Smith normal form*, J can be decomposed as

$$J = T_1 \cdot \mathrm{diag}(j_1, \ldots, j_n) \cdot T_2, \tag{8.7}$$

where T_1, T_2 are $n \times n$ unimodular (unit absolute determinant integer) matrices, and the j_i are positive integers [44, p. 311]. The unimodularity of T_1 and T_2 implies, by Proposition 2.1.1, that the matrices $G_1' = G_1 \cdot T_1$ and $G_2' = G_2 \cdot T_2^{-1}$ are also generator matrices of Λ_1 and Λ_2, respectively. By (8.7), these generator matrices are related as $G_2' = G_1' \cdot \mathrm{diag}(j_1, \ldots, j_n)$, which is the desired diagonal form (8.6). $\square$

Note that the transformation by T_1 and T_2^{-1} corresponds, by Proposition 2.1.1, to a change of basis before and after the scaling by $j_1, \ldots, j_n$. [3]

8.2 Cosets and Voronoi codebooks

Recall the definition (2.7) of a lattice coset Λ_x. For λ in the fine lattice Λ_1, we call

$$\Lambda_{2,\lambda} \overset{\Delta}{=} \lambda + \Lambda_2 \tag{8.8}$$

a *coset of Λ_2 relative to Λ_1*. Due to the nesting relation between the two lattices, each relative coset $\Lambda_{2,\lambda}$ belongs to the fine lattice Λ_1.

As in (2.33), the *set of relative cosets* is denoted by

$$\Lambda_1/\Lambda_2 = \{\Lambda_{2,\lambda} : \lambda \in \Lambda_1\}, \tag{8.9}$$

---

[2] The definition of self-similarity in the literature sometimes allows for rotation, i.e., the coarse lattice is a rotated and scaled version of the fine lattice.

[3] When the lattices are not self-similar, diagonal nesting may come at the cost of "bad" (i.e., long/far-from-orthogonal) bases; see the notion of basis goodness in Section 2.1.1.

and is called a *quotient group*. The quotient group (8.9) relates to the fine lattice Λ_1 like the "continuous" quotient $\mathbb{R}^n/\Lambda_x$ relates to $\mathbb{R}^n$: its union covers Λ_1.

Proposition 8.2.1 (Number of relative cosets) *The number of distinct relative cosets, i.e., the size of the quotient group Λ_1/Λ_2, is given by*

$$|\Lambda_1/\Lambda_2| = \frac{V(\Lambda_2)}{V(\Lambda_1)} = \det(J) = \Gamma^n, \tag{8.10}$$

where J is the nesting matrix in (8.2), and Γ is the nesting ratio (8.4).

This result is very plausible, as the point density $\gamma_1 = 1/V(\Lambda_1)$ of the fine lattice is $\det(J)$ times larger than the point density $\gamma_2 = 1/V(\Lambda_2)$ of the coarse lattice. Yet, to prove Proposition 8.2.1 rigorously (see Corollary 8.2.1 below), we need a notion paralleling that of a "fundamental cell" for nested lattices. As in (2.8) and (2.33), we ask for a *minimal* set of *coset representatives* $S \subset \Lambda_1$, such that

$$\bigcup_{\lambda \in S} \Lambda_{2,\lambda} = \Lambda_1. \tag{8.11}$$

This question leads us to the notion of a *lattice-shaped codebook*, and the special case of a *Voronoi codebook*.

Definition 8.2.1 (Lattice-shaped (Voronoi) codebook) *For a pair of nested lattices $\Lambda_1 \subset \Lambda_2$, a lattice-shaped codebook is the intersection of the fine lattice with some fundamental cell $\mathcal{P}_0 = \mathcal{P}_0(\Lambda_2)$ of the coarse lattice:*

$$\mathcal{C}_{\Lambda_1,\mathcal{P}_0} = \Lambda_1 \cap \mathcal{P}_0. \tag{8.12}$$

The fundamental cell $\mathcal{P}_0$ is called the "shaping region." If $\mathcal{P}_0$ is the fundamental Voronoi cell $\mathcal{V}_0(\Lambda_2)$ of the coarse lattice, then (8.12) is called a Voronoi codebook. [4]

See Figure 8.4 and the following example.

Example 8.2.1 (Parallelepiped shaping and code enumeration) *For the diagonal form $G_2 = G_1 \cdot \mathrm{diag}(j_1, \ldots, j_n)$ of (8.6), if the shaping region $\mathcal{P}_0$ is the fundamental parallelepiped induced by the coarse generating matrix G_2 (see (2.14)), then the codebook is given by*

$$\mathcal{C}_{\Lambda_1,\mathcal{P}_0} = \{G_1 \cdot [k_1, \ldots, k_n]^t : \text{for } 0 \leq k_i \leq j_i - 1\}. \tag{8.13}$$

The codebook size in this case is clearly

$$|\mathcal{C}_{\Lambda_1,\mathcal{P}_0}| = j_1 \cdot \ldots \cdot j_n = \det(J). \tag{8.14}$$

---

[4] Note that by the definition of a fundamental cell, the association of border points to the codebook (i.e., points of Λ_1 which fall on the boundary of $\mathcal{P}_{0,2}$) is such that the cosets $\{\Lambda_{2,\lambda_1} : \lambda_1 \in (\Lambda_1 \mod \Lambda_2)\}$ are disjoint. In the case of a Voronoi fundamental cell, this is guaranteed by systematic breaking of ties in the Voronoi partition; see (2.10).

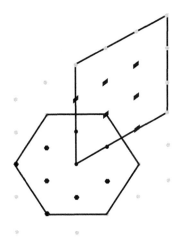

Figure 8.4 Lattice-shaped codebooks based on nested hexagonal lattices with two possible shaping regions: parallelepiped and Voronoi.

Example 8.2.1 would prove Proposition 8.2.1, if we show that $\mathcal{C}_{\Lambda_1, \mathcal{P}_0}$ is a complete set of coset representatives for any $\mathcal{P}_0(\Lambda_2)$. Indeed, as we shall see below, $\mathcal{C}_{\Lambda_1, \mathcal{P}_0}$ *plays the role of a fundamental cell* for partitioning Λ_1 with respect to Λ_2.

Proposition 8.2.2 (Partition by a lattice codebook) *Each point λ in the fine lattice Λ_1 can be written as a* unique *sum*

$$\lambda = \lambda_q + \lambda_e, \quad \lambda_q \in \Lambda_2 \text{ and } \lambda_e \in \mathcal{C}_{\Lambda_1, \mathcal{P}_0}. \tag{8.15}$$

Specifically, λ_q is the (unique) point in Λ_2 such that $\lambda \in \lambda_q + \mathcal{P}_0$, while $\lambda_e = \lambda \bmod_{\mathcal{P}_0} \Lambda_2$.

As in (2.12), we can think of λ_q as the quantization of λ and of λ_e as the resulting error, where here the quantizer is $Q_{\Lambda_2}(\cdot) = \{\Lambda_2, \mathcal{P}_0\}$. See (3.19) and (4.2).

Proof Since $\mathcal{P}_0$ induces a partition $\mathcal{P}_\lambda$, $\lambda \in \Lambda_2$, of $\mathbb{R}^n$, any point $\mathbf{x}$ can be written uniquely as the sum $Q_{\Lambda_2}(\mathbf{x}) + \mathbf{x}_e$, where the quantization point $Q_{\Lambda_2}(\mathbf{x})$ is in Λ_2, and the error $\mathbf{x}_e = \mathbf{x} \bmod_{\mathcal{P}_0} \Lambda_2$ is in $\mathcal{P}_0$. See (2.12). Thus, we only need to show that if $\mathbf{x}$ is in Λ_1, then the error $\mathbf{x}_e$ must be in Λ_1 as well, hence in the codebook $\Lambda_1 \cap \mathcal{P}_0$. This fact follows from the nesting relation between the two lattices: every point in Λ_2 is also in Λ_1, and therefore $Q_{\Lambda_2}(\mathbf{x})$ is in Λ_1. Thus for $\mathbf{x}$ in Λ_1, the error $\mathbf{x}_e$ is a difference between two vectors in Λ_1, so it must be in Λ_1 too. $\qquad \square$

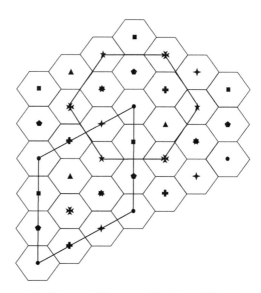

Figure 8.5 Relative cosets for the nested hexagonal lattices of Figure 8.3 (there are nine of these, each denoted by a different symbol). Any shaping region contains one member from each coset.

It follows that on the one hand, Λ_1 is covered *without overlap* by shifts of the lattice codebook by the coarse lattice:

$$\Lambda_1 = \dot{\bigcup}_{\lambda \in \Lambda_2} (\lambda + \mathcal{C}_{\Lambda_1, \mathcal{P}_0}) \tag{8.16}$$

where $\dot{\cup}$ denotes disjoint union. On the other hand, Λ_1 can be *decomposed into relative cosets* in Λ_1 / Λ_2 with coset representatives in $\mathcal{C}_{\Lambda_1, \mathcal{P}_0}$:

$$\Lambda_1 = \dot{\bigcup}_{\lambda \in \mathcal{C}_{\Lambda_1, \mathcal{P}_0}} (\lambda + \Lambda_2). \tag{8.17}$$

It follows from (8.17) that a lattice-shaped codebook is a *complete set of coset representatives* for the quotient group Λ_1 / Λ_2, as desired in (8.11); see Figure 8.5. Therefore we have the following result.

Corollary 8.2.1 (Codebook size) *The size of a lattice-shaped codebook is equal to the number of relative cosets, $|\mathcal{C}_{\Lambda_1, \mathcal{P}_0}| = |\Lambda_1 / \Lambda_2|$, independent of the choice of the shaping region $\mathcal{P}_0$. Thus (by the parallelepiped-shaping case in Example 8.2.1), for any shaping region $\mathcal{P}_0$,*

$$|\mathcal{C}_{\Lambda_1, \mathcal{P}_0}| = |\Lambda_1 / \Lambda_2| = \det(J) = \Gamma^n, \tag{8.18}$$

which proves Proposition 8.2.1.

Since after a modulo $(\mathcal{P}_0)$ operation every relative coset collapses into a single point (which is its intersection with the fundamental cell $\mathcal{P}_0$), we also have (see (2.31)) the following.

Corollary 8.2.2 (Alternative form) *The lattice codebook (8.12) can also be written as*

$$\mathcal{C}_{\Lambda_1, \mathcal{P}_0} = \Lambda_1 \ \mathrm{mod}_{\mathcal{P}_0} \ \Lambda_2. \tag{8.19}$$

The property that a codebook is a complete set of relative coset representatives, and the definition of a coset leader (2.32), provide an alternative view of a *Voronoi* codebook.

Corollary 8.2.3 (Least energy codebook) *The Voronoi codebook consists of the* leaders *of all relative cosets:*

$$\mathcal{C}_{\Lambda_1, \mathcal{V}_0} = \{\text{Coset Leader } (\Lambda_{2, \lambda}): \ \lambda \in \Lambda_1\}. \tag{8.20}$$

Thus, $\mathcal{V}_0$ minimizes the codebook second moment over all possible shaping regions in (8.12).

Choice of a shaping region What would be a good choice for the codebook shaping region $\mathcal{P}_0$ in Definition 8.2.1, for a given nested lattice pair? Codeword (and coset) enumeration is clearly easier for a parallelepiped codebook, as in Example 8.2.1. Other aspects of the coding problem (power constraint, source distribution), however, favor taking the *shortest* vector of each relative coset, hence a *Voronoi* shaped cell. Luckily, we can enjoy the benefit of both. By the *modulo equivalence* of fundamental cells (2.23b), we have

$$(\mathbf{x} \ \mathrm{mod}_{\mathcal{P}_0} \ \Lambda_2) \ \mathrm{mod}_{\mathcal{Q}_0} \ \Lambda_2 = \mathbf{x} \ \mathrm{mod}_{\mathcal{Q}_0} \ \Lambda_2 \tag{8.21}$$

for any fundamental cells $\mathcal{P}_0$ and $\mathcal{Q}_0$ of Λ_2. Thus, it is possible to start with a parallelepiped cell $\mathcal{P}_0$ (for simple codeword enumeration) and, later, determine the final shape of the codebook by the Voronoi region $\mathcal{Q}_0 = \mathcal{V}_0(\Lambda_2)$.

8.3 Nested linear, lattice and trellis codes

Let us consider a few examples for constructing nested codes and nested lattices.

Consider first two binary linear codes C_1 and C_2 in $\{0, 1\}^n$, where, as in (2.48), each code C_i is determined by an $(n - k_i) \times n$ parity-check matrix H_i, for $i = 1, 2$.

Suppose that $k_2 < k_1 < n$, and that the parity-check matrices are related as

$$
H_2 = \begin{bmatrix} \overbrace{H_1}^{(n-k_1)\times n} \\ --- \\ \underbrace{\Delta H}_{(k_1-k_2)\times n} \end{bmatrix} , \tag{8.22}
$$

where all the row vectors are linearly independent. Since each row vector of the parity-check matrix corresponds to one constraint, it follows that each codeword of C_2 satisfies the $n - k_1$ parity checks of C_1 plus $k_1 - k_2$ additional checks defined by the matrix ΔH. Hence, C_2 must be a sub-code of C_1, i.e,

$$
C_2 \subset C_1. \tag{8.23}
$$

The quotient group C_1/C_2 contains $2^{k_1-k_2}$ different cosets of C_2 relative to C_1:

$$
|C_1/C_2| = 2^{k_1-k_2}.
$$

Each coset in C_1/C_2 can be represented by the *relative syndrome*

$$
\Delta \mathbf{s} = \Delta H \cdot \mathbf{c} \tag{8.24}
$$

associated with any codeword $\mathbf{c}$ in the coset.

The codes C_1 and C_2 are also defined by their $n \times k_i$ generator matrices G_i, $i = 1, 2$ (2.54). The nesting relation (8.23) implies that these matrices are related as

$$
G_1 = [\underbrace{G_2}_{n\times k_2} \mid \underbrace{\Delta G}_{n\times(k_1-k_2)}]. \tag{8.25}
$$

The nesting relation (8.23) can be extended to q-ary linear codes over the alphabet $\mathbb{Z}_q = \{0, 1, \ldots, q - 1\}$. We thus have $C_2 \subset C_1 \subset \mathbb{Z}_q^n$, where the rightmost term is called an *extended alphabet*.

8.3.1   Nested construction A

By lifting the nested codes above to $\mathbb{R}^n$, using construction A, we generate *nested construction A lattices* $\Lambda_1 = \{\mathbf{x} \in \mathbb{R}^n : \mathbf{x} \bmod q \in C_1\}$ and $\Lambda_2 = \{\mathbf{x} \in \mathbb{R}^n : \mathbf{x} \bmod q \in C_2\}$ (Section 2.5).

Example 8.3.1 (Coded BPSK) *Coded binary phase-shift keying (BPSK) consists of binary codewords mapped to ± 1 symbols. It amounts to a scaled and shifted modulo-2 lattice with cubical shaping. Specifically, let*

$$
\Lambda_1 = \{2\mathbf{x} : \mathbf{x} \bmod 2 \in C_1\}
$$

for some n-dimensional linear binary code C_1, and $\Lambda_2 = 4\mathbb{Z}^n$. Then, the dithered codebook

$$C = (\mathbf{u} + \Lambda_1) \cap \mathcal{V}_0(\Lambda_2),$$

with $\mathbf{u} = [-1, \ldots, -1]$, is a coded BPSK constellation.

Note that Λ_2 in this example can be written as $\{2\mathbf{x} : \mathbf{x} \bmod 2 = \mathbf{0}\}$; so C_2 in this case is a degenerate code, containing just the all-zero codeword. For a more general situation, consider the following example.

Example 8.3.2 (Construction A for $E_8 \subset D_8$) *D_8 and E_8 are two construction A lattices in $\mathbb{R}^8$. See Example 2.4.2 and Problem 2.12 (where D_8 is constructed from C_1 which is a single parity-check code in $\mathbb{Z}_2^8$), and Example 2.5.2 (where E_8 is constructed from C_2 which is the $(8, 4, 4)$ Hamming code). Since all vectors in C_2 have an even number of 1s, we have $C_2 \subset C_1$, so $E_8 \subset D_8$.*

Due to the nesting relation (8.23) between the underlying (q-ary) codes, the fine and coarse lattices satisfy a chain of nesting relations:

$$q\mathbb{Z}^n \subset \Lambda_2 \subset \Lambda_1 \subset \mathbb{Z}^n. \tag{8.26}$$

Similarly, the quotient sub-groups of relative cosets satisfy the chain

$$\mathbb{Z}^n / \Lambda_1 / \Lambda_2 / q\mathbb{Z}^n, \tag{8.27}$$

corresponding to the relation $\mathbb{Z}_q^n / C_1 / C_2 / \mathbf{0}$ of the generating linear codes. By (2.57) and Proposition 8.2.1, for a prime alphabet size p the codebook size (number of relative cosets) is p^{k_1}/p^{k_2}, so the nesting ratio (8.4) is given by

$$\Gamma = p^{(k_1 - k_2)/n}. \tag{8.28}$$

8.3.2   Syndrome dilution scheme

In light of the four-lattice nesting chain (8.26), the fine grid $\mathbb{Z}^n$ can be viewed as a "simple" unbounded constellation, which is *diluted* by C_1 (through the parity-check matrix H_1 and construction A) to obtain the fine lattice Λ_1, and further *diluted* by C_2 (through the additional checks of ΔH) to obtain the coarse lattice Λ_2. This viewpoint leads to a simple practical implementation.

The idea is to represent each point $\mathbf{z}$ in $\mathbb{Z}_q^n$ by its associated coarse lattice point and a syndrome, and then to dilute the set of possible syndromes by an (outer) linear code. Every syndrome thus corresponds to a coset of the coarse lattice relative to $\mathbb{Z}^n$; the zero syndrome (which is always included, by the linearity of the outer code) corresponds to the coarse lattice Λ_2 itself, while the union over all possible (diluted) syndromes gives the fine lattice Λ_1.

We explain the scheme for the binary alphabet ($q = 2$) case; the extension to any q should be straightforward. We start with the coarse lattice $\Lambda_2 = \{\mathbf{x} \in \mathbb{R}^n :$

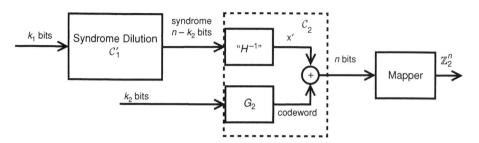

Figure 8.6 Nested construction A by syndrome dilution.

$\mathbf{x} \bmod 2 \in \mathcal{C}_2\}$, where $\mathcal{C}_2$ is defined by an $n \times k_2$ binary generator matrix G_2. Each integer vector $\mathbf{z}$ can be written as a sum of its even part $2 \cdot \lfloor \mathbf{z}/2 \rfloor$ and a reminder part, where the reminder part is defined by a pair of a k_2 bit vector (which together with the even part defines a coarse lattice point $\lambda \in \Lambda_2$), and an $n - k_2$ bit syndrome vector $\mathbf{s}$ (representing the coset of $\mathbf{z}$ relative to Λ_2).

To obtain the fine lattice $\Lambda_1 = \{\mathbf{x} \in \mathbb{R}^n : \mathbf{x} \bmod 2 \in \mathcal{C}_1\}$, the syndrome bits are diluted by an *outer code* $\mathcal{C}_1'$, which is an $(n - k_2, k_1)$ linear binary code (where $n - k_2 > k_1$). The fine lattice Λ_1 thus corresponds to a subset of 2^{k_1} syndromes of $\mathcal{C}_2$; hence there are 2^{k_1} cosets of Λ_2 relative to Λ_1 (out of the total of 2^{n-k_2} integer points in a fundamental cell of Λ_2). See Figure 8.6.

8.3.3   Mixed nesting dimensions

In many applications of nested lattices, one of the component lattices should be a good quantizer, while the other should be a good channel code. The target levels of goodness (and complexity) may, however, differ. Usually, the *equivalent dimension* of the channel coding component lattice needs to be much higher than that of the quantizer component. (We encountered a similar situation in separating shaping from coding in the VRDM system (Section 6.5.4)).

Consider, for example, the case where the fine lattice acts as a channel code, while the coarse lattice acts as a quantizer. In this case, the coarse lattice $\Lambda_{2,n}$ is a Cartesian product (3.40)

$$\Lambda_{2,n} = \underbrace{\Lambda_{2,k} \times \cdots \times \Lambda_{2,k}}_{m \text{ times}} \tag{8.29}$$

of some component lattice $\Lambda_{2,k}$, and it is a sublattice of a general (non-product) $n = mk$ dimensional fine lattice. (Cubical shaping amounts to the case $k = 1$.) Thus, while the coding complexity is determined by the full fine lattice dimension n as before, the shaping complexity (and gain) is determined by the (smaller) dimension k of the component lattice $\Lambda_{2,k}$.

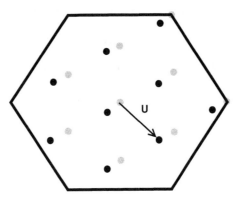

Figure 8.7 Dithered constellation.

Nesting of trellis codes The application of this idea to trellis codes is simple. As discussed in Section 2.5.2, an unbounded trellis code can be represented either in a real-valued filter representation (2.62a), or in a discrete-alphabet representation (2.62b). In the former, nesting can be obtained by letting the coarse filter be an integer scaling of the fine filter. In the latter, nesting is generated by picking two nested convolutional codes, similarly to nested linear codes (8.22) and (8.25), and lifting them to $\mathbb{R}^n$ via construction A.

8.4 Dithered codebook

Dither has already proven to be useful in randomizing an *unbounded* lattice code. Dither is also useful in the context of lattice shaping. In particular, randomizing the relative position of the nested lattice pair (Λ_1, Λ_2) enhances the symmetry of a lattice-shaped codebook, a symmetry that is broken by the arbitrary association of border points (points of Λ_1 which fall on the boundary of $\mathcal{P}_0(\Lambda_2)$).

Definition 8.4.1 (Dithered codebook) *For a given value of the dither vector $\mathbf{u}$ in $\mathbb{R}^n$, a* dithered codebook *consists of all shifted fine lattice points $\mathbf{u} + \lambda_1$, $\lambda_1 \in \Lambda_1$, inside some fundamental cell $\mathcal{P}_0$ of the coarse lattice Λ_2, i.e.,*

$$\mathcal{C}_{\mathbf{u}, \Lambda_1, \mathcal{P}_0} \stackrel{\Delta}{=} (\mathbf{u} + \Lambda_1) \cap \mathcal{P}_0 = (\mathbf{u} + \Lambda_1) \bmod \Lambda_2, \qquad (8.30)$$

where, as in the definition of a lattice-shaped (Voronoi) codebook (Definition 8.2.1), mod Λ_2 is understood as mod $\mathcal{P}_0$.

For $\mathbf{u} = 0$, this reduces to the earlier definition of a lattice codebook (Definition 8.2.1). See Figure 8.7.

Since a shifted fundamental cell $\tilde{\mathcal{P}}_0 = \mathcal{P}_0 - \mathbf{u}$ is a valid fundamental cell in Definition 2.2.1 (and since intersecting with a shift is the same as first shifting backward, then intersecting, then shifting forward), we can rewrite the dithered codebook (8.30) as a shifted non-dithered codebook:

$$\mathbf{u} + (\Lambda_1 \bmod \tilde{\mathcal{P}}_0).$$ (8.31)

It follows that the previous properties of *disjoint covering* (8.16) and codebook size (8.18) apply also to the dithered case, now with the fine lattice replaced by its coset $\Lambda_{1,\mathbf{u}} = \mathbf{u} + \Lambda_1$. As a corollary, we obtain the following.

Proposition 8.4.1 (Size invariant to dither) *The size of the dithered codebook* (8.30) *is invariant of the value of the dither vector* $\mathbf{u}$, *and is given by* $|\Lambda_1/\Lambda_2| = \det(J) = \Gamma^n$ (8.18), *where* Γ *is the nesting ratio* (8.4). *Hence, the coding rate* (6.1) *is given by*

$$R(\Lambda_1/\Lambda_2) = \frac{1}{n} \log |\text{codebook}| = \log \Gamma$$ (8.32)

bit per dimension. By (6.5), *this is the rate-per-unit-volume difference:* $R(\Lambda_1/\Lambda_2) = R_\infty(\Lambda_1) - R_\infty(\Lambda_2)$. *For construction A nesting, this is the difference between the rates of the component linear codes* $R(\Lambda_1/\Lambda_2) = R(\mathcal{C}_1) - R(\mathcal{C}_2)$, *which is upper bounded by the logarithm of the alphabet size.*

Dither distribution What should be the distribution of the dither vector $\mathbf{u}$ in (8.30)? In view of the use of dither in Chapters 4 through 6, there are two possible randomization schemes for nested lattices: the dither can be uniform over a fundamental cell of the fine lattice, or uniform over a fundamental cell of the coarse lattice. [5] As we shall see in the next section, the actual choice depends on which lattice determines the average cost: power in channel coding or distortion in source coding.

Another interesting observation is that we can use a *generalized dither*, i.e., a uniform distribution on *any* fundamental cell of the *coarse* lattice, not necessarily on the shaping region $\mathcal{P}_0$ (Section 4.2). This follows from the *nested partition* property discussed below.

8.4.1   Nested partition and generalized dither

Recall the parallelepiped codebook

$$\mathcal{C}_{\Lambda_1, \mathcal{P}_0} = \{G_1 \cdot [k_1, \dots, k_n]^t : \text{ for } 0 \le k_i \le j_i - 1\}$$

of Example 8.2.1, where the j_i are the elements of the diagonal nesting matrix. By dividing the ith edge of $\mathcal{P}_0$ into j_i equal intervals, for $i = 1, \dots, n$, the parallelepiped shaping region $\mathcal{P}_0$ is partitioned into $\prod_{i=1}^n j_i = \det(J)$ shifted copies of $\mathcal{P}_{0,1}$, the fine fundamental parallelepiped induced by the generator matrix G_1.

---

[5] When the nesting ratio is not too low, a discrete approximation of a continuous uniform dither over the coarse lattice cell (see Section 4.1.1) can be achieved using a dither which is uniform over the fine lattice points inside the codebook.

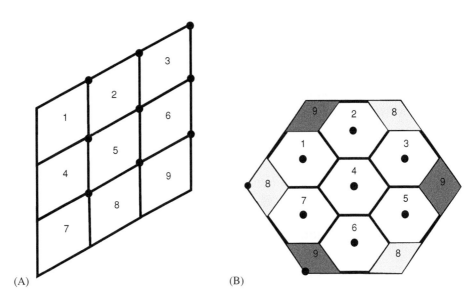

Figure 8.8 Nested partitions: (A) parallelepiped cells based on a diagonal form; (B) Voronoi cells.

As illustrated in Figure 8.8, this nice nested partition property holds, in fact, for *any* fundamental cells – not necessarily parallelepipeds – provided that each point is reduced modulo Λ_2.

Proposition 8.4.2 (Nested partition) *For any fundamental cells* $\mathcal{P}_{0,1} = \mathcal{P}_0(\Lambda_1)$ *and* $\mathcal{P}_0 = \mathcal{P}_0(\Lambda_2)$ *of the nested lattices* Λ_1 *and* Λ_2, *respectively,*

$$\mathcal{P}_0 = \bigcup_{\lambda \in \mathbb{C}_{\Lambda_1, \mathcal{P}_0}} [(\lambda + \mathcal{P}_{0,1}) \bmod_{\mathcal{P}_0} \Lambda_2], \tag{8.33}$$

where the union is disjoint.

Proof See Problem 8.1. $\square$

The nesting of fundamental cells (Proposition 8.4.2) implies a simple way to generate dither for the fine lattice.

Proposition 8.4.3 (Generalized dither) *For nested lattices* Λ_1 *and* Λ_2, *if* **u** *is uniform over a coarse fundamental cell* $\mathcal{P}_0$, *and* $\mathcal{P}_{0,1}$ *is a fine fundamental cell, then*

$$\mathbf{U} \bmod_{\mathcal{P}_{0,1}} \Lambda_1 \quad \text{is uniform over } \mathcal{P}_{0,1}; \tag{8.34}$$

i.e., **u** *is modulo-uniform with respect to* Λ_1. *Hence, by Lemma 4.2.1,* **u** *is a generalized dither for* Λ_1.

Proof It follows from (8.33) that every point **x** in $\mathcal{P}_{0,1}$ has exactly $\det(J)$ points **y** in $\mathcal{P}_0$ which satisfy **y** $\bmod_{\mathcal{P}_{0,1}} \Lambda_1 = \mathbf{x}$. This implies (8.34). $\square$

Example 8.4.1 (Dither for nested construction A) *Consider the nested construction A*

$$q\mathbb{Z}^n / \Lambda_2 / \Lambda_1 / \mathbb{Z}^n$$

described in Section 8.3. Since $q\mathbb{Z}^n$ is nested in both Λ_2 and Λ_1, it follows from Proposition 8.4.3 that

$$\mathbf{u} \sim \text{Unif(CUBE)}$$

is a generalized dither for both Λ_1 and Λ_2, where $\text{CUBE} = [0, q)^n$ is the cubic fundamental cell of the lattice $q\mathbb{Z}^n$.

8.5   Good nested lattices

The shaping and side-information coding schemes in the following chapters require good pairs of nested lattices, where one of the lattices (the fine one or the coarse one) is good for AWGN channel coding, while the other lattice is good for quantization under mean-squared distortion. The key to proving the existence of such lattices is to consider an appropriate *ensemble* of lattices.

Self-similar nested lattices (8.5), where the coarse lattice is an integer-scaled version of the fine lattice, can provide arbitrarily good nested lattice pairs for *integer nesting ratios* $\Gamma = 2, 3, \ldots$; this follows from the existence of lattices which are *simultaneously good* (Theorem 7.9.1). They cover the coding rates $R = \log(2)$, $\log(3), \log(4), \log(5), \ldots$ (see (8.32)), which are relatively dense for the high rates, but sparse for the low rates (say, in the range $0 < R < 2$ bit).

To see the difficulty of constructing good nested lattices at low rates, consider a nesting ratio between one and two, say $\Gamma \approx \sqrt{2}$, i.e., a rate of $R \approx 1/2$ bit per dimension. In the diagonal form (8.6), about half of the basis vectors of both lattices are the same, while half of the coarse basis vectors are twice the corresponding fine basis vectors. Intuitively, if one of the Voronoi cells tends to be "spherical," then the other cell tends to be "ellipsoidal" (although the actual shape depends also on the angles between the basis vectors).

In Chapter 7 we used the MHS ensemble as a "black box" to show that good lattices exist. Construction of good *nested* lattices, however, requires a more intimate knowledge of the ensemble structure. We develop two approaches below, both based on a random construction A (Sections 2.5 and 7.9).

8.5.1   Transformed construction A

The first approach is to apply a linear transformation T to a construction A lattice Λ_C; see Sections 2.4 and 2.5. A construction A lattice satisfies the nesting relation $q\mathbb{Z}^n \subset \Lambda_C$ (Proposition 2.5.1), with a nesting ratio $\Gamma = q^{k/n}$, where q and k are the alphabet size and the information dimension of the underlying code C, respectively. Thus, the transformed lattice satisfies

$$qT\mathbb{Z}^n \subset T\Lambda_C,$$

with the same nesting ratio, independent of T. Similarly to the $\mathbb{L}_{n,k,p,\gamma}$ (Loeliger) ensemble of Section 7.9.3, if the alphabet size is a *large prime* p, and the underlying code C is *random*, then, with an appropriate scaling $a = a(p, k, n)$, the points of $aT\Lambda_C$ tend to be uniform over the parallelepiped $aT[0, q)^n$. For a suitable choice of the transformation T (for example, the generator matrix of a good quantizer or channel code), we thus obtain a good nested lattice pair, though with a large nesting ratio $p^{k/n}$.

8.5.2   Random nested construction A

In the second approach, we extend the random construction A ensemble (Definition 7.9.2), to an ensemble $\mathbb{L}_{n,k_1,p,\Gamma} = \{\Lambda_2 \subset \Lambda_1\}$ of nested construction A lattices with an arbitrary fixed nesting ratio Γ; see Section 8.3. The Voronoi codebooks generated by these nested pairs have a fixed coding rate $R = \log \Gamma$.

Both fine and coarse components of each member in the ensemble are modulo-p lattices, given by

$$\Lambda_i = a \cdot \Lambda_{C_i} \quad i = 1, 2, \tag{8.35}$$

where $C_2 \subset C_1$ are nested p-ary linear codes. The linear codes C_2 and C_1 are generated by $n \times k_2$ and $n \times k_1$ matrices G_2 and G_1, respectively, which are related as $G_1 = [G_2 \mid \Delta G]$ (8.25). By (8.28), the nesting ratio Γ determines the coarse code information dimension, as

$$k_2 = k_1 - n \frac{\log \Gamma}{\log p} \tag{8.36}$$

which is assumed to be a positive integer. [6] The scaling coefficient a in (8.35) guarantees that the point density of the fine lattice is γ_1,

$$a = \frac{1}{\gamma_1^{1/n}} \cdot \frac{1}{p^{1-k_1/n}}; \tag{8.37}$$

see (7.90); hence, the point density of the coarse lattice is $\gamma_2 = \gamma_1/\Gamma^n$.

Like in the random construction A (Definition 7.9.2), the elements of G_1 are drawn independently and uniformly over $\mathbb{Z}_p$. Thus, by Theorem 7.9.1, for a suitable growth of p and k_1 with the dimension n (while keeping Γ fixed), both fine and coarse components of a random nested lattice pair in the ensemble $\mathbb{L}_{n,k_1,p,\Gamma}$ are simultaneously good for quantization and modulation in the presence of AWGN. [7]

---

[6] For large p and n, a slight modification of Γ will round k_2 to an integer.

[7] Note that if $k_1 = rn$ is linear in n, hence $\log(p)$ is logarithmic in n (see (7.96) and the discussion immediately after), then $\Delta k = k_1 - k_2$ grows *sub-linear* with n, implying also that $k_2 \approx rn$.

By Theorem 7.8.2, this statement is also true for a semi-spherical noise. We thus conclude the following.

Theorem 8.5.1 (Good nested lattices) *For any desired nesting ratio Γ, there exists a sequence of nested lattices $(\Lambda_1^{(n)}, \Lambda_2^{(n)})$ of increasing dimension n, such that each component lattice is a good quantizer (Definition 7.3.2) and a good constellation (Definition 7.7.1), i.e., $G(\Lambda_n^{(i)}) \to 1/2\pi e$, and $\mu(\Lambda_n^{(i)}, P_e) \to 2\pi e$ for all $P_e > 0$, as $n \to \infty$, for $i = 1, 2$. Moreover, $\mu_{\text{euclid}}(\Lambda_n^{(i)}, \mathbf{Z}_n, P_e) \to 2\pi e$ for any semi-spherical noise sequence $\mathbf{Z}_n$, for $i = 1, 2$ (Definition 7.8.3).*

8.5.3   Nesting with a fixed alphabet

In Section 7.9.5 we discussed the asymptotic near optimality of construction A lattice quantizers with a fixed alphabet, that is, a lattice sequence where the alphabet p is not growing with the dimension n. In particular, random modulo-2 lattice quantizers (i.e., $p = 2$) are nearly optimal, provided that the coding rate $R = k/n$ is in the neighborhood of $R \approx 0.4$ [146].

As for goodness for modulation in the presence of AWGN, the *error-floor* phenomenon of unbounded modulo-p lattices disappears in the case of bounded (lattice-shaped) codebooks. The reason is that, as we shall see in the next chapter, we do not count transitions inside the coarse cubic coset $\mathbf{c} + p\mathbb{Z}^n$ as error events. [8] Under this less restrictive decoding criterion, modulo-2 lattices are asymptotically nearly optimal also as AWGN channel codes. Thus, for rates up to about 0.4 bit per dimension, there exist asymptotically simultaneously good nested modulo-2 lattice codes.

Summary of Chapter 8

Nested lattices $\Lambda(G_2)$ is a sublattice of $\Lambda(G_1)$ if

$$G_2 = J \cdot G_1,$$

for some integer matrix J.

Lattice-shaped codebook For a shaping region equal to a fundamental cell $\mathcal{P}_0(\Lambda_2)$ of the coarse lattice,

$$\mathcal{C}_{\Lambda_1, \mathcal{P}_0(\Lambda_2)} = \Lambda_1 \cap \mathcal{P}_0(\Lambda_2) = \Lambda_1 \bmod {}_{\mathcal{P}_0}\Lambda_2.$$

Dithered lattice-shaped codebook

$$\mathcal{C}_{\Lambda_1, \mathcal{P}_0(\Lambda_2), \mathbf{u}} = [\mathbf{u} + \Lambda_1] \cap \mathcal{P}_0(\Lambda_2) = [\mathbf{u} + \Lambda_1] \bmod {}_{\mathcal{P}_0}\Lambda_2.$$

Voronoi codebook The case where the shaping region is $\mathcal{V}_0(\Lambda_2)$.

---

[8]   In Voronoi modulation, each message is represented by a whole coset. See Chapter 9.

> **Codebook size** For any shaping region $\mathcal{P}_0(\Lambda_2)$ and dither $\mathbf{u}$,
>
> $$|\mathcal{C}_{\Lambda_1,\mathcal{P}_0(\Lambda_2),\mathbf{u}}| = |\Lambda_1/\Lambda_2| = |\det(J)|.$$
>
> **Coding rate**
>
> $$R(\Lambda_1/\Lambda_2) = \frac{1}{n} \log_2 |\Lambda_1/\Lambda_2| = \log_2(\Gamma) \text{ bit per dimension,}$$
>
> where $\Gamma = |\Lambda_1/\Lambda_2|^{1/n}$ is the nesting ratio.
>
> **Good nested lattices** There exists a sequence of nested lattice pairs $\{\Lambda_2^{(n)} \subset \Lambda_1^{(n)}\}$ of a fixed nesting ratio Γ, such that $\Lambda_2^{(n)}$ and $\Lambda_1^{(n)}$ are asymptotically good for both quantization and channel coding.

Problems

P.8.1 (Nested partition; proof of Proposition 8.4.2) Show that for any fundamental cells $\mathcal{P}_{0,1}$ and $\mathcal{P}_0$ of the nested lattices Λ_1 and Λ_2, respectively,

$$\mathcal{P}_0 = \dot{\bigcup_{\lambda \in \mathbb{C}_{\Lambda_1.\mathcal{P}_0}}} \left[(\lambda + \mathcal{P}_{0,1}) \bmod_{\mathcal{P}_0} \Lambda_2 \right]$$

and that the union is *disjoint*. Guidance: the proof is based on two observations. (1) Even without a modulo-Λ_2 reduction, the union above is a valid fundamental cell of Λ_2. (2) Reducing one fundamental cell modulo another fundamental cell is a one-to-one mapping.

P.8.2 (Nested Voronoi cells) Show that if Λ_2 is a sublattice of Λ_1, then their Euclidean fundamental Voronoi cells $\mathcal{V}_{0,2}$ and $\mathcal{V}_{0,1}$ satisfy

$$\mathcal{V}_{0,1} \subset \mathcal{V}_{0,2}$$

for some systematic association of border points. Conclude that for a nested construction A with alphabet size q:

$$\text{CUBE} \subset \mathcal{V}_{0,1} \subset \mathcal{V}_{0,2} \subset q\text{CUBE}, \tag{8.38}$$

where $\text{CUBE} = [-1/2, 1/2)^n$ is the fundamental Voronoi cell of $\mathbb{Z}^n$, and $q\text{CUBE} = [-q/2, q/2)^n$ is the fundamental Voronoi cell of $q\mathbb{Z}^n$.

P.8.3 (Properties of coset representatives set) Show that a lattice-shaped codebook $\mathbb{C}_{\Lambda_1,\mathcal{P}_0}$ is a minimal subset S of Λ_1 satisfying (8.11), and a *maximal* subset S of Λ_1 such that for all λ_1, λ_1' in S the difference vector $\lambda_1 - \lambda_1'$ is not in Λ_2. Vice versa, any minimal/maximal subset of Λ_1 satisfying these properties is a lattice-shaped codebook with respect to some fundamental cell of Λ_2. Guidance: use the same reasoning as in Lemmas 2.3.2 and 2.3.3.

Historical notes

Quotient groups and chains of nested linear codes are very common in the coding literature [23]. In coded modulation, Ungerboeck's set-partition scheme [259] generates a simple nesting of an uncoded (cubic) constellation in a coded one. Multi-level codes extend this idea to higher levels of nesting in the Euclidean space; see Imai and Hirakawa [123], Forney [83] and Calderbank [29]. In the lattice framework, construction D generates a chain of nested lattices; see [49] and [84].

Voronoi codes were introduced by Conway and Sloane [47]; they showed that a lattice codebook can be shaped by the Voronoi region of an integer multiple of the same lattice (i.e., a self-similar sublattice). Forney [85] extended this idea to general nesting configurations, and analyzed the performance in the high rate (high SNR) regime; see also Forney and Wei [94]. Erez and Zamir [71] introduced randomized (dithered) Voronoi codes for general rates and SNR.

The existence of good construction A lattices (cubic coarse, general fine) follows from Loeliger's ensemble [175]. The existence of good pairs of nested lattice codes (for the Wyner–Ziv problem) is addressed by Zamir and Shamai [292]; see also Servetto [235]. Erez and Zamir [71] proposed the idea of generating good nested lattices (for dirty-paper coding) by transforming a random construction A; see also [66, 293], and Krithivasan and Pradhan [142]. In Section 8.5 we follow Ordentlich and Erez [206], who gave a direct proof for the existence of good pairs (under both goodness measures), based on first nesting linear codes over a prime alphabet and then lifting them to the Euclidean space.

9 Lattice shaping

As information theory shows, Gaussian sources and channels should be encoded using "Gaussian codebooks." A Gaussian variable maximizes the entropy for a given second moment. For source coding, this implies that a Gaussian codebook optimizes the volume-distortion and volume-overload trade-offs; for channel coding, it optimizes the volume-error and volume-power trade-offs. A Gaussian codebook has a Gaussian – or equivalently, *spherical* – shape, with roughly evenly spaced points as codewords. Can a lattice code replace a Gaussian codebook?

In variable-rate (entropy-coded) quantization (Chapter 5) and in non-uniform signaling (Chapter 6), the codebook is the whole (unbounded) lattice, and is *not truncated* to fit the source variance or the transmission power constraint. The lack of shaping is compensated for by variable-rate coding, which amounts to probabilistic ("soft") shaping: remote lattice points are rarely used so their effect on the *average* coding rate and power is negligible. Fixed-rate (or pick-amplitude constrained) coding, however, requires "hard" shaping, i.e., a *bounded* codebook.

In this chapter we examine the performance of a codebook (or constellation) whose codewords and shaping region both have a lattice structure, as described in Chapter 8. We saw earlier in Chapter 7 that for a large lattice dimension, the fundamental Voronoi region of a good lattice can *approximate a ball*; or equivalently, a uniform distribution on this region can approximate a white-Gaussian distribution. A Voronoi codebook can thus serve as a structured alternative to the Gaussian codebooks of information theory.

Preview

Let us first describe informally how to design nested lattices for Voronoi quantization (source coding) and modulation (channel coding); see Figure 9.1.

Voronoi quantization As for an unbounded quantizer, the fine lattice has to be a "good quantizer"; it should provide a good compromise between lattice point density and average MSE, meaning a small NSM (3.21). To compress the source at the lowest rate, the coarse lattice, on the other hand, should *minimize* the cell volume $V(\Lambda_2)$ while keeping the *overload probability* (the probability a source vector falls

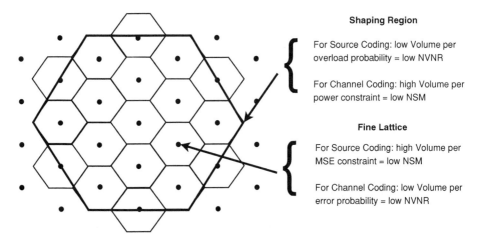

Figure 9.1 Voronoi codes for source and channel coding.

outside the shaping region) low. The coarse lattice should thus act like a "good channel code," meaning a small NVNR (3.37). Translating cell volumes into second moments, the resulting coding rate (8.32) can be written roughly as

$$R(\Lambda_1/\Lambda_2) = \frac{1}{n}\log\left(\frac{V(\Lambda_2)}{V(\Lambda_1)}\right) \approx \frac{1}{2}\log\left(\frac{\text{source variance}}{\text{quadratic distortion}}\right) \qquad (9.1)$$

bit per sample, up to a loss term depending on the NSM of the fine lattice and the NVNR of the coarse lattice.

Voronoi modulation In the channel coding counterpart, the roles of the two lattices are reversed. The fine lattice now has to be a "good channel code," i.e., dense yet noise resistant, as in the case of an infinite constellation. Whereas for maximizing the coding rate, the coarse lattice should *maximize* the cell volume $V(\Lambda_2)$ while keeping the average transmit power constraint. Thus, the coarse lattice now acts as a "good quantizer," and the resulting coding rate is roughly

$$R(\Lambda_1/\Lambda_2) = \frac{1}{n}\log\left(\frac{V(\Lambda_2)}{V(\Lambda_1)}\right) \approx \frac{1}{2}\log\left(\frac{\text{transmit power}}{\text{noise variance}}\right) \qquad (9.2)$$

bit per channel use, up to a loss term depending on the NVNR of the fine lattice and the NSM of the coarse lattice.

This characterization is limited, however, because it relies on high SNR approximations. It also does not tell us much about the advantage of the lattice structure, and how it affects the decoding complexity. Sections 9.1–9.3 describe lattice encoding and decoding in detail, and analyze the gap to capacity in the *high SNR* regime.

In the *low SNR* regime, the noise plays a dominant role, and "Gaussian" shaping is, in fact, not necessary to approach the optimum performance. The *medium* range of SNR is still interesting to explore, and we study the effect of three elements.

- **Dither,** which symmetrizes the error probability, power and distortion, and simplifies the analysis.
- **Lattice encoding and decoding,** which decouple coding and shaping, hence reduce the complexity of quantization and demodulation.
- **Wiener estimation,** which enhances the system performance within the lattice encoding and decoding framework.

Sections 9.4–9.6 reveal the role of these three elements in approaching Shannon's capacity when the SNR is not necessarily high. Section 9.10 shows how, with the same approach, Voronoi quantization can approach the quadratic-Gaussian rate-distortion function for any resolution.

9.1 Voronoi modulation

Recall the AWGN channel $\mathbf{Y} = \mathbf{X} + \mathbf{Z}$, (6.6), under the power constraint $\frac{1}{n} E\{\|\mathbf{X}\|^2\} \leq P$, (6.16). A Voronoi modulation system is based on a pair of nested lattices $\Lambda_2 \subset \Lambda_1$, where the coarse lattice Λ_2 is used for *shaping* (under the power constraint P) while the fine lattice Λ_1 is used for *coding* (against the noise $\mathbf{Z}$).

9.1.1 Encoding

The encoding part of the system consists of a mapping from messages to representation vectors, dithering, and shaping, as illustrated in Figure 9.2.

1. **Message representation** Each message $m = 1, \ldots, M$ amounts to a relative coset in the quotient group (8.9) Λ_1/Λ_2. Message m is represented by its coset representative $\mathbf{v}_m$ in some (arbitrary) fundamental cell $\mathcal{P}_0$ of Λ_2, i.e., in the lattice codebook

$$\mathbb{C}_{\Lambda_1, \mathcal{P}_0} = \Lambda_1 \cap \mathcal{P}_0 = \{\mathbf{v}_1, \ldots, \mathbf{v}_M\} \qquad (9.3)$$

called the *enumeration codebook*. A good choice for $\mathcal{P}_0$ – which simplifies the mapping – is a *parallelepiped* fundamental cell (Example 8.2.1). The coding rate R is thus given by (8.32),

$$R = \frac{1}{n} \log M = \frac{1}{n} \log |\Lambda_1/\Lambda_2| = \log \Gamma. \qquad (9.4)$$

2. **Dithering** The encoder then adds a vector $\mathbf{u}$ to the message representative vector $\mathbf{v}_m$. We shall assume that $\mathbf{u}$ is a realization of a random dither $\mathbf{U}$, which is uniform over $\mathcal{P}_0$ (or a generalized dither as in Section 4.2 and Proposition 8.4.3), independent of the transmitted message $\mathbf{v}_m$, and known to both the transmitter (encoder) and receiver (decoder).

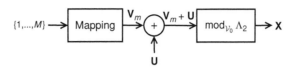

Figure 9.2 Voronoi modulation: mapping to coset representative, dithering, and shaping.

3. **Shaping** Finally, the encoder reduces the sum $\mathbf{v}_m + \mathbf{u}$ modulo the coarse fundamental (Euclidean) *Voronoi* cell $\mathcal{V}_0 = \mathcal{V}_0(\Lambda_2)$, so the channel input is the *least energy* vector in the coarse coset $\mathbf{v}_m + \mathbf{u} + \Lambda_2$:

$$\mathbf{x}(\mathbf{v}_m, \mathbf{u}) = (\mathbf{v}_m + \mathbf{u}) \bmod_{\mathcal{V}_0} \Lambda_2 \tag{9.5a}$$

$$= \underset{\mathbf{x} \in \mathbf{v}_m + \mathbf{u} + \Lambda_2}{\arg\min} \|\mathbf{x}\|. \tag{9.5b}$$

The set of possible output vectors $\mathbf{x}(\mathbf{v}_1, \mathbf{u}), \ldots, \mathbf{x}(\mathbf{v}_M, \mathbf{u})$ is thus a dithered Voronoi codebook (Definition 8.4.1). As discussed in Corollary 8.2.3, the choice of $\mathcal{V}_0$ in (9.5a) minimizes the average second moment – the *transmitted power* – of the dithered codebook over all possible shaping regions in (9.5a).

For a random message W, the encoder output $\mathbf{X} = \mathbf{x}(\mathbf{v}_W, \mathbf{u})$ becomes random. The actual distribution of $\mathbf{X}$, and in particular its power, depends on the chance of selecting each message and on the value of the dither $\mathbf{u}$. Nevertheless, for a *random* dither $\mathbf{U}$, the "crypto" lemma guarantees a strong property of the encoder output (4.10): $\mathbf{X}$ is uniformly distributed over $\mathcal{V}_0$ for *any* message $W = m$. In particular, the output power is equal to the second moment of the coarse lattice.

Proposition 9.1.1 (Lattice codebook power) *For a uniform (or modulo-uniform) dither with respect to Λ_2 (Lemma 4.2.1) and any message $W = m$, the channel input power (averaged over the dither $\mathbf{U}$) is equal to the second moment of the coarse lattice:*

$$E\{\|\mathbf{x}(\mathbf{v}_m, \mathbf{U})\|^2\} = \sigma^2(\Lambda_2). \tag{9.6}$$

Thus, the average transmit power $E\|\mathbf{X}\|^2$ is $P = \sigma^2(\Lambda_2)$, independent of the distribution of the message W.

Another benefit of the uniform distribution of the dithered codeword, is that it *symmetrizes* the mapping from messages to codewords. In particular, the *decoding error probability* (averaged over the dither) is *identical* for all the messages.

9.1.2   Decoding

We turn to the decoding of the message $W = m$ from the received vector $\mathbf{Y} = \mathbf{x}(\mathbf{v}_m, \mathbf{u}) + \mathbf{Z}$, given a (specific) dither vector $\mathbf{U} = \mathbf{u}$. We shall consider three possible decoders: a maximum likelihood decoder, a lattice decoder (combined with an

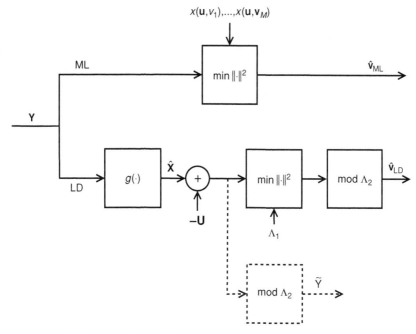

Figure 9.3 Voronoi demodulation: estimation, undithering, fine-lattice decoding and mapping to coset. The upper branch shows ML decoding, for comparison. The dashed line shows the generation of the decision vector $\hat{\mathbf{Y}}$ see (9.11f).

estimator), and a coset decoder. Figure 9.3 shows a block diagram of the first two decoders.

Maximum likelihood (ML) decoding

The ML decoder minimizes the average error probability

$$P_e = \frac{1}{M} \sum_{m=1}^{M} \Pr\{\hat{\mathbf{v}} \neq \mathbf{v}_m \mid \text{message } m \text{ was transmitted}\} \tag{9.7}$$

by maximizing the probability of the noise $P_Z(\mathbf{Y} - \mathbf{x})$ over all input vectors $\mathbf{x}$ in (9.5a); see the discussion near (6.7). For AWGN, this amounts to the dithered lattice point *closest* to $\mathbf{y}$:

$$\hat{\mathbf{v}}_{\text{ML}} = \arg \min_{\mathbf{v}_1,\ldots,\mathbf{v}_M} \|\mathbf{Y} - \mathbf{x}(\mathbf{v}_m, \mathbf{u})\| . \tag{9.8}$$

The ML decoder is *optimal*. Under ML decoding, the role of the (deterministic) dither $\mathbf{u}$ is merely to minimize the average transmit power $P(\mathbf{u}) = 1/M \sum_{m=1}^{M} \|\mathbf{x}(\mathbf{v}_m, \mathbf{u})\|^2$, while its effect on the error probability is small. [1]

---

[1] For coded BPSK, for example, the ML decoding error probability is *invariant* of the dither value $\mathbf{u}$, while the power is minimized by $\mathbf{u} = [1, \ldots, 1]$ (a symmetric constellation); see Example 8.3.1.

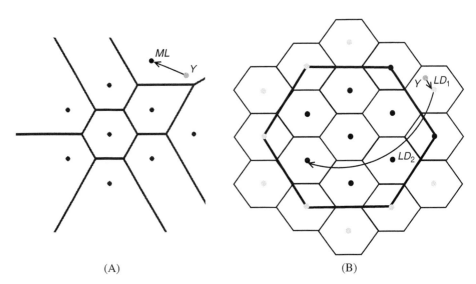

(A)   (B)

Figure 9.4 ML decoding versus lattice decoding. The former projects received vectors outside the shaping region into the codebook. The latter first quantizes the received vector to the nearest fine lattice point (step LD_1), and then reduces the result modulo Λ_2 (step LD_2).

ML decoding may, however, be *complex*. It suffers from the fact that the minimization in (9.8) must take into account the shaping region $\mathcal{V}_0 = \mathcal{V}_0(\Lambda_2)$. Thus, unless Λ_2 is a simple cubic lattice (where the shaping constraints factor across dimensions), ML decoding of a Voronoi constellation is complicated by the *joint* structure of the two nested lattices.

Another weakness of ML decoding is that the error probability and the search complexity (for a given dither value) *vary* among the codewords. Codewords near the edge of the shaping region have fewer neighbors compared to the "inner" codewords, hence their decoding error probability and search complexity are usually smaller.

Estimation and lattice decoding

Lattice decoding is a simpler approach, which ignores the "complicated" shaping region $\mathcal{V}_0(\Lambda_2)$. In this approach, after removing the dither, the output is quantized to the fine lattice Λ_1 (for AWGN the quantization is to the nearest lattice point) and the associated relative coset is identified. Figure 9.4 illustrates this operation.

Since this is, in general, a suboptimal decoder, we can enhance its performance by pre-processing the channel output – with an appropriate *estimator* – prior to decoding.

Definition 9.1.1 (Euclidean lattice decoder) *Estimation and Euclidean lattice decoding of the message* $\mathbf{v}_m$ *is defined by*

$$\hat{\mathbf{v}}_{\mathrm{LD}} = \left(\arg\min_{\lambda \in \Lambda_1} \| \hat{\mathbf{X}} - \mathbf{u} - \lambda \| \right) \bmod \Lambda_2, \tag{9.9a}$$

where the estimator

$$\hat{\mathbf{X}} = g(\mathbf{Y}) \tag{9.9b}$$

is some function of the channel output $\mathbf{Y}$*, and where the modulo-*Λ_2 *operation maps the minimizing* $\lambda \in \Lambda_1$ *to its coset representative in the enumeration codebook* (9.3).[2]

In contrast to the ML decoder (9.8), the *complexity* of the lattice decoder (9.9a) is *identical* for all messages, and is independent of the dither value. Furthermore, the complexity of the entire system is now more *balanced* between the encoder and the decoder: while the encoder performs an NN search with respect to the coarse lattice (in the Voronoi shaping stage (9.5a)), the decoder performs an NN search to find the nearest fine lattice point, but only a "simple" modulo-Λ_2 operation to identify the relative coset.

We shall discuss later the choice of the estimator function $g(\cdot)$ and the role of dither *randomness* at the decoder.

Coset decoding

The decoding rule (9.9a) maps all points $\mathbf{v} + \lambda'$ in the relative coset $\mathbf{v} + \Lambda_2$ to the same message $\mathbf{v}$. It can thus be improved by searching for the most likely *relative coset*, i.e., the coset $\lambda + \Lambda_2$ which maximizes the *sum* (over $\lambda' \in \Lambda_2$) of the likelihoods of its points $(\lambda + \lambda')$. For $\hat{\mathbf{X}} = \mathbf{Y}$, and AWGN of variance N, the likelihood of a point is inverse proportional to the exponent of its square distance from the received vector normalized by $2N$. Hence, the *coset decoder* is defined as

$$\hat{\mathbf{v}}_{\mathrm{CD}} = \left(\arg\max_{\lambda \in \mathbb{C}_{\Lambda_1, \mathcal{P}_0}} \sum_{\lambda' \in \Lambda_2} \exp\left\{ -\frac{\|\hat{\mathbf{X}} - \mathbf{u} - \lambda - \lambda'\|^2}{2N} \right\} \right) \bmod \Lambda_2, \tag{9.10}$$

where again the $\bmod \Lambda_2$ operation is with respect to the enumeration codebook $\mathbb{C}_{\Lambda_1, \mathcal{P}_0}$ (9.3). Like the lattice decoder, this decoder ignores the "complicated" shaping region $\mathcal{V}_0$.

The maximization in (9.10) is equivalent to ML decoding (6.11) of the fine lattice Λ_1, with respect to a *folded* (modulo-Λ_2) AWGN distribution. Hence, it can be rewritten as a noise-matched lattice decoder $Q_{\Lambda_1}(\cdot)$.

The coset decoder (9.10) is better than the lattice decoder, and slightly more complicated (due to the summation in (9.10)). Nevertheless, as we shall see, its advantage is negligible for high SNR or a large lattice dimension.

---

[2] The subtraction of $\mathbf{u}$ prior to quantizing to Λ_1 amounts to decoding to the shifted lattice constellation $\mathbf{u} + \Lambda_1$. The decoded fine lattice point λ may, however, be outside the codebook (even if the enumeration and the shaping cells are identical), hence the mod Λ_2 operation is necessary.

9.1.3   Full system: pre-emphasis, decision vector and modulo loss

We can summarize the lattice encoding and decoding scheme (9.5), (9.9), for a given message representative $\mathbf{v}_m$ and dither vector $\mathbf{u}$, by

$$\textbf{Encoding} \quad \mathbf{x} = (\mathbf{v}_m + \mathbf{u}) \bmod_{\mathcal{V}_0} \Lambda_2 \qquad (9.11\text{a})$$

$$\textbf{Decoding} \quad \hat{\mathbf{X}} = g(\mathbf{Y}) \qquad (9.11\text{b})$$

$$\hat{\mathbf{v}} = Q_{\Lambda_1}(\hat{\mathbf{X}} - \mathbf{u}) \bmod \Lambda_2, \qquad (9.11\text{c})$$

where we make the following remarks.

1. **Modulo operations** The encoder modulo operation (9.11a) is taken with respect to the (Euclidean) Voronoi cell $\mathcal{V}_0(\Lambda_2)$, while the decoder modulo operation (9.11c) can be taken with respect to *any* "enumeration" cell $\mathcal{P}_0(\Lambda_2)$.
2. **Lattice decoding cells** The fine lattice quantizer $Q_{\Lambda_1}(\cdot)$ may correspond to a nearest-neighbor search as in (9.9a), a coset decoder as in (9.10), or a *noise-matched* lattice decoder (6.11), matched to the true distribution of the estimation error $\hat{\mathbf{X}} - \mathbf{X}$ modulo the coarse lattice.
3. **Pre-emphasis** The encoding stage can be modified to

$$\mathbf{x} = f([\mathbf{v}_m + \mathbf{u}] \bmod_{\mathcal{V}_0} \Lambda_2); \qquad (9.11\text{d})$$

i.e., a non-linear mapping $f(\cdot)$ is applied before the transmission. The role of such a *pre-emphasis* mapping is to match the input $\mathbf{x}$ to a *non*-additive-noise channel; see Problem 9.4 and footnote 11.
4. **Decision vector** For future analysis, it is useful to rewrite the lattice decoder (9.11c) in terms of a *decision vector* $\tilde{\mathbf{Y}}$ (see Problem 9.1):

$$\hat{\mathbf{v}} = Q_{\Lambda_1}(\tilde{\mathbf{Y}}) \bmod \Lambda_2, \qquad (9.11\text{e})$$

where

$$\tilde{\mathbf{Y}} = [\hat{\mathbf{X}} - \mathbf{u}] \bmod \Lambda_2. \qquad (9.11\text{f})$$

We see that in effect, both the lattice decoder (9.9a) and coset decoder (9.10) amount to first "folding" the estimated vector into the coarse fundamental cell – the mapping from $\mathbf{Y}$ to $\tilde{\mathbf{Y}}$ – and only then making a decision. Since the mapping from $\mathbf{Y}$ to $\tilde{\mathbf{Y}}$ is many-to-one, it incurs some loss of information. We shall discuss this *modulo loss* of lattice decoding later.

9.2   Syndrome dilution scheme

The simplest Voronoi constellation has a cubic shape, for example, $\Lambda_2 = 4\mathbb{Z}^n$ for coded BPSK (Example 8.3.1). In terms of construction A, this constellation can be defined easily using an $n \times k$ generator matrix of some underlying linear code: $\{\mathbf{x} = G\mathbf{w} : \mathbf{w} \in \mathbb{Z}_2^k\}$, see (2.54). To define more efficient (and complex) shaping

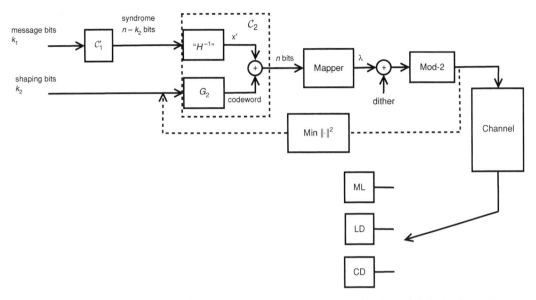

Figure 9.5 Voronoi modulation based on a nested construction A, with ML, lattice and coset decoders. The k_1 input bits carry the message W. The $n - k_2$ redundant bits define a syndrome of C_2. Shaping amounts to finding k_2 "control" bits, that select the least-energy transmit vector $\lambda + \mathbf{u}$ in the coset corresponding to the message syndrome.

regions, however, we shall use the syndrome representation of some underlying shaping linear code.

Recall the nesting of construction A lattices by *syndrome dilution* (Section 8.3). Each integer point $\mathbf{z}$ in $\mathbb{Z}^n$ is defined by three elements: its even part $\lfloor \mathbf{z}/2 \rfloor$, and a k_2 bit vector associated with a codeword in C_2 (defining together a coarse lattice point $\lambda \in \Lambda_2$), and an $n - k_2$ bit syndrome vector $\mathbf{s}$ (representing the coset of $\mathbf{z}$ relative to Λ_2). The $n - k_2$ syndrome bits are diluted by a $(n - k_2, k_1)$ linear code C_1' (where $n - k_2 > k_1$), which defines the fine lattice Λ_1, so that $2\mathbb{Z}^n \subset \Lambda_2 \subset \Lambda_1 \subset \mathbb{Z}^n$.

As illustrated in Figure 9.5, the Voronoi modulator sends information over the k_1 information bits of C_1'. These bits select a fine lattice point $\mathbf{x}' = f(b^{k_1})$, inside some enumeration cell of Λ_2. Shaping of the dithered vector $\mathbf{x}' + \mathbf{u}$ is done by finding the *leader* (i.e., the least-energy vector) of the coset $\Lambda_2 + \mathbf{x}' + \mathbf{u}$; this leader is represented by the k_2 bits of the corresponding coarse codeword in C_2.

The process above defines a set of 2^{k_1} possible transmitted vectors $\mathbf{x}$ (depending on the dither $\mathbf{u}$) that lies inside the fundamental Voronoi region of Λ_2. An ML decoder (9.8) runs over all elements in this constrained set, and finds the nearest vector to the received $\mathbf{Y}$. A lattice decoder (9.9a) runs over all pairs of k_2 bits (defining a coarse lattice point) and k_1 bits (defining a valid syndrome vector), and finds the nearest *fine* lattice point to $\mathbf{Y}$. A coset decoder (9.10) runs over all k_1 bits, and for each associated syndrome it sums the likelihoods of its coset (by running over all vectors of k_2 bits).

In all cases, the information is decoded as the k_1 information bits of the selected point or coset.

Example 9.2.1 (Trellis shaping [87]) *A well-known practical coding and shaping technique, due to Forney, uses trellis codes in place of the two linear codes C_1' and C_2. The encoder then uses a Viterbi search algorithm [82] to find the least-energy sequence within the message coset.*

Note that unlike lattice decoding, coset decoding can provide "soft information," which is essential for iterative decoding schemes; see [69]. Its advantage becomes more transparent for a modulo-2 (or any finite alphabet) lattice, because the distances between points in the cubic (sub-)cosets are small.

9.3   The high SNR case

The case of a trivial estimator in (9.9b)

$$\hat{\mathbf{X}} = \mathbf{Y} \tag{9.12}$$

is simple to analyze. We shall see that under high SNR conditions, the corresponding decision vector (9.11f) is a *sufficient statistic* for optimal decoding (Section 9.3.4). Moreover, for a sequence of good nested lattices, a trivial estimator followed by a lattice decoder achieves the high SNR capacity of the AWGN channel (Section 9.3.5), as well as of more general additive-noise channels (Section 9.3.6).

9.3.1   Lattice decoding with no estimation

When $\hat{\mathbf{X}} = \mathbf{Y}$, the decision vector (9.11f) becomes

$$\tilde{\mathbf{Y}} = [\mathbf{Y} - \mathbf{u}] \bmod \Lambda_2. \tag{9.13}$$

The mapping from the message representative $\mathbf{v}_m$ to $\tilde{\mathbf{Y}}$ then obeys the following simple relation.

Lemma 9.3.1 (Equivalent mod Λ channel: non-estimation case) *For an additive-noise channel $\mathbf{Y} = \mathbf{x} + \mathbf{Z}$, the channel from the message $\mathbf{v}_m \in \mathbb{C}_{\Lambda_1, \mathcal{P}_0}$ to the decision vector $\tilde{\mathbf{Y}}$ in (9.13) is given by*

$$\tilde{\mathbf{Y}} = [\mathbf{v}_m + \mathbf{Z}] \bmod \Lambda_2, \tag{9.14}$$

where mod Λ_2 is with respect to $\mathcal{P}_0$, as in (9.13).

Note that the equivalent channel (9.14) is independent of the dither vector $\mathbf{u}$.

Proof By the distributive law of the modulo operation (2.22b) (see also (8.21)),

$$\tilde{\mathbf{Y}} = [(\mathbf{v}_m + \mathbf{u}) \bmod \Lambda_2 + \mathbf{Z} - \mathbf{u}] \bmod \Lambda_2 \tag{9.15}$$

$$= [\mathbf{v}_m + \mathbf{u} + \mathbf{Z} - \mathbf{u}] \bmod \Lambda_2 \tag{9.16}$$

$$= [\mathbf{v}_m + \mathbf{Z}] \bmod \Lambda_2. \tag{9.17}$$

$$\square$$

Since the relation between the message representative $\mathbf{v}_m$ and the decision vector $\tilde{\mathbf{Y}}$ is *independent* of the dither vector, no further loss of information is incurred (beyond that already incurred by the modulo operation in (9.13)) if the decoder ignores the dither value when decoding $\mathbf{v}_m$ from $\tilde{\mathbf{Y}}$.

The following theorem makes precise the relation in (9.2) between the coding rate $R(\Lambda_1/\Lambda_2)$ (equivalently, nesting ratio Γ) and SNR $= P/N$, by taking into account the NSM and NVNR of the two lattices.

Theorem 9.3.1 (Minimum required SNR) *Assume Voronoi modulation with a uniform (or modulo-uniform) dither with respect to Λ_2, and lattice decoding with the trivial estimator (9.12), in the presence of AWGN of variance N. Then, a lattice decoding error probability $P_e = \Pr\{\hat{\mathbf{v}} \neq \mathbf{v}_m\}$ is achievable with a signal to noise ratio*

$$\text{SNR} = G(\Lambda_2) \cdot \mu(\Lambda_1, P_e) \cdot \Gamma^2 \tag{9.18}$$

or higher, where $G(\Lambda_2)$ and $\mu(\Lambda_1, P_e)$ are the NSM (3.21) and the NVNR (3.37) of the coarse and fine lattices, respectively, and $\Gamma = |\Lambda_1/\Lambda_2|^{1/n}$ is their nesting ratio (8.4). This condition becomes necessary in the limit of a large Γ (equivalently, large SNR or small P_e).

The only role of random dither in this theorem is to ensure that the transmission power is $P = \sigma^2(\Lambda_2)$. This effect becomes insignificant in the limit of high SNR, in which case (9.18) holds for *any* value of the dither; see the discussion in Section 9.9.

Proof In view of the equivalent mod Λ channel of Lemma 9.3.1, the lattice decoder (9.11e) is correct whenever the noise vector $\mathbf{Z}$ falls in the Voronoi cell $\mathcal{V}_0(\Lambda_1)$ of the fine lattice, or in one of its shifts by the coarse lattice Λ_2.[3] Ignoring the latter cases, the lattice decoding error probability is thus upper bounded by

$$P_e = \Pr\{\mathbf{Z} \notin \mathcal{V}_0(\Lambda_1) \bmod \Lambda_2\} \tag{9.19a}$$

$$\leq \Pr\{\mathbf{Z} \notin \mathcal{V}_0(\Lambda_1)\} \tag{9.19b}$$

$$= P_e(\Lambda_1, N) \tag{9.19c}$$

$$\overset{\triangle}{=} \tilde{P}_e \tag{9.19d}$$

(Definition 3.3.2), where the notation $\{\mathbf{a} \in \mathcal{B} \bmod \Lambda\}$ is short for $\{[\mathbf{a} \bmod \Lambda] \in [\mathcal{B} \bmod \Lambda]\}$. By the definition of the NVNR, the cell volume of the fine lattice is

---

[3] One can think of $\mathcal{V}_0(\Lambda_1) + \Lambda_2$ as the decoding region of the "zero relative coset" in Λ_1/Λ_2.

$V(\Lambda_1) = [\mu(\Lambda_1, \tilde{P}_e) \cdot N]^{n/2}$ (or larger). By the definition of the NSM, and using the second-moment to power-constraint relation (9.6), the volume of the coarse cell is $V(\Lambda_2) = [P/G(\Lambda_2)]^{n/2}$ (or smaller). Thus, at the optimal operation point

$$\Gamma^2 = \left(\frac{V(\Lambda_2)}{V(\Lambda_1)}\right)^{2/n} \tag{9.19e}$$

$$= \frac{P/G(\Lambda_2)}{\mu(\Lambda_1, \tilde{P}_e) \cdot N} \tag{9.19f}$$

$$\geq \frac{\text{SNR}}{G(\Lambda_2) \cdot \mu(\Lambda_1, P_e)}, \tag{9.19g}$$

where the last inequality follows by setting $P/N = \text{SNR}$, and because $\tilde{P}_e \geq P_e$ from (9.19) and $\mu(\Lambda_1, P_e)$ is monotonically decreasing with P_e. Thus, the SNR in the right-hand side of (9.18) is sufficient to achieve an error probability P_e. The "only if" part under high SNR conditions is left as an exercise (Problem 9.2). □

Recall from (8.32) that the coding rate $R(\Lambda_1/\Lambda_2)$ is equal to $\log \Gamma = \frac{1}{n} \log |\Lambda_2/\Lambda_1|$. Substituting in (9.18), we see that at the optimum operation point, the SNR, error probability P_e and coding rate $R(\Lambda_1/\Lambda_2)$, are related as

$$R(\Lambda_1/\Lambda_2) \geq \frac{1}{2} \log(\text{SNR}) - \frac{1}{2} \log(G(\Lambda_2) \cdot \mu(\Lambda_1, P_e)), \tag{9.20}$$

with equality in the limit of high SNR.

9.3.2   Gap to capacity

The analysis above holds for *all* SNR. In view of the high SNR AWGN channel capacity (Proposition 6.3.2), from (9.20) we have the following.

Corollary 9.3.1 (Capacity loss at high SNR)

$$\Delta_{\text{HSNR}} \overset{\Delta}{=} \lim_{\text{SNR} \to \infty} [C_{\text{AWGN}} - R(\Lambda_1/\Lambda_2)] = \frac{1}{2} \log(G(\Lambda_2) \cdot \mu(\Lambda_1, P_e)), \tag{9.21}$$

where C_{AWGN} is the AWGN channel capacity (6.17).

To minimize the capacity loss in (9.21), the fine lattice Λ_1 should have a low NVNR, i.e., be a *good channel code*, thus minimizing the cell volume per target error probability. At the same time, the coarse lattice Λ_2 should have a low NSM, i.e., be a *good quantizer*, to maximize the codebook volume for a given power constraint. It is therefore convenient to break (9.21) into two separate contributions.

Definition 9.3.1 (Shaping and coding losses) *The capacity loss of Voronoi modulation with lattice decoding at high SNR is the sum of the*

$$\text{Shaping loss} \overset{\Delta}{=} \frac{1}{2} \log(2\pi e G(\Lambda_2)) \tag{9.22}$$

of the coarse lattice, and the

$$\text{Coding loss} \triangleq \frac{1}{2} \log \left(\frac{\mu(\Lambda_1, P_e)}{2\pi e} \right) \tag{9.23}$$

of the fine lattice.

Note that the shaping loss is non-negative, and it can be interpreted as the divergence of the dither from white Gaussianity (Theorem 7.3.3). The coding loss is non-negative for a sufficiently small P_e (Section 3.3), and it coincides with the capacity loss of an infinite constellation (Chapter 6).

Example 9.3.1 (Cubic shaping) *Suppose a coded PAM or QAM system. This scheme can be represented by a coarse lattice $\Lambda_2 = a\mathbb{Z}^n$, nested in some fine lattice. Since $G(\mathbb{Z}^n) = 1/12$, the shaping loss is*

$$\text{Shaping loss} = \frac{1}{2} \log \left(\frac{2\pi e}{12} \right) \approx 0.254 \text{ bit}, \tag{9.24}$$

corresponding to a power loss of $10 \log_{10}(2\pi e/12) \approx 1.5$ dB.

We discuss below some implications and extensions of the high SNR analysis of Voronoi modulation with (estimation-free) lattice decoding.

9.3.3 Shaping gain

In the context of Voronoi modulation, the NSM ratio

$$\Gamma_s(\Lambda) \triangleq \frac{G(\mathbb{Z})}{G(\Lambda)} = \frac{1/12}{G(\Lambda)} \tag{9.25}$$

is known as the *shaping gain* relative to a cubic lattice. This quantity is the same as the *vector-quantizer gain* of Definition 3.2.3. It can also be expressed in logarithmic terms, as $\frac{1}{2} \log_2(1/12/G(\Lambda))$ [bits], or $10 \log_{10}(1/12/G(\Lambda))$ [dB]. In view of Example 9.3.1, the logarithmic shaping loss satisfies

$$\text{Shaping loss}(\Lambda) + \text{Shaping gain}(\Lambda) = \text{Shaping loss}(\mathbb{Z}^n). \tag{9.26}$$

The right-hand side, 0.254 bit or 1.5 dB, thus represents the *maximal achievable shaping gain* in Voronoi modulation.

Spherical shaping Instead of shaping the codebook using the coarse Voronoi cell, it can be shaped by intersecting the fine lattice with an n-dimensional ball. Although by doing so we lose the elegant nesting structure, the resulting shaping gain is slightly higher:

$$\Gamma_s(n\text{-dimensional sphere}) = 1/12G_n^* > \Gamma_s(n\text{-dimensional lattice}),$$

where G_n^* is the NSM of a sphere (7.15).

Mixed nesting dimensions As Tables 7.1 and 7.2 show, the potential shaping gain (of either a lattice or a spherical shaping scheme) is at most ≈ 1.5 dB, while the potential coding gain is larger (e.g., 6.6 dB at SER $= 10^{-5}$). It is thus practically more efficient to combine a low-dimensional shaping lattice with a high-dimensional coding lattice, as described in Section 8.3.3. Such a mixed-dimensional nesting is realized by a product coarse lattice: $\Lambda_2 = \Lambda_2' \times \ldots \times \Lambda_2'$. Since $G(\Lambda_2) = G(\Lambda_2')$ (see (3.23)), the shaping loss (or gain) is determined by the component lattice Λ_2'.

PAPR Another practical (negative) consequence of shaping is the increase in the *peak to average power ratio* (PAPR). For lattice shaping, the PAPR is defined as

$$\text{PAPR} = \frac{\left(\max_{\mathbf{x} \in \mathcal{V}_0(\Lambda_2)} \|\mathbf{x}\|_\infty\right)^2}{\sigma^2(\Lambda_2)}, \tag{9.27}$$

where $\|\mathbf{x}\|_\infty \stackrel{\Delta}{=} \max\{|x_1|, \ldots, |x_n|\}$ is the ℓ_∞ norm of $\mathbf{x} = (x_1, \ldots, x_n)$. For cubic shaping PAPR $= 3$, while for spherical and good high-dimensional lattice shaping it grows linearly with the dimension; see Problem 9.3.

9.3.4 Lattice versus ML decoding

We turn to inspect the *coding loss* term (9.23) of the Voronoi modulation scheme with lattice decoding. Is this part of the *modulo loss* of the lattice decoder with respect to the ML decoder? We shall see that at high SNR the answer is negative.

As discussed in Section 9.1, the advantage of ML decoding is due to codewords near the surface of the shaping region, because their decision cell is larger than a lattice Voronoi cell. In fact, a border decision cell is unbounded outward; see Figure 9.4. Nevertheless, although in high dimensions most of the volume concentrates on the outer shell (3.4), in the high SNR limit the fraction of cells which are *exactly* at the border becomes negligible:

$$\frac{\text{non-border points}}{\text{total points}} \leq \frac{V((\sqrt{\text{SNR}} - d) \cdot \mathcal{V}_0(\Lambda_2))}{V(\sqrt{\text{SNR}} \cdot \mathcal{V}_0(\Lambda_2))} \tag{9.28}$$

$$= \left(\frac{\sqrt{\text{SNR}} - d}{\sqrt{\text{SNR}}}\right)^n \tag{9.29}$$

$$\to 1, \quad \text{as SNR} \to \infty, \tag{9.30}$$

where d is the diameter of the fine lattice Voronoi cell. Thus, the *average* error probability (9.7) in ML decoding is asymptotically dominated by the inner (lattice-shaped) cells, and coincides with the lattice decoding error probability. [4]

---

[4] This analysis hints that locating *all* codewords on the surface of a sphere may be advantageous in terms of error probability. Nevertheless, as shown in Section 13.8, the error exponent of Voronoi modulation with lattice decoding coincides at high rate and SNR conditions with the Poltyrev sphere packing error exponent, which is an upper bound on the performance of *any* coding and decoding scheme.

The diminishing advantage of ML decoding over lattice decoding can also be measured in information theoretic terms. Let $\mathbf{U}$ represent the channel input, which is uniform over the shaping region $\mathcal{V}_0(\Lambda_2)$. It can be shown that the residual information contained in the non-modulo channel output $\mathbf{Y} = \mathbf{U} + \mathbf{Z}$, given the modulo channel output $[\mathbf{Y}]$ mod Λ_2, becomes negligible as the SNR increases:

$$I(\mathbf{U}; \mathbf{Y} \mid [\mathbf{Y}] \bmod \Lambda_2) \to 0, \quad \text{as } \sigma_z^2 \to 0. \tag{9.31}$$

Thus, $[\mathbf{Y}]$ mod Λ_2 is asymptotically a sufficient statistic for the decoding of $\mathbf{U}$; see Problem 9.5.

9.3.5 Good nested lattices

As we learned in Chapter 7, the ultimate shaping and coding gains are achievable by high-dimensional lattices.

Definition 9.3.2 (Good Voronoi modulation: high SNR) *We say that a sequence $(\Lambda_{2,n} \subset \Lambda_{1,n})$ of nested lattice pairs is "good for Voronoi modulation at high SNR," if $G(\Lambda_{2,n}) \to 1/2\pi e$, and $\mu(\Lambda_{1,n}, P_e) \to 2\pi e$ for all $P_e > 0$.*

The existence of such good nested lattice pairs was proved in Section 8.5. Combining this fact and Corollary 9.3.1, we conclude the following.

Corollary 9.3.2 (Achieving capacity) *For a sequence $(\Lambda_{1,n}, \Lambda_{2,n})$ of nested lattices which are good for Voronoi modulation, the high SNR capacity loss vanishes asymptotically:*

$$\Delta_{\text{HSNR}} = \frac{1}{2} \log(G(\Lambda_{2,n}) \cdot \mu(\Lambda_{1,n}, P_e)) \to 0, \quad \text{as } n \to \infty, \tag{9.32}$$

for all $P_e > 0$.

Thus, the Voronoi modulation system, with the (suboptimal) *lattice decoder* (9.9a), approaches the Shannon AWGN channel capacity in the limit of high SNR. As we shall see, the same decoder can close the gap to Shannon also for a *finite* SNR, using an appropriate estimator $\hat{\mathbf{X}} = g(\mathbf{Y})$ at the receiver front end.

9.3.6 Non-AWGN

Before moving to the non-high SNR case, let us consider how the system and results above change if $\mathbf{Z}$ is a *general* (not necessarily white-Gaussian) additive noise. In the general case, we can either keep the Euclidean lattice decoder (9.9a) and suffer a mismatch decoding loss (Section 6.4.1), or replace it by a noise-matched lattice quantizer Q_{Λ_1} (induced by the noise distribution $P_Z(\cdot)$), i.e., $\hat{\mathbf{v}}_{\text{LD}} = Q_{\Lambda_1}(\mathbf{Y} - \mathbf{u})$ mod Λ_2 (Definition 6.2.3). The latter is correct whenever the noise $\mathbf{Z}$ falls inside $\mathcal{V}_0^{\mathbf{Z}}(\Lambda_1)$, the generalized fundamental Voronoi cell (6.12) of Λ_1, or in one of its shifts by Λ_2.

As we have seen before in (6.27), a general noise is characterized by its *entropy power* $P_E(\mathbf{Z})$, rather than by its power. Thus, we define the *signal to entropy power ratio* as

$$\text{SENR} \triangleq \frac{P}{P_E(\mathbf{Z})}. \tag{9.33}$$

In these terms, the threshold SNR result of Theorem 9.3.1 becomes

$$\text{SENR} \geq G(\Lambda_2) \cdot \mu_{\text{matched}}(\Lambda_1, \mathbf{Z}, P_e) \cdot \Gamma^2, \tag{9.34}$$

where $\mu_{\text{matched}}(\Lambda_1, \mathbf{Z}, P_e)$ is the generalized NVNR (6.28). And the capacity loss at high SNR conditions becomes

$$\Delta_{\text{HSNR}} = \frac{1}{2} \log(G(\Lambda_2) \cdot \mu_{\text{matched}}(\Lambda_1, \mathbf{Z}, P_e)); \tag{9.35}$$

see (6.21). For noise satisfying the generalized AEP and a sequence of simultaneously good nested lattices, Δ_{HSNR} vanishes asymptotically as the dimension goes to infinity; see Section 8.5.

9.4   Shannon meets Wiener (at medium SNR)

Even if both component lattices are "good" (the shaping (9.22) and coding (9.23) losses are small), still the Voronoi modulation system with lattice decoding above does not meet the Shannon capacity limit for a *finite* SNR. In fact, for good lattices the rate approaches (see (9.20))

$$R(\Lambda_1/\Lambda_2) = \frac{1}{2} \log (\text{SNR}), \tag{9.36}$$

provided the SNR $= P/N$ is greater than or equal to 1, and 0 otherwise.[5] This rate corresponds to a *loss of* "1" inside the log in the AWGN channel capacity formula

$$C_{\text{AWGN}} = \frac{1}{2} \log (1 + \text{SNR}). \tag{9.37}$$

Although this "1" is negligible at high SNR conditions (because $\log(1 + x) - \log(x) \to 0$ as $x \to \infty$), it becomes critical when the SNR is smaller than 1, in which case the capacity is smaller than half a bit per channel use.

This capacity loss could be avoided if we used *ML decoding* (9.8). ML decoding means, however, breaking away from the lattice partition structure of the decision cells (Figure 9.4), hence increasing the decoding complexity.

---

[5] For small P_e and any pair of nested lattices, the minimum required SNR in (9.18) is *greater than* 1. This is because the nesting ratio Γ is always greater than 1, while $G(\Lambda_2) > 1/2\pi e$, and $\mu(\Lambda_1, P_e) > 2\pi e$ for sufficiently small P_e (Sections 3.3 and 7.1).

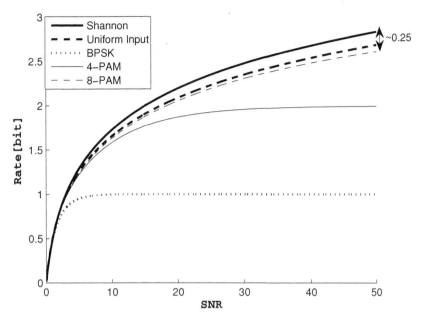

Figure 9.6 Capacity curves (in bits as a function of the SNR) for a discrete-symmetric input AWGN channel, corresponding to coded unshaped digital modulation. Shown also are the mutual information curve for a continuously uniform input, and the power-constrained capacity curve, $C_{\text{AWGN}} = \frac{1}{2}\log(1 + \text{SNR})$, corresponding to the mutual information for a Gaussian input. We see that BPSK saturates at 1 bit and 4-PAM at 2 bits, while a uniform input keeps an asymptotic constant gap of ≈ 0.254 bit (the *cubic shaping loss*) from the C_{AWGN} curve.

We want to enjoy the benefit of both worlds: close the gap to capacity for all SNR, while still preserve the lattice structure at both the encoding and decoding stages. For that, we shall employ linear (Wiener) estimation before the lattice decoding. This section provides the motivation – from Wiener's and Shannon's viewpoints – while an explicit scheme will be presented in the next sections.

9.4.1   The interesting regime of SNR

When deviating from the high-rate/high-SNR regime, we need to re-examine some of our basic assumptions. Is spherical ("Gaussian") shaping still necessary? Is lattice decoding still simpler than ML decoding?

Figure 9.6 shows the mutual information $I(X; X + Z)$ when the input X is uniform over some one-dimensional constellations.[6] At high SNR, the mutual information saturates, as expected, at the logarithm of the constellation size. At low SNR, on the other hand, all curves (including BPSK) have the same rate versus

---

[6] This mutual information corresponds to the maximum reliable rate achievable by "coding" these input symbols, i.e., using a high-dimensional constellation whose alphabet consists of these symbols.

SNR slope, and converge to the optimum Shannon capacity curve. Indeed, it can be shown that if X has power P but otherwise it is arbitrary, and $Z \sim N(0, N)$, then

$$\lim_{N \to \infty} \frac{I(X; X + Z)}{\frac{1}{2} \log(1 + P/N)} = 1.$$

(The sum of X and a much stronger Gaussian is "almost" Gaussian [262].) Thus, at low SNR the input does not have to be Gaussian, and *cubic* shaping (as in Example 9.3.1) is optimal. We can interpret this phenomenon as if at low SNR the modulator enjoys a "*natural shaping*" due to the Gaussian noise.

Binary (BPSK) modulation, even if coded, lies on the corners of a $[-1, +1]^n$ cube. Due to this strong symmetry, decoding of a binary codeword is simple, *even under an ML rule*; see Example 8.3.1. Thus, in the low SNR regime, unshaped coded binary modulation can achieve capacity at a relatively low complexity.

The low SNR regime is also optimal energy-wise: the *capacity per unit cost* (i.e., the ratio of $C(\text{SNR})$ to the SNR) is maximum at low SNR, where we have

$$C_{\text{AWGN}} \approx \frac{\text{SNR}}{2} \log(e), \tag{9.38}$$

and it decreases as we go to a higher SNR. Yet, getting a significant bit rate at low SNR requires a large bandwidth, i.e., the *spectral efficiency is low*. [7]

The practical compromise between these two conflicting effects – energy versus spectral efficiency – is often to work in the *medium* SNR regime. In this regime, shaping gives an advantage of up to 6% in capacity; see Figure 9.7.

9.4.2   Wiener: estimate and decode

Suppose that the output of the channel $\mathbf{Y} = \mathbf{X} + \mathbf{Z}$, with an input power P, is multiplied by the Wiener coefficient (4.44)

$$\alpha^* = \frac{P}{P + N} = \frac{\text{SNR}}{1 + \text{SNR}}. \tag{9.39}$$

This will effectively reduce the noise variance from N to (4.45)

$$\text{MSE}(\alpha^*) = (\alpha^*)^2 N + (\alpha^* - 1)^2 P = \frac{PN}{P + N} = \frac{N}{1 + 1/\text{SNR}}. \tag{9.40}$$

Let us view the mapping

$$\mathbf{X} \to \alpha^* \mathbf{Y} \tag{9.41}$$

---

[7] In continuous-time AWGN channels we usually look at the inverse of the capacity per unit cost, known as the *energy per information bit* E_b/N_0, which is normalized by the noise spectral level N_0.

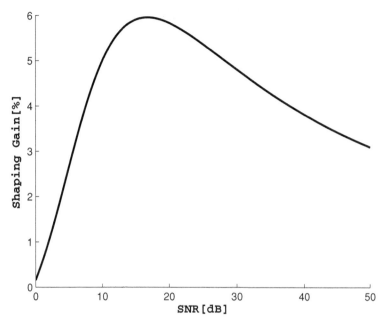

Figure 9.7 The relative shaping gain $[C_{\text{AWGN}} - I(U; U + Z)]/I(U; U + Z)$ in percent, as a function of the SNR.

as a *forward channel* with noise variance $\text{MSE}(\alpha^*)$. Plugging this value as the noise variance in the rate formula (9.36), gives

$$R(\Lambda_1/\Lambda_2) = \frac{1}{2} \log \left(\frac{P}{\text{MSE}(\alpha^*)} \right) = \frac{1}{2} \log (1 + \text{SNR}), \qquad (9.42)$$

which exactly meets the AWGN channel capacity.

This heuristic calculation ignores, however, the fact that for $\alpha^* < 1$, the equivalent noise $\alpha^* \mathbf{Y} - \mathbf{X} = (\alpha^* - 1)\mathbf{X} + \alpha^* \mathbf{Z}$ in the channel (9.41) is a *mixture* of the noise $\mathbf{Z}$ and input $\mathbf{X}$; so the equivalent noise is orthogonal to the output $\alpha^* \mathbf{Y}$ (by the orthogonality principle) but *correlated* with the input $\mathbf{X}$. Thus, the rate formula (9.37) (which assumes an *additive* noise) does not apply.[8]

The trick – as we saw for filtered dithered quantization in Sections 4.4 and 5.6, and as we shall see for Voronoi modulation in Section 9.5 below – is to use dither as a means for *decorrelation*, this time between the transmitted *message* and the channel *input* $\mathbf{X}$. But before we do that, let us take a Shannon view of the decoding problem.

---

[8] This phenomenon is known in the communication literature as *biased equalization*. Analysis that ignores the residual correlation leads to an "overoptimistic" capacity of $\frac{1}{2} \log(1 + P/\text{MSE}(\alpha^*)) = \frac{1}{2} \log(2 + P/N)$; see [237].

9.4.3   Shannon: joint and reversed typicality

Information theoretic measures enhance the insight into communication in the presence of noise. The mutual information $I(X; Y)$ between the channel input X and the channel output Y can be decomposed either in a *forward channel* form, as $h(Y) - h(Y|X)$, or in a *backward (or reverse) channel* form, as $h(X) - h(X|Y)$, where $h(Y|X)$ and $h(X|Y)$ are *conditional entropies*; see (A.6). The former expression corresponds to a partition of an output space of size $2^{nh(Y)}$ to decision cells of size $2^{nh(Y|X)}$ each. The latter expression corresponds to the proportion of a codeword region of size $2^{nh(X|Y)}$ in an input space of size $2^{nh(X)}$.

For an AWGN channel $Y = X + Z$ with Gaussian input $X \sim N(0, \sigma_x^2)$, these decompositions become

$$I(X; Y) = \frac{1}{2} \log(2\pi e \sigma_y^2) - \frac{1}{2} \log(2\pi e \sigma_z^2) \tag{9.43a}$$

$$= \frac{1}{2} \log(2\pi e \sigma_x^2) - \frac{1}{2} \log(2\pi e \sigma_{x|y}^2). \tag{9.43b}$$

Recall that for jointly Gaussian zero-mean vectors (X, Y), the conditional density $f(x|y)$ is Gaussian (see Proposition 4.4.2 and Problem 4.12):

$$f(x|y) = \frac{1}{\sqrt{2\pi \sigma_{x|y}^2}} \cdot e^{-\frac{(x - \alpha^* y)^2}{2\sigma_{x|y}^2}}, \tag{9.44}$$

where $\alpha^* = \rho \sigma_x / \sigma_y$ is the Wiener coefficient (4.36), and $\sigma_{x|y}^2 = (1 - \rho^2)\sigma_x^2$ is the conditional variance (or in this case, the LMMSE (4.37)). Specifically, for a power-constrained AWGN channel, we have $\sigma_x^2 = P, \sigma_z^2 = N, \sigma_y^2 = P + N, \rho = \sigma_x / \sigma_y = \sqrt{P/(P + N)}$, so $\sigma_{x|y}^2 = PN/(P + N)$; see (4.45). Thus,

$$h(X|Y) = \frac{1}{2} \log \left(2\pi e \frac{PN}{P + N} \right), \tag{9.45}$$

and (9.43) becomes [9]

$$I(X; Y) = \frac{1}{2} \log \left(\frac{P + N}{N} \right) \tag{9.46a}$$

$$= \frac{1}{2} \log \left(\frac{P}{\frac{PN}{P+N}} \right) \tag{9.46b}$$

$$= \frac{1}{2} \log \left(1 + \frac{P}{N} \right) = C_{\text{AWGN}}. \tag{9.46c}$$

We see that the "1" inside the logarithm can be attributed either to output power boost ($P + N$ instead of P) in the forward channel form (9.46a), or to noise

---

[9] Note that the Pythagorean relation of Figure 4.8 implies that the power sum of the input and equivalent noise in the reverse channel is equal to the input power of the forward channel: $\text{Var}(\alpha^* Y) + PN/(P + N) = P$.

reduction $(PN/(P + N)$ instead of $N)$ in the reverse channel form (9.46b). Does this information theoretic interpretation have any operational meaning?

Shannon's *joint-typicality decoding* approach (which we encountered in proving the existence of "good lattice constellations" in Section 7.7) links the reverse channel form (9.46b) to Wiener's estimate-and-decode paradigm. Recall from the Gaussian AEP (Theorem 7.2.1) that the typical set of an n-dimensional white-Gaussian vector of variance N is a sphere (or equivalently a ball (3.4)) of variance $\sqrt{nN}$. A similar notion of *joint typicality* holds for a pair of correlated vectors. Suppose X and Y are jointly Gaussian with zero mean and covariance matrix

$$\Sigma = \begin{pmatrix} \sigma_x^2 & \rho\sigma_x\sigma_y \\ \rho\sigma_x\sigma_y & \sigma_y^2 \end{pmatrix}$$

where ρ is the correlation coefficient. The joint p.d.f. of X and Y is given by

$$f(x, y) = \frac{1}{\sqrt{2\pi \det(\Sigma)}} \cdot e^{-\frac{1}{2}(x,y)\Sigma^{-1}(x,y)^t} \tag{9.47}$$

implying a joint differential entropy (A.5) of

$$h(X, Y) = \frac{1}{2}\log(2\pi e\sqrt{\det(\Sigma)}) \tag{9.48}$$

with marginal entropies $h(X) = \frac{1}{2}\log(2\pi e\sigma_x^2)$ and $h(Y) = \frac{1}{2}\log(2\pi e\sigma_y^2)$.

We say that vectors $\mathbf{x}$ and $\mathbf{y}$ are (ϵ-) *jointly typical* with respect to a joint p.d.f. $f(x, y)$, if each vector is *individually typical* with respect to its own (marginal) entropy (Definition 7.2.1), and as a pair they satisfy

$$\left| -\frac{1}{n}\log(f(\mathbf{x}, \mathbf{y})) - h(X, Y) \right| < \epsilon, \tag{9.49}$$

where $f(\mathbf{x}, \mathbf{y}) = \prod_{i=1}^{n} f(x_i, y_i)$. In the jointly Gaussian case, the marginal conditions induce the usual spherical typical set (7.25) per vector, while condition (9.49) induces an ellipsoidal set in $\mathbb{R}^n \times \mathbb{R}^n$. The *joint AEP* asserts that if $\mathbf{X}$ and $\mathbf{Y}$ are generated by a sequence of n i.i.d. copies of the pair (X, Y), then they are jointly typical with high probability for any $\epsilon > 0$ and sufficiently large n; see [53]. This forms a basis for *joint-typicality decoding*: given an output vector $\mathbf{Y}$, search for a codeword $\mathbf{X}(i)$ such that the pair $(\mathbf{X}(i), \mathbf{Y})$ is jointly typical.

An alternative form of joint-typicality decoding is based on the *reverse channel* from $\mathbf{Y}$ to $\mathbf{X}$. By the chain rule for the joint entropy (A.5), $h(X, Y) = h(X) + h(Y|X) = h(Y) + h(X|Y)$. Since $f(x, y) = f(y) \cdot f(x|y)$, we can replace condition (9.49) above by a constraint on the distance between $\frac{1}{n}\log f(\mathbf{x}|\mathbf{y})$ and the conditional entropy $h(X|Y)$.

In the Gaussian case, we have $h(X|Y) = \frac{1}{2}\log(2\pi e\sigma_{x|y}^2)$ (9.43b). Thus, in view of (9.44) the "reversed" joint AEP takes the following form, which is illustrated in Figure 9.8.

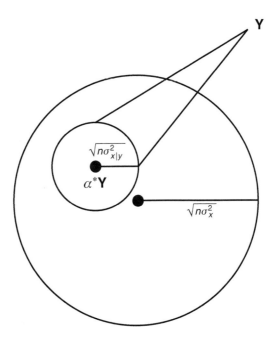

Figure 9.8 Geometry of reversed AEP.

Proposition 9.4.1 (Reversed joint AEP) *If the vector pair $(\mathbf{X}, \mathbf{Y})$ consists of i.i.d. copies of a zero-mean jointly Gaussian pair (9.47), then for any $\epsilon > 0$ and sufficiently large n we have with a large probability*

$$\mathbf{Y} \in \mathcal{B}\left(\mathbf{0},\ (1+\epsilon)\sqrt{n\sigma_y^2}\right) \tag{9.50a}$$

and

$$\mathbf{X} \in \mathcal{B}\left(\alpha^*\mathbf{Y},\ (1+\epsilon)\sqrt{n\sigma_{x|y}^2}\right). \tag{9.50b}$$

The reversed joint AEP in the Gaussian case implies a simple structure for joint-typicality decoding of a randomly generated codebook. If the input $\mathbf{X}(i)$ to an AWGN channel is a codeword from a white-Gaussian random code $\{\mathbf{X}(1), \ldots, \mathbf{X}(M)\}$, and $\mathbf{Y}$ is the received output, then a joint-typicality decoder will search for the transmitted codeword within a radius $\sqrt{n\sigma_{x|y}^2}$ around the point $\alpha^*\mathbf{Y}$. This search procedure is in line with the structure of *estimate-and-decode* discussed earlier in Section 9.4.2: multiply the channel output by the Wiener coefficient α^*, and look for a unique codeword within the typical radius. [10]

---

[10] This latter step corresponds to a threshold decoder (7.60). Alternatively, the decoder can look for the codeword *nearest* to $\alpha^*\mathbf{Y}$.

Achieving capacity with reversed joint-typicality decoding

It is easy to verify that this scheme achieves the *full* AWGN channel capacity. Suppose that the M codewords are independent and uniformly distributed over an n-ball of radius $\sqrt{n\sigma_x^2}$. By the AEP, this is asymptotically the same as a white-Gaussian codebook (Theorem 7.2.1); so for large n, the true codeword falls with high probability inside the search ball (9.50b). On the other hand, the probability p_{compete} that a competing (uniformly random) codeword $\mathbf{X}(j)$ falls inside the search ball is equal to the proportion of the intersection of a ball of radius $\sqrt{n\sigma_{x|y}^2}$ with a ball of radius $\sqrt{n\sigma_x^2}$, which is upper bounded by the ball volume ratio $(\sigma_{x|y}/\sigma_x)^n$. Hence, by the *success-threshold exponent* (Lemma A.2.1), the probability of finding any of the $M - 1$ wrong codewords in the search ball (9.50b) vanishes as n goes to infinity, provided that the coding rate $R = \frac{1}{n}\log(M)$ is smaller than

$$R < \frac{1}{n}\log\left(\frac{1}{p_{\text{compete}}}\right) \le \log\left(\frac{\sigma_x}{\sigma_{x|y}}\right) = C_{\text{AWGN}}, \tag{9.51}$$

where the last equality follows by setting $\sigma_x^2 = P$ and $\sigma_{x|y}^2 = PN/(P + N)$.

We know that direct NN decoding of $\mathbf{X}(i)$ from $\mathbf{Y}$ is optimal (for each dimension n). Does estimate-and-decode have any advantage?

We may visualize lattice decoding of a Voronoi constellation (Definition 9.1.1) as a situation where the original codebook has 2^{nR_1} random codewords with average power P_1, but the decoder knows that the true codeword has a much *smaller* power $P \ll P_1$. If the number of power P codewords is close to the maximum of $\approx 2^{nC} = (1 + P/N)^{n/2}$, then an NN search over the *entire* codebook is likely to fail, because $\mathbf{Y}$ will be closer to farther codewords of power $P + N$. To avoid this competition, we can limit the NN search to codewords of power P (corresponding to ML decoding). Or alternatively, first attenuate $\mathbf{Y}$ by the Wiener coefficient $\alpha^* = P/(P + N)$ and then perform unlimited NN search over the entire codebook. The reversed AEP (Proposition 9.4.1) guarantees that the latter procedure is correct with high probability for large n. See Problem 9.8.

9.5   The mod Λ channel

We return to analyze the Voronoi demodulator of Figure 9.3, now with a general estimator $\hat{\mathbf{X}} = g(\mathbf{Y})$ at the receiver front end. We shall assess the potential improvement due to the estimator g at general SNR conditions, by calculating the *Shannon capacity* of the equivalent channel from the message representative $\mathbf{v}$ to the decision vector $\hat{\mathbf{Y}}$.

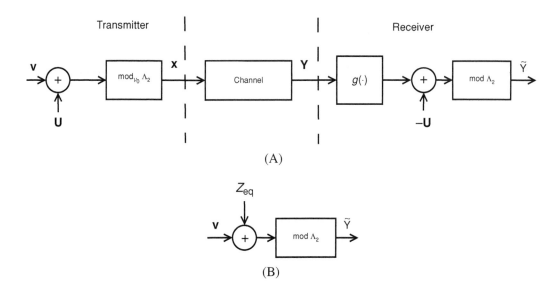

Figure 9.9 (A) Mapping from the modulator input vector **v** to the decision vector $\tilde{\mathbf{Y}}$; and (B) the equivalent mod Λ channel.

Figure 9.9 illustrates this channel and its equivalent form. Recall from (9.11) that $\tilde{\mathbf{Y}} = [g(\mathbf{Y}) - \mathbf{u}] \bmod \Lambda_2$, where $\mathbf{Y}$ is the output of the physical channel in response to an input $\mathbf{x} = [\mathbf{v} + \mathbf{u}] \bmod_{v_0} \Lambda_2$ and where the former modulo operation is with respect to any enumeration cell. Lemma 9.3.1 described the properties of the channel from the message representative $\mathbf{v}$ to $\tilde{\mathbf{Y}}$ for the *no*-estimation case $\hat{\mathbf{X}} = g(\mathbf{Y}) = \mathbf{Y}$. The case of a general estimator g is, however, fundamentally different, as now a *random* dither plays a crucial role.

Lemma 9.5.1 (Equivalent mod Λ channel: general case) *For a random dither* **U** *which is uniform (or modulo-uniform) over a fundamental cell of Λ_2, and for any input* $\mathbf{v} \in \mathbb{R}^n$ *(not necessarily in Λ_1) and an estimator* $\hat{\mathbf{X}} = g(\mathbf{Y})$, *the channel from* **v** *to* $\tilde{\mathbf{Y}}$ *is equivalent in distribution to the modulo-additive-noise channel shown in Figure 9.9(B):*

$$\tilde{\mathbf{Y}} = [\mathbf{v} + \mathbf{Z}_{eq}] \bmod \Lambda_2 \tag{9.52a}$$

with equivalent noise

$$\mathbf{Z}_{eq} = [g(\mathbf{Y}) - \mathbf{X}] \bmod \Lambda_2, \tag{9.52b}$$

where **X** *is uniform over* $\mathcal{V}_0(\Lambda_2)$, **Y** *is the corresponding channel output, and the pair* $(\mathbf{X}, \mathbf{Y})$ *is independent of* **v**.

Note that Lemma 9.5.1 holds for a *general* channel $p(\mathbf{y}|\mathbf{x})$ (not necessarily an additive-noise channel), as well as for Voronoi modulation with *pre-emphasis* $f(\cdot)$; see (9.11d). [11]

As Lemma 9.5.1 shows, *without conditioning* on the random dither $\mathbf{U}$, the channel output $\mathbf{Y}$ is independent of the message $\mathbf{v}$. The decision vector $\tilde{\mathbf{Y}}$ takes $\mathbf{U}$ into account, hence it *does depend* on $\mathbf{v}$. And the resulting equivalent channel (9.52a) is modulo additive.

Proof We add and subtract $\mathbf{x}$ in the argument of the modulo-Λ_2 operation in (9.11f), and write the first $\mathbf{x}$ explicitly as $\mathbf{x} = [\mathbf{v} + \mathbf{u}] \bmod_{\mathcal{V}_0} \Lambda_2$ from (9.5a):

$$\tilde{\mathbf{Y}} = \left[\underbrace{[\mathbf{v} + \mathbf{u}] \bmod_{\mathcal{V}_0} \Lambda_2}_{\mathbf{x}} + \hat{\mathbf{X}} - \mathbf{u} - \mathbf{x} \right] \bmod \Lambda_2. \tag{9.53}$$

By the distributive law (2.22b) (see also (8.21)), we can omit the inner modulo operation; hence the dither $\mathbf{u}$ cancels out, and we get

$$\tilde{\mathbf{Y}} = [\mathbf{v} + [g(\mathbf{Y}) - \mathbf{x}]] \bmod \Lambda_2, \tag{9.54}$$

where $g(\mathbf{Y}) = \hat{\mathbf{X}}$ is the input estimator. For a specific value $\mathbf{u}$ of the dither, the channel input $\mathbf{x}$ (and hence $g(\mathbf{Y}) - \mathbf{x}$) still depends on $\mathbf{v}$ by (9.5a). [12] Nevertheless, for a *random* uniform dither $\mathbf{U}$, the crypto lemma (Lemma 4.1.1) implies that $\mathbf{X} = [\mathbf{v} + \mathbf{U}] \bmod_{\mathcal{V}_0} \Lambda_2$ is uniform over $\mathcal{V}_0$, *independent* of $\mathbf{v}$. Hence, also the pair $(\mathbf{X}, \mathbf{Y})$ is independent of $\mathbf{v}$ (because $\mathbf{v} \leftrightarrow \mathbf{X} \leftrightarrow \mathbf{Y}$ form a Markov chain), and the lemma follows. □

The argument of the modulo-Λ_2 operation in (9.52b) is the *estimation error* $\hat{\mathbf{X}} - \mathbf{X}$, i.e., the error of $g(\mathbf{Y})$ in estimating $\mathbf{X}$. The modulo-Λ_2 operation "folds" the tails of the distribution of the estimation error into the enumeration fundamental cell $\mathcal{P}_0$.

Let us examine the information theoretic properties of the equivalent noise and the mod Λ channel. We shall start with the case of general channel $p(\mathbf{y}|\mathbf{x})$ and estimator g, and then move to an additive-noise channel $\mathbf{Y} = \mathbf{X} + \mathbf{Z}$ with a linear estimator $g(\mathbf{y}) = \alpha \mathbf{y}$.

9.5.1 Properties of the equivalent noise

1. The (folded) distribution of the equivalent noise $\mathbf{Z}_{eq}$ (9.52b) amounts to a distribution over the cosets of Λ_2, induced by the estimation error $\hat{\mathbf{X}} - \mathbf{X}$; hence it is invariant to the choice of the enumeration cell $\mathcal{P}_0$, up to shuffling of pieces of its support (see Figure 9.10).

---

[11] In the latter case, the equivalent noise is $\mathbf{Z}_{eq} = [g(\mathbf{Y}) - \tilde{\mathbf{X}}] \bmod \Lambda_2$, where now $\tilde{\mathbf{X}}$ is uniform over $\mathcal{V}_0$, and $\mathbf{Y}$ is the output of the composite channel $f(\tilde{\mathbf{X}}) \to p(\mathbf{y}|\mathbf{x}) \to \mathbf{Y}$. Note that $g(\mathbf{Y})$ can be viewed as an estimate for $\tilde{\mathbf{X}}$ rather than $\mathbf{X}$.

[12] Unless for an additive-noise channel with $g(\mathbf{Y}) = \mathbf{Y}$, in which case $g(\mathbf{Y}) - \mathbf{x} = \mathbf{Z}$ is independent of $\mathbf{v}$.

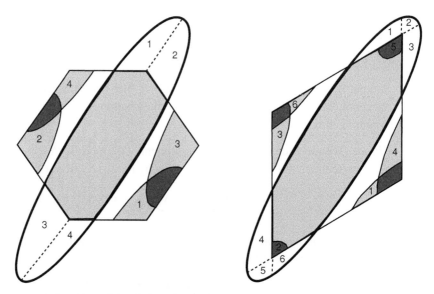

Figure 9.10 The effect of the choice of modulo cell on the folded distribution. Unlike the "perfect" picture (modulo equivalence of cells) in Section 2.3, the folded distribution may create overlaps or leave holes. But the total volume of holes or overlaps is invariant to the choice of the modulo cell.

2. In particular, the entropy of the equivalent noise, $h(\mathbf{Z}_{eq})$, is equal to the *coset entropy* $h(\hat{\mathbf{X}} - \mathbf{X} \bmod \Lambda_2)$ of the estimation error with respect to Λ_2 (A.23), which is invariant of the choice of $\mathcal{P}_0$; see Appendix A.3.

3. Plugging in the fundamental Voronoi cell as $\mathcal{P}_0$, we obtain

$$h(\mathbf{Z}_{eq}) = h(\hat{\mathbf{X}} - \mathbf{X} \bmod {}_{\mathcal{V}_0} \Lambda_2) \tag{9.55a}$$

$$\leq \frac{n}{2} \log \left(\frac{2\pi e}{n} E \| \hat{\mathbf{X}} - \mathbf{X} \bmod_{\mathcal{V}_0} \Lambda_2 \|^2 \right) \tag{9.55b}$$

$$\leq \frac{n}{2} \log \left(\frac{2\pi e}{n} E \| \hat{\mathbf{X}} - \mathbf{X} \|^2 \right), \tag{9.55c}$$

where (9.55b) follows from the maximum entropy property (A.13) of the Gaussian distribution; (9.55c) follows since the modulo-Voronoi operation does not increase the second moment (though it may increase the variance; see Problem 9.6). For an unbiased estimator (where $E\{\hat{\mathbf{X}} - \mathbf{X}\} = 0$) the second moment can be replaced by the variance.

4. Alternatively, by the properties of the coset entropy (Lemma A.3.2), $h(\hat{\mathbf{X}} - \mathbf{X} \bmod \Lambda_2)$ is smaller than or equal to the (unfolded) estimation-error entropy $h(\hat{\mathbf{X}} - \mathbf{X})$ (see Problem 9.7), which by the maximum entropy property implies (9.55c).

9.5.2   Capacity of the mod Λ channel

1. The Shannon capacity (6.15) of the mod Λ channel, $\tilde{\mathbf{Y}} = [\mathbf{V} + \mathbf{Z}_{eq}] \bmod \Lambda_2$ (9.52a), is equal to [13]

$$C_{\text{mod }\Lambda} = \frac{1}{n}[\log(V(\Lambda_2)) - h(\mathbf{Z}_{eq})], \tag{9.56}$$

and it is achieved by an input $\mathbf{V}$ which is uniform over a coarse fundamental cell $\mathcal{P}_0(\Lambda_2)$ (A.14).

2. $C_{\text{mod }\Lambda}$ *upper bounds* the discrete input capacity, where the channel input $\mathbf{V}$ is restricted to the fine lattice Λ_1 (as in the Voronoi modulation system).

3. $C_{\text{mod }\Lambda}$ is maximized by an estimator g^{**} that *minimizes* the equivalent noise *entropy*.

4. An alternative (simpler) approach to optimizing the estimator is to relate the capacity to the estimator's MSE. Using the MSE bound on the equivalent noise entropy (9.55),

$$C_{\text{mod }\Lambda} \geq \frac{1}{2}\log\left(\frac{P}{\frac{1}{n}E\|\hat{\mathbf{X}} - \mathbf{X}\|^2}\right) - \frac{1}{2}\log(2\pi e G(\Lambda_2)), \tag{9.57}$$

where we replaced the volume of the coarse lattice in (9.56) by its NSM and the power constraint P.

5. This lower bound is maximized by the MMSE estimator g^* of $\mathbf{X}$ from $\mathbf{Y}$ (4.40), which is, in general, non-linear, and it achieves $\frac{1}{n}E\|\hat{\mathbf{X}} - \mathbf{X}\|^2 = \text{Var}(\mathbf{X}|\mathbf{Y})$.

So far the analysis holds for a general channel $p(\mathbf{y}|\mathbf{x})$ and estimator g. We shall now restrict our attention to the AWGN channel, and see how a *linear* (scalar) estimator is sufficient to approach capacity for a *general* SNR.

9.5.3   Additive noise and linear estimation

1. For an additive-noise channel $\mathbf{Y} = \mathbf{X} + \mathbf{Z}$ and a linear estimator $\hat{\mathbf{X}} = \alpha\mathbf{Y}$, the estimation error is distributed as

$$\hat{\mathbf{X}} - \mathbf{X} \overset{\text{dist}}{=} \alpha\mathbf{Z} + (\alpha - 1)\mathbf{U}, \tag{9.58}$$

where $\mathbf{U} \sim \text{Unif}(\mathcal{V}_0)$. (On the right-hand side we use $\mathbf{U}$ instead of $\mathbf{X}$ to emphasize its uniformity.)

2. The equivalent noise (9.52b) is then given by

$$\mathbf{Z}_{eq} = \Big(\underbrace{\alpha\mathbf{Z}}_{\text{channel-noise}} + \underbrace{(\alpha - 1)\mathbf{U}}_{\text{self-noise}} \Big) \bmod \Lambda_2. \tag{9.59}$$

---

[13] $C_{\text{mod }\Lambda}$ is the maximum mutual information over all possible inputs $\mathbf{V}$; see (A.14).

3. As in Section 4.4, for $0 < \alpha < 1$, the estimation error (9.58) is a linear "mixture" of two components: channel noise weighted by α, and dither weighted by $\alpha - 1$. We call it a *mixture noise*. The weighted dither component is called *self-noise*, because it is a noise induced by the modulation system itself.

4. The *mixture ratio* is defined as the ratio between the variances of the channel noise component and the dither component:

$$\xi \triangleq \frac{\text{noise}}{\text{dither}} = \frac{\alpha^2 N}{(1 - \alpha)^2 P}, \tag{9.60}$$

where N is the variance of the noise $\mathbf{Z}$ (so far $\mathbf{Z}$ is not necessarily Gaussian).

5. The mean-squared estimation error $\frac{1}{n} E \|\hat{\mathbf{X}} - \mathbf{X}\|^2$ becomes in this case

$$\text{MSE}(\alpha) = \alpha^2 N + (1 - \alpha)^2 P. \tag{9.61}$$

6. And the lower bound (9.57) on the mod Λ channel capacity becomes

$$C_{\text{mod}\,\Lambda} \geq \frac{1}{2} \log \left(\frac{\text{SNR}}{\alpha^2 + (\alpha - 1)^2 \cdot \text{SNR}} \right) \tag{9.62a}$$

$$\triangleq C(\text{SNR}, \alpha), \tag{9.62b}$$

up to the $\frac{1}{2} \log(2\pi e G(\Lambda_2))$ shaping loss.

7. The lower bound (9.62) is maximized by the Wiener coefficient $\alpha^* = \text{SNR}/(1 + \text{SNR})$, so $C(\text{SNR}, \alpha)$ becomes the AWGN channel capacity C_{AWGN} (6.17).

Since for an AWGN channel, $C_{\text{mod}\,\Lambda}$ is always upper bounded by C_{AWGN} (due to the data-processing inequality (A.17)), we obtain the following.

Proposition 9.5.1 (Gap to C_{AWGN}) *For an AWGN channel with Wiener estimation ($\alpha = \alpha^* = \text{SNR}/(1 + \text{SNR})$), the capacity (9.56) of the* mod Λ *channel satisfies*

$$C_{\text{AWGN}} \geq C_{\text{mod}\,\Lambda} \geq C_{\text{AWGN}} - \frac{1}{2} \log(2\pi e G(\Lambda_2)), \tag{9.63}$$

where $C_{\text{AWGN}} = \frac{1}{2} \log(1 + \text{SNR})$ (6.17).

9.5.4   High-dimensional shaping

The negative term on the right-hand side of (9.63) is the *shaping loss* of the coarse lattice (9.22). This loss vanishes for a high-dimensional coarse lattice which is "good for quantization," i.e., for $G(\Lambda_2) \approx 1/2\pi e$ (Definition 7.3.2). Thus, the gap to the AWGN channel capacity can be closed at all SNR.

Observe that this high-dimensional Wiener-estimated mod Λ channel bridges between the Shannon and Wiener viewpoints discussed in Section 9.4. It converts the *reverse channel* (stemming from the orthogonality principle and the joint AEP) into a *forward channel*: in both these channels the output power is P and the noise power is $PN/(P + N)$.

The closing of the gap to the AWGN capacity happens in spite of several additional losses encountered in our analysis above: (i) using an *MMSE* criterion for estimation (rather than minimum-noise entropy which maximizes the capacity (9.56)); (ii) restriction to *linear scalar* estimation; and (iii) neglecting the noise entropy reduction due to folding, i.e., aliasing (Section 9.5.1).[14] How can that be?

Intuitively, for $G(\Lambda_2) \approx 1/2\pi e$ the dither $\mathbf{U}$ becomes white Gaussian (Theorem 7.2.3), thus the mixture $(\alpha - 1)\mathbf{U} + \alpha \mathbf{Z}$ of dither and AWGN in (9.59) becomes white Gaussian as well. Since the channel input is also white Gaussian, minimizing error entropy amounts to MMSE, which is achieved by a linear scalar estimator.

As for the third loss, the divergence data-processing inequality (A.19) implies that the entropy reduction due to folding is smaller for the noise than for the channel output. Thus, both these potential entropy reductions become negligible as the coarse lattice becomes a good quantizer.

9.5.5 Scalar shaping at low SNR

At low SNR conditions and small shaping dimensions, the lower bound (9.63) on the mod Λ channel capacity may be loose. In fact, it may even be negative if C_{AWGN} is small. To assess the low-SNR/low-dimension behavior of the mod Λ channel, consider its capacity (9.56) for a scalar shaping lattice $\Lambda_2 = \Delta \cdot \mathbb{Z}$:

$$C_{\mathrm{mod}\,\Delta\,\mathbb{Z}} = \log(\Delta) - h([\alpha Z + (\alpha - 1)U] \bmod \Delta), \qquad (9.64)$$

where $U \sim \mathrm{Unif}(0, \Delta)$, and Δ is the step size of the coarse lattice, so the transmission power is $P = \Delta^2/12$. Note that unlike for the case of a good shaping lattice, here the optimal value of the estimation coefficient α is *not* necessarily the Wiener coefficient.

Figure 9.11(A) shows $C_{\mathrm{mod}\,\mathbb{Z}}$ as a function of the SNR, for several values of α. These capacity curves are compared to the mutual information $I(U; U + Z)$, corresponding to a *non-modulo* uniform-input AWGN channel, and to the AWGN channel capacity C_{AWGN}. As the figure shows, while $I(U; U + Z)$ coincides with C_{AWGN} at low SNR, $C_{\mathrm{mod}\,\mathbb{Z}}$ grows much slower than C_{AWGN} for any value of α. In particular, $C_{\mathrm{mod}\,\mathbb{Z}}(\mathrm{SNR})$ is convex downward (∪) near the origin. Thus, if there is no strict amplitude constraint on the channel input, better performance in this range can be obtained by time sharing between the zero-rate/zero-SNR point, and some positive-rate/SNR point.

To see this behavior more clearly, Figure 9.11(B) shows the *capacity per unit cost* $C_{\mathrm{mod}\,\mathbb{Z}}(\mathrm{SNR})/\mathrm{SNR}$ of the mod $\mathbb{Z}$ channel. As we can see, the maximum value (of ≈ 0.28 [bit/unit energy]) is attained around an SNR equal to 1, for $\alpha \approx 1/2$. In

---

[14] We do not require Λ_2 to be a good AWGN channel code; thus, the probability that the mixture noise leaves the fundamental cell of Λ_2 (and possibly folds into itself) is not necessarily small.

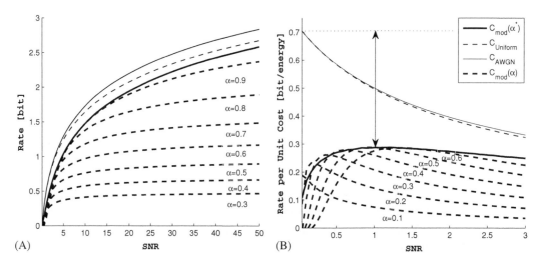

Figure 9.11 (A) The capacity $C_{\text{mod}\,\mathbb{Z}}(\text{SNR}, \alpha)$ of the mod $\mathbb{Z}$ channel as a function of the SNR for several values of α, compared to the Shannon capacity $C_{\text{AWGN}} = \frac{1}{2}\log(1 + \text{SNR})$, and the uniform-input (non-modulo) capacity $I(U; U + Z)$. (B) A zoom into the low SNR regime, showing the corresponding graphs of capacity per unit cost, $C_{\text{mod}\,\mathbb{Z}}(\text{SNR}, \alpha)/\text{SNR}$. For very low SNR (SNR < 0.5), the optimum linear coefficient α^{**} is, somewhat surprisingly, slightly larger than the Wiener coefficient α^*. Nevertheless, the best operation point in terms of energy per bit is around SNR = 1, where $\alpha^{**} \approx \alpha^* = 1/2$.

view of the slope $\log_2(e)/2 \approx 0.72$ [bit/unit energy] of $C_{\text{AWGN}}(\text{SNR})$ at SNR = 0 (see (9.38)), the mod $\mathbb{Z}$ channel loses about $10\log_{10}(0.72/0.28) \approx 4$ dB in *energy per bit* at the low SNR regime.

The gap between $C_{\text{mod}\,\mathbb{Z}}$ and $I(U; U + Z)$ reflects the *modulo loss*, i.e., the performance loss of lattice decoding compared to ML decoding for a uniform input. In Section 9.3.4 we observed that this loss vanishes in the high SNR regime. In contrast, as Figure 9.11 shows, lattice decoding is strictly inferior to ML decoding in the low SNR regime. Intuitively, while ML decoding enjoys a "natural shaping" at low SNR, lattice decoding does not (unless scalar shaping is replaced by a good high-dimensional shaping lattice).

9.6   Achieving C_{AWGN} for all SNR

The properties of the mod Λ channel indicate that from an information theoretic viewpoint, Voronoi modulation with good lattices can approach the AWGN channel capacity at *all* SNR. However, this promise does not apply directly to our system, because it ignores several factors.

1. **Granularity** The mod Λ channel capacity (9.56) ignores the restriction of the input **V** to be a point in the fine lattice Λ_1.

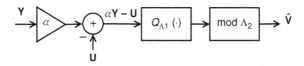

Figure 9.12 Linear Voronoi demodulation: linear estimation, undithering, lattice decoding and mapping to coset.

2. **One-shot coding** Achieving capacity requires, at least in theory, encoding across *many* channel inputs; for the mod Λ channel this means encoding across a super block of size $N = mn$ (n being the lattice dimension), where m is large. In contrast, the Voronoi modulator uses a *single* (n-dimensional) input vector in each coding session. [15]

3. **Mismatched decoding** Achieving capacity assumes, again in theory, that the decoder is *matched* to the distribution of the equivalent noise $\mathbf{Z}_{\text{eq}}$. Since the mixture $\alpha \mathbf{Z} + (\alpha - 1)\mathbf{U}$ is not quite Gaussian (for $\alpha < 1$), the decoder assumed by the capacity result is *not* the desired (simple) NN lattice decoder. [16]

Nevertheless, Theorem 9.6.1 below states that if the lattice dimension is large, then the Voronoi modulation and demodulation system (9.11) can approach the AWGN channel capacity *in spite of* the limitations above; i.e., using only points of Λ_1 as codewords, and with a Wiener-estimated NN lattice decoder:

$$\textbf{Encoding} \qquad \mathbf{x} = (\mathbf{v}_m + \mathbf{u}) \bmod_{\mathcal{V}_0} \Lambda_2 \tag{9.65a}$$

$$\textbf{Decoding} \qquad \hat{\mathbf{v}}_{\text{LD}} = \left(\underset{\lambda \in \Lambda_1}{\arg\min} \, \|\alpha \mathbf{Y} - \mathbf{u} - \lambda\| \right) \bmod \Lambda_2 \tag{9.65b}$$

where the vector $\mathbf{v}_m$ represents the message $1 \leq m \leq M$ to be transmitted, and $\mathbf{u}$ is the dither; see Figure 9.12. Note that for α equal to the Wiener coefficient, this scheme mimics the Wiener–Shannon estimate and decode idea discussed in Sections 9.4.2 and 9.4.3.

Theorem 9.6.1 (Asymptotic optimality) *The Voronoi modulation system (9.65), with good nested lattices, uniform (or modulo-uniform) dither with respect to Λ_2, Wiener estimation and Euclidean lattice decoding, achieves the AWGN channel capacity C_{AWGN}, with an arbitrarily small error probability.*

The precise definition of good nested lattices (Definition 9.6.2), as well as the proof of the theorem, appear towards the end of this section. Before that, we examine the decoding error probability and achievable rate for a general nested lattice pair.

---

[15] This is not an issue if $\Lambda_2 = (\Lambda'_2)^m$ is a product lattice, where m is large; see Sections 8.3.3 and 9.2.

[16] A Euclidean coset decoder (9.10) is partially matched to $\mathbf{Z}_{\text{eq}}$; it takes into account the folding into $\mathcal{P}_0(\Lambda_2)$, but not the uniform distribution of the self-noise (dither) component.

9.6.1   Error probability for a random dither

While for a specific dither value the decoding error probability depends on the transmitted message, a random uniform dither symmetrizes the error probability among the messages. Moreover, it greatly simplifies the analysis.

If the dither $\mathbf{U}$ is uniform (or modulo-uniform) over the coarse fundamental cell $\mathcal{P}_0$, then the statistical relation between the message representative $\mathbf{v}_m$ and the decision vector $\tilde{\mathbf{Y}} = [\alpha \mathbf{Y} - \mathbf{u}] \mod \Lambda_2$, is equivalent to the input/output relation in the mod Λ channel of Figure 9.9. Thus, by Lemma 9.5.1, the probability of error in NN lattice decoding is given by

$$P_e = \Pr\{\mathbf{Z}_{\text{eq}} \notin \mathcal{V}_0(\Lambda_1) \mod \Lambda_2\} \qquad (9.66)$$

for *any* transmitted message $\mathbf{v}_m$, where the notation $\{\mathbf{a} \in \mathcal{B} \mod \Lambda\}$ is short for $\{[\mathbf{a} \mod \Lambda] \in [\mathcal{B} \mod \Lambda]\}$. As in (9.19), this probability can be upper bounded by omitting the modulo-lattice reduction, i.e.,

$$P_e \leq \Pr\{\alpha \mathbf{Z} + (\alpha - 1)\mathbf{U} \notin \mathcal{V}_0(\Lambda_1)\} \triangleq \tilde{P}_e, \qquad (9.67)$$

where $\mathbf{U} \sim \text{Unif}(\mathcal{V}_0(\Lambda_2))$. Note that the upper bound $\tilde{P}_e$ is the NN (mismatched-decoding) error probability (6.32) of the fine lattice with respect to the mixture noise (9.58).

To define the NVNR in the presence of mixture noise, observe that as the noise variance N decreases from infinity to zero, the mixture error probability (9.66) decreases from 1 to a minimum value

$$P_{\min} \triangleq \Pr\{(\alpha - 1)\mathbf{U} \notin \mathcal{V}_0(\Lambda_1)\}, \qquad (9.68)$$

which is the fraction of the scaled coarse Voronoi cell $(\alpha - 1)\mathcal{V}_0(\Lambda_2)$ outside the fine Voronoi cell $\mathcal{V}_0(\Lambda_1)$.

Definition 9.6.1 (NVNR with mixture noise) *For P_e in the range $P_{\min} < P_e < 1$, the equivalent mixture NVNR is defined as*

$$\mu_{\text{mix}}(\Lambda_1, P_e; \Lambda_2, \alpha) \triangleq \mu_{\text{euclid}}(\Lambda_1, \alpha \mathbf{Z} + (\alpha - 1)\mathbf{U}, P_e), \qquad (9.69a)$$

where μ_{euclid} is the Euclidean ("mismatched") NVNR (6.33) of the fine lattice with respect to the mixture noise (9.58), i.e.,

$$\mu_{\text{euclid}}(\Lambda_1, \alpha \mathbf{Z} + (\alpha - 1)\mathbf{U}, P_e) = \frac{V^{2/n}(a\Lambda_1)}{\frac{1}{n} E \|\alpha \mathbf{Z} + (\alpha - 1)\mathbf{U}\|^2}, \qquad (9.69b)$$

where the scaling parameter a is such that $\Pr\{\alpha \mathbf{Z} + (\alpha - 1)\mathbf{U} \notin \mathcal{V}_0(a\Lambda_1)\} = P_e$. An alternative notation for the same quantity replaces the linear estimation coefficient α by the equivalent Gaussian to dither mixture ratio $\xi = \alpha^2 N / (1 - \alpha)^2 P$ (9.60):

$$\mu_{\text{mix}}(\Lambda_1, P_e; \Lambda_2, \xi). \qquad (9.69c)$$

Note that this definition is oblivious to the coupling between the nested lattices.

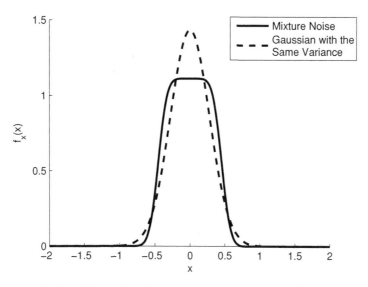

Figure 9.13 Distribution of the scalar mixture $\alpha Z + (\alpha - 1)U$, with $\sigma^2 = 1$, $U \sim$ Unif$(-1/2, 1/2)$, and $\alpha = 0.1$, compared to that of a pure Gaussian with the same variance. The tail of the mixture is clearly lighter.

Comparison to a pure Gaussian noise

The distribution of the mixture $\alpha \mathbf{Z} + (\alpha - 1)\mathbf{U}$ is more *concentrated* than that of a white-Gaussian distribution with the same variance σ^2. Thus, the resulting probability of error (9.67) tends to be larger than $P_e(\Lambda_1, \sigma^2)$ of Definition 3.3.2 for small VNR, and smaller than $P_e(\Lambda_1, \sigma^2)$ for large VNR; see Figure 9.13.

It is interesting to examine the implication of this behavior with respect to the relation between the mixture noise NVNR and that of a pure Gaussian. Since the mixture is more concentrated than a Gaussian distribution with the same variance, we have

$$\mu_{\mathrm{euclid}}(\Lambda_1, \alpha \mathbf{Z} + (\alpha - 1)\mathbf{U}, P_e) < \mu(\Lambda_1, P_e) \tag{9.70}$$

for $\alpha > 0$, provided that P_e is sufficiently small. See Appendix A.5. The condition of a small P_e is meaningful, however, only if $P_{\min}$ is close to zero, i.e., $(\alpha - 1)\mathcal{V}_0(\Lambda_2)$ is mostly contained in $\mathcal{V}_0(\Lambda_1)$.

What would be the best choice for the estimation coefficient α? While the mixture variance MSE(α) is always minimized by the Wiener coefficient $\alpha = \alpha^* = P/(P + N)$, the *tail* of the mixture distribution is *lighter* for a smaller value of α. The tail of the mixture is significant if the VNR of the fine lattice with respect to the AWGN is large, i.e., $V^{2/n}(\Lambda_1) \gg N$, in which case P_e is minimized by some $\alpha^{**} < \alpha^*$; see Figure 9.14.

A mixture noise is more favorable than pure AWGN also in terms of the *error exponent* of a random fine lattice. Section 13.8 shows that the decoding error

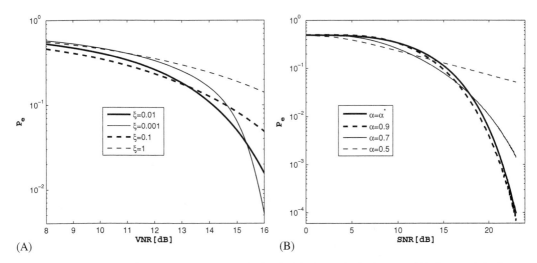

Figure 9.14 The mixture error probability (9.67) for scalar lattices. In (A) the x-axis is the VNR $\Delta_1^2/\mathrm{Var}(Z_{\mathrm{eq}})$, where the equivalent noise Z_{eq} is parameterized by the Gaussian to dither ratio ξ (we do not restrict the coarse and fine lattice to be nested). In (B) the x-axis is SNR $= (\Delta_2^2/12)/N$, where the nesting ratio $\Gamma = \Delta_2/\Delta_1 = 2$ is held fixed, and the equivalent noise is parameterized by the estimation coefficient α. In both figures we see that error-probability-wise, a Gaussian dominated mixture is better at low VNR or SNR, while a dither dominated mixture is better at high VNR or SNR.

probability (9.66) decays exponentially with the fine lattice dimension n:

$$P_e \approx e^{-nE(\mu,\alpha,\mathrm{SNR})}, \tag{9.71}$$

where $\mu = V^{2/n}(\Lambda_1)/N$ is the VNR of the fine lattice (which implicitly depends on the gap to capacity $C_{\mathrm{AWGN}} - R$), and where the SNR is given by $\mathrm{Var}(\mathbf{U})/\mathrm{Var}(\mathbf{Z}) = P/N$. The optimum value α^{**}, that maximizes the error exponent $E(\mu, \alpha, \mathrm{SNR})$, depends on both the VNR and SNR; for high SNR $\alpha^{**} = 1$ (a pure Gaussian noise), while for low SNR α^{**} vanishes (a dither dominated mixture). [17] As it turns out, for a fixed VNR, $E(\mu, \alpha^{**}, \mathrm{SNR})$ *increases* as the SNR decreases, i.e., the error probability becomes smaller as the mixture becomes "less Gaussian."

Interestingly, this behavior holds even if Λ_2 is a good shaping lattice, i.e., even if $\mathbf{U}$ tends to be closer to Gaussian as n goes to infinity.

9.6.2 The gap to capacity: general lattices

Equipped with the definition of the mixture noise NVNR (Definition 9.6.1), we turn to examine the gap to capacity of a given n-dimensional Voronoi modulation system, as we did for the high SNR case in Section 9.3. We begin with a lemma which

---

[17] In either case, α^{**} tends to be smaller than the Wiener coefficient $\alpha^* = \mathrm{SNR}/(1 + \mathrm{SNR})$ as μ grows.

characterizes the coding rate versus error probability of a general (not necessarily linear) Voronoi modulation system.

Lemma 9.6.1 (Achievable rate in Voronoi modulation and demodulation with a general estimator) *The coding rate $R(\Lambda_1/\Lambda_2)$ of the Voronoi modulation and demodulation system (9.11), with a uniform (or modulo-uniform) dither with respect to Λ_2, a general estimator $g(\cdot)$ and NN lattice decoding, is lower bounded by*

$$\frac{1}{2} \log\left(\frac{P}{\frac{1}{n} E \|\tilde{\mathbf{Z}}\|^2} \right) - \frac{1}{2} \log(G(\Lambda_2) \cdot \mu_{\text{euclid}}(\Lambda_1, \tilde{\mathbf{Z}}, P_e)), \tag{9.72}$$

for any random variable $\tilde{\mathbf{Z}}$ which is equal to $\mathbf{Z}_{\text{eq}}$ modulo the coarse lattice, i.e., $\tilde{\mathbf{Z}} \bmod \Lambda_2 = \mathbf{Z}_{\text{eq}}$, where $\mathbf{Z}_{\text{eq}}$ is the equivalent noise in the mod Λ channel (9.52b). Here, P is the transmission power, μ_{euclid} is the Euclidean ("mismatched") NVNR (6.33), and P_e is the decoding error probability (9.66). In particular, the lemma holds for $\tilde{\mathbf{Z}}$ which is equal to the estimation error $\hat{\mathbf{X}} - \mathbf{X} = g(\mathbf{Y}) - \mathbf{X}$, or to the equivalent noise $\mathbf{Z}_{\text{eq}}$ itself.

Like the equivalent mod Λ channel (Lemma 9.5.1), this lemma holds for a general channel $p(\mathbf{y}|\mathbf{x})$ (not necessarily additive). Note that for $\tilde{\mathbf{Z}} = \hat{\mathbf{X}} - \mathbf{X}$, (9.72) is equal to the lower bound on the mod Λ channel capacity (9.57), up to a coding-loss term of $\frac{1}{2} \log[\mu_{\text{euclid}}(\Lambda_1, \hat{\mathbf{X}} - \mathbf{X}, P_e)/2\pi e]$.

Proof By Lemma 9.5.1, the equivalent channel seen by the decoder is $\tilde{\mathbf{Y}} = [\mathbf{v} + \mathbf{Z}_{\text{eq}}] \bmod \Lambda_2$. Following the line of the proof in the non-estimation case (Theorem 9.3.1 and (9.20)), we have

$$R(\Lambda_1/\Lambda_2) = \frac{1}{n} \log\left(\frac{V(\Lambda_2)}{V(\Lambda_1)} \right) \tag{9.73a}$$

$$= \frac{1}{2} \log\left(\frac{P/G(\Lambda_2)}{\mu_{\text{euclid}}(\Lambda_1, \tilde{\mathbf{Z}}, \tilde{P}_e) \cdot \frac{1}{n} E \|\tilde{\mathbf{Z}}\|^2} \right), \tag{9.73b}$$

where in the denominator of (9.73b) we used the definition of the Euclidean NVNR (6.33), and where $\tilde{P}_e \triangleq \Pr\{\tilde{\mathbf{Z}} \notin \mathcal{V}_0(\Lambda_1)\}$ is the (unfolded) error probability of the fine lattice with respect to $\tilde{\mathbf{Z}}$. The lemma follows from (9.73b) since (i) the error probability P_e is equal to $\Pr\{\tilde{\mathbf{Z}} \notin \mathcal{V}_0(\Lambda_1) \bmod \Lambda_2\}$ for all $\tilde{\mathbf{Z}}$ satisfying the condition, (ii) $P_e \leq \tilde{P}_e$, and (iii) the NVNR is monotonic in the error probability. $\square$

Using the relation (9.69) between the Euclidean NVNR and the mixture NVNR in the linear estimation case, and substituting the linear estimation error $\alpha \mathbf{Z} + (\alpha - 1)\mathbf{U}_{\text{eq}}$ (the mixture noise) as $\tilde{\mathbf{Z}}$ in Lemma 9.6.1, we conclude the following theorem.

Theorem 9.6.2 (Achievable rate with linear estimation) *The Voronoi modulation and demodulation system (9.65), with uniform (or modulo-uniform) dither with respect to Λ_2, linear estimation $g(\mathbf{y}) = \alpha \mathbf{y}$ and NN lattice decoding, can achieve a* *coding rate of*

$$R(\Lambda_1/\Lambda_2) \geq \frac{1}{2}\log\left(\frac{P}{\mathrm{MSE}(\alpha)}\right) - \frac{1}{2}\log(G(\Lambda_2) \cdot \mu_{\mathrm{mix}}(\Lambda_1, P_e; \Lambda_2, \alpha)), \quad (9.74)$$

with a decoding error probability P_e (9.66), where P and $\mathrm{MSE}(\alpha)$ are the transmission power and the estimation mean-squared error (9.61), respectively, and μ_{mix} is the mixture noise NVNR (Definition 9.6.1).

For a small error probability, μ_{mix} can be replaced by the Gaussian NVNR $\mu(\Lambda_1, P_e)$; see (9.70).

Theorem 9.6.2 holds for *any* linear coefficient α. For $\alpha = 1$, it reduces to the non-estimation case of Section 9.3.1 (where $\mathrm{MSE} = N$, and the mixture NVNR becomes the usual Gaussian NVNR). Nevertheless, having in mind the asymptotic (large-dimensional) case, we plug in the MSE minimizer α^* (which achieves MSE of $PN/(P + N)$), and conclude the following.

Corollary 9.6.1 (Gap to capacity with Wiener estimation) *The gap to the AWGN channel capacity for $\alpha = \alpha^* = P/P + N$ is upper bounded by*

$$\Delta \triangleq C_{\mathrm{AWGN}} - R(\Lambda_1/\Lambda_2) \leq \frac{1}{2}\log(G(\Lambda_2) \cdot \mu_{\mathrm{mix}}(\Lambda_1, P_e; \Lambda_2, \alpha^*)) \quad (9.75)$$

for all $\mathrm{SNR} = P/N$. For a small error probability P_e (see (9.70)),

$$\Delta \leq \frac{1}{2}\log(G(\Lambda_2) \cdot \mu(\Lambda_1, P_e)), \quad (9.76)$$

i.e., the gap to capacity is at most the sum of shaping and coding losses of the high SNR case (Definition 9.3.1).

To close the gap to capacity, we look for nested lattice pairs for which both shaping and coding losses are small.

9.6.3 The gap to capacity: good lattices

We first observe that for a good shaping lattice, the mixture noise seen by the coding lattice is roughly AWGN.

Proposition 9.6.1 (Good mixture noise is semi-spherical) *If Λ_n is a sequence of good lattice quantizers (Definition 7.3.2), then, for any $0 \leq \alpha \leq 1$, the mixture $\mathbf{Z}_n = \alpha\mathbf{Z} + (\alpha - 1)\mathbf{U}_n$ of AWGN and uniform dither $\mathbf{U}_n \sim \mathrm{Unif}(\mathcal{V}_0(\Lambda_n))$ is a semi-spherical noise (Definition 7.8.2), i.e.,*

$$\frac{1}{n}h(\mathbf{Z}_n) \to \frac{1}{2}\log(2\pi e\mathrm{MSE}(\alpha)) \quad (9.77)$$

and

$$\Pr\left\{\mathbf{Z}_n \in \mathcal{B}\left(0, (1+\epsilon)\sqrt{n\mathrm{MSE}(\alpha)}\right)\right\} \to 1 \quad (9.78)$$

as $n \to \infty$ for any $\epsilon > 0$, where $\text{MSE}(\alpha) = \alpha^2 N + (\alpha - 1)^2 P$ *(9.61), and where N and P are the second moments of the AWGN $\mathbf{Z}$ and the lattice Λ_n, respectively.*

Proof The proof follows since the dither of good lattice quantizers is a semi-spherical noise (see discussion after Definition 7.8.2), and since the sum of a semi-spherical noise and AWGN is a semi-spherical noise (Problem 7.12). See also Ordentlich and Erez [206]. □

Based on this fact, we extend Definition 9.3.2 of good Voronoi modulation lattices, to the general SNR case.

Definition 9.6.2 (Good Voronoi modulation: general SNR) *We say that a sequence $(\Lambda_1^{(n)}, \Lambda_2^{(n)})$ of nested lattices of a growing dimension n is "good for Voronoi modulation," if*

(i) the coarse lattice is a good quantizer, i.e., $G(\Lambda_2^{(n)}) \to 1/2\pi e$ (Definition 7.3.2); and

(ii) the fine lattice is a good channel code under NN decoding against the mixture noise $\mathbf{Z}_n = \alpha \mathbf{Z} + (\alpha - 1)\mathbf{U}_n$, where $\mathbf{U}_n \sim \text{Unif}(\mathcal{V}_0(\Lambda_2^{(n)}))$, i.e.,

$$\mu_{\text{mix}}\left(\Lambda_1^{(n)}, P_e; \Lambda_2^{(n)}, \alpha\right) \to 2\pi e \qquad (9.79)$$

as $n \to \infty$, for all $P_e > 0$ and $0 < \alpha \leq 1$ (Definition 7.8.3).

The existence of nested lattices good for Voronoi modulation follows from Theorem 8.5.1, and the tendency of the mixture noise (for good shaping lattices) to AWGN (Proposition 9.6.1). Using such lattices in Corollary 9.6.1 proves Theorem 9.6.1.

9.7 Geometric interpretation

The linear estimation lattice decoder (9.65b) first attenuates the channel output, and then decodes to the nearest codeword in the codebook (9.3). An equivalent solution is to take the channel output *as is* (unattenuated), and then decode to the nearest codeword in an *inflated* codebook. Other equivalent solutions can be obtained by combining partial attenuation and inflation.

Specifically, the linear estimation lattice decoder (9.65b) is equivalent to

$$\hat{\mathbf{v}}_{\text{LD}} = \alpha \left[\left(\arg\min_{\lambda' \in (\Lambda_1/\alpha)} \left\| \mathbf{Y} - \frac{\mathbf{u}}{\alpha} - \lambda' \right\| \right) \mod \frac{\Lambda_2}{\alpha} \right]. \qquad (9.80)$$

That is, the decoded codeword λ' belongs to an *inflated codebook*

$$\mathcal{C}_{\Lambda_1/\alpha, \mathcal{P}_0/\alpha} = \frac{\Lambda_1}{\alpha} \mod \frac{\Lambda_2}{\alpha}. \qquad (9.81)$$

See identities (2.43)–(2.44) and Figure 9.15.

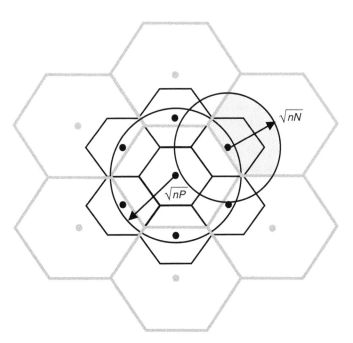

Figure 9.15 A lattice inflation interpretation of lattice decoding after linear estimation with $\alpha < 1$, versus ML decoding. It is seen from the picture that the inflated decision cells capture a higher portion of the output probability mass than the original decision cells.

This kind of *mismatch* (or bias) between the encoder and the decoder is not common. In fact, biased estimation is *sub*optimal under NN decoding to the shaped codebook. Nevertheless, interestingly, the optimality of the linear estimation lattice decoder (9.65b) at high dimensions implies that, under the restriction to *lattice decoding*, bias may be beneficial.

We can think of codebook inflation as if the attenuation is performed at the *encoder*. That is, the encoder attenuates the codebook (by α) before the transmission; the codebook power is initially P/α^2, so the power constraint is satisfied after the attenuation. The decoder then searches for the nearest point to $\mathbf{Y}$ in the (unattenuated) codebook.

Figure 9.16 shows a generalization of this scheme, where the codebook is attenuated by α_1 before the transmission, while the channel output is attenuated by α_2 before the decoding, where $\alpha_1\alpha_2 = \alpha$. In a *symmetric* system, both the encoder and decoder attenuate by the same factor $\alpha_1 = \alpha_2 = \sqrt{\alpha}$.

9.8   Noise-matched decoding

In Section 9.6 we restricted our attention to NN (Euclidean) decoding (9.65b). Luckily, this was sufficient to achieve capacity, because the dither – and hence the equivalent mixture noise $\mathbf{Z}_{\text{eq}}$ – become closer to AWGN for a sequence of good

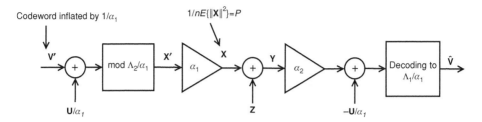

Figure 9.16 Attenuation divided between the encoder and decoder. It can be verified that the decoder coefficient α_2 is the optimal linear estimator of the "inflated signal" $\mathbf{X}'$.

(high-dimensional) coarse lattices. In return, the error probability in NN decoding of a good fine lattice tends to zero as the dimension goes to infinity.

Practical coding schemes, however, tend to separate shaping from coding, and give up the shaping gain, i.e., use a *simple cubic* coarse lattice; see the discussion on mixed-nesting dimensions in Section 9.3.3. In this case, the uniform dither U, and hence the mixture noise $\alpha Z + (\alpha - 1)U$, are not quite Gaussian (in particular at *low SNR*; see Figure 4.9).

Another feature of practical coding schemes is that they are based on modulo-q (construction A) lattices. This means that the fine lattice – due either to cubical shaping or to construction A – contains a cubic sub-lattice. Hence, it cannot be "good for AWGN channel coding"; see Section 7.9.5. While this is not a real problem for a bounded (Voronoi) lattice constellation (because all the points in a cubic (sub-) coset carry the same message, so intra-coset transitions are not counted as error events), the NN decoder fails to take advantage of this structure.

To resolve the two problems above, Gaussian mixture noise and cubic-coset messages, the lattice decoder needs to *match* the noise true statistics, i.e., replace the NN rule in (9.65b) by a fine lattice quantizer $Q_{\Lambda_1}(\cdot)$ that performs ML decoding with respect to the equivalent noise $\mathbf{Z}_{eq}$; see (6.11) and Section 9.1.3. Such noise-matched lattice decoding means taking into account two factors:

1. the folding (mod Λ_2) of the noise (which amounts to performing *coset decoding* (9.10)); and
2. the uniformity (non-Gaussianity) of the dither component $\mathbf{U}$.

Note that this still amounts to lattice decoding because Q_{Λ_1} ignores the shape $\mathcal{V}_0(\Lambda_2)$ of the Voronoi codebook.

To see the improvement due to noise-matched decoding, we shall find a finer characterization for the performance of *Euclidean decoding* over the mod Λ channel. For that, consider the following (tighter) lower bound on the mod Λ channel capacity, due to (9.55b):

$$C_{\text{mod }\Lambda} \geq \underline{C}_{\text{mod }\Lambda}^{(\text{euclid-th})} \triangleq \left[\frac{1}{n} \log[V(\Lambda_2)] - \frac{1}{2} \log\left(2\pi e\sigma_{eq}^2\right)\right]^+, \qquad (9.82)$$

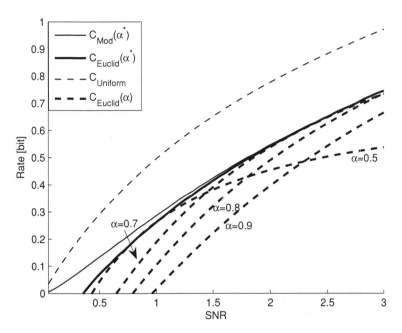

Figure 9.17 Capacity curves for scalar Voronoi modulation with noise-matched lattice decoding and Euclidean-threshold lattice decoding. For comparison, the non-modulo capacity $I(U, U + Z)$ is also shown.

where

$$\sigma_{eq}^2 \triangleq \frac{1}{n} E \| [\hat{\mathbf{X}} - \mathbf{X}] \bmod_{V_0} \Lambda_2 \|^2 \tag{9.83}$$

is the second moment per dimension of the equivalent noise under a Voronoi modulo operation, and $[x]^+ = \max\{0, x\}$. A *Euclidean-threshold decoder* searches for a unique codeword within a sphere (modulo Λ_2) of radius r_D around the received vector, where r_D is slightly larger than $\sqrt{n\sigma_{eq}^2}$; see (6.35) and (7.74). The right-hand side of (9.82) is a lower bound on the *mismatched* Euclidean-threshold decoder *capacity* of the mod Λ channel, denoted $\underline{C}_{\bmod \Lambda}^{(euclid-th)}$; see Section 6.4.1 and Problems 9.9–9.10.

Figure 9.17 shows the capacities $C_{\bmod \mathbb{Z}}$ and $\underline{C}_{\bmod \mathbb{Z}}^{(euclid-th)}$ as a function of the SNR, for a cubic shaping lattice. For the noise-matched capacity we use formula (9.64), while for the Euclidean-threshold capacity we use

$$\underline{C}_{\bmod \mathbb{Z}}^{(euclid-th)} = \left[\log(\Delta) - \frac{1}{2} \log \left(2\pi e E \left([\alpha Z + (\alpha - 1)U] \bmod \Delta \right)^2 \right) \right]^+, \tag{9.84}$$

where mod Δ here is to the interval $(-\Delta/2, \Delta/2)$. As in Figure 9.11, at the low SNR regime both curves are strictly below $I(U, U + Z)$, the information rate of a non-modulo AWGN channel; but $\underline{C}_{\bmod \mathbb{Z}}^{(euclid-th)}$ is even lower, and it vanishes below some positive value of SNR.

Note that the capacity curves above do not reflect the coding loss, i.e., they correspond to a good (high-dimensional) fine lattice. For a general fine lattice, a coding loss term ($\frac{1}{2}\log(\mu_{\text{matched}}/2\pi e)$ or $\frac{1}{2}\log(\mu_{\text{euclid}}/2\pi e)$) should be subtracted from the corresponding capacity; see the definitions of noise-matched and Euclidean NVNR in Section 6.4.

The gap between the noise-matched capacity (9.64) and the lower Euclidean-threshold capacity (9.84) (provided that the latter is positive) can be interpreted as the divergence of the equivalent noise Z_{eq} from Gaussianity (7.38):

$$\text{Noise-matching gain} = D(Z_{\text{eq}}; Z_{\text{eq}}^*). \tag{9.85}$$

This gain is upper bounded by the divergence of the dither from Gaussianity $\frac{1}{2}\log(2\pi e/12)$; and it vanishes at the high SNR regime, where Z_{eq} becomes Gaussian, so NN (or Euclidean-threshold) decoding achieves capacity.

9.9 Is the dither really necessary?

We return to the question we addressed in the context of quantization (the ECDQ in Section 5.7), now in the context of Voronoi modulation with lattice decoding: is the dither only an analytical tool, and does it have to be random?

The following example shows that for binary modulation, i.e., an un-shaped (cubic) constellation, dithering is wasteful in terms of power, and is unnecessary for ML decoding.

Example 9.9.1 (Randomized BPSK) *Consider a dithered and coded binary constellation, as described in Example 8.3.1, where the fine lattice is a (scaled) modulo-2 lattice, and the coarse lattice is $4\mathbb{Z}$. The transmit power with a random uniform dither $U \sim \text{Unif}(-2, 2)$ is $P = \sigma^2(4\mathbb{Z}) = 4^2/12 \approx 1.33$. In contrast, for a standard (non-randomized) BPSK constellation (which corresponds to a fixed dither value $u = -1$) the power is only $P = 1$. For ML decoding, since the error probability is invariant to the dither value, this 33% extra power of a randomized constellation does not give us any benefit. The Shannon capacity of a standard BPSK constellation at $\text{SNR} = 1$ (i.e., noise power of $N = 1$), for example, is $\approx C_{\text{AWGN}} = 1/2$ bit (see Figure 9.6), and randomization cannot increase it. In contrast, the mod Λ channel capacity is $C_{\text{mod4}} = \log(4) - h(Z_{\text{eq}}) \approx 0.38$ bit for a noise-matched decoder (9.56), and $C_{\text{mod4}}^{(\text{euclid-th})} = \log(4) - \frac{1}{2}\log(2\pi e \frac{4/3}{4/3+1}) \approx 0.36$ bit for a Euclidean-threshold decoder (9.84); see Figures 9.11 and 9.17. Thus, the restriction to lattice decoding reduces the capacity, in spite of the $\approx 33\%$ extra power of the modulator.*

Binary modulation, however, cannot achieve capacity (unless the SNR is low; see Figure 9.6). Let us then turn to examine the role of the dither in a general Voronoi-shaped modulation system. Since the coding rate R is invariant to the

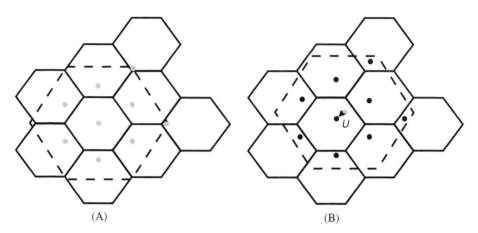

(A)   (B)

Figure 9.18 Effect of scaling and dither on codewords and decision cells. (A) undithered scaled case ($\alpha = 2/3$); (B) dithered and scaled case.

dither (Proposition 8.4.1), we examine the effect of the dither on the two other system parameters: error probability and power.

Dithering is insignificant at high-SNR/high-rate conditions. Since the constellation points are very dense inside the Voronoi shaping region, the transmit power is almost unaffected by shifting these points (modulo the shaping region). Also, for the no-estimation case ($\hat{\mathbf{X}} = \mathbf{Y}$ or $\alpha = 1$), which is optimal at high SNR, the error probability in lattice decoding is invariant to the transmitted message and the value of the dither (see Lemma 9.3.1).

In contrast, at *non*-high-SNR conditions, where the optimal estimation coefficient α is smaller than 1, *random* dither plays a crucial role in equalizing the power and error probability among the messages. Let us first examine the case of a *specific* (non-random) dither vector $\mathbf{u}$ and transmitted message $\mathbf{v}_m$, as illustrated in Figure 9.18. In light of (9.65b), the error probability is the probability that the attenuated output $\alpha \mathbf{Y}$ leaves the mth fine Voronoi cell, shifted by the dither $\mathbf{u}$:

$$P_e(\mathbf{v}_m, \mathbf{u}) = \Pr\{\alpha \mathbf{Y} \notin [\mathbf{u} + \mathbf{v}_m + \mathcal{V}_0(\Lambda_1)] \mod \Lambda_2\}. \tag{9.86}$$

Using the same steps as in the analysis of the mod Λ channel (Lemma 9.5.1), we rewrite (9.86) as

$$P_e(\mathbf{v}_m, \mathbf{u}) = \Pr\{[\alpha \mathbf{Z} + (\alpha - 1)\mathbf{x}] \notin \mathcal{V}_0(\Lambda_1) \mod \Lambda_2\} \tag{9.87}$$

where $\mathbf{x} = \mathbf{x}(\mathbf{v}_m, \mathbf{u})$ (9.65a); hence the "self-noise" component $(\alpha - 1)\mathbf{x}$ here is a deterministic (message dependent) bias.

Observe that if $\alpha \neq 1$, then $P_e(\mathbf{v}_m, \mathbf{u})$ in (9.87) depends on the transmitted message $\mathbf{v}_m$. Figure 9.19 demonstrates this dependence by a numerical example. To appreciate the variation, suppose that the dither exactly "matches" one of the message representative vectors, i.e., $\mathbf{u} = -\mathbf{v}_j$ for some j, for example, assume a zero dither

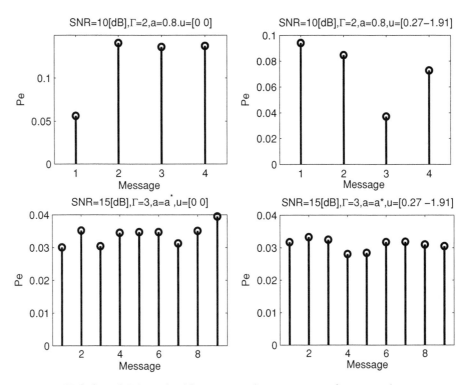

Figure 9.19 Variation of $P_e(\mathbf{v}_m, \mathbf{u})$ with respect to the messages m, for zero and non-zero dither. (Hexagonal constellations with nesting ratios $\Gamma = 2$ and 3. The linear coefficient is denoted by a.)

$\mathbf{u} = 0$ for the zero codeword $\mathbf{v}_0$. In this case, the self-noise component $(\alpha - 1)\mathbf{x}$ is zero if $\mathbf{v}_0$ is transmitted, and non-zero for all other messages (see Figure 9.18), implying that the error probability is minimum if message $\mathbf{v}_0$ is transmitted:

$$P_e(\mathbf{v}_m, \mathbf{u} = 0) > P_e(\mathbf{v}_0, \mathbf{u} = 0) = \Pr\{\alpha\mathbf{Z} \notin \mathcal{V}_0(\Lambda_1) \mod \Lambda_2\}. \qquad (9.88)$$

See Figure 9.19 and Problem 9.11. More generally, a dither vector $\mathbf{u}$ tends to favor a message $\mathbf{v}_m$ where $\mathbf{u} + \mathbf{v}_m$ is close to the origin.

Figure 9.20 shows that even the *average* (power and error probability) over the *messages* may still vary with the value of the dither. (Here the average power is $\bar{P}(\mathbf{u}) = \frac{1}{nM} \sum_m \|\mathbf{x}(\mathbf{v}_m, \mathbf{u})\|^2$ and the average error probability is $\bar{P}_e(\mathbf{u}) = 1/M \sum_m P_e(\mathbf{v}_m, \mathbf{u})$.) It follows that if one only cares about the message-average performance (i.e., can tolerate power and error-probability variation among the messages), then each desired operation point (i.e., a pair of average power and error probability) is optimized by some *specific* (non-random) dither value.

What happens to this variation for good (high-dimensional) lattices? We know from the discussion earlier in this chapter that Voronoi modulation with random dither and Wiener-estimated lattice decoding asymptotically achieves the Shannon

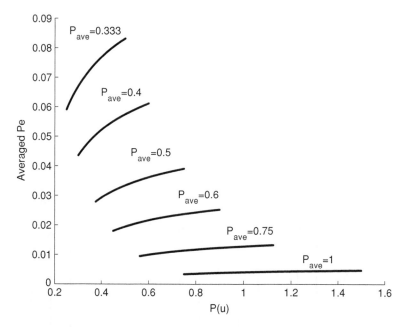

Figure 9.20 Variation of the message-average power $\bar{P}(u)$ and error probability $\bar{P}_e(u)$, as a function of the dither value u, for scalar Voronoi modulation (PAM) at a rate $R = 1$ bit. Each curve is for a different constellation amplitude Δ, where $P_{\text{ave}} \triangleq \Delta^2/12$ is the average power for a random dither.

capacity, i.e., the best attainable performance. Thus, the variation (around the uniform dither case) of the message-average power $\bar{P}(\mathbf{u})$ and error probability $\bar{P}_e(\mathbf{u})$ must vanish as the dimension goes to infinity – otherwise we could find a dither value for which the Shannon bound is exceeded.

Indeed, for a large lattice dimension, $\mathbf{u}$ is typically *orthogonal* to the message vector $\mathbf{v}_m$, and $\|\mathbf{x}(\mathbf{v}_m, \mathbf{u})\|$ is typically $\approx \sqrt{nP}$. Hence, asymptotically, *almost all* dither values (though not necessarily the zero dither) are good in terms of the message-average power and error probability.

9.10   Voronoi quantization

The duality of source coding and channel coding (Chapter 2) suggests that the encoder and decoder in quantization are the counterparts, in reversed order, of the encoder and decoder in modulation. This is indeed a useful view in the dithered Voronoi codebook setting, with some small modifications.

Consider the quantization problem of Chapter 4, where a vector source $\mathbf{S}$ is encoded under some per letter difference distortion measure $d(\hat{s} - s)$, and reconstructed as $\hat{\mathbf{S}}$. Let $\Lambda_2 \subset \Lambda_1$ be a pair of n-dimensional nested lattices, with relative cosets Λ_1/Λ_2; let $\mathcal{P}_0$ and $\mathcal{V}_0$ denote *enumeration* and *shaping* fundamental cells, respectively, of the coarse lattice Λ_2; finally let $\mathcal{C}_{\Lambda_1, \mathcal{P}_0} = \{\mathbf{v}_1, \dots, \mathbf{v}_M\}$ denote the corresponding codebook (Chapter 8 and Section 9.1).

Given a source vector $\mathbf{S} = \mathbf{s}$, the encoding and decoding operations are defined as follows:

$$\text{Encoding} \quad \mathbf{v} = Q_{\Lambda_1}(\alpha \mathbf{s} + \mathbf{u}) \bmod \Lambda_2 \tag{9.89}$$

$$\text{Decoding} \quad \hat{\mathbf{s}} = \beta \cdot [(\mathbf{v} - \mathbf{u}) \bmod \nu_0 \Lambda_2], \tag{9.90}$$

where $\mathbf{u}$ is the dither, and α and β are scalar "estimation" coefficients. The quantizer $Q_{\Lambda_1}(\cdot)$ is matched to the distortion measure $d(\cdot)$; in the squared-distortion case, its output is given by the NN rule:

$$Q_{\Lambda_1}(\alpha \mathbf{s} + \mathbf{u}) = \arg \min_{\lambda \in \Lambda_1} \|\alpha \mathbf{s} + \mathbf{u} - \lambda\|. \tag{9.91}$$

Correct decoding is the event where $\hat{\mathbf{S}} = \beta[Q_{\Lambda_1}(\alpha \mathbf{s} + \mathbf{u}) - \mathbf{u}]$, i.e., the modulo-$\Lambda_2$ operation is not effective. The error, or *overload*, probability (for random source $\mathbf{S}$ and dither $\mathbf{U}$), is defined as the probability of the complementary event, $P_e = \Pr\{\hat{\mathbf{S}} \neq \beta[Q_{\Lambda_1}(\alpha \mathbf{s} + \mathbf{u}) - \mathbf{u}]\}$.

This system is similar to Voronoi modulation (Figure 9.16), only the encoder and decoder switch roles. At *high-resolution* quantization we set $\alpha = \beta = 1$, so "correct decoding" amounts to the source $\mathbf{S}$ falling in the coarse Voronoi region $\mathcal{V}_0$. The coding rate $R(\Lambda_1/\Lambda_2) = \frac{1}{n} \log |\Lambda_1/\Lambda_2|$ becomes in this case

$$R(\Lambda_1/\Lambda_2) = R^*(D_{cd}) + \frac{1}{2} \log(G(\Lambda_1) \cdot \mu(\Lambda_2, P_e)), \tag{9.92}$$

where $R^*(D) = \frac{1}{2} \log(\sigma_s^2/D)$ is the quadratic-Gaussian rate-distortion function (5.33), and $D_{cd} = \sigma^2(\Lambda_1)$ is the "granular distortion," i.e., the distortion conditional on correct decoding.

The second term in (9.92) is the *shaping and overload redundancy*, paralleling the shaping and coding loss in Voronoi modulation at high SNR (9.21), but with the coarse and fine lattices switching roles. For a large lattice dimension and "good" nested lattices, this redundancy term goes to zero.

Definition 9.10.1 (Good Voronoi quantization: high resolution) *We say that a sequence $(\Lambda_{2,n} \subset \Lambda_{1,n})$ of nested lattice pairs of increasing dimension n is "good for Voronoi quantization at high resolution," if $G(\Lambda_{1,n}) \to 1/2\pi e$, and $\mu(\Lambda_{2,n}, P_e) \to 2\pi e$ for all $P_e > 0$.*

Note that we can let $P_e \to 0$, implying that $D_{cd} \to D$, where $D = E\|\hat{\mathbf{S}} - \mathbf{S}\|^2/n$ is the *total distortion*, so the Voronoi quantizer approaches $R^*(D)$. [18]

For a *general resolution*, we observe that as $P_e \to 0$, the system becomes a *filtered dithered quantizer* $\hat{\mathbf{s}} = \beta \cdot [Q_{\Lambda_1}(\alpha \mathbf{s} + \mathbf{u}) - \mathbf{u}]$ (see Sections 4.5 and 5.6.2).

---

[18] Since the overload distortion (i.e., the conditional distortion given incorrect decoding) is bounded (by $\sigma^2(\Lambda_2)$), its contribution to the average distortion D becomes negligible as $P_e \to 0$.

Hence, for a uniform (or modulo-uniform) dither with respect to Λ_1, the optimal coefficients α and β follow from Wiener estimation principles. However, we shall not make a precise analysis of this case, for two reasons: (i) Voronoi quantization (with a high-dimensional shaping lattice) is not common in practical point-to-point coding systems; (ii) Voronoi quantization is a useful concept in source coding with side information (the *Wyner–Ziv problem*), which we address in Chapter 10.

Summary of Chapter 9

Voronoi modulation

$$\mathbf{x}(\mathbf{v}_m, \mathbf{u}) = [\mathbf{v}_m + \mathbf{u}] \bmod {}_{\mathcal{V}_0}\Lambda_2,$$

where $1 \leq m \leq M = |\Lambda_1/\Lambda_2|$ is the message, $\mathbf{v}_m$ is the message representative vector and $\mathbf{u}$ is a dither.

Coding rate $R(\Lambda_1/\Lambda_2) = \frac{1}{n}\log_2 |\Lambda_1/\Lambda_2|$ bit per channel use.

Transmit power If the dither $\mathbf{U}$ is modulo-uniform with respect to Λ_2, then

$$\frac{1}{n}E\|\mathbf{X}\|^2 = \sigma^2(\Lambda_2), \quad \text{for all } 1 \leq m \leq M.$$

Lattice decoding $\hat{\mathbf{v}} = Q_{\Lambda_1}[g(\mathbf{Y}) - \mathbf{u}] \bmod \Lambda_2$, where $g(\mathbf{Y})$ estimates $\mathbf{X}$.

Error probability For NN decoding ($Q_{\Lambda_1} = Q_{\Lambda_1}^{(NN)}$) and no estimation ($g(\mathbf{y}) = \mathbf{y}$),

$$P_e \leq \Pr(\mathbf{Z} \notin \mathcal{V}_0(\Lambda_1)), \quad \text{for all } 1 \leq m \leq M.$$

Gap to capacity at high SNR

$$C_{\text{AWGN}} - R(\Lambda_1/\Lambda_2) = \underbrace{\frac{1}{2}\log(2\pi e G(\Lambda_2))}_{\text{Shaping loss}} + \underbrace{\frac{1}{2}\log\left(\frac{\mu(\Lambda_1, P_e)}{2\pi e}\right)}_{\text{Coding loss}}.$$

Equivalent mod Λ channel

$$\tilde{\mathbf{Y}} = [\mathbf{v} + \mathbf{Z}_{\text{eq}}] \bmod \Lambda_2,$$

where $\mathbf{Z}_{\text{eq}}$ is the estimation error $g(\mathbf{Y}) - \mathbf{X}$ modulo Λ_2.

Coding rate with noise-matched lattice decoding

$$R(\Lambda_1/\Lambda_2) \geq C_{\text{mod }\Lambda} - \underbrace{\frac{1}{2}\log\left(\frac{\mu_{\text{matched}}(\Lambda_1, \mathbf{Z}_{\text{eq}}, P_e)}{2\pi e}\right)}_{\text{Coding loss}}.$$

Error probability with linear estimation $g(\mathbf{y}) = \alpha\mathbf{y}$

$$P_e = \Pr(\mathbf{Z}_{eq} \notin \mathcal{V}_0(\Lambda_1) \mod \Lambda_2), \quad \text{for all } 1 \le m \le M,$$

where

$$\mathbf{Z}_{eq} = [\alpha\mathbf{Z} + (\alpha - 1)\mathbf{U}] \mod \Lambda_2.$$

For *noise-matched* lattice decoding, $\mathcal{V}_0(\Lambda_1)$ is replaced by $\mathcal{V}_0^{\mathbf{Z}_{eq}}(\Lambda_1)$.

Gap to capacity with Wiener estimation ($\alpha^* = P/P + N$)

$$C_{AWGN} - R(\Lambda_1/\Lambda_2) \le \text{Shaping loss}(\Lambda_2) + \text{Coding loss}(\Lambda_1).$$

Good lattices and deterministic dither If Λ_1 and Λ_2 are good for shaping and coding, respectively, then $R(\Lambda_1/\Lambda_2)$ approaches C_{AWGN} for all SNR, and for almost any dither value $\mathbf{U} = \mathbf{u}$.

Problems

P.9.1 (Decision vector) Prove that the lattice decoder (9.11c) is equivalent to (9.11e), i.e., calculate the decision vector $\tilde{\mathbf{Y}}$ (undither and reduce modulo Λ_2), then quantize and reduce again modulo Λ_2.

P.9.2 (High SNR performance) Prove the "only if" part of Theorem 9.3.1 in the limit of small noise power. (Guidance: show that the probability of the event that (9.19b) ignores is upper bounded by the probability that the noise exceeds the shaping region.)

P.9.3 (PAPR) (a) Show that the PAPR (9.27) of cubical shaping $\Lambda_2 = \mathbb{Z}^n$ is equal to three. (b) Show that for a general Λ_2, the PAPR is upper bounded by $n\rho_{cov}^2(\Lambda_2)/2\pi eG(\Lambda_2) < n\rho_{cov}^2(\Lambda_2)$. Conclude that for Rogers-good lattices, the PAPR increases linearly with the dimension.

P.9.4 (Pre-emphasis) Consider the channel $Y = (1 + X)(1 + Z)$, where $Z \sim N(0, N)$, and where the input X is non-negative and is subject to the constraint $E[\log(1 + X)]^2 \le P$. (a) Prove that the high SNR capacity of this channel is $C_{HSNR} = \frac{1}{2}\log(P/N)$. (b) Find a suitable pre-emphasis $X = f(\tilde{X})$ (see (9.11d)) and an estimator $\hat{X} = g(Y)$ in (9.11), so that Voronoi modulation with lattice decoding achieves C_{HSNR} at high SNR. (Guidance: for both parts use the identity $h(\log(X)) = h(X) - E\{\log(X)\}$.)

P.9.5 (Modulo loss at high SNR) Prove that the modulo loss vanishes at high SNR (9.31). Hint: use the chain rule and the fact that $\mathbf{U} \leftrightarrow \mathbf{U} + \mathbf{Z} \leftrightarrow [\mathbf{U} + \mathbf{Z}] \mod \Lambda_2$

form a Markov chain, to show that

$$I(\mathbf{U}; \mathbf{U} + \mathbf{Z} \mid [\mathbf{U} + \mathbf{Z}] \bmod \Lambda_2) = I(\mathbf{U}; \mathbf{U} + \mathbf{Z}) - I(\mathbf{U}; [\mathbf{U} + \mathbf{Z}] \bmod \Lambda_2);$$
(9.93)

see Appendix A.1. Then, show that $[\mathbf{U} + \mathbf{Z}] \bmod \Lambda_2$ converges to $\mathbf{U} + \mathbf{Z}$ in distribution as $\sigma_z^2 \to 0$, and use the semi-continuity of the divergence [164, 219].

P.9.6 (Modulo-Voronoi) Show that a modulo-Voronoi operation does not increase the second moment. (Hint: see Problem 2.7.) Give an example of a non-zero-mean variable whose *variance* increases by a modulo-Voronoi operation.

P.9.7 (Lemma A.3.2: modulo reduces entropy)) For a random variable X with a differential entropy $h(X)$ and a lattice Λ, prove that:

$$h(X) \geq h(X \bmod \Lambda)$$

with equality if and only if X is contained with probability 1 in some fundamental cell of Λ. (Hint: use the data-processing inequality: $I(Q; X \bmod \Lambda) \leq I(Q; X)$, where $Q = Q_{\Lambda, P_0}(X)$ is the quantizer associated with the modulo operation.)

P.9.8 (Random coding interpretation of lattice decoding [2]) Let $\mathcal{C}_1$ be a "large" random Gaussian codebook of dimension n, power P_1 and with 2^{nR_1} codewords. Let $\mathcal{C}$ be the truncation of $\mathcal{C}_1$ to a ball of power P (i.e., of radius $\sqrt{nP}$), where $P < P_1$, and let 2^{nR} be the number of codewords in $\mathcal{C}$. The encoder uses the "small" codebook $\mathcal{C}$ for transmission over an AWGN channel $Y = X + Z$ with noise power N. The decoder is a reverse-typicality decoder matched to the *large* codebook, i.e., it uses the rule (9.50b), with $X \sim N(0, P_1)$ and $Y = X + Z$. Let SNR $= P/N$. Show that reliable decoding is possible for R smaller than the α-capacity $C(\text{SNR}, \alpha)$, (9.62b), with $\alpha = P_1/(P_1 + N)$. Guidance: show that the true codeword is within a distance $\sqrt{nN(\alpha^2 + (1 - \alpha)^2 \text{SNR})}$ from αY, and follow the argument in (9.51).

P.9.9 (Mismatched-decoding capacity) Let $\mathcal{Z}_{d,a}$ denote the family of random variables Z in the interval $[0, a)$ satisfying some d-moment constraint $E\{d(Z)\} \leq d$, and let $Z_{\text{ME}} = \max_{Z \in \mathcal{Z}_d} h(Z)$ denote the maximum-entropy variable in the family. The d-threshold capacity is the maximum achievable rate with a decoder of the form:

$$\hat{\mathbf{c}} = \begin{cases} \mathbf{c}, & \text{if } d([\mathbf{Y} - \mathbf{c}] \bmod a) < d_{\text{th}} \text{ for a unique } \mathbf{c} \in \mathcal{C}, \\ ?, & \text{otherwise}, \end{cases}$$

see (6.35) for the case where d is the squared-Euclidean distance. Show that the d-threshold capacity of a memoryless modulo-additive noise channel $Y = X + Z \bmod a$, with noise $Z \in \mathcal{Z}_{d,a}$, is at least the regular capacity with $Z = Z_{\text{ME}}$, i.e., $C^{d\text{-th}} \geq \log(a) - h(Z_{\text{ME}})$.

Guidance: use Lemma 7.8.1 with a decoding region equal to a d-ball, i.e., $\mathcal{S} = \{\mathbf{x} : d(\mathbf{x}) \leq d\}$. By the law of large numbers, argue that the probability that $\mathbf{Z} \in \mathcal{S}$

goes to 1 as the dimension goes to infinity. Finally, for a uniform code ensemble, argue that the competing codeword term in the lemma vanishes for all rates smaller than $\log(a) - h(Z_{\mathrm{ME}})$.

P.9.10 (Euclidean-threshold decoding) (1) Show that for any body S and random variable $\mathbf{Z}$, the probability $\mathrm{Pr}([\mathbf{Z} \in S] \bmod \Lambda)$ is invariant of the fundamental cell $\mathcal{P}_0$ with respect to which the modulo-Λ operation is taken (where $\{[\mathbf{Z} \in S] \bmod \Lambda\}$ means $\{[\mathbf{Z}] \bmod \Lambda \in [S] \bmod \Lambda\}$). (2) For a ball $S = \mathrm{Ball}(0, r)$, show that if $r \leq r_{\mathrm{pack}}(\Lambda)$, then this probability is equal to $\mathrm{Pr}([\mathbf{Z}] \bmod_{\mathcal{V}_0} \Lambda \in S)$, where $\mathcal{V}_0$ is the fundamental Voronoi region. (3) Use these facts and Problem 9.9 to prove the formula (9.82) for the Euclidean-threshold capacity of the mod Λ channel.

P.9.11 (Biased decoding) Prove that if a body S is convex and symmetric about the origin (e.g., the Euclidean Voronoi cell of a lattice), $\mathbf{Z}$ is AWGN, and $\mathbf{x}$ is a non-zero vector, then

$$\mathrm{Pr}\{\mathbf{Z} \notin [t \cdot \mathbf{x} + S]\}$$

is monotonically increasing with $|t|$. Conclude that the error probability in the presence of AWGN is minimized when the Voronoi cells are unbiased.

Guidance: (i) show that the complementary probability $p(\mathbf{x}) = \mathrm{Pr}\{\mathbf{Z} \in [\mathbf{x} + S]\}$ is the convolution $f_W(\mathbf{x}) * 1_S(\mathbf{x})$ of the AWGN density f_W and the indicator function 1_S of the body S; (ii) show that each of these functions is symmetric about the origin and log-concave (see Appendix A.4 for the definition); (iii) conclude that also $p(\mathbf{x})$ is symmetric about the origin and log-concave (use Lemma A.4.1); (iv) conclude that $p(\mathbf{x})$ has a maximum at the origin, and is monotonically non-increasing along rays from the origin; (v) show the desired result.

Historical notes

Shannon [240] showed that the capacity-achieving input for a power-constrained AWGN channel is Gaussian, and that – by the AEP – this input corresponds to a spherically shaped codebook; see the book by Cover and Thomas [53]. Shannon also gave the insightful illustration of duality between the forward and backward (reverse) channels, which, in the Gaussian case, links between the information and estimation viewpoints. For the role of spherical shaping in the context of error exponents, see the book by Gallager [97].

Zamir and Erez [286] showed that a Gaussian input is robust, in the sense that it loses at most half a bit in capacity for any (power-constrained) additive-noise channel.

The connection to lattices was made in the work of de Buda [57, 58], who showed that Shannon's Gaussian codebook can be replaced by a high-dimensional spherically shaped lattice codebook. Conway and Sloane [47] and Forney [85] introduced the idea of a Voronoi-shaped codebook. Calderbank and Sloane [34] proposed the coding gain as the total power advantage of a coded spherically shaped constellation over a cubically shaped uncoded constellation. Forney [83] proposed separating the shaping and coding gains (as done in Section 9.3). Forney and Wei [94] used "continuous approximation" to determine the "shaping gain" of a Voronoi code in terms of the NSM of the shaping lattice.

Trellis shaping was introduced by Forney [87]. He combined Marcellin and Fischer's trellis-coded quantizer [178] (as a shaping scheme) with a convolutional code (as a coding scheme). See also Eyuboglu and Forney [75].

A different approach to shaping follows the non-equiprobable signaling route of Chapter 6, but at a fixed coding rate. Calderbank and Ozarow [33] proposed the shell mapping technique, which was extended by Khandani and Kabal [133] and Laroia *et al.* [150]; see the tutorial paper by Calderbank [30] and the book by Tretter [256].

The theoretical performance limits of lattice codes over a power-constrained AWGN channel were studied by several authors [57, 58, 163, 175, 261]. Urbanke and Rimoldi [261] distinguished between a spherical shell and a solid shell shaping region, and between lattice decoding (which ignores the shaping region) and maximum likelihood decoding. They showed that for both shaping regions, maximum likelihood decoding can approach the full Shannon capacity $\frac{1}{2}\log(1 + \text{SNR})$ of the AWGN channel. They left open the question (implied by de Buda's original work [57] and Loeliger [175]) whether the suboptimal lattice decoder is limited by the smaller rate of $\frac{1}{2}\log(\text{SNR})$. Erez and Zamir [71] closed this gap by showing that with a Wiener estimation receiver, a dithered Voronoi codebook with lattice decoding can achieve the full AWGN channel capacity. Interestingly, this result was inspired by the solution for a side-information problem [68, 293]: a lattice precoding scheme for Costa's dirty-paper channel [50]. Forney [89, 90] showed that this "Wiener–Shannon connection" has interesting implications on coding, modulation and decoding for Gaussian channels (see also the paper by Cioffi *et al.* [41]).

The modulo-lattice channel (continuous and nested versions) was proposed by Forney *et al.* [92] to model successive decoding of a multi-level constellation. Erez, Zamir and Shamai [68, 71] showed that a dithered Voronoi code with a general estimator and a modulo receiver (termed a "modulo-lattice transformation") is equivalent to a modulo-lattice channel; see also [72] and Philosof, Erez, Khisti and Zamir [214, 217].

Forney [89] argued that almost all dither values must be asymptotically optimal (as the dimension goes to infinity) for the Wiener estimated modulo-lattice channel. Yona and Haim [280] observed that for $\alpha < 1$, a zero dither favors the zero codeword, and hence cannot be optimal. Ling and Belfiore [166] showed that an undithered lattice constellation (with probabilistic shaping) can approach the AWGN channel capacity above some minimum SNR value.

10 Side-information problems

Classical information theory deals with *point-to-point* communication, where a single source is transmitted over a channel to a single destination. In a *distributed scenario*, there may be more than one source or more than one channel and destination. The simplest setup, which captures much of the essence of the problem, is that of sources and channels with *side information*. The idea of coding with side information appeared for the first time in the seminal work of Slepian and Wolf from 1973 [247]. Let us illustrate this idea with a couple of examples.

Predictive coding of temperature

Suppose I wish to communicate tomorrow's temperature to my friend, after hearing the weather forecast. If the relevant range is 21–28 °C, then I need three bits of information to tell her my number. But suppose that in this season of the year the temperature changes daily by *exactly one* degree. If today's temperature is known to both of us, then clearly one bit of information is sufficient: "0" means -1 °C and "1" means $+1$ °C. But what if today's temperature is known only to my friend?

In the general case, X and Y are two correlated memoryless sources, where X is known to the encoder and Y is known to the decoder. By the principles of lossless compression, communicating X requires a coding rate of about the entropy of X in the absence of the side information Y; and about the conditional entropy of X given Y, if Y is available at *both* encoder and decoder. The surprising result of Slepian and Wolf is that the latter (smaller) rate is sufficient even if Y is available *only* at the decoder.

Returning to our temperature story, if the number X is even (my friend knows that, since her number Y is odd), then telling whether $X/2$ is even or odd (i.e., if X divides by four or not) requires just one bit, and it provides enough information to know whether X is $Y - 1$ or $Y + 1$. And in general, the parity of $\lfloor X/2 \rfloor$ provides enough information to decode X given Y.

The idea is thus to divide the source elements into *sets*, and encode an element by saying to which set it belongs. In the temperature story, the sets are the even and odd numbers.

This protocol is known as a *binning* or a *coloring* scheme. In the general case of encoding n-dimensional source vectors $\{\mathbf{X}\}$, we use 2^{nR} colors. The source vectors are divided into 2^{nR} bins, and each bin is given a different color. Our goal is to form the bins such that for a *typical* $(\mathbf{X}, \mathbf{Y})$ pair, the combination of side information (which is an n-dimensional vector $\mathbf{Y}$) and a color will uniquely identify the source vector $\mathbf{X}$; see Figure 10.1(A). Not very surprisingly, asymptotically (as n goes to infinity) we can form these bins at *random*. And the price we pay is small: the probability of a reconstruction error (where the identification is not unique or wrong) goes to zero as n goes to infinity. [1]

The Slepian–Wolf source coding problem has a *channel coding dual*, known as the *Gelfand–Pinsker problem*, where this time the side information is available only at the *encoder*.

Embedding a hidden message

Suppose I wish to hide a message in a sequence of integer numbers given to my friend; to keep the message unnoticeable, I am allowed either to leave a number unchanged or to increase it by 1. If my friend knew the original (unaltered) sequence, then each integer number could carry one bit of information: unchanged means "0," and increased by 1 means "1." But what if she does *not* know the original sequence?

In the literature, this problem bears the picturesque name *writing on a dirty paper*, or in different contexts *digital watermarking* and *writing to a memory with defects*. From the temperature example above, it is easy to guess what the protocol should be: let an even number mean "0" and odd number mean "1," and alter the integer sequence accordingly.

The underlying principle is again to divide signals into sets (call it coloring or binning), and identify a message with a choice of set. The goal here is to construct the sets of possible transmitted signals, such that every set will contain at least one element X which meets the channel *state* S and the *constraint*; see Figure 10.1(B). In the hidden message story, the sets are the even and odd numbers, the channel state S is the original sequence of integers, and the constraint is the altering rule.

A general Gelfand–Pinsker setup consists of a (possibly constrained) channel from (X, S) to Y, where the encoder knows S and controls X. As in the Slepian–Wolf problem, a *random binning* solution is asymptotically optimal in the limit of a large coding block length (where reliable encoding and decoding is guaranteed with a high probability). The similarity is even stronger in the deterministic (noiseless) channel case, where $Y = f(X, S)$: the capacity with side information S at the encoder is equal to the conditional entropy $H(Y|S)$ (maximized over all possible

---

[1] Note that for a general source and side information pair, a small probability of error is usually unavoidable (just like in coding for a noisy channel); see [8, 141, 296].

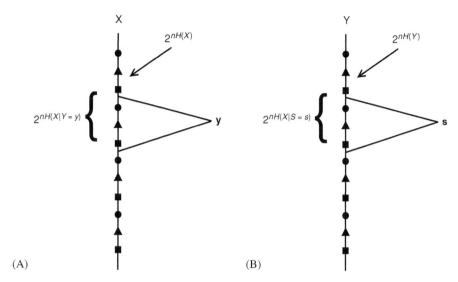

Figure 10.1 Random binning schemes: (A) the Slepian–Wolf problem; (B) the Gelfand–Pinsker problem. For the former, throw each **x** vector into one of 2^{nR} bins, with a uniform probability. If $R > H(X|Y)$, then with a *high probability*, **X** will be the only vector in its bin which is *jointly typical* with the side information **Y**.

inputs X depending on S), which is the same as the capacity when the state S is available at *both* the encoder and decoder.

In this chapter we are interested in "noisy" extensions of the two problems above: lossy source coding with side information at the decoder, and noisy channel coding with side information at the encoder. The former is known as the *Wyner–Ziv problem*, while the latter is known as (a general form of) the *writing on a dirty paper* problem. We shall mainly be interested in the quadratic-Gaussian versions of the two problems.

As of today, unfortunately, most practical communication systems do not utilize side information; the main references for these setups are thus (still) the information theoretic non-constructive random coding solutions. Our goal is to show that with just a slight modification, the Voronoi modulation and quantization schemes of Chapter 9 efficiently solve the dirty-paper channel and the Wyner–Ziv source coding problems. Thus, these simple lattice coding schemes give us, almost "for free," the benefit of structure in the basic side-information setups.

We shall first motivate the use of structured side-information codes in Section 10.1, by solving *binary-symmetric* versions of the Slepian–Wolf and the Gelfand–Pinsker problems, using the syndromes of a linear code. In Section 10.2, we shall extend the discussion to Gaussian multi-terminal settings, where side information plays a major role. Sections 10.3 and 10.4 define the rate-distortion function with side information, and show how to approach it with a nested lattice

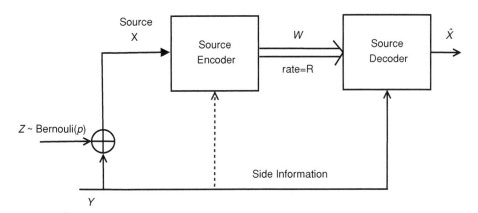

Figure 10.2 Binary Slepian–Wolf problem: source coding with side information for a doubly symmetric binary source.

coding scheme. Finally, Sections 10.5 and 10.6 do the same for capacity with side information and the dirty-paper channel.

10.1   Syndrome coding

Consider a memoryless *doubly symmetric binary source* (DSBS) (X_1, Y_1), $(X_2, Y_2), \ldots$. Each X_i or Y_i is symmetric (i.e., $\Pr(X_i = 1) = \Pr(Y_i = 1) = 1/2$), and the pair (X_i, Y_i) is connected via a BSC with a crossover probability $p < 1/2$:

$$Y_i = X_i \oplus Z_i, \tag{10.1}$$

where $\oplus$ is modulo-2 addition, i.e., $\Pr(X_i \neq Y_i) = \Pr(Z_i = 1) = p$.

In the setup shown in Figure 10.2, the **X**-component is the "source," while the **Y**-component is the "side information." The encoder maps the vector $\mathbf{X} = (X_1, \ldots, X_n)$, into a message $W = f(\mathbf{X})$, where $W \in \{1, \ldots, 2^{nR}\}$. The decoder receives W, and the correlated vector $\mathbf{Y} = (Y_1, \ldots, Y_n)$, and reconstructs the source as $\hat{\mathbf{X}} = g(W, \mathbf{Y})$. In the lossless coding case, the goal is to make the (block) error probability $\Pr(\hat{\mathbf{X}} \neq \mathbf{X})$ small.

Clearly, without access to the side information **Y**, the source **X** is "un-compressible"; hence, the coding rate R must be 1 bit per source sample. On the other hand, if *both* the encoder and decoder know the correlated source **Y**, then they can reduce the coding rate to the conditional entropy $R = H(X|Y) = H_B(p)$, where $H_B(p) = -p \log(p) - (1 - p) \log(1 - p)$ is the binary entropy of the crossover probability p. (Recall that $H_B(p) \leq 1$ bit, with equality if and only if $p = 1/2$.)

As discussed earlier, the Slepian–Wolf theory guarantees that a *random binning* scheme can achieve the conditional entropy $H(X|Y) = H_B(p)$, even if only the decoder has access to $\mathbf{Y}$. Here we shall show that the same rate is achievable by a *linear* coding scheme. The solution is based on a "good" linear binary (n, k) code, paralleling the notion of "good lattice codes" in Chapter 7.

10.1.1   Good linear codes

Recall from Section 2.5, that a linear binary (n, k) code $\mathcal{C}$ is defined by an $(n - k) \times n$ parity-check matrix H, or an $n \times k$ generator matrix G. Letting $\mathbf{s} = H\mathbf{x}$ denote the *syndrome* of a vector $\mathbf{x}$, the code $\mathcal{C}$ is the set of all zero-syndrome vectors in $\{0, 1\}^n \triangleq \mathbb{Z}_2^n$, which is also the column space of the matrix G. More generally, the binary space can be partitioned into 2^{n-k} distinct *cosets* $\mathcal{C}_{\mathbf{s}}$, where $\mathcal{C}_{\mathbf{s}} = \{\mathbf{x} : H\mathbf{x} = \mathbf{s}\}$ is the set of all n-vectors with a fixed syndrome $\mathbf{s}$. Each coset is a shift of the code, $\mathcal{C}_{\mathbf{s}} = \mathcal{C} \oplus \mathbf{v}_{\mathbf{s}}$, where $\mathbf{v}_{\mathbf{s}}$ can be any member of $\mathcal{C}_{\mathbf{s}}$. Thus, $\mathcal{C}_{\mathbf{s}}$ can be written as the set of n-vectors $\mathbf{x}$ of the form $G\mathbf{w} \oplus \mathbf{v}_{\mathbf{s}}$, where $\mathbf{w}$ runs over all k-vectors.

A set of *coset representatives* contains a unique member $\mathbf{v}$ from every coset. Let us restrict attention to the set of coset *leaders*, i.e., minimum Hamming-weight representatives, paralleling the shortest representative (2.32) in the case of a lattice. Define a *leader function* $\mathbb{Z}_2^{n-k} \to \mathbb{Z}_2^n$, which maps a syndrome $\mathbf{s}$ to the leader of its coset:

$$\text{leader}(\mathbf{s}) = \arg \min_{\mathbf{x} \in \mathcal{C}_{\mathbf{s}}} w_H(\mathbf{x}) \tag{10.2}$$

(where ties are broken arbitrarily). Using this function, define the *modulo code* operation on a vector $\mathbf{x} \in \mathbb{Z}_2^n$ with respect to the linear code $\mathcal{C}$, as the leader of the coset of $\mathbf{x}$:

$$\mathbf{x} \bmod \mathcal{C} \triangleq \text{leader}(H\mathbf{x}) \tag{10.3a}$$

$$\triangleq \text{error}(\mathbf{x}, \mathcal{C}). \tag{10.3b}$$

The definition (10.3a) parallels the modulo-lattice operation (2.21): it satisfies the *distributive law* (2.22b), and it amounts to the *error* vector to the closest codeword in the code $\mathcal{C}$ (where again ties are broken arbitrarily), as evident from the second notation (10.3b). More generally, the error vector to the closest codeword in a coset $\mathcal{C}_{\mathbf{s}}$, is given by

$$\text{error}(\mathbf{x}, \mathcal{C}_{\mathbf{s}}) = \text{leader}(H\mathbf{x} \oplus \mathbf{s}) \tag{10.4a}$$

$$= [\mathbf{x} \oplus \text{leader}(\mathbf{s})] \bmod \mathcal{C} \tag{10.4b}$$

which coincides with (10.3) for $\mathbf{s} = 0$. See Problems 10.1, 10.2 and 10.3.

ML decoding over a BSC (10.1) with a crossover probability $<1/2$, amounts to finding the codeword $\mathbf{c}$ with the *minimum Hamming distance* to the received vector

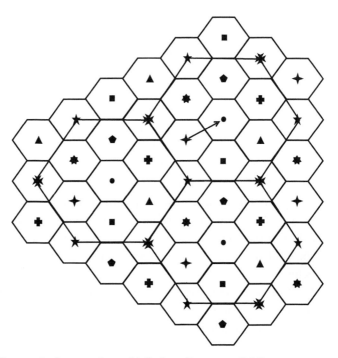

Figure 10.3 Cosets of a linear code, and ML decoding over a BSC.

Y, i.e., the $\mathbf{c} \in \mathcal{C}$ that achieves error($\mathbf{Y}, \mathcal{C}$). In light of (10.3), this amounts to

$$\hat{\mathbf{c}} = \mathbf{Y} \oplus \hat{\mathbf{Z}}, \quad \text{where } \hat{\mathbf{Z}} = \text{error}(\mathbf{Y}, \mathcal{C}), \tag{10.5}$$

as illustrated in Figure 10.3.

An error event $\hat{\mathbf{c}} \neq \mathbf{c}$ in the procedure above amounts to $\hat{\mathbf{Z}} \neq \mathbf{Z}$. Since $H\mathbf{c} = 0$, hence $H\mathbf{Y} = H(\mathbf{c} + \mathbf{Z}) = H\mathbf{Z}$, an error event amounts to

$$\mathbf{Z} \bmod \mathcal{C} \neq \mathbf{Z}. \tag{10.6}$$

A linear code is *good for coding* over a BSC with parameter p, if it achieves a small error probability $P_e = \Pr\{\hat{\mathbf{c}} \neq \mathbf{c}\} < \epsilon$, at a rate $R = k/n$ close to the BSC capacity $C = 1 - H_B(p)$, for some $\epsilon > 0$. Thus for such a code, the syndrome length $|\mathbf{s}| = n - k$ and the error probability P_e satisfy

$$n - k \approx n H_B(p), \quad \text{and} \quad \Pr\{\mathbf{Z} \bmod \mathcal{C} \neq \mathbf{Z}\} < \epsilon \tag{10.7}$$

respectively.

Another notion of goodness is for *source coding under Hamming distortion*. Following (10.5)–(10.7), the minimum Hamming distance between a source vector $\mathbf{x}$ and a codeword $\mathbf{c} \in \mathcal{C}$, is given by $w_H(\text{error}(\mathbf{x}, \mathcal{C}))$. The rate-distortion function (5.31) of a Bernoulli(1/2) source $\mathbf{X}$, under an average Hamming distortion D, is given by $R(D) = 1 - H_B(D)$. A linear (n, k) code is *good for source coding* with

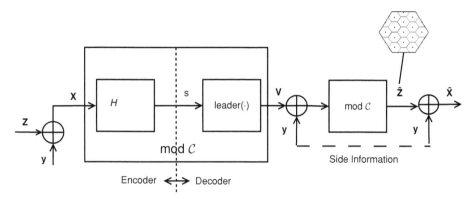

Figure 10.4 A linear code solution for the binary Slepian–Wolf problem, which demonstrates the derivation in (10.11b). The picture of the hexagonal codebook is a heuristic interpretation for the modulo-code operation.

parameter D, if at a rate $R = k/n$ close to $R(D)$ it achieves an average Hamming distortion D:

$$n - k \approx n H_B(D), \quad \text{and} \quad E\{w_H(\text{error}(\mathbf{X}, \mathcal{C}))\} = n D. \tag{10.8}$$

It follows from the *maximum-entropy principle* (A.13), that the quantization error vector $\text{error}(\mathbf{X}, \mathcal{C})$ is nearly (divergence-wise) a Bernoulli(D) process.

We see that the number of cosets of a good code with a parameter θ (where θ is either the crossover probability for channel coding, or the distortion for source coding), is roughly

$$2^{n-k} \approx 2^{n H_B(\theta)}.$$

10.1.2 Source with side information

We return to the linear code solution for the binary Slepian–Wolf problem in Figure 10.2. The syndrome encoder identifies the coset of the source $\mathbf{X}$ by sending its syndrome, i.e., a *linear* function $\mathbb{Z}_2^n \to \mathbb{Z}_2^{n-k}$ of $\mathbf{X}$. Hence, the message W contains $|\mathbf{s}| = n - k$ bits, and the coding rate is

$$R = 1 - k/n \tag{10.9}$$

bit per source sample. The decoder estimates the noise $\mathbf{Z}$ as the error between the side information $\mathbf{Y}$ and the coset $\mathcal{C}_\mathbf{s}$, (10.4), and adds it to $\mathbf{Y}$. The encoding and decoding operations are thus summarized by:

$$\textbf{Encoding} \quad \mathbf{s} = H\mathbf{X} \tag{10.10a}$$

$$\textbf{Decoding} \quad \hat{\mathbf{Z}} = \text{error}(\mathbf{Y}, \mathcal{C}_\mathbf{s}) \tag{10.10b}$$

$$\hat{\mathbf{X}} = \mathbf{Y} \oplus \hat{\mathbf{Z}}. \tag{10.10c}$$

Since $\mathbf{Y} = \mathbf{X} \oplus \mathbf{Z}$, and error$(\mathbf{Y}, C_s) = $ leader$(H\mathbf{Y} \oplus \mathbf{s})$, where $\mathbf{s} = H\mathbf{X}$, it follows from linearity that

$$
\begin{aligned}
\hat{\mathbf{Z}} &= \text{leader}(H\mathbf{Y} \oplus H\mathbf{X}) \\
&= \text{leader}(H(\mathbf{Y} \oplus \mathbf{X})) \\
&= \text{leader}(H\mathbf{Z}).
\end{aligned}
\tag{10.11a}
$$

An equivalent representation of the system, shown in Figure 10.4, represents the message by the leader vector $\mathbf{v} = \text{leader}(\mathbf{s}) = \text{leader}(H\mathbf{X}) = \mathbf{X} \bmod C$; see (10.3). Hence, we have from (10.4)

$$
\begin{aligned}
\hat{\mathbf{Z}} &= [\mathbf{Y} \oplus \mathbf{v}] \bmod C \\
&= [\mathbf{Y} \oplus (\mathbf{X} \bmod C)] \bmod C \\
&= [\mathbf{Y} \oplus \mathbf{X}] \bmod C \\
&= \mathbf{Z} \bmod C,
\end{aligned}
\tag{10.11b}
$$

where the one before last equality follows from the *distributive law* of the modulo operation.

If the code C is a *good* BSC code in the sense of (10.7), then the result in (10.11) is equal to $\mathbf{Z}$, and hence the reconstruction $\hat{\mathbf{X}}$ (10.10c) is correct, with probability of at least $1 - \epsilon$. And from (10.9), the coding rate is $R \approx H_B(p)$, which is the desired conditional entropy $H(X|Y)$.

Observe that in the linear coding scheme above, the *random bins* of Figure 10.1 are replaced by the *cosets* $\{C_s\}$ of the linear code C. This structure is, in fact, common to all the linear (and later, lattice) side-information coding schemes in this chapter.

10.1.3   Channel with known noise

Let us consider now a binary version of the Gelfand–Pinsker problem, paralleling the *information embedding* problem at the beginning of this chapter. Suppose a BSC $Y = X \oplus Z$, where the input X is subject to an average Hamming-weight constraint, i.e., every vector $\mathbf{X}$ of n channel inputs must satisfy $w_H(\mathbf{X}) \le qn$, for some $0 < q < 1/2$. In the context of information embedding, the random variable Z is viewed as a source signal, in which we want to embed a "watermark" W, such that the Hamming distortion between the watermarked signal Y and the source Z is at most q.

Similarly to the additive-noise channel capacity formula (6.15), it is easy to show that the capacity of this channel is given by $C = H_B(q \otimes p) - H_B(p)$, where p is the BSC crossover probability, and $q \otimes p = q(1 - p) + (1 - q)p$ is the *binary convolution*. In contrast, if the noise Z is known at the decoder, and hence can

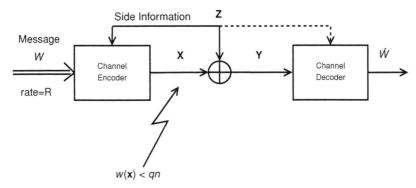

Figure 10.5 BSC coding subject to an input constraint with side information (channel noise) at the encoder.

be XOR-ed with **Y** and *canceled*, then the larger capacity of $C = H_B(q)$ can be achieved, *independent* of the crossover probability p.

The more interesting case, shown in Figure 10.5, is when Z is known *only to the encoder*.[2] Unlike the decoder, the encoder cannot simply cancel the noise, i.e., XOR the desired information with Z, because this would imply a cost in Hamming weight, and may even violate the input constraint. Nevertheless, the Gelfand–Pinsker theory implies that if the noise Z is available at the encoder *non-causally* (i.e., the whole noise vector $\mathbf{Z} = Z_1, \ldots, Z_n$ is known prior to the selection of the input vector $\mathbf{X} = X_1, \ldots, X_n$), then the full capacity of $H_B(q)$ can be achieved.[3] See Figure 10.6. We show below a *linear* coding solution that achieves this capacity.

This time, the scheme is based on a code $\mathcal{C}$ which is *good for source coding* under an average Hamming distortion q (10.8). Each message W contains $n - k$ bits, and is represented by a syndrome **s**. By (10.8), the rate $R = (n - k)/n$ is $\approx H_B(q)$ bit per channel use, as desired. Given the side information **Z**, the encoding and decoding operations are:

$$\textbf{Encoding} \quad \mathbf{X} = \text{error}(\mathbf{Z}, \mathcal{C}_\mathbf{s}) \qquad (10.12a)$$
$$\textbf{Decoding} \quad \hat{\mathbf{s}} = H \cdot \mathbf{Y}, \qquad (10.12b)$$

where error$(\mathbf{Z}, \mathcal{C}_\mathbf{s})$ is the error (10.4) between **Z** and the coset $\mathcal{C}_\mathbf{s}$.[4]

---

[2] In the context of *copyright protection*, this setup is known as *public* or *blind watermarking*, where only the encoder has access to the original source.

[3] In the *causal side-information* setup, due to Shannon [243], the input X_i at time instance i depends only on the past and present noise values $Z_1, \ldots, Z_i$. The capacity obviously can only be smaller than the non-causal capacity, as is the case here. See Problem 10.5.

[4] Contrast that with the conventional representation (2.54) of a k-bit information message **w** by the n-bit vector $\mathbf{x} = G\mathbf{w}$, in channel coding with no side information.

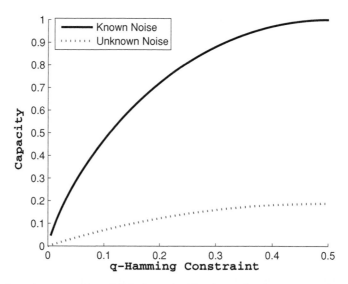

Figure 10.6 Capacity curves for a BSC channel with a Hamming input constraint: noise known at the receiver or at the transmitter, and unknown noise.

Assuming that $\mathbf{Z}$ is Bernoulli($1/2$) (a "worst case" noise) then, by (10.8), the error (10.12a) satisfies an average input constraint

$$E\{w_H(\mathbf{X})\} = qn \tag{10.13}$$

as desired. (Otherwise $\mathbf{X}$ can be symmetrized by dithering.) We are left to show that the estimate $\hat{\mathbf{s}}$ in (10.12b) is indeed correct. By the linearity of the channel, $H\mathbf{Y} = H\mathbf{X} \oplus H\mathbf{Z}$. Now, from (10.4), $\mathbf{X} = \mathrm{leader}(\mathbf{s} \oplus H\mathbf{Z})$; hence $H\mathbf{X} = \mathbf{s} \oplus H\mathbf{Z}$ (by the definition of a coset and the leader function (10.2)), so

$$\hat{\mathbf{s}} = (\mathbf{s} \oplus H\mathbf{Z}) \oplus H\mathbf{Z} = \mathbf{s} \tag{10.14}$$

since the two $H\mathbf{Z}$ terms cancel out.

An equivalent (and more insightful) analysis represents the message $\mathbf{s}$ by its leader vector $\mathbf{v}$, and uses modulo-$\mathcal{C}$ operations (Figure 10.7). We rewrite the encoding (10.12a) and decoding (10.12b) as

$$\mathbf{X} = (\mathbf{v} \oplus \mathbf{Z}) \bmod \mathcal{C}, \quad \text{where } \mathbf{v} = \mathrm{leader}(\mathbf{s}) \tag{10.15a}$$
$$\hat{\mathbf{v}} = \mathbf{Y} \bmod \mathcal{C}. \tag{10.15b}$$

By the distributive law of the modulo operation, we then have

$$\hat{\mathbf{v}} = [(\mathbf{v} \oplus \mathbf{Z} \bmod \mathcal{C}) \oplus \mathbf{Z}] \bmod \mathcal{C} = \mathbf{v} \tag{10.16}$$

since the two noise terms cancel each other, and $\mathbf{v} \bmod \mathcal{C} = \mathbf{v}$.

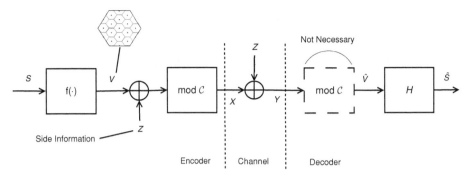

Figure 10.7 Coset-based scheme for channel coding with perfect side information.

10.1.4 Other settings and extensions

Writing to a memory with defects

The original motivation for channel coding with side information came from writing to a memory with defective cells. In this problem, k out of n cells are stuck at some fixed unchangeable values. The channel states $(S_1, \ldots, S_n)$ indicate the "stuck-at" values of these cells, as well as the location of the $n - k$ writable cells. These states are known to the encoder before writing, but are unknown to the decoder. The idea of the coding protocol is to associate with each message a distinct set of length-n codewords $\{(X_1, \ldots, X_n)\}$; the set should be sufficiently large that for any vector of channel states $(S_1, \ldots, S_n)$, i.e., a pattern of k (or less) defective cells, there exists at least one codeword in the set that coincides with that pattern, and hence can be written to the memory. It can be shown that a scheme based on a linear *erasure-correction* code achieves asymptotically a capacity of $n - k$ out of n cells, which is the same as the capacity if the side information was available also at the decoder.

Multi-terminal settings

The side-information problems above can be used as building blocks for more general "multi-terminal" setups.

Distributed compression The general form of the Slepian–Wolf setup consists of two separate encoders. Each encoder encodes one of two correlated sources X and Y, which are to be reconstructed by a *joint* decoder. The Slepian–Wolf *rate region* defines the set of coding rates R_1 and R_2, which allow X and Y to be reconstructed losslessly (with an arbitrarily small probability of error):

$$R_1 \geq H(X|Y) \tag{10.17a}$$

$$R_2 \geq H(Y|X) \tag{10.17b}$$

$$R_1 + R_2 \geq H(X, Y). \tag{10.17c}$$

Note that by the chain rule $H(Y) + H(X|Y) = H(X, Y)$ (which is the rate when the sources are encoded *jointly*). It follows that the rate pair $\{R_1 = H(X|Y), R_2 = H(Y)\}$ is a "corner point" of the pentagon rate region above. Furthermore, we can achieve this point via coding with side information: the Y-encoder compresses Y to its entropy $H(Y)$ (so the joint decoder can reproduce Y independent of the message sent by the X-encoder), while the X-encoder uses binning to compress X at a rate of $H(X|Y)$, given the side information Y at the joint decoder.

Problem 10.6 shows a linear coding implementation of this idea, in the case where (X, Y) is a DSBS.

Deterministic broadcast The channel counterpart of this setup is a two-user deterministic broadcast channel. This channel has two inputs X_1 and X_2, controlled by a *joint* encoder, and two outputs Y_1 and Y_2, received by two *separate* decoders, where each output is a (deterministic) function of the two inputs: $Y_i = f_i(X_1, X_2)$, $i = 1, 2$. The *sum capacity* of this broadcast problem is the same as the capacity of a *point-to-point* setup (i.e., with *joint* decoding from (Y_1, Y_2)). It is the maximum of the output joint entropy $H(Y_1, Y_2)$, over all possible (constrained) random input pairs (X_1, X_2) [220].

Assume that there exists a transformation of the pair (X_1, X_2) to a pair $(\tilde{X}_1, \tilde{X}_2)$, such that $Y_1 = \tilde{X}_1$ and $Y_2 = \tilde{f}(\tilde{X}_1, \tilde{X}_2)$. The joint encoder can then use the following simple protocol: transmit to decoder 1 at a rate $H(Y_1)$ using the input $\tilde{X}_1$, and to decoder 2 at a rate $H(Y_2|Y_1)$ using the input $\tilde{X}_2$ and knowledge (at the encoder alone) of the side information Y_1. The sum rate is then $H(Y_1, Y_2)$, as desired.

Problem 10.7 shows a linear coding implementation of this idea, for a binary symmetric broadcast channel under a Hamming input constraint.

Nested syndrome coding

The source and channel coding problems we discussed so far can be considered as "noiseless." In the Slepian–Wolf DSBS setup (Figure 10.2), the target was a *lossless* reconstruction of the source, while in the binary Gelfand–Pinsker setup (Figure 10.1), the channel was effectively noiseless.

The *noisy* versions of these two problems are more challenging, and closer in spirit to the lattice coding framework. In lossy Slepian–Wolf coding (lossy source coding with side information), the source must be *quantized* prior to binning, hence its correlation with the side-information signal is slightly weakened. In noisy Gelfand–Pinsker problems, i.e., setups where only *part* of the noise is known to the encoder, the bins must be slightly *diluted*, to allow their decoding in the presence of the residual noise.

The linear coding solutions of these problems are based on *nested binary codes* $C_1(n, k_1)$ and $C_2(n, k_2)$, where $C_2 \subset C_1$, and on their *relative syndromes* (Section 8.3). These nested binary side-information codes are studied in [293]. Our focus below shifts back to the continuous world and to lattices.

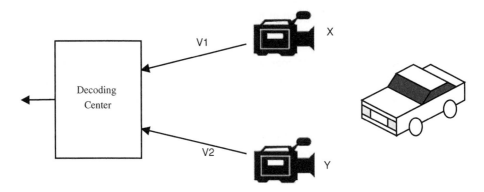

Figure 10.8 Distributed video compression.

10.2   Gaussian multi-terminal problems

Communication with continuous signals provides "real world" examples for coding with side information, as a part of a multi-terminal communication system. We describe below two such examples, one for source coding, the other for channel coding. These examples motivate the setups and lattice coding schemes in the next sections.

10.2.1   Distributed video compression

Imagine two digital video cameras observing the same scene from different locations, as illustrated in Figure 10.8. Each camera compresses its video stream and transmits it to a data fusion center. This *distributed compression* setup reminds us of the two-terminal Slepian–Wolf problem in Section 10.1.4, but with a lossy encoding.

   A fundamental principle in standard video compression algorithms is *motion compensation*. The encoder predicts (or interpolates) consecutive video frames based on the *temporal* correlation in the scene, hence it can reduce the coding rate for the same reconstruction quality.[5] Can distributed encoders take advantage of the *spatial correlation* between their video streams? Indeed, the Slepian–Wolf and Wyner–Ziv theories promise (nearly) optimal compression performance, even in the absence of encoder cooperation, provided that *decoding* is done *jointly*. More precisely, due to the spatial correlation, the joint decoder can use the encoding of one video stream as side information for the decoding of the other video stream.

---

[5] For example, MPEG or V264 reduces the bit rate of a raw video stream from about 1 gigabit per second to 5–10 megabit per second, with almost no significant reduction in video quality.

Motion compensation at the decoder

An interesting variation on distributed video compression, in a *point-to-point* setup, is called "Wyner–Ziv video coding." The idea is to encode each frame *separately*, but at the (low) rate of *joint* encoding. The motivation is to reduce the complexity of the video encoder (say, in a "light" device such as a cellular phone), and move the complexity to the decoder. The encoder bins (or "hashes") each video frame separately, and avoids the (hard) motion compensation step. The decoder incorporates this binning information, and possibly some individually encoded "I" frames, into the video correlation model, and reconstructs the video stream.

We can model these distributed video coding scenarios quite well within the *quadratic-Gaussian* framework, i.e., jointly Gaussian sources in time and in space, under a squared-error fidelity criterion. Theoretically, under the quadratic-Gaussian model there is no loss in rate-distortion performance for having the side information only at the decoder. (See more on the loss in general models later.) As we shall see in the following sections, a coding ("binning") scheme based on nested lattices $\Lambda_2 \subset \Lambda_1$ approaches the quadratic-Gaussian Wyner–Ziv rate-distortion function, in the limit of a large lattice dimension. For a finite dimension, the rate loss at high-resolution conditions is the same as that of Voronoi quantization (Section 9.10): $\frac{1}{2} \log[G(\Lambda_1)\mu(\Lambda_2, P_e)]$ bit per sample, where P_e is the equivalent overload probability. Furthermore, this scheme with *probabilistic decoding* works quite well even under more general continuous (non-quadratic/non-Gaussian) models.

10.2.2   MIMO broadcast

The quadratic-Gaussian model also describes well some applications of the Gelfand–Pinsker theory, and again motivates the use of nested lattice codes.

Imagine a cellular base station with two antennas, transmitting to two mobile users, with a single antenna each. This "down-link" channel is characterized by a 2×2 channel matrix H (describing the path gains from antennas X_1 and X_2 at the base station to the receiver antennas of users A and B), followed by independent AWGN at each receiver. Since the receivers do not cooperate, this setup is a noisy version of the deterministic broadcast channel of Section 10.1.4.

One simple broadcast strategy is *time-sharing*, where the base station transmits (with both antennas and full power) to one user at a time. A more efficient strategy, though, is to transmit *simultaneously* to both users in a *coordinated* manner. As it

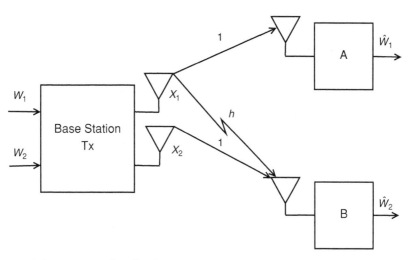

Figure 10.9 Dirty-paper coding for the MIMO broadcast channel. The interfering signal hX_1 is viewed as side information known at the transmitter.

turns out, this *MIMO broadcast* strategy almost *doubles* the capacity of each user at high SNR conditions. [6]

The idea of MIMO broadcast is similar to deterministic broadcast (Section 10.1.4). Figure 10.9 illustrates a simplified setup, where the matrix H is *triangular*, i.e., the first base station antenna sees *both* users A and B, while the second antenna only sees user B: [7]

$$H = \begin{pmatrix} 1 & 0 \\ h & 1 \end{pmatrix} \quad \Rightarrow \quad \begin{aligned} A &= X_1 \\ B &= \underbrace{hX_1}_{I} + X_2. \end{aligned} \tag{10.18}$$

The base station uses its first antenna (X_1) to transmit information to user A. Since both users see the first antenna, user B will get an undesired replica ($I = hX_1$) of this signal. Nevertheless, the base station *knows* this interference (assuming it knows the matrix H); hence, by the Gelfand–Pinsker theory, it can take it into account while transmitting to user B (channel coding with side information at the encoder). Specifically, the base station uses its second antenna (X_2) to transmit a

---

[6] The time-sharing strategy corresponds to time-division multiple access (TDMA), or equivalently, to frequency-division multiple access (FDMA). At high SNR, the time-sharing capacity in a K-user setup is $\approx \frac{W}{K} C(K \, \mathrm{SNR}_i)$ bit per second per Hz for user i, where $C(x) \overset{\Delta}{=} \log_2(1 + x)$, and $\mathrm{SNR}_i = \|\mathbf{h}_i\|^2 P / W N_0$, where P is the power of each transmit antenna, W is the channel bandwidth, N_0 is the AWGN spectral density, and $\|\mathbf{h}_i\|^2$ is the norm of the row of H corresponding to user i. In contrast, the MIMO broadcast (also called space-division multiple access (SDMA)) capacity of user i is $\approx W \, C(\mathrm{SNR}_i)$ bit per second per Hz. Since at high SNR conditions the capacity is *logarithmic* in P, the latter is roughly K times larger.

[7] QR-factorization is a method to triangularize a general matrix H by a unitary "pre-coding" matrix, at no cost in the total transmission power; see [27, 105].

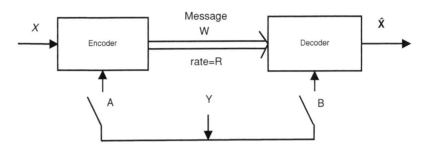

Figure 10.10 Case of a source with side information.

signal that cancels the interference $I = hX_1$ from the first antenna, and at the same time carries the information intended to user B. This method, where the effect of a known interference I is eliminated at the transmitter with no cost in power, is known as *pre-coding* or *dirty-paper coding*.

Section 10.6 describes a lattice dirty-paper coding scheme, based on nested lattices ($\Lambda_2 \subset \Lambda_1$), which is similar to the Voronoi modulator of Chapter 9. Under AWGN (i.e., a channel of the form $Y = X + I + Z$, where Z is AWGN and I is interference), this scheme achieves the zero-interference capacity $C = \frac{1}{2}\log(1 + \text{SNR})$, up to the "usual" loss of $\frac{1}{2}\log[G(\Lambda_2)\mu(\Lambda_1, P_e)]$ bit per channel use, where P_e is the decoding error probability.

10.3   Rate distortion with side information

We now turn our attention to noisy side-information problems, keeping in mind their application to the multi-terminal setups we saw above.

Consider the lossy extension of the configuration in Figure 10.2, of encoding a source $\mathbf{X}$ with side information $\mathbf{Y}$, subject to some distortion constraint $E\{d(\mathbf{X}, \hat{\mathbf{X}})\} \le D$. Unlike in the lossless coding case (where $D = 0$), the theoretical performance for a general source and side-information pair $(\mathbf{X}, \mathbf{Y})$ and distortion measure $d(x, \hat{x})$ depends on whether the side information $\mathbf{Y}$ is available at both the encoder and decoder, or only at the decoder. See Figure 10.10 and Table 10.1.

In the case where $\mathbf{Y}$ is available at both, optimal performance for memoryless source and side information is given by the *conditional rate-distortion function*:

$$R_{x|y}(D) = \inf_{\hat{X}:\, d(X,\hat{X})\le D} I(X; \hat{X}|Y) \tag{10.19}$$

which is the natural extension of the "no-SI" rate-distortion function (5.31). Note that here the conditional output distribution $p(\hat{x}|x, y)$ can take a general form, which amounts to the fact that the encoder completely determines $\hat{\mathbf{X}}$.

When $\mathbf{Y}$ is available only at the decoder, the encoder sends *some* information about $\mathbf{X}$, which the decoder combines with $\mathbf{Y}$ to obtain $\hat{\mathbf{X}}$. (Note that the uninformed

Table 10.1 Cases of source coding with side information

Note that encoder-only side information is useless, unless the distortion measure depends on the side information, i.e., $d = d(x, \hat{x}, y)$; see [179].

	Switch position		Encoding	Decoding	R, D functions
	A	B			
no SI	open	open	$m = f(\mathbf{x})$	$\hat{\mathbf{x}} = g(m)$	$R_x(D)$
SI at both	close	close	$m = f(\mathbf{x}, \mathbf{y})$	$\hat{\mathbf{x}} = g(m, \mathbf{y})$	$R_{x|y}(D)$
SI at decoder	open	close	$m = f(\mathbf{x})$	$\hat{\mathbf{x}} = g(m, \mathbf{y})$	$R_{WZ}(D)$
SI at encoder	close	open	$m = f(\mathbf{x}, \mathbf{y})$	$\hat{\mathbf{x}} = g(m)$	$R_x(D)$

encoder does *not* completely determine $\hat{\mathbf{X}}$.) The optimal performance is given in this case by the Wyner–Ziv function:

$$R_{\text{WZ}}(D) = \inf_U I(X; U|Y), \qquad (10.20a)$$

where the infimum is taken over all *auxiliary* random variables U, such that

- $U \leftrightarrow X \leftrightarrow Y$ form a Markov chain;
- there exists a function $g(u, y)$, such that $E\{d(X, \hat{X})\} \leq D$ for $\hat{X} = g(U, Y)$.

Due to the Markov chain condition in (10.20a), the minimization argument can also be written as

$$H(U|Y) - H(U|X) = I(U; X) - I(U; Y). \qquad (10.20b)$$

The Wyner–Ziv formula is a non-explicit single-letter characterization. Like the rate-distortion function (5.31), it has a closed-form expression only in a few special cases. [8]

Clearly, not having the side information $\mathbf{y}$ at the encoder can only make performance worse. And indeed, the Markov chain constraint on the minimization (10.20a) implies that $R_{\text{WZ}}(D) \geq R_{x|y}(D)$. This inequality is, in fact, *strict* in the DSBS/Hamming case: the conditional rate-distortion function is given by $R_{x|y}(D) = H_B(p) - H_B(D), 0 \leq D \leq p$, while the Wyner–Ziv function is given by

$$R_{\text{WZ}}(D) = \ell.\text{c.e.} \left\{ H_B(p \otimes D) - H_B(D), \quad (p, 0) \right\}, \ 0 \leq D \leq p, \quad (10.21)$$

where $\otimes$ denotes binary convolution, and $\ell.\text{c.e.}$ stands for lower convex envelop (see Figure 10.11).

---

[8] The Blahut–Arimoto algorithm [10, 22, 53] is an iterative method to compute the rate-distortion function, which was extended to the Wyner–Ziv function by Willems [266].

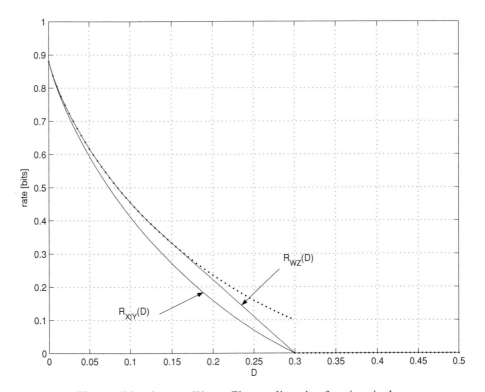

Figure 10.11 The conditional versus Wyner–Ziv rate-distortion functions in the binary-Hamming case.

Fortunately, the two rate-distortion functions coincide under the widely used quadratic-Gaussian model:

$$R_{\mathrm{WZ}}(D) = R_{x|y}(D) = \frac{1}{2} \log \left(\frac{\sigma_{x|y}^2}{D} \right), \quad 0 \le D \le \sigma_{x|y}^2, \qquad (10.22)$$

where $\sigma_{x|y}^2$ is the conditional variance of X given Y. We can realize $R_{x|y}(D)$ and $R_{\mathrm{WZ}}(D)$ in this case using pre/post-scaled AWGN "test channels," similar to the channel realizing the *un*conditional rate-distortion function (5.33). Assuming for simplicity $EX = EY = 0$, these test channels are given by

$$\hat{X} = a^* Y + \beta [\alpha(X - a^* Y) + N] \qquad (10.23a)$$

in the conditional case, and

$$U = \alpha X + N \qquad (10.23b)$$

$$g(u, y) = a^* y + \beta [u - \alpha a^* y] \qquad (10.23c)$$

in the Wyner–Ziv case, where $a^* = \rho \sigma_x / \sigma_y$ is the Wiener coefficient for estimating X from Y (4.36), and $N \sim N(0, \sigma_n^2)$ is independent of (X, Y), and where the scaling coefficients α and β and the noise variance σ_n^2 satisfy

$$\alpha\beta = 1 - \frac{D}{\sigma_{x|y}^2} \quad \text{and} \quad \sigma_n^2 = \frac{\alpha}{\beta}D, \tag{10.23d}$$

for an arbitrary positive scalar α. The conditional distributions of U induced by (10.23b) and (10.23d), are

$$U|X \sim N(\alpha X, \sigma_n^2) \quad \text{and} \quad U|Y \sim N(\alpha a^* Y, \alpha^2 \sigma_{x|y}^2 + \sigma_n^2) \tag{10.23e}$$

respectively.

Note the similarity between (10.23) and the (unconditional) test channel (5.35); here the estimation error $X - a^* Y$ plays the role of a source, with the estimation mean-squared error $E(X - a^* Y)^2 = \sigma_{x|y}^2$ being the source variance, and β plays the role of a Wiener estimator against the noise N. Clearly, all three cases – unconditional, conditional and Wyner–Ziv – coincide when X and Y are uncorrelated ($a^* = \rho = 0$).

Achieving $R_{WZ}(D)$ via random binning [53]

In the quadratic-Gaussian case, the random coding scheme has a simple geometric interpretation, as illustrated in Figure 10.12. For simplicity of the exposition, let $\alpha = 1$, and $\beta = 1 - D/\sigma_{x|y}^2$; thus, $U = X + N$, where $\sigma_n^2 = D/\beta$, and $\sigma_u^2 = \sigma_x^2 + \sigma_n^2$. To generate the code, draw 2^{nR_1} independent codewords $\mathbf{u}$, where each codeword is uniform on a ball of radius slightly larger than $r_u = \sqrt{n\sigma_u^2}$ about the origin; then assign the codewords at random to 2^{nR} bins. The encoder quantizes the source vector $\mathbf{X}$ to the closest codeword $\mathbf{u}$, and sends the identity $m \in \{1, \ldots, 2^{nR}\}$ of the bin containing $\mathbf{u}$ to the decoder. If

$$R_1 > I(X; U) = \tfrac{1}{2} \log(1 + \sigma_x^2/\sigma_n^2),$$

then (for a large n, by the Gaussian joint AEP, see Section 9.4.3) $\mathbf{u}$ is typically inside a ball of radius $r_n = \sqrt{n\sigma_n^2}$ around $\mathbf{X}$. The decoder looks for the codeword $\hat{\mathbf{u}}$ in bin m which is closest to the source estimate $a^* \mathbf{Y}$, and reconstructs the source as $\hat{\mathbf{X}} = a^* \mathbf{Y} + \beta[\hat{\mathbf{u}} - a^* \mathbf{Y}]$. If the decoder is correct, i.e., $\hat{\mathbf{u}} = \mathbf{u}$, then by (10.23) this reconstruction satisfies the desired distortion constraint D. To find a condition on R for correct decoding, note that by (10.23e) $\mathbf{u}$ is typically within a distance $r^* = \sqrt{n(\sigma_{x|y}^2 + \sigma_n^2)}$ from the estimator $a^* \mathbf{Y}$. The probability that any other $\mathbf{u}$ in bin i is closer to $a^* \mathbf{Y}$ is upper bounded by $2^{nR_1} \cdot (r^*/r_u)^n / 2^{nR}$ (because the probability to fall within a radius r^* is $(r^*/r_u)^n$, while the probability to fall in bin i is 2^{-nR}); thus, the product probability goes to zero as $n \to \infty$ if

$$R_1 < I(U; Y) + R$$

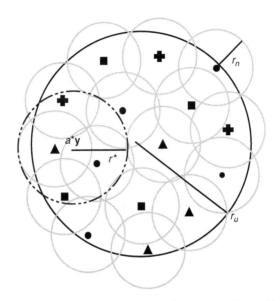

Figure 10.12 Random binning in the quadratic-Gaussian Wyner–Ziv problem: quantization spheres (r_n) and the detection sphere (r^*) inside the source sphere (r_u).

(because $I(U;Y) = \log(r_u/r^*)$). We see that both requirements for typical encoding and correct decoding are satisfied if R is larger than $I(X;U) - I(U;Y) = R_{\mathrm{WZ}}(D)$.

Although for general sources the functions $R_{\mathrm{WZ}}(D)$ and $R_{x|y}(D)$ are different, they coincide as $D \to 0$, corresponding to the zero rate loss in the discrete lossless Slepian–Wolf problem discussed earlier.[9] Moreover, the gap between the two functions is bounded by a universal constant.

Lemma 10.3.1 (Wyner–Ziv rate loss [284]) *Under a quadratic distortion measure,*

$$R_{\mathrm{WZ}}(D) - R_{X|Y}(D) \le 1/2 \text{ bit} \qquad (10.24)$$

for any *joint distribution of X and Y.*

This bound and the proof technique are similar to the universal bound on the ECDQ redundancy in Section 5.9.

The Berger–Tung rate region

The following set of achievable rates (R_1, R_2) extends the Wyner–Ziv formula to two-terminal lossy source coding (Section 10.2), paralleling the Slepian–Wolf rate

---

[9] $R_{\mathrm{WZ}}(D) - R_{x|y}(D) \to 0$ as $D \to 0$, for smooth sources under difference distortion measures [284].

region (10.17) in the lossless case:

$$R_1 \geq I(X; U|Y) \tag{10.25a}$$

$$R_2 \geq I(Y; V|X) \tag{10.25b}$$

$$R_1 + R_2 \geq I(X, Y; U, V) \tag{10.25c}$$

for some auxiliary random variables U and V forming a "long" Markov chain $U \leftrightarrow X \leftrightarrow Y \leftrightarrow V$, and satisfying the distortion constraints as in the Wyner–Ziv case (10.20). As we shall see in Section 12.1, the *Berger–Tung rate region* (10.25) does not necessarily characterize the *best* achievable rates (R_1, R_2) (even for the best U and V). In fact, multi-terminal lossy source coding is still an open problem in *network information theory*; see the book by El Gamal and Kim [64].

One important exception is the quadratic-Gaussian case. The Berger–Tung rate region (10.25) (with additive-Gaussian U and V) is tight for correlated Gaussian sources under a squared-error distortion [264].

Another aspect of lossy multi-terminal source coding is that, unlike the lossless coding case (Section 10.1.4), it suffers a rate loss with respect to the joint-encoding case. Nevertheless, as in the Wyner–Ziv problem (Lemma 10.3.1), this rate loss is *universally upper bounded*. For quadratic distortion and *any* joint distribution of X and Y, the redundancy of the Berger–Tung sum rate (10.25c) above the *joint-encoding rate-distortion function* is upper bounded by 1 bit:

$$\inf\{R_1 + R_2\} - R_{\text{coordinate}}(D_1, D_2) \leq 1 \text{ bit}, \tag{10.26}$$

where $R_{\text{coordinate}}(D_1, D_2) = \inf I(X, Y; \hat{X}, \hat{Y})$, and where the infimum in (10.26) is over all (R_1, R_2) and (U, V) in (10.25). This half-a-bit per source bound extends to *any* number of sources. It implies that the Berger–Tung rate region is *nearly optimal*, provided that the rates can be *freely* allocated among the encoders; see [285].

10.4   Lattice Wyner–Ziv coding

We turn to show how to achieve the Wyner–Ziv function $R_{\text{WZ}}(D)$ in the quadratic-Gaussian case using a nested lattice coding scheme.

It is convenient to assume that X and Y are related as

$$X = Y + Z, \tag{10.27}$$

where Z, called the *innovation* (or *unknown*) source component, is a zero-mean Gaussian with variance σ_z^2, independent of the value of Y, and Y is an arbitrary (not necessarily Gaussian) random variable. [10] In this case, the Wiener coefficient for

---

[10] By the orthogonality principle (4.38), any jointly Gaussian pair $(X, \tilde{Y})$ can be described in the form $X = a^* \tilde{Y} + Z$, where a^* is the Wiener coefficient (4.36), and $Z \sim N(0, \sigma_{x|y}^2)$ is independent of $\tilde{Y}$. The relation (10.27) corresponds to $Y = a\tilde{Y}$ and $\sigma_z^2 = \sigma_{x|y}^2$.

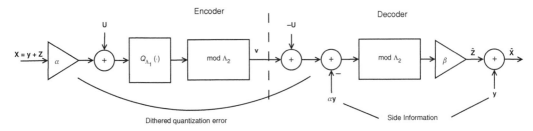

Figure 10.13 Lattice Wyner–Ziv encoding.

estimating X from $Y = X + Z$ is $a^* = 1$, and the conditional variance of X given Y is $\sigma_{x|y}^2 = \sigma_z^2$; hence the Wyner–Ziv rate-distortion function (10.22) is $R_{\text{WZ}}(D) = \frac{1}{2}\log(\sigma_z^2/D)$.

The task of the Wyner–Ziv encoder is to encode X in a way that conveys the innovation component Z, without wasting bits on the Y component which is known to the decoder. It is insightful to assume that the Y component is much *stronger* than the innovation Z, so "naive" encoding of X is very wasteful.

The idea of the coding scheme is similar to *syndrome coding* in the binary Slepian–Wolf problem (Section 10.1.2). The encoder quantizes the source vector $\mathbf{x}$ to a point λ in the fine lattice Λ_1, and sends its *relative coset* λ/Λ_2 to the decoder. This message contains $\log|\Lambda_2/\Lambda_1|$ bits. The decoder knows that λ is in the vicinity of the side-information vector $\mathbf{y}$, at a distance determined by the fine lattice resolution and the strength of the innovation $\mathbf{Z}$. Hence, if the coarse lattice Λ_2 is *sparse* enough, then λ would be the *only* member in its coset within this distance from $\mathbf{y}$, so the decoder can uniquely decode λ.

Note that no statistical assumptions are made about the side-information vector $\mathbf{y}$. If the quantization resolution is high, i.e., $D \approx \sigma^2(\Lambda_1) \ll \sigma_z^2$, then, to guarantee the coset sparseness condition, the number of relative cosets $|\Lambda_2/\Lambda_1|$ must be at least $\approx (\sigma_z/\sqrt{D})^n$. This corresponds to a coding rate of $R \approx \frac{1}{2}\log(\sigma_z^2/D)$, as desired.

The qualitative description above implicitly assumes "good" (high-dimensional) lattices. The lattice Wyner–Ziv coding scheme below uses general nested lattices, and incorporates *dither* (to randomize the quantization error) and *Wiener estimation* (to optimize the rate-distortion trade-off). Using these elements, we can characterize the gap to the quadratic-Gaussian Wyner–Ziv rate-distortion function at *any* quantization resolution, and show that this gap vanishes for good nested lattices.

10.4.1   Nested lattice binning scheme

The scheme, shown in Figure 10.13, slightly modifies the Voronoi quantizer of Section 9.10. It is based on a nested lattice pair $\Lambda_2 \subset \Lambda_1$, which generates a lattice-shaped codebook $\mathcal{C}_{\Lambda_1,\mathcal{P}_0} = \{\mathbf{v}_1, \ldots, \mathbf{v}_M\}$. Given a source vector $\mathbf{x} = \mathbf{y} + \mathbf{z}$, and a

Table 10.2 Resemblance between lattice Wyner–Ziv coding and the random binning solution (box in Section 10.3)

The resemblance becomes more apparent in the limit where the known part of the source is strong ($\sigma_y^2 \to \infty$ for a fixed σ_z^2), so $R_1 \to \infty$. Note that here the Wiener coefficient a^* is 1, because $X = Y + Z$.

Random binning scheme	Lattice coding scheme
rate R_1 codebook	fine lattice Λ_1
2^{nR} bins	$\lvert \Lambda_1/\Lambda_2 \rvert$ relative cosets
$R_1 > I(X; U)$ (successful encoding)	Λ_1 is a good quantizer
$R_1 < I(U; Y) + R$ (reliable decoding)	Λ_2 is good for channel coding

dither vector $\mathbf{u}$, the encoding and decoding operations are as follows:

$$\textbf{Encoding} \quad \mathbf{v} = Q_{\Lambda_1}(\alpha\mathbf{x} + \mathbf{u}) \bmod \Lambda_2 \tag{10.28a}$$

$$\textbf{Decoding} \quad \hat{\mathbf{z}} = \beta\,[(\mathbf{v} - \mathbf{u} - \alpha\mathbf{y}) \bmod \nu_0\Lambda_2] \tag{10.28b}$$

$$\hat{\mathbf{x}} = \mathbf{y} + \hat{\mathbf{z}}, \tag{10.28c}$$

where the scalar coefficients α and β are defined later.

Observe the following features of the lattice Wyner–Ziv coding system.

1. **Voronoi quantization** The only difference with respect to Voronoi quantization without side information (Section 9.10) is the subtraction and addition of the (scaled) side-information vector $\mathbf{y}$, before and after the modulo operation at the decoder.
2. **Lattice binning** Each message $\mathbf{v}$ represents an entire relative coset $\{\lambda \in \Lambda_1 : \lambda \bmod \Lambda_2 = \mathbf{v}\}$ in Λ_1/Λ_2. Hence, the scheme mimics the random coding solution for the quadratic-Gaussian Wyner–Ziv problem, with dithered relative cosets playing the role of "bins"; see Table 10.2 and Figure 10.14.
3. **Scaling coefficients** The role of the "pre-scaling" coefficient α is to *match* the source to the fine lattice resolution, and is equivalent to scaling the lattice by $1/\alpha$; see (2.43). At high-resolution quantization ($D \ll \sigma_z^2$), the role of the "post-scaling" coefficient β is simply to undo the pre-scaling, i.e., $\beta = 1/\alpha$. For a general resolution, β plays the role of a *Wiener coefficient*, that estimates the innovation from its dithered-quantized version, as in the random coding solution (10.23). [11]
4. **Symmetric complexity** The modulo-Λ_2 operation at the encoder can be taken with respect to a simple enumeration cell of the coarse lattice. In contrast, the modulo-Λ_2 operation at the decoder – that extracts the (quantized) Gaussian innovation – must be taken with respect to a (Euclidean) Voronoi shaping cell, in

---

[11] The multiplication by β at the decoder can be done *before* the modulo-Λ_2 operation, provided that Λ_2 is scaled by β.

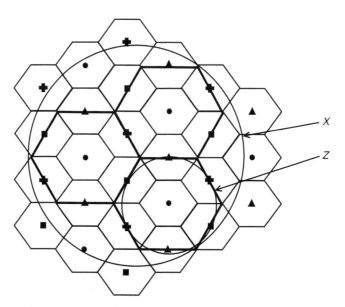

Figure 10.14 Lattice binning in the quadratic-Gaussian case.

order to minimize the error probability. Thus, the complex NN search is *evenly divided* between the encoder (the quantization step Q_{Λ_1}) and the decoder (the shaping step $\mathrm{mod}\ _{\nu_0} \Lambda_2$).

10.4.2 System performance

It would be convenient to characterize performance by the distortion when the system operates "properly" (as it should for a typical innovation vector $\mathbf{Z}$), and the probability of "improper" system operation. For that, let us define an "ideal" reconstruction vector $\tilde{\mathbf{x}}$, corresponding to an *unbounded* fine lattice quantizer (i.e., with the modulo-Λ_2 operations in (10.28) eliminated):

$$\tilde{\mathbf{x}} = \mathbf{y} + \beta \left[Q_{\Lambda_1}(\alpha \mathbf{x} + \mathbf{u}) - \mathbf{u} - \alpha \mathbf{y} \right]. \tag{10.29}$$

The event $\mathcal{E}$ where $\hat{\mathbf{X}} \neq \tilde{\mathbf{X}}$, is called a *decoding error* (or "overload"), and its probability is denoted

$$P_e = \Pr\{\mathcal{E}\} = \Pr\{\hat{\mathbf{X}} \neq \tilde{\mathbf{X}}\}. \tag{10.30}$$

Based on this event, the *total distortion* between $\mathbf{X}$ and $\hat{\mathbf{X}}$ can be broken into the sum of two terms:

$$D_t \triangleq \frac{1}{n} E\|\hat{\mathbf{X}} - \mathbf{X}\|^2 = D_{\mathrm{cd}} + D_{\mathrm{o}}, \tag{10.31}$$

where the *correct-decoding distortion* is defined as

$$D_{\text{cd}} = \frac{1}{n} E\{\|\hat{\mathbf{X}} - \mathbf{X}\|^2 \cdot 1_{\bar{\mathcal{E}}}\} \tag{10.32}$$

while the *overload distortion* is defined as

$$D_{\text{o}} = \frac{1}{n} E\{\|\hat{\mathbf{X}} - \mathbf{X}\|^2 \cdot 1_{\mathcal{E}}\}, \tag{10.33}$$

where

$$1_{\mathcal{E}} = \begin{cases} 1, & \text{if } \hat{\mathbf{X}} \neq \tilde{\mathbf{X}} \\ 0, & \text{if } \hat{\mathbf{X}} = \tilde{\mathbf{X}} \end{cases} \tag{10.34}$$

denotes the *error indicator*, and $1_{\bar{\mathcal{E}}} = 1 - 1_{\mathcal{E}}$ denotes the *correct-decoding indicator*. [12]

It should not surprise us that for good nested lattices $\Lambda_2 \subset \Lambda_1$, optimum performance can be achieved using the parameters of the "test channel" realization of the Wyner–Ziv function (10.23):

$$\alpha\beta = 1 - \frac{D}{\sigma_z^2} \quad \text{and} \quad \sigma^2(\Lambda_1) = \frac{\alpha}{\beta} D; \tag{10.35}$$

for example, the symmetric pre/post-scaling solution (5.51) is $\alpha = \beta = \sqrt{1 - D/\sigma_z^2}$ and $\sigma^2(\Lambda_1) = D$.

Theorem 10.4.1 (Performance characterization) *For the lattice Wyner–Ziv coding scheme (10.28), with uniform (or modulo-uniform) dither $\mathbf{U}$ with respect to Λ_1, and scaling parameters as specified in (10.35), the correct-decoding distortion (10.32) satisfies*

$$D_{\text{cd}} \leq D, \tag{10.36a}$$

the coding rate $R(\Lambda_1/\Lambda_2) = \frac{1}{n} \log |\Lambda_1/\Lambda_2|$ satisfies:

$$R(\Lambda_1/\Lambda_2) = R_{\text{WZ}}(D) + \frac{1}{2} \log(G(\Lambda_1) \cdot \mu_{\text{mix}}(\Lambda_2, P_e; \Lambda_1, \xi)) \tag{10.36b}$$

at Gaussian to dither mixture ratio (9.60)

$$\xi = \sigma_z^2/D - 1, \tag{10.36c}$$

and the decoding error probability (10.30) is equal to

$$P_e = \Pr\{\alpha\mathbf{Z} + \mathbf{U}_{\text{eq}} \notin \mathcal{V}_0(\Lambda_2)\}, \tag{10.36d}$$

---

[12] An equivalent writing of (10.32) is

$$(1 - P_e) \cdot \frac{1}{n} E\{\|\hat{\mathbf{X}} - \mathbf{X}\|^2 | \bar{\mathcal{E}}\}$$

where $E\{\cdot | \bar{\mathcal{E}}\}$ denotes conditional expectation given correct decoding, and similarly for (10.33).

where $R_{\mathrm{WZ}}(D) = \frac{1}{2}\log(\sigma_z^2/D)$ is the Wyner–Ziv function (10.22), $G(\Lambda_1)$ is the NSM of the fine lattice, $\mu_{\mathrm{mix}}(\Lambda_2, P_e; \Lambda_1, \xi)$ is the mixture NVNR (9.69) of the coarse lattice, and $\mathbf{U}_{\mathrm{eq}}$ is the equivalent dither (4.8), which is uniform over $\mathcal{V}_0(\Lambda_1)$ and independent of the innovation $\mathbf{Z}$.

The proof is given in Section 10.4.3 below. We make a few remarks regarding Theorem 10.4.1.

1. **Side information** The system performance is independent of the statistics (or even the value) of the side information $\mathbf{Y}$. Specifically, the redundancy term in (10.36b) is the same as in Voronoi quantization with *no* side information (Section 9.10).

2. **Equivalent mixture source** As evident from (10.36d), the equivalent source seen by the coarse (shaping) lattice is the sum

$$\alpha\mathbf{Z} + \mathbf{U}_{\mathrm{eq}} \triangleq \mathbf{Z}_{\mathrm{eq}} \qquad (10.37)$$

composed of a Gaussian component $\alpha\mathbf{Z}$ and a dither component $\mathbf{U}_{\mathrm{eq}}$. Substituting the system parameters (10.35), the variances of the equivalent source and the dither are given by

$$\mathrm{Var}(\alpha\mathbf{Z} + \mathbf{U}_{\mathrm{eq}}) = \frac{\alpha}{\beta}\sigma_z^2, \quad \text{and} \quad \mathrm{Var}(\mathbf{U}_{\mathrm{eq}}) = \frac{\alpha}{\beta}D \qquad (10.38)$$

respectively, thus the Gaussian to dither mixture ratio (9.60) is indeed given by (10.36c).

3. **Redundancy for a small error probability** By (9.70), the mixture NVNR is upper bounded by the Gaussian NVNR for a small P_e; the *redundancy term* in (10.36b) is thus upper bounded by

$$\frac{1}{2}\log(G(\Lambda_1)\,\mu(\Lambda_2, P_e)), \qquad (10.39)$$

i.e., by the sum of the *quantization loss* of the fine lattice and the *coding loss* of the coarse lattice.

4. **High-resolution quantization** When $D \ll \sigma_z^2$, hence $\xi \gg 1$, the equivalent (mixture) source is dominated by the Gaussian component $\mathbf{Z}$. Thus, the error probability approaches $P_e(\Lambda_2, \sigma_z^2)$, the mixture NVNR approaches $\mu(\Lambda_2, P_e)$, and the redundancy term in (10.36b) becomes (10.39) (for all P_e). We conclude that in both this and the previous cases, Λ_1 plays the role of a (good) quantizer, while Λ_2 plays the role of a (good) channel code for an AWGN channel.

The system designer may care about the *total* distortion $D_t = D_{\mathrm{cd}} + D_{\mathrm{o}}$ (10.31), rather than the correct-decoding distortion D_{cd} treated in Theorem 10.4.1.

Theorem 10.4.2 (Vanishing overload) *For a sequence $(\Lambda_1^{(m)}, \Lambda_2^{(m)})$ of nested lattices, possibly with a varying dimension $n(m)$, let $P_e(m)$ and $D_{\mathrm{o}}(m)$ denote the* *decoding error probability and overload distortion, respectively, of the corresponding Wyner–Ziv coding system. If $P_e(m)$ and $P_e(m)\, r_{\text{cov}}^2(\Lambda_2^{(m)})/n(m)$ go to zero as m goes to infinity (where r_{cov} denotes the lattice-covering radius), then $D_o(m)$ also goes to zero.*

The proof is given in Section 10.4.3 below.

To make P_e and $P_e\, r_{\text{cov}}^2(\Lambda_2)/n$ small for a *fixed* dimension n, the coarse lattice should be *sparse* (i.e., with a large VNR) with respect to the equivalent source (10.37); see Lemma 10.4.1 below. But this comes at the cost of a large redundancy (10.39) above $R_{\text{WZ}}(D)$. Alternatively, *both* the redundancy and overload distortion D_o can be arbitrarily small for a sequence of *good* nested lattices. The following notion of goodness extends Definition 9.10.1.

Definition 10.4.1 (Good Voronoi quantization: general resolution) *We say that a sequence $(\Lambda_1^{(n)}, \Lambda_2^{(n)})$ of nested lattices of a growing dimension n is "good for Voronoi quantization," if (i) the fine lattice is a good quantizer (Definition 7.3.2), i.e., $G(\Lambda_1^{(n)}) \to 1/2\pi e$, (ii) the coarse lattice is a good channel code under NN decoding against a "noise" being the equivalent source (10.37), i.e.,*

$$\mu_{\text{euclid}}(\Lambda_2^{(n)}, \mathbf{Z}_{\text{eq }n}, P_e) \to 2\pi e \quad \text{for all } P_e > 0 \tag{10.40}$$

(Theorem 7.8.1(B)), and (iii) the coarse lattice has a non-vanishing covering efficiency $\rho_{\text{cov}}(\Lambda_2^{(n)}) \geq \delta$, for some $\delta > 0$, for all n.

The existence of nested lattices which are good for Voronoi quantization follows from Theorem 8.5.1, since the equivalent source is semi-spherical; see Proposition 9.6.1.

Corollary 10.4.1 (Optimality for good nested lattices) *For a sequence of nested lattices good for Voronoi quantization (Definition 10.4.1), the lattice coding scheme (10.28) achieves the quadratic-Gaussian Wyner–Ziv rate-distortion function $R_{\text{WZ}}(D_t)$ with respect to the total distortion (10.31).*

Specifically, for *symmetric* pre/post-scaling ($\alpha = \beta = \sqrt{1 - D/\sigma_z^2}$ in (10.35)), this coding scheme satisfies $\sigma^2(\Lambda_1^{(n)}) = D$ for all n, and then $\sigma^2(\Lambda_2^{(n)}) \to \sigma_z^2$, $P_e(n) \to 0$, $D_o(n) \to 0$, $D_t(n) \to D$, and $R \to R_{\text{WZ}}(D)$, as $n \to \infty$.

10.4.3 Proof of Theorem 10.4.1 and Theorem 10.4.2

A: Proof of (10.36d) Let $\mathbf{e}_q$ denote the error (4.4) in dithered quantization of $\alpha\mathbf{x}$ by the fine lattice:

$$\mathbf{e}_q = Q_{\Lambda_1}(\alpha\mathbf{x} + \mathbf{u}) - (\alpha\mathbf{x} + \mathbf{u}). \tag{10.41}$$

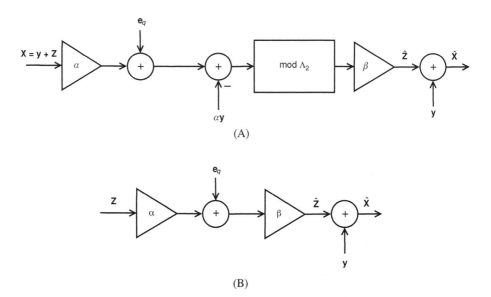

Figure 10.15 (A) Equivalent channel for the Wyner–Ziv lattice coding scheme of Figure 10.13. (B) Equivalent channel under the correct-decoding assumption.

Observing that the argument of the modulo-Λ_2 operation at the encoder (10.28a) is $\alpha \mathbf{x} + \mathbf{u} + \mathbf{e}_q$, we rewrite the final reconstruction as

$$\hat{\mathbf{x}} = \mathbf{y} + \beta \left[((\alpha \mathbf{x} + \mathbf{u} + \mathbf{e}_q) \bmod \Lambda_2 - \mathbf{u} - \alpha \mathbf{y}) \bmod \nu_0 \Lambda_2 \right] \tag{10.42a}$$

$$= \mathbf{y} + \beta \left[(\alpha \mathbf{z} + \mathbf{e}_q) \bmod \nu_0 \Lambda_2 \right] \tag{10.42b}$$

$$\stackrel{\text{c.d.}}{=} \mathbf{y} + \beta \left[\alpha \mathbf{z} + \mathbf{e}_q \right] \tag{10.42c}$$

$$= \tilde{\mathbf{x}}, \tag{10.42d}$$

where $\stackrel{\text{c.d.}}{=}$ denotes *equality conditional on correct decoding* $\tilde{\mathcal{E}}$, and $\tilde{\mathbf{x}}$ is the ideal reconstruction (10.29). Here, (10.42b) follows from the distributive law of the modulo operation (8.21), and since $\mathbf{y} + \mathbf{z} = \mathbf{x}$; while (10.42c)–(10.42d) follow from the definition of $\tilde{\mathbf{x}}$. See Figure 10.15. We conclude that the decoding error probability (10.30) is given by

$$P_e = \Pr\{(\alpha \mathbf{Z} + \mathbf{E}_q) \notin \mathcal{V}_0(\Lambda_2)\}. \tag{10.43}$$

This proves (10.36d), since by the crypto lemma (Lemma 4.1.1 and Theorem 4.1.1), $(\mathbf{Z}, \mathbf{E}_q)$ are distributed as $(\mathbf{Z}, \mathbf{U}_{\text{eq}})$.

B: Proof of (10.36b) Now, using the definition of the mixture NVNR (9.69), the formula (10.36d) for the decoding error probability, and the variance of the equivalent source (10.38), and similarly, by the definition of the NSM, and the system parameters (10.35), the volumes of the coarse and fine lattices satisfy

$$V^{2/n}(\Lambda_2) = \mu_{\text{mix}}(\Lambda_2, P_e; \Lambda_1, \xi) \cdot \frac{\alpha}{\beta} \sigma_z^2 \qquad (10.44a)$$

$$V^{2/n}(\Lambda_1) = \frac{1}{G(\Lambda_1)} \cdot \frac{\alpha}{\beta} D. \qquad (10.44b)$$

Hence, the coding rate $R(\Lambda_1/\Lambda_2) = \frac{1}{n} \log |\Lambda_1/\Lambda_2| = \frac{1}{n} \log(V(\Lambda_2)/V(\Lambda_1))$ satisfies (10.36b).

C: Proof of (10.36a) From (10.42c)–(10.42d), conditional on correct decoding $\bar{\mathcal{E}} = \{\tilde{\mathbf{x}} = \hat{\mathbf{x}}\}$, the equivalent error vector is

$$\hat{\mathbf{x}} - \mathbf{x} = \tilde{\mathbf{x}} - \mathbf{x} = \beta \mathbf{e}_q - (1 - \alpha\beta)\mathbf{z}. \qquad (10.45)$$

Thus, the correct-decoding distortion (10.32) of $\hat{\mathbf{X}}$ satisfies:

$$n D_{\text{cd}} = E\{\|\hat{\mathbf{X}} - \mathbf{X}\|^2 \cdot 1_{\bar{\mathcal{E}}}\} \qquad (10.46a)$$

$$= E\{\|\tilde{\mathbf{X}} - \mathbf{X}\|^2 \cdot 1_{\bar{\mathcal{E}}}\} \qquad (10.46b)$$

$$\leq E\{\|\tilde{\mathbf{X}} - \mathbf{X}\|^2\} \qquad (10.46c)$$

$$= E\|\beta\mathbf{E}_q - (1 - \alpha\beta)\mathbf{Z}\|^2 \qquad (10.46d)$$

$$= E\|\beta\mathbf{U}_{\text{eq}}\|^2 + E\|(1 - \alpha\beta)\mathbf{Z}\|^2 \qquad (10.46e)$$

$$= \beta^2 \sigma^2(\Lambda_1) + (1 - \alpha\beta)^2 \sigma_z^2 \qquad (10.46f)$$

$$= n D \qquad (10.46g)$$

for any $\mathbf{y}$, where in (10.46e) we used the crypto lemma, and in (10.46g) we substituted the system parameters (10.35). This completes the proof of Theorem 10.4.1. □

D: Proof of Theorem 10.4.2 We first derive a bound on the error norm $\|\hat{\mathbf{x}} - \mathbf{x}\|$ which holds under either correct or incorrect decoding. For any source vector $\mathbf{x} = \mathbf{y} + \mathbf{z}$, we have $\mathbf{x} - \hat{\mathbf{x}} = \mathbf{z} - \hat{\mathbf{z}}$ from (10.28), and therefore

$$\|\mathbf{x} - \hat{\mathbf{x}}\|^2 = \|\mathbf{z} - \hat{\mathbf{z}}\|^2 \qquad (10.47a)$$

$$\leq 2\|\mathbf{z}\|^2 + 2\|\hat{\mathbf{z}}\|^2 \qquad (10.47b)$$

$$\leq 2\|\mathbf{z}\|^2 + 2 r_{\text{cov}}^2(\beta\Lambda_2), \qquad (10.47c)$$

where (10.47b) follows from the triangle inequality, and (10.47c) follows since, by (10.28b), $\hat{\mathbf{z}}$ always falls inside $\mathcal{V}_0(\beta\Lambda_2)$, the coarse lattice cell scaled by β. It follows that the overload distortion (10.33) can be upper bounded by

$$D_{\text{o}} \leq \frac{2}{n} E\{\|\mathbf{Z}\|^2 \cdot 1_{\mathcal{E}}\} + P_e \frac{2}{n} \beta^2 r_{\text{cov}}^2(\Lambda_2). \qquad (10.48)$$

Since $\frac{1}{n} E\|\mathbf{Z}\|^2 = \sigma_z^2$ is finite, the first term in (10.48) vanishes if $P_e = \Pr\{1_{\mathcal{E}}\}$ goes to zero. And the second term vanishes if $P_e r_{\text{cov}}^2(\Lambda_2)/n$ goes to zero. □

E: Sparse coarse lattice A *sparse coarse lattice* amounts to a large VNR, the "noise" being the equivalent source (10.37):

$$\mu = \frac{V^{2/n}(\Lambda_2)}{\text{Var}(\alpha \mathbf{Z} + \mathbf{U}_{\text{eq}})} \gg 1. \tag{10.49}$$

Lemma 10.4.1 below shows that we can make D_{o} (as well as P_e) as small as desired by taking a sufficiently sparse coarse lattice, hence $D_{\text{t}} \approx D_{\text{cd}}$. This will come, however, at the cost of increasing the redundancy (10.39).

For a fixed dimension n and a fine lattice Λ_1, let $\Lambda_2^{(1)}, \Lambda_2^{(2)}, \ldots,$ be a sequence of lattices which are nested in Λ_1, and

$$\mathcal{V}_0(\Lambda_2^{(m)}) \supset k_m \cdot \mathcal{V}_0(\Lambda_2), \quad \text{where } k_m \to \infty \text{ as } m \to \infty. \tag{10.50}$$

That is, the coarse Voronoi cell grows without bound, at least linearly with k_m. It follows that the VNR μ of Λ_2 grows as k_m^2, hence the redundancy above $R_{\text{WZ}}(D)$ grows as $\log(k_m)$. On the good side, both the decoding error probability (10.30) and the overload distortion (10.33) vanish.

Lemma 10.4.1 (Sparse coarse lattice) *For the sequence $\Lambda_2^{(m)} \subset \Lambda_1$ defined above, we have $P_e(m) \to 0$ and $P_e(m) \, r_{\text{cov}}^2(\Lambda_2^{(m)}) \to 0$, hence (by Theorem 10.4.2) $D_{\text{o}}(m) \to 0$, as $m \to \infty$.*

Proof Since the Voronoi cell of the coarse lattice $\Lambda_2^{(m)}$ grows without bound, $r_{\text{cov}}^2(\Lambda_2^{(m)}) \sim k_m^2$, and $P_e(m) = \text{Pr}(\mathcal{E}) = E\{1_{\mathcal{E}}\}$ goes to zero. Furthermore, due to the "Gaussian tail" of the equivalent source $\alpha\mathbf{Z} + \mathbf{U}_{\text{eq}}$ in (10.36d), the error probability decays as $P_e(m) \sim e^{-k_m^2}$. Hence, $P_e(m) \cdot r_{\text{cov}}^2(\Lambda_2^{(m)}) \sim k_m^2 \, e^{-k_m^2}$, which goes to zero as m (and hence k_m) go to infinity; so the second term vanishes too. $\square$

10.4.4   Source-matched decoding

At low-resolution scalar quantization, i.e., when D is close to σ_z^2, and $\Lambda_1 = a\mathbb{Z}^n$, we observe a phenomenon similar to scalar shaping at low SNR (Section 9.8). Since the uniform dither component U in the equivalent source (10.37) is not negligible, the equivalent source $\alpha Z + U$ is not quite Gaussian. Hence, a *Euclidean* Voronoi cell (i.e., an NN rule) in the decoder modulo operation (10.28b) is not optimal, because it does not minimize the error probability (10.36d). [13]

As discussed in Section 9.8, we can improve the system performance in this case by *matching the decoder* to the equivalent source. That is, we keep the *same* coarse lattice Λ_2, but use in the modulo-Λ_2 operation in (10.28b) an *ML rule* with respect

---

[13] The loss is evident from the positive redundancy of the system at the maximum-distortion point $D = \sigma_z^2$. (Note that the log NSM-NVNR term in (10.36b) is constant with respect to D.) This loss can be easily avoided using a simple decoder, which reconstructs $\hat{\mathbf{X}} = \mathbf{Y}$, and thus achieves a distortion $D = \sigma_z^2$ with a *zero* rate.

to the equivalent source $\alpha Z + U$; see (6.12), where the equivalent source is viewed as "noise."

The characterization of the system performance becomes simple when the coarse lattice is a *good* (high-dimensional) channel code, while the fine lattice is arbitrary; see the discussion about mixed nesting dimensions in Section 8.3.3. To simplify the exposition, we shall restrict our attention to *scalar quantization*, i.e., a cubic fine lattice.

Lemma 10.4.2 (Source-matched decoding) *Assume a cubic fine lattice $\Lambda_{1,n} = \Delta_1 \cdot \mathbb{Z}^n$, and system parameters satisfying a distortion D in (10.35); i.e., $\alpha\beta = 1 - D/\sigma_z^2$ and $\Delta_1^2/12 = \alpha D/\beta$. Let the coarse lattice $\Lambda_{2,n}$ belong to a sequence of lattices of a growing dimension n, such that: (a) $\Lambda_{2,n}$ is nested in $\Lambda_{2,n}$ for all n; and (b) $\Lambda_{2,n}$ is a good channel code for the noise $\alpha Z + U$ (see Theorem 7.8.1); i.e., the noise-matched NVNR $\mu_{\mathrm{matched}}(\Lambda_{2,n}, \alpha Z + U, P_e)$ goes to $2\pi e$ as n goes to infinity, where $U \sim \mathrm{Unif}(-\Delta_1/2, \Delta_1/2)$ is the dither. Then, as $n \to \infty$,*

$$R(\Lambda_{1,n}/\Lambda_{2,n}) \to I(Z; \alpha Z + U). \tag{10.51}$$

Proof The asymptotic channel coding goodness of the coarse lattice implies that $\frac{1}{n}\log(V(\Lambda_{2,n})) \to h(\alpha Z + U)$; see the definition of noise-matched NVNR (6.28) and entropy power (A.15). Since $\frac{1}{n}\log(V(\Lambda_{1,n})) = \log(\Delta_1) = h(U)$, we obtain $R(\Lambda_1/\Lambda_2) = \frac{1}{n}\log V(\Lambda_{2,n})/V(\Lambda_{1,n}) = h(\alpha Z + U) - h(U)$, which can be rewritten as the mutual information formula (10.51). $\square$

Note that the rate formula (10.51) coincides with that of (pre/post-filtered) ECDQ of the innovation source $\mathbf{Z}$ (Section 5.6.2). It holds also for a *non-white-Gaussian* $\mathbf{Z}$, provided the asymptotic channel coding goodness condition of the coarse lattice holds.

As Figure 10.16 shows, scalar lattice Wyner–Ziv coding features a phenomenon similar to Figure 9.17 (scalar Voronoi modulation with noise-matched decoding): at high distortion (D close to σ_z^2), the systems's rate-distortion curve has a *reversed convexity* ($\cap$ instead of $\cup$). Thus, better performance can be obtained in this range by time sharing between the zero-rate/maximum-distortion point, and some positive rate point.

If Λ_2 is *not* a good (high-dimensional) coding lattice, then we cannot neglect the effect of overload on the total distortion (10.31). (In particular, if Λ_2 is a construction A lattice with a fixed alphabet, then the symbol error rate cannot vanish with the dimension; see Section 7.9.5.) To keep the overload distortion small, we must use a sparser coarse lattice (see Lemma 10.4.1), implying a cost in rate. Another implication of the overload distortion is that it increases the equivalent variance of the source. As Figure 10.16(C) shows, we can compensate for that by using an estimation coefficient β that is slightly *smaller* than the Wiener solution $\beta^* = \alpha^2\sigma_z^2\sigma^2(\Lambda_1)/[\alpha^2\sigma_z^2 + \sigma^2(\Lambda_1)]$.

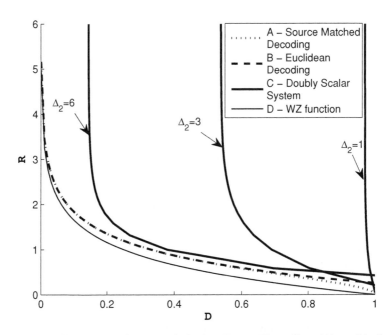

Figure 10.16 Rate-distortion performance in lattice Wyner–Ziv coding: (A) a cubic fine lattice and a good coarse lattice with source-matched decoding (Theorem 10.4.2); (B) the same lattices with Euclidean decoding; (C) doubly scalar system (where both coarse and fine lattices are scalar) with several values of the coarse step size Δ_2; and (D) the Wyner–Ziv function. Note that the rate in (C) is given by $R = \log(\Delta_2/\Delta_1)$, and the distortion is approximated by $D \cong (1 - P_e)[1/\sigma_z^2 + 12/\Delta_1^2]^{-1} + D_o$.

10.4.5   Universal lattice coding

We saw that the operation and performance of the lattice Wyner–Ziv coding scheme is *independent* of the statistics of the side information Y. As for the innovation source Z, the scheme rate loss is bounded by a *universal constant*, provided that the coarse lattice is "good," and the decoder is matched to the innovation statistics. [14]

Non-white-Gaussian innovation As in Lemma 10.3.1, we can compare the coding rate $R(\Lambda_1/\Lambda_2)$ with the conditional rate-distortion function $R_{X|Y}(D)$. In the case of an *additive-innovation* model $\mathbf{Y} = \mathbf{X} + \mathbf{Z}$, $R_{X|Y}(D)$ is equal to the rate-distortion function $R_Z(D)$ of the innovation Z, while $R(\Lambda_1/\Lambda_2) = I(Z; Z + U)$ by Lemma 10.4.2 (where for simplicity we set $\alpha = 1$). Observe that the rate loss $I(Z; Z + U) - R_Z(D)$ is the same as in ECDQ of the innovation Z (Chapter 5); thus, by Theorems 5.5.1 and 5.9.1, this loss is universally upper bounded by Ziv's constant $C_{Ziv}(\Lambda_1)$ (5.55) for all D, and it approaches $\frac{1}{2} \log(2\pi e G(\Lambda_1))$ as D goes to zero. In particular, for a *scalar* fine lattice, the loss is at most ≈ 0.754 bit, and it is ≈ 0.254 bit at high-resolution quantization. For a *good* fine lattice, these numbers are

---

[14] Note that a random ensemble of lattices can be universally good for a family of noise distributions; see Section 7.9.

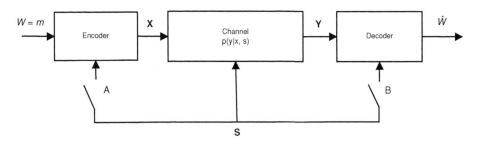

Figure 10.17 A channel with side information.

replaced by 1/2 bit and zero, respectively. Thus, for an additive-innovation, lattice Wyner–Ziv coding achieves the universal half-a-bit bound of Lemma 10.3.1.

Multi-terminal lattice coding The universal property of lattice Wyner–Ziv coding extends to the multi-terminal case. Recall the discussion in Section 10.2, on how to achieve a corner point of the multi-terminal rate region by successive application of Wyner–Ziv coding. Indeed, the sum rate of a multi-terminal dithered lattice coding scheme (with good coarse lattices and source-matched decoding) is given by the mutual information in a doubly additive-noise channel $I(X, Y; X + U_1, Y + U_2)$, where U_1 and U_2 are the equivalent dithers. This sum rate is at most ≈ 0.754 bit per terminal above the joint-coding rate for scalar fine lattices, and $1/2$ bit per terminal for good fine lattices. Thus, this scheme achieves (asymptotically) the same universal $1/2$ bit bound as the Berger–Tung solution; see (10.26).

See Problem 10.13 regarding achieving a non-corner rate point of the Berger–Tung region.

10.5 Channels with side information

A state-dependent noisy memoryless channel consists of an input X, an output Y and a state S. The input X is controlled by the encoder, the state S is drawn by nature (according to some distribution $p(s)$), and the output Y is governed by some conditional distribution $p(y|x, s)$, depending on both the input and state. The channel state S plays the role of "side information." As in the Wyner–Ziv setting, performance depends on whether S is available at the encoder, the decoder or at both.

Figure 10.17 and Table 10.3 illustrate the four possible cases of side-information availability. A delicate issue in the fourth case (side information at the encoder) is whether the encoder knows the state sequence $s_1, \ldots, s_n$ causally or non-causally. In the causal case, studied by Shannon [243], the current channel input x_t depends only on states up to this time, i.e., $x_t = f_t(m, s_1, \ldots, s_t)$, for $t = 1, \ldots, n$. Table 10.3 shows the non-causal case: the Gelfand–Pinsker setting, where each component of the transmission vector $\mathbf{x} = (x_1, \ldots, x_n)$ may depend on the entire state

Table 10.3 Cases of channel coding with side information

m denotes the transmitted message, x, y and s are the channel input, output and state, respectively, while $\hat{m}$ is the decoded message.

	Switch position		Encoding	Decoding	Capacity functions
	A	B			
no SI	open	open	$\mathbf{x} = f(m)$	$\hat{m} = g(\mathbf{y})$	$C_{\text{nosi}} = C(p(y\vert x))$
SI at both	closed	closed	$\mathbf{x} = f(m, \mathbf{s})$	$\hat{m} = g(\mathbf{y}, \mathbf{s})$	$C_{\text{si both}} = C(p(ys\vert xs))$
SI at decoder	open	close	$\mathbf{x} = f(m)$	$\hat{m} = g(\mathbf{y}, \mathbf{s})$	$C_{\text{si dec}} = C(p(ys\vert x))$
SI at encoder	closed	open	$\mathbf{x} = f(m, \mathbf{s})$	$\hat{m} = g(\mathbf{y})$	$C_{\text{si enc}} = C_{\text{GP}}$

vector $\mathbf{s} = (s_1, \ldots, s_n)$. Although at first sight the causal case seems more natural, many interesting practical applications are, in fact, non-causal, for example, MIMO broadcast, memory with defective cells and digital watermarking.

As mentioned in the beginning of this chapter, in the noiseless (or deterministic) Gelfand–Pinsker problem, capacity does not increase if the side information is also available at the decoder. Specifically, C_{GP} is equal to $C_{\text{si both}}$, and both capacities are equal to the conditional entropy $H(Y\vert S)$, maximized over all conditional input distributions $p(x\vert s)$. In contrast, in the noisy channel case, the situation is similar to rate distortion with side information (Section 10.3): C_{GP} is generally *smaller* than $C_{\text{si both}}$, with equality in the important quadratic-Gaussian case. [15]

Formulas for capacity with side information

In the complete side-information case, the capacity is a straightforward extension of (6.15):

$$C_{\text{si both}} = \max I(X; Y\vert S), \qquad (10.53)$$

where the maximization is over all valid conditional inputs $p(x\vert s)$. The encoder-only side-information capacity is given by

$$C_{\text{GP}} = \max[I(U; Y) - I(U; S)] \qquad (10.54a)$$
$$= \max[H(U\vert S) - H(U\vert Y)], \qquad (10.54b)$$

---

[15] As an example of a case where $C_{\text{GP}} < C_{\text{si both}}$, consider the binary channel case, where the known interference S is a binary symmetric source, the unknown noise Z is an independent Bernoulli-p source, and the channel input satisfies an input Hamming constraint $\frac{1}{n} E w_H(\mathbf{X}) \leq \delta$. The capacity with side information at the transmitter is given in this case by

$$C_{\text{GP}} = C(\delta) = \text{u.c.e.}\{H(\delta) - H(p), \ (0, 0)\} \quad 0 \leq \delta \leq 0.5, \qquad (10.52)$$

where u.c.e.$\{\cdot\}$ denotes upper convex envelope as a function of δ. Thus, we lose in capacity for not knowing S at the receiver, because with an informed receiver we could achieve a capacity $C = H(\delta * p) - H(p)$, which is larger for any $0 < \delta < 1/2$.

> where the maximization is over all auxiliary random variables U, and inputs X which are a function $X = g(U, S)$ (possibly under some input-cost constraint), such that $U \leftrightarrow (X, S) \leftrightarrow Y$ form a Markov chain. Note the resemblance between (10.53)–(10.54) and the side-information rate-distortion functions (10.19)–(10.20). See [64] for more details on side-information capacities.

Our main focus in this section will be the dirty-paper channel:

$$Y = X + S + Z, \tag{10.55}$$

where Z is an unknown additive noise, and S is an interference signal known to the transmitter but not to the receiver. [16] We shall assume that the channel input must satisfy an average power constraint $\frac{1}{n} E \parallel \mathbf{X} \parallel^2 \leq P$ (6.16).

This setting extends the configuration of channel coding with perfect side information discussed in Section 10.1.3. Here, there is an additional noise component (Z) which is unknown to both the transmitter and receiver. We can also view this setup as the digital watermarking problem illustrated in Figure 10.18: a watermark W is embedded into a source S, while keeping the mean-squared distortion between S and the watermarked signal $S + X$ below P. In this point of view, the noise Z represents a possible "attack" attempting to "remove" the watermark.

Costa [50], adhering to the Gelfand–Pinsker setting, showed that if S and Z are statistically independent Gaussian variables, then the maximization in the Gelfand–Pinsker formula (10.54) is achieved (for any distribution of S) by

$$U = \alpha S + X, \quad \text{where } X \sim N(0, P) \text{ is independent of } S, \tag{10.56a}$$

where $\alpha = \alpha^* = P/(P + N)$ is the Wiener coefficient, and N is the variance of Z. For this choice, the conditional distribution of U given S, and the conditional distribution of U given Y, are given by [17]

$$U|S \sim N(\alpha S, P) \quad \text{and} \quad U|Y \sim N(\alpha Y, PN/(P + N)) \tag{10.56b}$$

respectively. The Gelfand–Pinsker capacity (10.54) then becomes

$$C_{\text{GP}} = C_{\text{AWGN}} = \frac{1}{2} \log\left(1 + \frac{P}{N}\right). \tag{10.57}$$

We see that the effect of the interference S is canceled out *completely*, as if it were

---

[16] Note that any channel $Y = X + Z$ with side information S, where Z and S are jointly Gaussian, can be reduced to the form (10.55).

[17] The Wiener estimation error of U from Y is $(\alpha - 1)X + \alpha Z$, which is orthogonal to Y, just as in the zero interference case (Proposition 4.4.3).

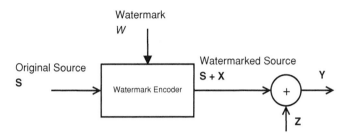

Figure 10.18 Dirty-paper coding viewed as digital watermarking.

zero or were available also at the receiver. The proof is based on the general random binning solution for channels with side information [100].

Achieving the Costa capacity via random binning

The random coding scheme that approaches the Costa capacity (10.57) has a simple geometric interpretation, *dual* to the quadratic-Gaussian Wyner–Ziv random coding scheme. For simplicity we shall assume a Gaussian $S \sim N(0, \sigma_s^2)$, so U is Gaussian $N(0, \sigma_u^2)$, with $\sigma_u^2 = \alpha^2 \sigma_s^2 + P$ by (10.56a). The code is generated by drawing 2^{nR_1} independent codewords $\mathbf{u}$, where each codeword is uniform over a ball of radius slightly larger than $r_u = \sqrt{n\sigma_u^2}$ about the origin, and then assigning the codewords at random to 2^{nR} bins. Each bin represents a possible message $m = 1, \ldots, 2^{nR}$. The encoder looks in bin m for the closest codeword $\mathbf{u}$ to the scaled interference αS, and transmits the difference vector $\mathbf{X} = \mathbf{u} - \alpha S$. The decoder multiplies the channel output $\mathbf{Y}$ by α, quantizes it to the closest codeword $\hat{\mathbf{u}}$, and decodes the message m as the bin of $\hat{\mathbf{u}}$. By (10.56b) $\mathbf{u}$ is typically within a radius $r^* = \sqrt{nPN/(P+N)}$ around $\alpha \mathbf{Y}$; hence the decoder is correct with high probability provided that

$$R_1 < I(U; Y)$$

(recall the reversed AEP of Proposition 9.4.1). As for the encoder, if

$$R_1 > I(U; S) + R,$$

then with high probability there will be in bin m at least one (random) codeword $\mathbf{u}$ which is jointly typical with S (the chance of typicality is $\approx 2^{-nI} = (r_x/r_u)^n$, and the chance of falling in bin m is 2^{-nR}, see the success-exponent threshold (Lemma A.2.1)). By (10.56b), such a typical $\mathbf{u}$ is within a radius $r_x = \sqrt{nP}$ around αS; so the "error" vector $\mathbf{X}$ to this $\mathbf{u}$ (and certainly to the *closest* $\mathbf{u}$) satisfies the power constraint P. Both requirements for typical encoding and correct decoding above can be satisfied, as long as R is smaller than $I(U; Y) - I(U; S) = C_{GP}$ of (10.54).

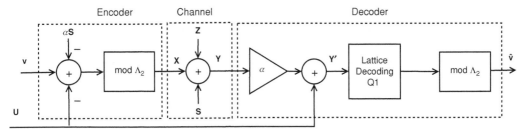

Figure 10.19 Lattice dirty-paper coding.

10.6 Lattice dirty-paper coding

The main problem we face in coding for the dirty-paper channel $Y = X + S + Z$ (10.55) is how to subtract the interference S without increasing the power of the transmitted signal X. In fact, it would be insightful to think about S as a signal with an *infinite* power: a simple subtraction of such a signal is impossible.

The trick is to subtract the interference vector *modulo a coarse Voronoi cell* tuned to meet the power constraint P. This is similar to the idea of "hiding a secret message in an integer sequence" (the problem described in the beginning of this chapter), or to canceling the known noise in a Hamming-constrained BSC (Section 10.1.3). Each message $1 \leq m \leq M$ is represented by a (relative) coset of the coarse lattice; being aware of the interference vector $\mathbf{s}$, the encoder looks for a vector $\mathbf{x}$ in the coarse Voronoi cell $\mathcal{V}_0(\Lambda_2)$, such that $\mathbf{x} + \mathbf{s}$ (the known part of the channel output) is a point in the desired coset. Indeed, for any representative member λ of the mth coset, taking $\mathbf{x}$ to be $(\lambda - \mathbf{s}) \bmod \Lambda_2$ implies that $\mathbf{x} + \mathbf{s}$ is a point λ' in the mth coset. And with an appropriate dithering, $\mathbf{x}$ meets – on the average – the desired transmit power constraint P. From the receiver side, if the fine lattice is sparse enough with respect to the (unknown) noise $\mathbf{Z}$, then the decoder will be able to decode λ' from the noisy channel output $\mathbf{Y} = \lambda' + \mathbf{Z}$, and identify the transmitted coset, which represents the message m.

Figure 10.19 shows the full lattice dirty-paper coding scheme, which also includes multiplication by a linear estimation coefficient α, in a way that *mimics* Costa's random binning solution of Section 10.5 (see Table 10.4):

$$\textbf{Encoding} \quad \mathbf{x} = (\mathbf{v}_m + \mathbf{u} - \alpha\mathbf{s}) \bmod_{\mathcal{V}_0} \Lambda_2 \qquad (10.58a)$$

$$\textbf{Decoding} \quad \hat{\mathbf{v}} = \left(\arg\min_{\lambda \in \Lambda_1} \|\alpha\mathbf{Y} - \mathbf{u} - \lambda\| \right) \bmod \Lambda_2, \qquad (10.58b)$$

where as usual $\mathbf{v}_m \in \mathcal{C}_{\Lambda_1, \mathcal{P}_0}$ is the message (coset) representative, for $1 \leq m \leq M$, $\hat{\mathbf{v}}$ is the decoded message representative, and $\mathbf{u}$ is the dither.

Observe that the lattice dirty-paper coding scheme (10.58) is just a simple variation on the linear Voronoi modulation scheme (9.65) of Chapter 9: the only difference is the subtraction of the (scaled) interference $\alpha\mathbf{s}$ at the encoder. In particular, for

Table 10.4 Resemblance between lattice dirty-paper coding and Costa's random binning

The resemblance is enhanced in the limit of a strong interference ($\sigma_s^2 \to \infty$), where $R_1 \to \infty$.

Costa's random binning scheme	Lattice dirty-paper coding
rate R_1 codebook	fine lattice Λ_1
2^{nR} bins	$\|\Lambda_1/\Lambda_2\|$ relative cosets
$R_1 < I(U; Y)$ (reliable decoding)	Λ_1 is good for channel coding
$R_1 > I(U; S) + R$ (successful encoding)	Λ_2 is a good quantizer
	(distortion equal to the power constraint P)

a uniform (or modulo-uniform) dither U with respect to Λ_2, the average transmit power satisfies Proposition 9.1.1, i.e., $P = \sigma^2(\Lambda_2)$ independent of the message and interference. At high SNR conditions ($P \gg N$), we can set $\alpha = 1$, so the encoder and decoder simplify to the description given in the beginning of the section. For a general α and SNR conditions, this scheme has the same equivalent mod Λ channel as in the zero interference case.

Lemma 10.6.1 (Equivalent mod Λ channel) *Assume lattice dirty-paper coding (10.58) in the channel $Y = X + S + Z$ (10.55), with a uniform (or modulo-uniform) dither U with respect to Λ_2, and an estimation coefficient α. For any value $S = s$ of the interference vector, the relation between the input v and the decision vector $\tilde{Y} = (\alpha Y - U) \bmod \Lambda_2$ is equivalent to the (zero interference) mod Λ channel of Lemma 9.5.1:*

$$\tilde{Y} = [v + Z_{eq}] \bmod \Lambda_2, \tag{10.59}$$

where the equivalent noise $Z_{eq} = (\alpha Z + (\alpha - 1)U) \bmod \Lambda_2$ (see (9.59)) is independent of v.

Proof We follow the proof of Lemma 9.5.1, noting that for any input x and dither u,

$$\tilde{Y} = \Big[\underbrace{[v + u - \alpha s] \bmod_{\mathcal{V}_0} \Lambda_2}_{x} + \underbrace{\alpha(x + s + Z)}_{\alpha Y} - u - x \Big] \bmod \Lambda_2 \tag{10.60a}$$

$$= [v + (\alpha - 1)x + \alpha Z] \bmod \Lambda_2 \tag{10.60b}$$

where (10.60b) follows by the distributive law of the modulo-lattice operation (2.22b). The rest of the proof is the same as that of Lemma 9.5.1: for a uniform dither U, the crypto lemma (Lemma 4.1.1) implies that the randomized input X is uniform over $\mathcal{V}_0(\Lambda_2)$ independent of v, which proves (10.59). □

This lemma implies that the performance of the lattice dirty-paper coding scheme is independent of the interference s. Hence, the coding rate and capacity loss are the same as in the zero interference case (Section 9.6). In particular, the scheme can achieve the capacity C_{AWGN} for a sequence of good (high-dimensional) nested lattice pairs (Theorem 9.6.1). And noise-matched lattice decoding (Section 9.8) (possibly with *non*-Wiener estimation) can enhance the system performance for low dimensions and low SNR, or for construction A lattices.

The main conclusion is summarized in the following theorem.

Theorem 10.6.1 (System performance) *The coding rate* $R(\Lambda_1/\Lambda_2) = \frac{1}{n} \log |\Lambda_1/\Lambda_2|$ *of the lattice coding scheme* (10.58), *with a random dither and Wiener estimation coefficient* $\alpha^* = P/(P + N)$, *over the dirty-paper channel* (10.55), *satisfies*

$$R(\Lambda_1/\Lambda_2) \geq C_{\text{AWGN}} - \frac{1}{2} \log(G(\Lambda_2) \cdot \mu_{\text{mix}}(\Lambda_1, P_e; \Lambda_2, \alpha^*)), \qquad (10.61)$$

where $C_{\text{AWGN}} = \frac{1}{2} \log(1 + P/N)$ *is the zero interference AWGN channel capacity, and* P_e *is the decoding error probability. Thus, the scheme achieves* C_{AWGN} *asymptotically for nested lattice pairs that are good for Voronoi modulation over an AWGN channel (Definition 9.6.2).*

10.6.1   Sufficient statistics for strong interference

In spite of the strong resemblance between their lattice coding solutions, the dirty-paper and zero interference cases differ in a few aspects. One such aspect is the efficiency of lattice decoding compared to ML decoding. For the AWGN channel, lattice decoding reduces complexity at the price of a performance loss, called *modulo loss* (unless the SNR is high or the nested lattices are high dimensional and good, in which case lattice decoding is optimal; see the discussion in Sections 9.1.2, 9.3.4, 9.5.5 and 9.8). In contrast, in the case of the dirty-paper channel, the advantage of ML decoding over lattice decoding is smaller, and it vanishes completely when the interference S is "strong" compared to the transmitter power P.

To see why, consider a hypothetical situation, where the addition operation in the dirty-paper channel (10.55) is *modulo* some multiple of the coarse lattice cell, and the interference S is *uniform* over that extended cell; say, for simplicity, the coarse lattice is a step-Δ scalar lattice, the addition in (10.55) is modulo $N\Delta$ for some integer $N > 1$, and $S \sim \text{Unif}(0, N\Delta)$. In this case, the conditional channel-output distribution $p(y|\text{message} = m, \text{dither} = u)$ is cyclic-periodic with a period Δ, i.e., $p(y \oplus \Delta|m, u) = p(y|m, u)$, for all m and u, where $\oplus$ denotes addition modulo $N\Delta$. It follows that $Y_q = Q_\Delta(Y)$ is uniformly distributed over $\{0, \ldots, N - 1\}$, independent of (Y_e, m, u), where $Y_e = Y \bmod \Delta = Y - Y_q$:

$$p(y = y_q + y_e|m, u) = \frac{1}{N} p(y_e|m, u),$$

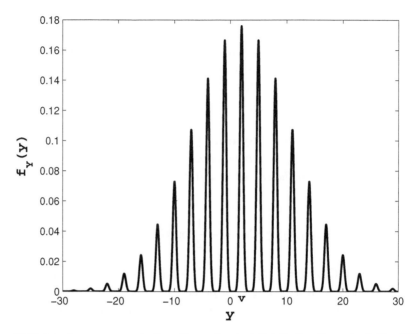

Figure 10.20 Interference concentration. Channel output distribution for scalar lattice DPC, for a specific message v in the presence of a strong Gaussian interference ($\sigma_s \gg \Delta \gg \sigma_z$). The figure assumes $u = 0$, so the coset shift is equal to v.

for all y_q in $\{0, \dots, N-1\}$. Thus, Y_e (which is equivalent to the decision vector $\tilde{Y} = [Y_e - u] \bmod \Delta$) is a *sufficient statistic* for decoding m from (Y, u), and the modulo loss is *zero*.

For a general interference S and regular addition in (10.55), however, the output distribution is not quite periodic in Δ, hence Y_q depends on (m, u), and ignoring it may result in loss of performance; see Figure 10.20. For example, in the zero interference ($S = 0$) case, Y_q identifies the larger ML Voronoi cell of a border codeword, and hence can reduce its decoding error probability; see Figure 9.4. Yet, the additional ambiguity induced by the interference S weakens the dependence of Y_q on the message m, and eliminates it completely as S becomes "strong."

The situation is similar to high-resolution quantization (Section 4.1.1), where the source tends to be uniform in the quantization cells, and the quantizer output tends to an "impulse train." To make a rigorous statement of this property, we shall use the mutual information as a measure for dependence.

Theorem 10.6.2 (Modulo output is a sufficient statistic) *For a lattice dirty-paper encoder* (10.58a) *and a Gaussian interference* $S \sim N(0, \sigma_s^2)$, *the mutual information between the message representative vector* **V** *and the channel output*

$\mathbf{Y} = [\mathbf{V} + \mathbf{u} - \alpha\mathbf{S}] \bmod \Lambda_2 + \mathbf{S} + \mathbf{Z}$ *satisfies*

$$\lim_{\sigma_s^2 \to \infty} I(\mathbf{V}; \mathbf{Y}|\mathbf{U} = \mathbf{u}) = I(\mathbf{V}; \tilde{\mathbf{Y}}) \tag{10.62}$$

for all $\mathbf{u}$, *or, equivalently,* $I(\mathbf{V}; \mathbf{Y}|\tilde{\mathbf{Y}}, \mathbf{U} = \mathbf{u}) \to 0$, *where* $\tilde{\mathbf{Y}} = [\alpha\mathbf{Y} - \mathbf{u}] \bmod \Lambda_2$ *is the decision vector.*

It follows that for a strong interference, $\mathbf{V}$ and $Q_{\Lambda_2}(\mathbf{Y})$ are conditionally independent given $\tilde{\mathbf{Y}}$. Thus, $\tilde{\mathbf{Y}}$ becomes a *sufficient statistic* for decoding the message from $\mathbf{Y}$ and $\mathbf{u}$, i.e., ML decoding is no better than lattice decoding.

Proof For simplicity, assume $\alpha = 1$, so $\mathbf{X} = [\mathbf{V} + \mathbf{u} - \mathbf{S}] \bmod \Lambda_2$ and $\tilde{\mathbf{Y}} = [\mathbf{Y} - \mathbf{u}] \bmod \Lambda_2$. See an extension to a general estimation coefficient $0 < \alpha < 1$ in Problem 10.12. Noting that for any vector $\mathbf{a}$,

$$[\mathbf{a} - \mathbf{S}] \bmod \Lambda_2 + \mathbf{S} = [\mathbf{a} - \mathbf{S}] - Q_{\Lambda_2}[\mathbf{a} - \mathbf{S}] + \mathbf{S} \tag{10.63}$$

$$= \mathbf{a} + Q_{\Lambda_2}[\mathbf{S} - \mathbf{a}], \tag{10.64}$$

we have

$$I(\mathbf{V}; \mathbf{Y}|\mathbf{U} = \mathbf{u}) = I(\mathbf{V}; \mathbf{V} + \mathbf{u} + \mathbf{Z} + Q_{\Lambda_2}[\mathbf{S} - \mathbf{V} - \mathbf{u}]) \tag{10.65}$$

$$\to I(\mathbf{V}; [\mathbf{V} + \mathbf{u} + \mathbf{Z}] \bmod \Lambda_2) \tag{10.66}$$

$$= I(\mathbf{V}; [\mathbf{Y}] \bmod \Lambda_2), \tag{10.67}$$

where the limit (10.66) follows from Lemma A.6.1 (Appendix A.6), which implies that a strong additive lattice-distributed noise is information-wise equivalent to a modulo-lattice reduction, and (10.67) follows from the distributive law of the modulo operation. The equivalent statement $I(\mathbf{V}; \mathbf{Y}|\tilde{\mathbf{Y}}) \to 0$ follows from the chain rule for mutual information (A.7). $\square$

Note that this theorem holds for any SNR and lattice dimension. A negative implication is that the "natural shaping" – enjoyed by ML decoding in the interference-free AWGN channel at the low SNR regime (Section 9.4.1) – disappears in the presence of a strong interference. Apparently, good high-dimensional shaping is a *necessary* feature of a dirty-paper coding scheme that approaches capacity, even at *low* SNR.

10.6.2 Causal side information˜

When the shaping lattice is cubical ($\Lambda_2 = a\mathbb{Z}^n$, while $\Lambda_1 \supset \Lambda_2$ is arbitrary), the modulo-Λ_2 operation at the dirty-paper encoder (10.58a) is scalar, i.e., the encoder cancels the interference $\mathbf{S} = (S_1, \ldots, S_n)$ component-by-component. This fits the definition of a channel with a *causal* side information, discussed in Section 10.5.

Example 10.6.1 (Tomlinson–Harashima pre-coding) *In the context of channels with* inter-symbol interference *(ISI), the causal dirty-paper coding technique is known as Tomlinson–Harashima pre-coding (THP)* [117, 254]. *The THP combines*

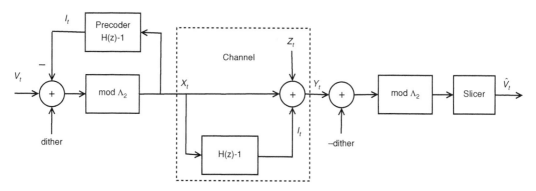

Figure 10.21 Lattice ISI pre-coding (high SNR version). $H(z)$ is the Z-transform of the channel filter.

coded modulation with ISI cancelation at the encoder. Assuming a monic-filter channel of the form $Y_t = X_t + \sum_{i=1}^{\infty} h_i X_{t-i} + Z_t$, the THP encoder outputs

$$X_t = [V_t - \underbrace{\sum_{i=1}^{\infty} h_i X_{t-i}}_{\text{interference}}] \bmod \Delta_2,$$

where $V_1, \ldots, V_n$ is a coded PAM or QAM vector, bounded in the range $(-\Delta_2/2, \Delta_2/2)$. The decoder reduces the channel output modulo Δ_2, to obtain the decision variable $\tilde{Y}_t = [V_t + Z_t] \bmod \Delta_2$, from which it decodes the symbol V_t. See Figure 10.21.

Erez *et al.* [67] showed that when the interference is strong and the SNR is high, dirty-paper coding is equivalent to entropy-constrained quantization with *high resolution*; see Section 5.5.3. It follows from high-resolution quantization theory [106] that lattice-based interference cancelation is optimal over *all* causal dirty-paper coding schemes, hence the shaping loss of $\frac{1}{2}\log(2\pi e/12) \approx 0.254$ bit is, in fact, unavoidable. [18] Is the higher loss (of ≈ 4 dB) of a scalar shaping lattice at the *low SNR* regime (see Sections 9.5.5 and 9.8) also unavoidable in causal dirty-paper coding with a strong interference?

The answer to the above question depends on which one is lower: the SNR or the SIR (signal to interference ratio). Liu and Viswanath [172] proposed an *opportunistic non-causal dirty-paper* coding scheme, which operates in the limit of low SNR, and achieves the *full* AWGN capacity per unit cost independent of the level of the interference; see Section 9.4.1. Borade and Zheng [24] showed a similar result in the causal side-information case, using a scheme that transmits energy

[18] It is tempting to extend this result to dirty-paper coding with *block anticipation*, i.e., the encoder sees the interference sequence in blocks of size n (at time instant $mn + 1 \le t \le (m+1)n$ the encoder sees $S_{mn+1}, \ldots, S_{(m+1)n}$). If Gersho's conjecture is true (see Section 3.2), then a lattice dirty-paper coding scheme (with the best n-dimensional lattice) is optimal, hence the shaping loss of $\frac{1}{2}\log(2\pi e G_n)$ is unavoidable.

in a message-dependent interval (as in pulse-position modulation), whenever the interference S exceeds a certain threshold that depends on the SNR and SIR. It follows that the order of limits matters: while shaping (i.e., non-causal dirty-paper coding) seems to be necessary to achieve capacity for any SNR in the limit of a strong interference, causal dirty-paper coding is sufficient to achieve capacity for any (fixed) interference level, in the limit of low SNR.

Summary of Chapter 10

Syndrome source coding To encode a binary source $X = Y \oplus Z$ with side information Y at the decoder, send the syndrome

$$\mathbf{s} = H \cdot \mathbf{X},$$

and reconstruct $\mathbf{X}$ as

$$\hat{\mathbf{X}} = \mathrm{error}(\mathbf{Y}, \mathcal{C}_\mathbf{s}) \oplus \mathbf{Y},$$

where the code $\mathcal{C}$ (with a parity-check matrix H) is good for a BSC with noise Z.

Syndrome channel coding To send an information vector $\mathbf{s}$ over a BSC $Y = X \oplus Z$ with a Hamming constraint q, and where the noise Z is known at the encoder, transmit

$$\mathbf{X} = \mathrm{error}(\mathbf{Z}, \mathcal{C}_\mathbf{s}),$$

and decode $\mathbf{s}$ as

$$\hat{\mathbf{s}} = H \cdot \mathbf{Y},$$

where the code $\mathcal{C}$ is a good quantizer with a Hamming distortion q.

Lattice Wyner–Ziv coding To quantize a source $X = Y + Z$ with side information Y at the decoder, send the relative coset

$$\mathbf{v} = Q_{\Lambda_1}(\mathbf{X} + \mathbf{U}) \bmod \Lambda_2$$

in Λ_1/Λ_2, and reconstruct $\mathbf{X}$ as

$$\hat{\mathbf{X}} = \mathbf{Y} + \beta^* \cdot [(\mathbf{v} - \mathbf{U} - \mathbf{Y}) \bmod \nu_0 \Lambda_2],$$

where the fine lattice Λ_1 is a good quantizer (matched to the target distortion), the coarse lattice Λ_2 is a good channel code for the noise Z, $\mathbf{U}$ is the dither, and β^* is the Wiener coefficient for estimating $\mathbf{Z}$ from $\mathbf{Z} + \mathbf{U}$.

Gap from the Wyner–Ziv function

$$R(\Lambda_1/\Lambda_2) - R_{\mathrm{WZ}}(D) \leq \text{Quantization loss}(\Lambda_1) + \text{Coding loss}(\Lambda_2),$$

for a sufficiently small overload probability P_e.

Lattice dirty-paper coding To send coset information $\mathbf{v}_m$ (for $1 \leq m \leq |\Lambda_1/\Lambda_2|$) over the channel $Y = X + S + Z$, where the interference S is available at the encoder, transmit the vector

$$\mathbf{X} = (\mathbf{v}_m + \mathbf{U} - \alpha^* \mathbf{S}) \bmod_{\mathcal{V}_0} \Lambda_2,$$

and decode $\mathbf{v}_m$ as

$$\hat{\mathbf{v}} = Q_{\Lambda_1}(\alpha^* \mathbf{Y} - \mathbf{U}) \bmod \Lambda_2,$$

where the fine lattice Λ_1 is a good channel code for the noise Z, the coarse lattice Λ_2 is a good quantizer (matched to the target transmit power), $\mathbf{U}$ is the dither, and α^* is the Wiener coefficient for estimating X from $X + Z$.

Equivalence to an interference-free channel The channel from $\mathbf{v}_m$ to $(\alpha^* \mathbf{Y} - \mathbf{U}) \bmod \Lambda_2$ is independent of the statistics of S.

Sufficient statistics for a strong interference For $S \sim N(0, \sigma_s^2)$ and $\sigma_s^2 \to \infty$, the decision vector $(\alpha^* \mathbf{Y} - \mathbf{U}) \bmod \Lambda_2$ becomes a sufficient statistic for decoding $\mathbf{v}_m$ from $\mathbf{Y}$.

Gap to the Gelfand–Pinsker capacity

$$C_{\mathrm{GP}} - R(\Lambda_1/\Lambda_2) \leq \text{Shaping loss}(\Lambda_2) + \text{Coding loss}(\Lambda_1),$$

for a sufficiently small decoding-error probability P_e.

Problems

P.10.1 (Modulo code operation) Prove that $\mathbf{x} \bmod \mathcal{C}$ as defined in (10.3) is the error vector to the closest codeword in the code $\mathcal{C}$. Hint: use (10.2).

P.10.2 (Error to coset) Prove (10.4). Guidance: observe that the error between a vector $\mathbf{x}$ and a coset $\mathcal{C}_\mathbf{s}$ is the minimum Hamming-weight vector in the shifted coset $\mathbf{x} \oplus \mathcal{C}_\mathbf{s}$ which, by linearity, is the coset representative of the syndrome $\mathbf{s}' = H\mathbf{x} \oplus \mathbf{s}$.

P.10.3 (Distributive law of the modulo-$\mathcal{C}$ operation) Prove that $(\mathbf{x} \bmod \mathcal{C}) \oplus \mathbf{y} \bmod \mathcal{C} = (\mathbf{x} \oplus \mathbf{y}) \bmod \mathcal{C}$.

P.10.4 (BSC with a Hamming input constraint) Show that the capacity of the channel in Figure 10.5 *without* side information is given by $C = H_B(q \otimes p) - H_B(p)$, where q is the Hamming input constraint, p is the BSC crossover probability, and $q \otimes p = q(1 - p) + (1 - q)p$ is the *binary convolution*. Show that if the noise Z is known at the decoder, then the larger capacity of $C = H_B(q)$ can be achieved, *independent* of the crossover probability p.

P.10.5 (Causal noise cancelation via "naive" pre-coding) For the Hamming-constrained BSC above, show that if the noise is known at the encoder, then "naive" noise cancelation can achieve a rate of $R = H_B(p'_x)$, where p'_x is the solution of $q = p'_x \otimes p$, provided that $q > p$. Draw the graph of $R(q)$ versus q for $0 \le q \le 1/2$. Time share it with the zero-rate/zero-cost point $(0, 0)$, to increase the rate at some range of input constraints $0 \le q \le q^*$. Find q^*.

P.10.6 (Linear distributed compression for DSBS) Let (X, Y) be a DSBS (10.1). Show how to achieve the corner point $R_X = H_B(p)$, $R_Y = 1$ bit, in the plane of rate pairs (R_X, R_Y), using syndrome coding for the X source.

P.10.7 (Linear deterministic broadcast) Show how to implement the scheme of Section 10.1.3 in the case where X_1, X_2, Y_1, Y_2 are all binary, $Y_1 = X_1 = \tilde{X}_1$, $Y_2 = \tilde{X}_1 \oplus \tilde{X}_2$, and where $\tilde{X}_2$ needs to satisfy an average Hamming constraint q_2.

P.10.8 (Rate loss in Wyner–Ziv and the orthogonality principle) Show that for any auxiliary random variable U in the Wyner–Ziv formula (10.20), the rate loss with respect to the conditional mutual information $I(X; \hat{X}|Y)$ (corresponding to side information everywhere) is given by $I(X; U|\hat{X}, Y)$. (Hint: use the chain rule for mutual information.) Show that for the optimal U in the Gaussian case (10.23b), the reconstruction error $\hat{X} - X$ is orthogonal to $(U, Y, \hat{X})$. Conclude that in this case $I(X; U|\hat{X}, Y) = 0$.

P.10.9 (Capacity loss in Gelfand–Pinsker and the orthogonality principle) Show that for any auxiliary random variable U in the Gelfand–Pinsker formula (10.54), the capacity loss with respect to the conditional mutual information $I(X; Y|S)$ (corresponding to side information everywhere) is given by $I(U; S|Y)$. (Hint: use the chain rule for mutual information.) Show that for the optimal U in the power-constrained Gaussian case (10.56a), the estimation error $U - \alpha Y$ is independent of (S, Y). Conclude that in this case $I(U; S|Y) = 0$.

P.10.10 (Broadcast with "onion peeling" at the transmitter) The two-user Gaussian broadcast channel is defined by $Y_1 = X + Z_1$ and $Y_2 = X + Z_2$, where $Z_i \sim N(0, N_i)$, and $N_2 > N_1$, i.e., receiver 2 is worse than receiver 1. The standard solution is by superposition coding: $X = X_1 + X_2$, where X_1 with power αP carries private information to the good receiver at rate $R_1 = c(\alpha P/N_1)$, and X_2 with power $(1 - \alpha)P$ carries common information for both receivers at rate $R_2 = c((1 - \alpha)P/(\alpha P + N_2))$, where $c(x) \triangleq \frac{1}{2}\log(1 + x)$. Decoding at the bad receiver treats the interference from X_1 as noise, while decoding at the good receiver is by "onion peeling": first X_2 is decoded and subtracted, and then X_1 is decoded interference free. Show that the "onion peeling" step can be replaced by lattice dirty-paper coding at the transmitter.

Guidance: for simplicity, assume first high SNR conditions. Consider transmission of the form $X = (X_1 - X_2) \bmod \Lambda_2 + X_2$, and find suitable receivers. Secondly, add dither and suitable Wiener estimation for the general SNR case.

P.10.11 (Differential pulse-code modulation (DPCM) with decoder-only prediction) Let $X_1, X_2, \ldots$ denote a stationary-Gaussian source, to be encoded under the MSE criterion $D = E(\hat{X} - X)^2$. DPCM is a sequential coding scheme, which at time instance n encodes the prediction error $Z_n = \tilde{X}_n - X_n$, where $\tilde{X}_n = f(\hat{X}_{n-1}, \hat{X}_{n-2}, \ldots)$ is a predictor of X_n given the past reconstructions. At the decoder, the source is reconstructed by adding the encoded prediction error $\hat{Z}_n$ to the prediction $\tilde{X}_n$. Show that the prediction step at the encoder can be avoided, and replaced by lattice Wyner–Ziv coding.

Guidance: for simplicity, assume high-resolution quantization and scalar lattices. Consider an encoder of the form $\mathbf{v} = Q_{\Lambda_1}(X) \bmod \Lambda_2$, and find suitable lattice parameters and a decoder system. Analyze the system rate-distortion performance. (For extension to general resolution with multi-dimensional lattices, see [291].)

P.10.12 (Lattice decoding is optimal for dirty-paper coding with strong interference) Extend the proof of Theorem 10.6.2 to a general estimation coefficient $0 \le \alpha \le 1$. Hint: show that $I(\mathbf{V}; \alpha\mathbf{Y} - \mathbf{u})$ is equal to $I(\mathbf{V}; [\mathbf{V} - \mathbf{u}] \bmod \Lambda_2 + \mathbf{Z}_{\text{mix}} + Q_{\Lambda_2}[\alpha\mathbf{S}])$, argue why $Q_{\Lambda_2}[\alpha\mathbf{S}]$ is asymptotically independent of $\mathbf{Z}_{\text{mix}}$ in the limit of strong interference (even for a *specific* $\mathbf{u}$), and use Lemma A.6.1.

P.10.13 (Non-corner rate point in multi-terminal source coding [Pradhan-Ramchandran 2005]) (1) Show how to achieve a general rate point (R_1, R_2), where $R_1 + R_2 = 1 + H_B(p)$, in distributed coding of a DSBS (i.e., the Slepian–Wolf problem of Section 10.1.2) using a linear code. Guidance: assume encoding of the form $(H\mathbf{x}, G_1\mathbf{x})$ and $(H\mathbf{y}, G_2\mathbf{y})$, for some binary matrices H, G_1 and G_2 with dimensions $k_0 \times n$, $k_1 \times n$ and $k_2 \times n$, respectively; and decoding of the form $\hat{\mathbf{z}} =$ leader $(H\mathbf{x} \oplus H\mathbf{y})$ and $\hat{\mathbf{y}} = \hat{\mathbf{x}} \oplus \hat{\mathbf{z}}$, where $\hat{\mathbf{x}} =$ inverse $(H\mathbf{x}, G_1\mathbf{x}, G_2\tilde{\mathbf{x}})$, and where $G_2\tilde{\mathbf{x}} = G_2\mathbf{y} \oplus G_2\hat{\mathbf{z}}$. Find conditions on the matrices that guarantee correct decoding with high probability for a large n.

(2) Propose a nested lattice extension of this scheme to lossy multi-terminal source coding (i.e., the quadratic-Gaussian Berger–Tung problem described in Section 10.3).

Historical notes

The main breakthroughs in network information theory happened in the 1970s and early 1980s; see the books by Cover and Thomas [53] and El Gamal and Kim [64].

Multi-terminal lossless source coding was introduced by Slepian and Wolf [247] in 1973, where they showed that the sum rate is not worse than in the coordinated encoding case. Cover [51] simplified their proof (and extended it to ergodic sources) by introducing the random binning technique. A special case of the Slepian–Wolf setup is lossless source coding with (uncompressed) side information at the decoder. A linear-code interpretation for coding with side information (the syndrome code of Section 10.1) was described by Wyner [274]. The rate-distortion function with side information was found in 1976 by Wyner and Ziv [276] and Wyner [275]. Berger and Tung [20] proposed an achievable rate region for multi-terminal source coding with distortion (the lossy Slepian–Wolf problem), and Wagner *et al.* [264] showed that this region is tight for the quadratic-Gaussian case.

Although channels with side information were already treated (for the causal side-information case) by Shannon in 1958 [243], it took some time until a setup paralleling the Wyner–Ziv problem was identified. In the mid-1970s, Kuznetsov and Tsybakov [147] found codes for writing to computer memories with defective-cell information. In 1980, Gelfand and Pinsker [100] gave a complete information theoretic solution for the capacity of channels with non-causal side information at the encoder. Their formula had several interesting consequences. In 1983, Heegard and El Gamal [120] re-examined the problem of memories with side information, and found the capacity for noiseless and noisy setups. Costa [50] used the Gelfand–Pinsker formula to confirm a conjecture of Cover about the capacity of Gaussian channels with side information, a setup which they called "writing on a dirty paper." Costa showed that a transmitter can eliminate the effect of a known interference without a power penalty.

Channel coding with side information was implicit already in research from the late 1970s on non-degraded broadcast channels. Marton's inner bound [180] builds upon viewing the transmission to one user as a side information for the transmission to another user; see the discussion in Section 10.2. From a different angle, the deterministic (noiseless) broadcast channel, introduced and solved by Pinsker [220], strongly resembles the Slepian–Wolf distributed lossless source coding setup. More than two decades later, Caire and Shamai [27] showed that Costa's dirty-paper coding can be used to achieve the optimum sum rate over the Gaussian MIMO broadcast channel.

An early version of nested linear codes as a structured binning scheme appears in Heegard's paper from 1983 on partitioned linear block codes for memories with noise and defects [119]. For lossy source coding with side information, Shamai *et al.* [238, 239] introduced in 1996 a nested linear code that achieves the binary-Hamming Wyner–Ziv rate-distortion function. An extension to the quadratic-Gaussian setup using nested lattice codes was proposed by Zamir and Shamai [292], with an optimality proof for the high-resolution regime. Erez *et al.* [68, 293] proposed a lattice-based approach to dirty-paper channel coding. They showed that a scalar lattice is asymptotically optimal for causal dirty-paper coding in the high SNR regime (pointing out the relation to Willems' scalar dirty-paper coding scheme [267, 268]), and that a good (high-dimensional) lattice quantizer can achieve the Costa capacity at all SNR. In [293], they unified the nested lattice binning approach, and showed its optimality for general multi-terminal problems, including – beyond the classical side-information setups – lossy distributed source coding, noisy memories, ISI pre-coding and broadcast channels.

An interesting observation made in [68, 293] is the relation between lattice dirty-paper coding and pre-coding for inter-symbol interference channels, as proposed in the early 1970s by Tomlinson [254] and Harashima and Miyakawa [117]. This viewpoint implies that a large body of literature on combining coding, pre-coding (equalization at the transmitter) and shaping applies to the dirty-paper channel, see, for example, the work by Eyuboglu and Forney [75] and Laroia *et al.* [149, 150, 151]. Trellis and turbo-coded versions of dirty-paper coding with dither and MMSE estimation were studied by Philosof *et al.* [214], Erez and ten Brink [69] and Fischer [80]. In the parallel area of linear MIMO channels, Ginis and Cioffi [105] developed pre-coding schemes for channel equalization at the transmitter; see the generalized decision-feedback equalizer (GDFE) interpretation of Cioffi and Forney [42], and the trellis pre-coder of Yu and Cioffi [281].

Another closely related setup is digital watermarking, where the encoder embeds a digital message in an analog signal under some distortion constraint. An information theoretic relation to the Gelfand–Pinsker problem was found by Moulin *et al.* [191]. The capacity as a game between a watermark embedder and attacker was established by Cohen and Lapidoth [43], and further studied by Somech-Baruch and Merhav [249]. Lattice-based schemes were proposed by Chen and Wornell [38] and by Eggers *et al.* [63].

In the source coding area, practical trellis and lattice-based schemes for side information and multi-terminal coding setups were proposed by Pradhan and Ramchandran [224], Servetto [235] and Liu *et al.* [173]. Another large body of work focused on "Wyner–Ziv video coding," as discussed in Section 10.2; see the work of Puri *et al.* [226] and Shemer [246], which was preceded by a patent of Witsenhausen and Wyner [271].

A central theme in this chapter is the duality between source coding and channel coding, with and without side information; see, for example, Cover and Chiang [52] and Barron *et al.* [15].

11 Modulo-lattice modulation

Co-written with Yuval Kochman

So far we have discussed source and channel coding problems separately. In this chapter we address the combination of the two, known as joint source-channel coding (JSCC). In a JSCC scheme, the encoder maps a source sequence into a channel input block, while the decoder produces a source estimate from the corresponding channel outputs. This mapping can be constructed using a concatenation of source and channel codes, in which case it is known as a separation-based scheme. While Shannon's separation principle states that this approach yields asymptotically the smallest possible average distortion, other schemes may have significant advantages as we will see in the following. For example, analog transmission (symbol-by-symbol mapping) has very desirable properties of low complexity and high robustness. However, except for some very special source-channel pairs, it is inferior to the digital (separation-based) approach in terms of average distortion. Therefore, finding schemes which are neither fully digital nor fully analog is of great interest.

In this context, lattices play a natural role: since they are defined directly in the Euclidean space, they do not require mapping of the source sequence into a digital representation and back, as done in a separation-based scheme. We shall present a "semi-analog" JSCC technique, based on lattice codes, called "modulo-lattice modulation" (MLM). This technique uses lattices only for the purpose of shaping, while not using any data-bearing code (that is, the "fine lattice" of nested lattice schemes is replaced by uncoded transmission). The MLM technique enjoys the best of both analog and digital worlds. It yields the optimal expected distortion for a large class of Gaussian-quadratic JSCC problems, while being robust to channel uncertainty and saving complexity with respect to the separation-based scheme.

After a formal presentation of the JSCC problem in Section 11.1, we shall define new lattice figures of merit for JSCC in Section 11.2. In Section 11.3 we shall introduce an analog version of the lattice side-information schemes of Chapter 10, which forms the basis for a robust joint source-channel lattice modulation. Finally, in Section 11.4 we shall use this scheme for analog bandwidth conversion.

11.1   Separation versus JSCC

We first extend the basic source and channel coding framework, introduced in Chapter 1, to include the joint source-channel setting. The source is a vector, $\mathbf{S} = S_1, \ldots, S_n$, that is reconstructed at the destination as $\hat{\mathbf{S}} = \hat{S}_1, \ldots, \hat{S}_n$. The goal of the scheme is to achieve a low distortion D. For that, the encoder produces a channel input vector $\mathbf{X} = X_1, \ldots, X_m$, and the corresponding output $\mathbf{Y} = Y_1, \ldots, Y_m$ is given to the decoder.

Information theory tells us that any scheme must satisfy

$$n \cdot R(D) \leq m \cdot C, \tag{11.1}$$

where D is the average distortion, and $R(\cdot)$ and C are the source rate-distortion function and channel capacity, respectively. Note that, although the last two quantities are measured in bits, the scheme is not required to contain any digital element. Rather, the rates appear just as a convenient way to lower bound the distortion. Remarkably, information theory also tells us that the bound (11.1) is asymptotically achievable in the limit of a large block length. More precisely, we take the source block n and channel block m to infinity together, where the *bandwidth expansion ratio*

$$\rho = \frac{\text{channel dimension}}{\text{source dimension}} = \frac{m}{n} \tag{11.2}$$

is held fixed. [1] Then, Shannon's celebrated separation theorem states that the optimal asymptotic average distortion D^* is given by

$$R(D^*) = \rho \cdot C. \tag{11.3}$$

As the name suggests, the optimal asymptotic performance is indeed achieved by separating the tasks of source and channel coding. In fact, the structure of a separation-based encoder was already demonstrated in Figure 1.4. The source vector $\mathbf{S}$ is first mapped into a bit sequence $b_1, \ldots, b_k$, which represents some quantized value $\tilde{\mathbf{S}}$. We know that there exist good quantizers of rate $k/n \to R(D)$, if D is the distortion between $\mathbf{S}$ and $\tilde{\mathbf{S}}$. Then the bits are sent over the channel using a good channel code with rate $k/m \to C$. The decoder will work in reverse order: first it will recover, with high probability, the bit sequence, and then it will map it to a reconstruction $\hat{\mathbf{S}} = \tilde{\mathbf{S}}$. If a channel decoder error occurs, then we do not know much about $\hat{\mathbf{S}}$. Still one can show that the effect of such errors becomes negligible, and the overall expected distortion approaches D.

Source-channel separation is very important, conceptually and practically. It can be seen as providing a theoretical justification for the reality of designing

---

[1]   Of course, for an integer n, the resulting m may not be an integer. This can be solved by using $m = \lfloor \rho \cdot n \rfloor$, but we shall ignore the rounding effect as it is insignificant asymptotically.

compression and communication schemes: these two tasks are usually performed by different people, who use distinctive tools and expertise. The engineer who designs a modem does not need to know whether the bits sent come from text or video; likewise, for quantizer design, one does not care whether the index will be sent over a wired or wireless medium. Yet, other JSCC schemes may be advantageous in terms of complexity and robustness, as the following shows.

Example 11.1.1 (Gaussian source and channel) *A Gaussian source with variance σ_s^2 is sent over an AWGN channel with power constraint P and noise variance N. The optimal distortion is given by (11.3). For MSE distortion, after substituting $R(D) = R^*(D)$ (5.33) and $C = C_{\text{AWGN}}$, we have:*

$$D^* = \frac{\sigma_s^2}{(1 + P/N)^\rho}.$$ (11.4)

Now we turn to the scalar case, i.e., $m = n = 1$. Consider the simple analog scheme

$$X = \sqrt{\frac{P}{\sigma_s^2}} \cdot S$$ (11.5)

$$\hat{S} = \sqrt{\frac{\sigma_s^2}{P}} \cdot \alpha^* Y$$ (11.6)

where $\alpha^* = P/(P + N)$ is the MMSE coefficient of the channel (4.44). It is easy to verify that the resulting MSE distortion is given by

$$D = \frac{\sigma_s^2}{1 + P/N}.$$ (11.7)

Remarkably, this is just (11.4) with $\rho = 1$.

This example shows that for some special cases, the optimal asymptotic performance is achieved already by a scalar analog scheme. While the separation-based scheme has the same average distortion as the scalar scheme, we can point out three major advantages of the latter.

1. **Delay** While the separation-based scheme requires joint processing of many samples, the analog scheme operates sample-by-sample.
2. **Complexity** In a separation-based scheme, the complexity grows with block-length. The analog scheme only performs two multiplications per sample.
3. **Robustness** The separation-based scheme requires knowledge of the noise power n at the decoder (in order to set the code rates). In contrast, (11.5) is independent of N.

What can be done in the Gaussian setting with MSE distortion, when $\rho \neq 1$? Unfortunately, it is not possible to achieve optimal performance using short block-lengths. Still, one may wish to use such a scheme for the advantages listed above.

At the end of the chapter, we will show how this is achieved using MLM. We first define the relevant lattice figures of merit. Then we consider the case where $\rho = 1$ but side information is present, and finally leverage the results to get back to the case $\rho \neq 1$.

11.2   Figures of merit for JSCC

In Chapter 3 we saw how the goodness of a lattice for source and channel coding can be characterized in terms of the ratio between a corresponding square-radius and the volume. The volume plays a major role, since it corresponds to the density of lattice points, hence the rate. As demonstrated above, in a JSCC scheme digital data may be replaced by analog signals. How should we characterize the performance of a lattice with respect to such signals?

It turns out that the question of interest is the following. Let $\mathbf{Z}$ be a random vector and Λ a given lattice. We would like to have

$$\Pr\{\mathbf{Z} \notin \mathcal{V}_0(\Lambda)\} = P_e \tag{11.8}$$

for some error probability $0 < P_e < 1$. Define

$$L(\Lambda, P_e) = \frac{\sigma^2(\Lambda)}{\sigma_Z^2} \tag{11.9}$$

to be the lattice Gaussian second-moment to noise ratio (SMNR), that is, the ratio between the lattice second moment $\sigma^2(\Lambda)$ and the noise per element variance σ_Z^2. As it turns out, we can characterize this quantity in terms of the NSM and NVNR defined in Chapter 3.

First, assume that $\mathbf{Z}$ is Gaussian i.i.d. On one hand (3.21)

$$V^{2/n}(\Lambda) = \frac{\sigma^2(\Lambda)}{G(\Lambda)}, \tag{11.10}$$

and on the other (3.37)

$$V^{2/n}(\Lambda) = \mu(\Lambda, P_e) \cdot \sigma_Z^2. \tag{11.11}$$

It follows that for an error probability P_e, the Gaussian SMNR is given by

$$L(\Lambda, P_e) = \mu(\Lambda, P_e) \cdot G(\Lambda). \tag{11.12}$$

Since $G(\Lambda) > 1/(2\pi e)$ (3.26) and for small enough P_e, $\mu(\Lambda, P_e) > 2\pi e$ (3.43), we have that for such P_e

$$L(\Lambda, P_e) > 1. \tag{11.13}$$

Luckily, the existence of lattices which are simultaneously good for source and channel coding (Theorem 7.9.1) implies that for any $P_e > 0$ there exists a sequence of lattices of increasing dimension Λ_n, such that

$$\lim_{n \to \infty} L(\Lambda_n, P_e) = 1. \tag{11.14}$$

That is, in the infinite-dimensional limit, a white-Gaussian vector with any variance smaller than the second moment of the lattice, falls with high probability inside the basic cell.

As in the case of lattice decoding in the digital channel setting (Section 9.6), it turns out that outside the high SNR limit, the case of interest is when the Gaussian noise $\mathbf{Z}$ is replaced by a mixture of Gaussian and "self" noises. Let $\mathbf{T} = \mathbf{Z} + \mathbf{U}$, where $\mathbf{Z}$ is i.i.d. Gaussian and $\mathbf{U}$ is uniformly distributed over a scaled version of the basic cell $\mathcal{V}_0(\Lambda)$. As in (9.60), we define ξ to be the ratio between the variances of the Gaussian and self-noise components. Denoting by P_e the probability that $\mathbf{T}$ leaves $\mathcal{V}_0(\Lambda)$, we replace (11.11) by the Euclidean NVNR expression (9.69c):

$$V^{2/n}(\Lambda) = \mu_{\text{mix}}(\Lambda, P_e; \Lambda, \xi) \cdot \sigma_T^2. \tag{11.15}$$

Consequently,

$$\frac{\sigma^2(\Lambda)}{\sigma_T^2} = L_{\text{mix}}(\Lambda, P_e; \xi), \tag{11.16}$$

where

$$L_{\text{mix}}(\Lambda, P_e; \xi) = \mu_{\text{mix}}(\Lambda, P_e; \Lambda, \xi) \cdot G(\Lambda) \tag{11.17}$$

is the mixture SMNR. Notice that, unlike in Chapters 9 and 10, here the lattice that generates the self-noise component is the very same lattice for which the error event is defined. As for the case of Gaussian noise, $L_{\text{mix}}(\Lambda, P_e; \xi) > 1$ for a sufficiently small P_e, for any lattice Λ and mixture ratio ξ; moreover, there exists a sequence of good lattices such that this quantity approaches 1. Furthermore, it is plausible that the mixture SMNR is close to the Gaussian SMNR, and for small enough P_e, $\mu(\Lambda, P_e) \geq \mu_{\text{mix}}(\Lambda, P_e; \Lambda, \xi)$ for all ξ; see Section 9.6.1. Thus, the Gaussian $L(\Lambda, P_e)$ provides achievable performance.

11.3 Joint Wyner–Ziv/dirty-paper coding

Suppose that the source with side information, discussed in Section 10.4, needs to be transmitted over the power-constrained interference channel of Section 10.6, as demonstrated in Figure 11.1. We denote the source as $\mathbf{S} = \mathbf{Q} + \mathbf{J}$, where $\mathbf{Q}$ is Gaussian with element variance σ_Q^2 and $\mathbf{J}$ is an arbitrary signal, known at the decoder.

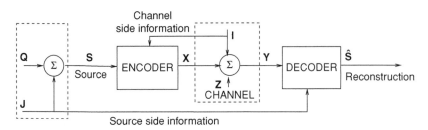

Figure 11.1 The joint Wyner–Ziv/dirty-paper coding problem.

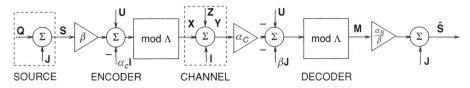

Figure 11.2 Analog Wyner–Ziv/dirty-paper coding scheme.

The channel is given by $\mathbf{Y} = \mathbf{X} + \mathbf{Z} + \mathbf{I}$, where the input power is constrained to P, $\mathbf{Z}$ is AWGN of power N and $\mathbf{I}$ is an arbitrary signal, known at the encoder. We restrict our attention to the case where the bandwidth expansion factor is $\rho = 1$ (thus all of the vectors are n-dimensional), and we are interested in the MSE distortion D.

It is not trivial that the separation principle should hold in the presence of side information. To see why it does hold in the problem of interest, recall that by (10.57), $C_{GP} = C_{AWGN}$, while by (10.22), $R_{WZ}(D) = R_{S|J}(D)$. Denoting the performance of the separation-based scheme by D^*, it follows that $C_{AWGN} = R_{S|J}(D^*)$, thus

$$D^* = \frac{\sigma_Q^2}{1 + P/N}. \tag{11.18}$$

Comparing with (11.7), we see that the separation-based scheme was able to cancel completely the signals $\mathbf{I}$ and $\mathbf{J}$; that is, it achieves the optimal performance of sending the source innovations $\mathbf{Q}$ through an interference-free AWGN channel with noise $\mathbf{Z}$. Clearly, we could not expect lower distortion.[2] Still, we seek a JSCC scheme that will have the same performance, while preserving some of the advantages of an analog scheme.

The MLM scheme which does that is closely related to the modulo-lattice side-information schemes of Sections 10.4 and 10.6. Directly combining the schemes using separation would result in two nested lattice pairs. The MLM scheme, depicted in Figure 11.2, however, drops the fine lattices altogether, replacing the quantization

---

[2] In fact, the separation principle holds in the general (non-Gaussian) combined Wyner–Ziv/dirty-paper setting, see [185].

lattice of the WZ system, and the lattice codebook of the DPC system, by linear mapping of the source vector. Further, it merges the two coarse lattices into one – which is used to handle both the side-information signals $\mathbf{I}$ and $\mathbf{J}$:

$$\textbf{Encoding} \quad \mathbf{X} = (\beta\mathbf{S} - \alpha_C\mathbf{I} + \mathbf{U}) \bmod \Lambda \qquad (11.19a)$$

$$\textbf{Decoding} \quad \mathbf{M} = [\alpha_C\mathbf{Y} - \beta\mathbf{J} - \mathbf{U}] \bmod \Lambda \qquad (11.19b)$$

$$\hat{\mathbf{S}} = \frac{\alpha_S}{\beta}\mathbf{M} + \mathbf{J}, \qquad (11.19c)$$

where $\mathbf{U}$ is uniform (or modulo-uniform) dither.

The analysis of the MLM scheme closely follows that of the separate source and channel side information schemes. First we can combine the source, channel, encoder and decoder equations, to arrive at a modulo-lattice equivalent channel, similar to that of Figure 10.15. We define the equivalent noise:

$$\mathbf{Z}_{\text{eq}} = \alpha_C\mathbf{Z} - (1 - \alpha_C)\mathbf{X} \qquad (11.20a)$$

and the signal

$$\mathbf{T} = \beta\mathbf{Q} + \mathbf{Z}_{\text{eq}}. \qquad (11.20b)$$

In terms of these, we have the following.

Lemma 11.3.1 (Equivalent modulo-Λ channel) *The decoder modulo output signal $\mathbf{M}$ of* (11.19b) *satisfies:*

$$\mathbf{M} = \mathbf{T} \bmod \Lambda.$$

The proof is similar to that of Theorem 10.4.1, using the distributive law of the modulo operation (8.21), and is left as an exercise.

What is the resulting performance? As in Section 10.4.2, we define a decoding error event, this time with respect to the signal $\mathbf{T}$, and a corresponding error probability:

$$P_e = \Pr\{\mathcal{E}\} = \Pr\{\mathbf{M} \neq \mathbf{T}\} = \Pr\{\mathbf{T} \notin \mathcal{V}_0(\Lambda)\}. \qquad (11.21)$$

Just as in (10.31)–(10.34) we write the total distortion as the sum of correct decoding and overload distortion, using the indicator function of $\mathcal{E}$.

We note that for correct decoding, the reconstruction satisfies, for any element $1 \le i \le n$,

$$\hat{S}_i = \frac{\alpha_S}{\beta} \cdot T_i + J_i \qquad (11.22a)$$

$$= \alpha_S \cdot \left(Q_i + \frac{Z_{i,\text{eq}}}{\beta} \right) + J_i. \qquad (11.22b)$$

We see that without errors, the source innovations pass through an equivalent linear channel, where β scales down the additive noise. From this point of view, we would like the scaling factor β to be as large as possible. However, the error probability P_e grows with β. Notice that the signal $\mathbf{T}$ is a combination noise, composed of a

Gaussian part

$$\mathbf{T}_G = \beta\mathbf{Q} + \alpha_C\mathbf{Z} \tag{11.23a}$$

and self-noise

$$\mathbf{T}_U = -(1 - \alpha_C)\mathbf{X}. \tag{11.23b}$$

Thus, for a given P_e, the statistics of $\mathbf{T}$ must satisfy (11.16), dictating the value of β.

Recalling Proposition 9.1.1 the channel power constraint can be satisfied by a lattice with second moment $\sigma^2(\Lambda) = P$. For any such given lattice Λ of blocklength n we can characterize the trade-off between the error-free distortion and error probability using the SMNR. We make the following choice of parameters. The channel MMSE coefficient is

$$\alpha_C = \frac{P}{P + N}. \tag{11.24a}$$

The encoder scaling coefficient β and the resulting Gaussian to uniform ratio ξ are given by the solution of:

$$\beta^2 = \left[\frac{1}{L_{\mathrm{mix}}(\Lambda, P_e; \xi)} - \frac{N}{P + N}\right]^+ \cdot \frac{P}{\sigma_Q^2} \tag{11.24b}$$

$$\xi = \frac{\beta^2\sigma_Q^2(P + N)^2 + P^2 N}{N^2 P}, \tag{11.24c}$$

where $[a]^+ = \max(a, 0)$, and the source MMSE coefficient is

$$\alpha_S = \frac{\beta^2\sigma_Q^2(P + N)}{\beta^2\sigma_Q^2(P + N) + PN}. \tag{11.24d}$$

Theorem 11.3.1 (Performance characterization) *For the MLM coding scheme* (11.19) *using a lattice Λ with second moment P, any $0 < P_e < 1$, and choice of parameters (and resulting mixture ratio)* (11.24), *the correct decoding distortion satisfies:*

$$D_{\mathrm{cd}} \le L_{\mathrm{mix}}(\Lambda, P_e; \xi) \cdot D^*, \tag{11.25}$$

and the error probability is P_e. Here, D^* is the optimal performance (11.18), *and $L_{\mathrm{mix}}(\Lambda, P_e; \xi)$ is the mixture SMNR* (11.17) *of the lattice. Furthermore, for small enough P_e,*

$$D_{\mathrm{cd}} \le L(\Lambda, P_e) \cdot D^*.$$

Before proving the theorem, we make some comments.

1. **High-dimensional limit** We can take the lattice Λ from a sequence Λ_n of JSCC-good lattices, such that $L_{\mathrm{mix}}(\Lambda_n, P_e; \xi) \to 1$, thus D_{cd} approaches the

optimum D^*. Indeed, in the limit, we have:

$$\beta^2 = \frac{P}{P+N} \cdot \frac{P}{\sigma_Q^2} \triangleq \beta_0^2 \tag{11.26a}$$

$$\alpha_S = \frac{P}{P+N} = \alpha_C, \tag{11.26b}$$

thus the linear channel (11.22) for the source innovations Q_i becomes identical to the channel that the optimal analog scheme of Example 11.1.1 yields for the source without side information.

But what about the overload distortion? Recall that Theorem 10.4.2 showed that the effect is vanishing. Looking at the proof, it is evident that the specific nature of the signal played no part, and it holds for correct decoding with respect to $\mathbf{T}$: as long as we take $P_e(n)$ and $P_e(n) r_{\text{cov}}^2 (\Lambda_n) / n$ to zero simultaneously, we will have $D(n) \to D_{\text{cd}}(n) \to D^*$.

2. **High SNR limit** When $P \gg N$, the MMSE factors α_C and α_S approach one, and the noise $\mathbf{Z}_{\text{eq}}$ becomes Gaussian. Thus, under correct decoding,

$$\hat{S}_i = S_i + \frac{Z_i}{\beta}, \tag{11.27a}$$

where

$$\beta^2 = \frac{1}{L(\Lambda, P_e)} \cdot \frac{P}{\sigma_Q^2}. \tag{11.27b}$$

3. **Non-trivial distortion** Whenever $\beta^2 > 0$, the distortion is strictly lower than the source variance. However, when

$$\frac{1}{L_{\text{mix}}(\Lambda, P_e; \xi)} \leq \frac{N}{P+N} \tag{11.28}$$

we have $\beta^2 = 0$, thus also $\alpha_S = 0$, and the scheme output is zero, yielding $D = \sigma_S^2$. This happens since in that case, the error probability we are seeking is too low with respect to the lattice quality, and $\mathbf{T}$ cannot be guaranteed to stay inside the basic cell of the lattice even if we take $\beta^2 = 0$ such that the signal component is zero!

4. **Optimal choice of parameters** The factor α_C chosen is the channel MMSE factor, which minimizes the variance of $\mathbf{Z}_{\text{eq}}$, thus also the variance of $\mathbf{T}$. However, it does not necessarily minimize the probability that correct decoding holds, i.e., $\mathbf{T} \in \mathcal{P}_0(\Lambda)$. This is because the choice of α_C also affects the mixture ratio ξ.

Proof of Theorem 11.3.1 In the case (11.28) the scheme output is zero, and the theorem is trivial. We assume, then, that $\beta^2 > 0$. Recall the definitions of $\mathbf{Z}_{\text{eq}}$ and $\mathbf{T}$ (11.20). Using the MMSE α_C, the per element variance of $\mathbf{Z}_{\text{eq}}$ is

$$\sigma_{\text{eff}}^2 = \frac{PN}{P+N}. \tag{11.29a}$$

By the additivity of the channel and the crypto lemma, $\mathbf{Z}_{\text{eq}}$ is independent of $\mathbf{Q}$, thus $\mathbf{T}$ has per element variance

$$\sigma_T^2 = \beta^2 \sigma_Q^2 + \sigma_{\text{eff}}^2. \tag{11.29b}$$

Furthermore, following the partition into Gaussian and uniform parts (11.23), the mixture ratio of $\mathbf{T}$ is given by ξ (11.24c). Combining (11.29) with (11.17) for second moment $\sigma^2(\Lambda) = P$, we get (11.24b). Since we assumed that (11.28) does not occur, there exists some β such that (11.24b) and (11.24c) are satisfied. Now assuming correct decoding, we have (11.22). Since α_S is the MMSE factor for the channel with input $\mathbf{Q}$ and additive noise $\mathbf{Z}_{\text{eq}}/\beta$, we have that the resulting distortion is:

$$D = \alpha_S \frac{\sigma_{\text{eff}}^2}{\beta^2} \tag{11.30a}$$

$$= \frac{\sigma_Q^2 P N}{\beta^2 \sigma_Q^2 (P + N) + P N} \tag{11.30b}$$

$$= L_{\text{mix}}(\Lambda, P_e; \xi) \cdot \frac{\sigma_Q^2 N}{P + N} \tag{11.30c}$$

as required.

11.3.1 Robustness to unknown noise variance

Recall that without SI, the simple analog scheme of Example 11.1.1 achieves perfect robustness to unknown noise variance N. We now show that the MLM scheme allows us to maintain some of this robustness even in the presence of SI.

We consider the same Gaussian-quadratic setting of Figure 11.1, except that now the noise variance N is unknown to the encoder. We denote by $D(N)$ and $D^*(N)$ the distortion achieved by a scheme and the optimum distortion (11.18), respectively, when the actual noise variance is N. Since we do not seek perfect robustness, we assume that this variance belongs to some set, and specifically that $N \leq N_0$ for some fixed $N_0 > 0$. Now suppose that we designed the scheme with some parameters $\alpha_C, \alpha_S, \beta^2$ such that correct decoding holds when $N = N_0$. It is easy to verify that it will also hold with high probability for any $N \leq N_0$.

We are interested in the high SNR regime, i.e., in the limit where the maximal noise variance N_0 is small. Thus, we consider a scheme with $\alpha_C = \alpha_S = 1$. Under these conditions, the maximal value of β^2 which allows correct decoding is given by

$$\beta^2 = \frac{P - N_0}{\sigma_Q^2}.$$

The distortion is then simply given by:

$$D(N) = \frac{N}{\beta^2} = \frac{N}{P - N_0} \sigma_Q^2.$$

Comparing to $D^*(N)$, we have the following.

Theorem 11.3.2 (Asymptotic robustness)   *For the joint Wyner–Ziv/dirty-paper problem, for any $\epsilon > 0$ there exists a scheme which is independent of the noise variance N, such that in the high-dimensional limit,*

$$D(N) \leq (1 + \epsilon)D^*(N) \tag{11.31}$$

for all sufficiently small N. That is, (11.31) holds for all $N \leq N_0(\epsilon)$, where $N_0(\epsilon)$ is positive for all $\epsilon > 0$.

11.4   Bandwidth conversion

We now abandon the SI scenario, and return to the basic quadratic-Gaussian source-channel problem as in Example 11.1.1, except that the number of channel uses m is different from the number of source samples n. We first describe a simple scheme for the scalar case $n = 1$, and then show how the MLM scheme provides an asymptotically optimal extension.

11.4.1   Iterative saw-tooth mapping

We start with the case $n = 1$, $m = 2$, i.e., each source sample S is translated into a pair of channel inputs X_1, X_2, and assume for simplicity that the source is uniform in the interval $[0, 1]$. Shannon has suggested using a "continuous constellation" that is a curve in the two-dimensional channel input space, parametrized by the source value: $(X_1(S), X_2(S))$. The choice of curve yields an inherent trade-off between small-scale and large-scale errors. This can be easily demonstrated using the following simple curve:

$$X_1(S) = S \tag{11.32a}$$
$$X_2(S) = [\beta S] \bmod \mathbb{Z}, \tag{11.32b}$$

where β is an integer. Suppose that we use a simple decoder, which maps the channel output Y_1, Y_2 into the closest point on the curve, and then sets $\hat{S}$ to be the unique S corresponding to that point according to (11.32). Note that the source interval from 0 to $1/\beta$ is mapped into the channel branch from $(0, 0)$ to $(1/\beta, 1)$. Thus, each source interval is stretched by a factor $\sqrt{\beta^2 + 1}$. We now distinguish between two kinds of reconstruction errors, similar to correct decoding and overload distortion in quantization (recall Section 10.4.2).

1. **Small errors** Suppose that

$$Q_{\mathbb{Z}}(\beta S) = Q_{\mathbb{Z}}(\beta \hat{S}), \tag{11.33}$$

that is, the channel noise is small enough such that $\hat{S}$ and S belong to the same branch of the curve. Then, thanks to the stretching, the noise is scaled down when the decoder maps back to $\hat{S}$. Thus, the distortion satisfies:

$$D \cong \frac{N}{\beta^2 + 1}. \tag{11.34}$$

2. **Large errors** If the noise is high enough such that the closest point to (Y_1, Y_2) is on a wrong branch, then the error is of the order of magnitude of the source variance. As β grows the branches become closer, thus the probability of such an event increases.

This low-delay, low-complexity scheme, while being suboptimal, does offer some robustness to unknown signal to noise ratio. Suppose that we are in the regime where the effect of large errors is already negligible, and then we keep decreasing the noise variance: the distortion will decrease in proportion. If we increase the noise variance, however, at some point large errors will dominate and performance will deteriorate rapidly. This is known as the "threshold effect" of non-linear systems.

The scheme can be extended to larger integer values of ρ by using the iterative map:

$$X_{i+1} = [\beta X_i] \bmod \mathbb{Z}, \ i = 1, \ldots, \rho - 1. \tag{11.35}$$

See Figure 11.3 for an illustration of the resulting transmission. The trade-off between small and large errors can be somewhat improved by replacing this "sawtooth map" with the continuous "tent map" of [37]. However, in order to approach zero large-error probability and optimal performance, even at a known noise variance N, we must replace $\mathbb{Z}$ with a high-dimensional lattice.

11.4.2   Optimal MLM-based scheme

We now present the asymptotically optimal MLM scheme for bandwidth expansion. That is, we take the source dimension n to infinity, while keeping a fixed bandwidth expansion factor ρ (11.2). Throughout this section we assume the high-dimensional limit, i.e., (11.26) applies, $D = D^*$ and P_e is arbitrarily small. Of course, finite-dimensional effects can be taken into account in order to find the achievable performance for a given lattice.

We treat the case of integer ρ, and start with $\rho = 2$. We use the first n channel inputs for analog transmission of the source, just like in the optimal scheme for

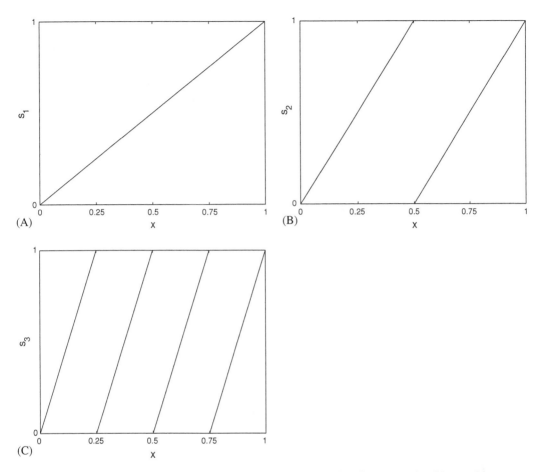

Figure 11.3 Saw-tooth mapping for bandwidth expansion factor $\rho = 3$, with stretching factor $\beta = 2$: (A) first transmission, (B) second transmission, (C) third transmission.

$\rho = 1$. After this stage, the decoder has an estimate $\hat{\mathbf{S}}^{(1)}$ of the source, which satisfies

$$\mathbf{S} = \hat{\mathbf{S}}^{(1)} + \mathbf{Q}^{(1)}, \tag{11.36}$$

where $\mathbf{Q}^{(1)}$ is the estimation error of the analog stage, independent of the estimate $\hat{\mathbf{S}}^{(1)}$ by the orthogonality principle. This error has per element variance

$$\sigma_Q^2 = D^{(1)} = \frac{N}{P+N}\sigma_S^2. \tag{11.37}$$

Now, we have n more channel uses. For these, we can view $\hat{\mathbf{S}}^{(1)}$ as source side information, available at the decoder. We have, then, a special case of the joint Wyner–Ziv/dirty-paper problem, with source innovations $\mathbf{Q}^{(1)}$, and without channel interference. For this stage we can now employ the MLM scheme, with resulting distortion:

$$D^{(2)} = \frac{N}{P+N} D^{(1)} \tag{11.38a}$$

$$= \left(\frac{N}{P+N}\right)^2 \sigma_S^2, \tag{11.38b}$$

which is indeed the optimal distortion. For larger integer ρ we can repeat the process, viewing the source estimation as side information. By induction, after $m = \rho n$ channel uses we will have

$$D^{(\rho)} = \frac{N}{P+N} D^{(\rho-1)} \tag{11.39a}$$

$$= \left(\frac{N}{P+N}\right)^\rho \sigma_S^2. \tag{11.39b}$$

In the exercises we present an MLM scheme for the dual case of bandwidth reduction, i.e., $1/\rho$ is an integer. For a rational ρ, one may combine the two; alternatively, for any ρ, and for the more general case of colored sources and channels, the MLM approach can be combined with time-domain processing to create the *analog matching* scheme; the interested reader is referred to [137].

11.4.3   Robustness to unknown noise variance

It is now natural to ask what can be achieved for bandwidth conversion when the noise variance is not known a priori at the encoder. For the case $\rho = 2$, we have the following.

Lemma 11.4.1 (Robustness for $\rho = 2$) *For the quadratic-Gaussian JSCC problem with bandwidth expansion factor $\rho = 2$, let $N_0 > 0$ be some noise variance, Then, in the infinite blocklength limit, the distortion*

$$D(N) = \frac{N}{P+N} \cdot \frac{N_0}{P+N_0} \cdot \sigma_S^2$$

is achievable for any noise variance $N \le N_0$ using the same encoder.

The proof is given as an exercise. Comparing to the optimum performance, we see that the distortion is the geometric mean between the distortion achieved by an optimal schemes with noise variances N_0 and N. One can give the following interpretation: the price of robustness is that the improvement with noise variance is only according to the source bandwidth. Yet, there is substantial gain compared to a separation-based scheme, where performance does not improve at all if $N < N_0$; see Figure 11.4.

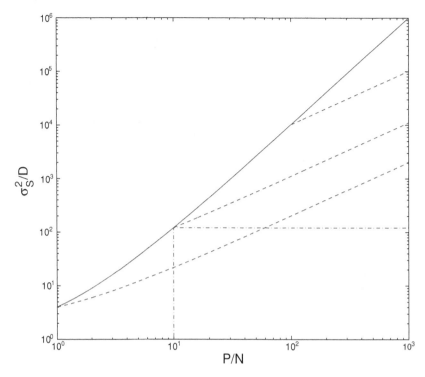

Figure 11.4 Robustness to unknown noise variance with $\rho = 2$. The solid curve is the ideal performance for known variance (11.4). Dashed curves represent the performance given by Lemma 11.4.1 for some values of N_0. For $P/N_0 = 10$, the performance of a separation-based scheme is also demonstrated using a dash-dotted curve.

Summary of Chapter 11

Gaussian second-moment to noise ratio

$$L(\Lambda, P_e) = \mu(\Lambda, P_e) \cdot G(\Lambda)$$

where μ is the Gaussian NVNR and G is the NSM.

Modulo-lattice modulation To send a source $\mathbf{S} = \mathbf{Q} + \mathbf{J}$, where $\mathbf{Q}$ is Gaussian with variance σ_Q^2 and $\mathbf{J}$ is known at the decoder, over the channel $\mathbf{Y} = \mathbf{X} + \mathbf{Z} + \mathbf{I}$, where $\mathbf{Z}$ is Gaussian with variance N and $\mathbf{I}$ is known at the encoder, transmit

$$\mathbf{X} = (\beta \mathbf{S} - \alpha_C \mathbf{I} + \mathbf{U}) \bmod \Lambda$$

where $\mathbf{U}$ is dither, and reconstruct as

$$\hat{\mathbf{S}} = \mathbf{J} + \frac{\alpha_S}{\beta} \cdot ([\alpha_C \mathbf{Y} - \beta \mathbf{J} - \mathbf{U}] \bmod \Lambda).$$

In the high-dimensional limit, MLM achieves the optimum:

$$D = \frac{\sigma_Q^2}{1 + \frac{P}{N}}.$$

Gaussian joint source-channel coding with BW mismatch The optimum

$$D = \frac{\sigma_S^2}{\left(1 + \frac{P}{N}\right)^\rho}$$

is achievable for integer ρ or integer $1/\rho$ by translation into an analog Wyner–Ziv or an analog dirty-paper problem, respectively.

Problems

P.11.1 (Equivalent modulo-lattice channel) Prove Lemma 11.3.1.

P.11.2 (MLM for BW reduction) In this exercise we prove the optimality of MLM for bandwidth reduction by an integer factor, i.e., $1/\rho$ is an integer. We start with the case $\rho = 1/2$. We see the source vector as the concatenation of two independent vectors $\mathbf{S}^{(1)}$ and $\mathbf{S}^{(2)}$, each of length m. We now allocate some channel power to each. $\mathbf{S}^{(1)}$ is sent using analog transmission of power $P^{(1)}$. Since this signal is known to the encoder, we can view the transmission of $\mathbf{S}^{(2)}$ as a special case of joint source-channel coding with side information, this time without Wyner–Ziv SI. Let $\mathbf{X}^{(2)}$ be the encoder output of an MLM scheme for that problem, then the total transmission is:

$$\mathbf{X} = \sqrt{\frac{P^{(1)}}{\sigma_S^2}} \mathbf{S}^{(1)} + \mathbf{X}^{(2)}.$$

1. What is $P^{(2)}$, the power of $\mathbf{X}^{(2)}$, such that the total transmitted power is P?
2. Show that $\mathbf{S}^{(1)}$ can be estimated with distortion

$$D^{(1)} = \frac{P^{(2)} + N}{P + N} \sigma_S^2.$$

3. Show that $\mathbf{S}^{(2)}$ can be estimated with distortion

$$D^{(2)} = \frac{N}{P^{(2)} + N} \sigma_S^2.$$

4. Show that there exists a choice of power allocation such that $D^{(1)} = D^{(2)} = D^*$.
5. Extend the results to any integer $1/\rho$.

P.11.3 (Robustness of MLM for BW expansion) In this exercise we prove Lemma 11.4.1. We note that (11.37) holds even when the encoder does not know the noise power N, thus we concentrate on the second stage of transmission.

1. Show that with

$$\beta^2 = \frac{P^2}{\sigma_S^2 N_0},$$

   correct decoding holds with high probability for any $N \leq N_0$.
2. Argue that, when there is no channel interference, the encoder and decoder only need to agree upon β (and not upon α_C).
3. Find the optimal decoder parameters α_C and α_S, and derive the performance.
4. Extend the result: what is $D^{(\rho)}(N)$ achieved by the MLM scheme for larger integer ρ for $N \leq N_0$, if we insist that $D^{(\rho)}(N_0)$ is optimal?

Historical notes

The joint source-channel coding problem dates back to Shannon [241], where the optimal performance for a white-Gaussian source-channel pair is derived (as a function of the bandwidth expansion factor), and a geometrical representation of source-channel maps (the "snake" diagram) is presented. Further study of "signal curves" was carried out by Kotel'nikov [140]. Several works have analyzed the performance of various modulation techniques to the optimum; these include a paper by Goblick [107] where it is shown that analog transmission is optimal in the white-Gaussian bandwidth-matched case; sufficient and necessary conditions for such matching between source-channel pairs were derived by Gastpar et al. [99]. In an important theoretical work, Ziv [298] showed that for the bandwidth-mismatched case, no single transmitter can be simultaneously optimal for two SNRs.

Various joint source-channel schemes have been proposed, some of them from a geometrical point of view, proposing low-dimensional source-channel maps that exhibit good robustness properties. An alternative is presented by hybrid digital-analog (HDA) methods, where the channel resources are split between a digital code and a simple linear analog component. Such schemes date back at least to the late 1960s, see, for example, [184]. Some later works took a more theoretical point of view, letting the digital element be a high-dimensional optimal code. Of these many works we mention Shamai et al. [239] and Mittal and Phamdo [189]. The former makes the connection between the bandwidth mismatch problem and a side-information problem.

A connection between the bandwidth expansion problem and *analog* Wyner–Ziv coding first appeared in Reznic et al. [227], and a modulo-lattice solution was proposed. Kochman and Zamir generalized the view to the joint Wyner–Ziv/dirty-paper problem [136], and combined the modulo-lattice schemes with filtering to propose an optimal scheme for any colored source and channel [137] (including bandwidth mismatch as a special case). In parallel, Wilson et al. [269] proposed a similar scheme that uses random codebooks rather than lattices. It is worth mentioning that the general (non-Gaussian) joint Wyner–Ziv/Gelfand–Pinsker problem was considered by Merhav and Shamai [185], who showed that the separation principle holds for a wide class of source-channel pairs.

The modulo-lattice modulation approach has been used in network information problems, such as computation over the multiple-access channel [193] and parallel relaying [135].

12 Gaussian networks

Co-written with Bobak Nazer

There are many ways in which we can use side-information paradigms as building blocks in general multi-terminal networks. Two such cases were discussed in Chapter 10: the broadcast channel (Section 10.1.4) and distributed compression of correlated sources (Section 10.2).

In these simple settings, the side information is concentrated in the "relevant" terminal in the network. In the broadcast channel, for example, the joint encoder may view the transmission to one terminal as side information for the transmission to another terminal. In multi-terminal source coding, the joint decoder may view the reconstruction of one source as side information for the reconstruction of another source. Lattice coding schemes, such as the lattice dirty-paper and Wyner–Ziv coding schemes of Chapter 10, can reduce the complexity (and perhaps offer some intuition) compared to random coding solutions. But they do not give us any performance advantage over the random i.i.d. coding solutions, which are known to be optimal for these settings.

A more interesting situation, however, occurs when *side information is distributed* among more than one terminal. Surprisingly, it turns out that in some distributed linear network topologies, structured solutions *outperform* random coding solutions. Moreover, in some cases they are, in fact, asymptotically optimal.

A common theme in the settings we discuss in this chapter is the distributed computation of a many-to-one function. The arguments of this function are information or side-information signals. The function reduces the effective dimension or entropy of these signals; it represents either some feature that we want to extract or utilize, or an intermediate decoding stage that will be completed at the final destination. Eventually we shall restrict our attention to *linear* functions.

We consider the following problems.

1. **The two-help-one (Körner–Marton) problem** Two encoders separately compress the sources X and Y so that the decoder can reconstruct the function $f(X, Y)$, either losslessly or subject to some distortion constraint. (The sources X and Y can be viewed as side information for the desired function $f(X, Y)$.)
2. **Multiple-access channel (MAC) with distributed state information** Communicate over a two-user MAC whose state S is a function $S = f(S_1, S_2)$ of side

information S_1 which is available at transmitter 1 only, and side information S_2 which is available at transmitter 2 only.

3. **Computation over a MAC** Given the output Y of a two-user MAC whose inputs X_1 and X_2 represent messages V_1 and V_2, decode a function $f(V_1, V_2)$ of the two messages. (The messages V_1 and V_2 can be viewed as side information for the function $f(V_1, V_2)$.) This setup is the basis for the *compute-and-forward* technique for relay networks: each relay decodes a different function of the messages, so that the final destination can recover all the messages.

4. **Interference alignment** Three or more transmitter-receiver pairs interfere with each other on a shared channel. Each receiver treats its channel observation Y_k as the output of an effective two-user MAC, comprising its desired signal X_k and a function of the interfering signals $f(X_1, \ldots, X_{k-1}, X_{k+1}, \ldots, X_K)$.

The settings above are part of the landscape of open problems in network information theory. Ideally, we would like to enjoy the benefit of *coordination*, in spite of the distributive nature of encoding and/or decoding in the network. In the two-help-one problem, for example, we would like to reduce the overall coding rate to the entropy of the function $f(X, Y)$, rather than the entropy of its arguments X and Y. The difficulty is, however, that the structure of a general function f (e.g., in terms of its distributivity or associativity properties) does not necessarily match the statistical structure of the sources and channels in the network.

We shall focus in this chapter on the (important) special case where both the desired function f and the channels between users in the network are *linear*. Interestingly, linear and lattice codes have a clear advantage in this case over the more unstructured random coding approach usually taken in network information theory. We demonstrate this advantage through canonical binary and Gaussian networks, featuring the four problems mentioned above.

12.1   The two-help-one problem

It is not hard to detect the few differences between the two faces in Figure 12.1. Once detected, it is also not too hard to describe them with just a few words. But would a few words be sufficient if the observers of the two faces were *separated*?

An information theoretic analog of this question is the two-help-one problem of Figure 12.2. This problem was proposed in a seminal paper from the late 1970s by Körner and Marton [139]. They showed that if one wishes to reconstruct the modulo-2 sum of two correlated binary sources from their independent encodings, then a sum rate of only *twice* that of joint encoding is sufficient. Interestingly, the only known scheme that can approach this performance is based on (random) linear coding, as opposed to random i.i.d. coding.

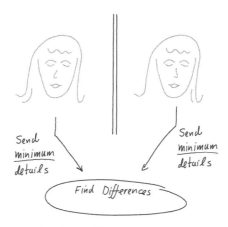

Figure 12.1 Find (and communicate) the differences.

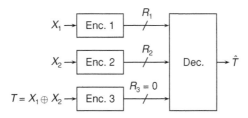

Figure 12.2 Körner–Marton problem: distributed source coding of the modulo sum $T = X_1 \oplus X_2$ of two dependent binary sources X_1 and X_2.

12.1.1 The Körner–Marton problem

Specifically, the Körner–Marton problem consists of a doubly symmetric binary source (X_1, X_2), as in the binary Slepian–Wolf problem (Section 10.1), and a "parity" variable $T = X_1 \oplus X_2$ indicating when X_1 and X_2 are different, i.e., $\Pr(T = 1) = \Pr(X_1 \neq X_2) = \theta$. The decoder needs only to reconstruct losslessly the parity source T. While direct compression of T requires a rate equal to the entropy $H(T) = H_B(\theta)$, the goal of the X_1 and X_2 encoders is to help reduce the rate of the T encoder further. The most interesting case is when the rate of the T encoder is *zero*, so coding is done entirely by the two helpers. See Figure 12.2 for a block diagram. If coordination between the X_1 and X_2 encoders were allowed, then they could compute the XOR sequence $T_1, \ldots, T_n$ and encode it at a rate of $H_B(\theta)$. Körner and Marton used a "genie-aided" argument to show that in the *un*coordinated case, the optimal rate region is no better than

$$R_1 \geq H_B(\theta) \qquad R_2 \geq H_B(\theta). \tag{12.1}$$

Specifically, if the receiver was given X_2 (or X_1) as genie-aided side information, it would still require $H_B(\theta)$ bits to compress X_1 (or X_2).

Furthermore, this rate region can be achieved by the *linear code* solution of the binary Slepian–Wolf problem (Section 10.1.2): each encoder transmits the syndrome of its observed source relative to a good linear binary code C for a BSC with crossover probability θ.

Let H denote the parity-check matrix of C. The encoding and decoding operations are:

$$\textbf{Encoder 1} \quad \mathbf{s}_1 = H\mathbf{x}_1 \tag{12.2a}$$

$$\textbf{Encoder 2} \quad \mathbf{s}_2 = H\mathbf{x}_2 \tag{12.2b}$$

$$\textbf{Decoding} \quad \hat{\mathbf{t}} = \text{leader}(\mathbf{s}_1 \oplus \mathbf{s}_2). \tag{12.2c}$$

By linearity, it follows that

$$\hat{\mathbf{t}} = \text{leader}(H(\mathbf{x}_1 \oplus \mathbf{x}_2)) \tag{12.3}$$

$$= \text{leader}(H\mathbf{t}). \tag{12.4}$$

Following the derivation in Section 10.1.2, since C is a good code for Bernoulli(θ) noise, we can recover $\mathbf{t}$ with probability at least $1 - \epsilon$ if the rate per user exceeds $H_B(\theta)$. Comparing this achievable region with the outer bound in (12.1), we see that this is the optimal rate region.

An alternative approach to the two-help-one problem is to use a random i.i.d. code induced by some *single-letter* formula. A natural choice for such a solution is the Berger–Tung rate region (10.25) for distributed compression of X_1 and X_2, with a single reconstruction $\hat{T}$, and a modified distortion measure, [1]

$$d(x_1, x_2, \hat{t}) = \begin{cases} 0, & \text{if } \hat{t} = x_1 \oplus x_2 \\ 1, & \text{otherwise.} \end{cases} \tag{12.5}$$

The resulting rate region is

$$R_1 \geq H_B(\theta) \qquad R_2 \geq H_B(\theta) \qquad R_1 + R_2 \geq 1 + H_B(\theta). \tag{12.6}$$

This rate region corresponds to lossless compression of X_1 and X_2, i.e., two-terminal Slepian–Wolf encoding (10.17). As Figure 12.3 shows, it is strictly contained in (12.1) since $H(X_1, X_2) = 1 + H_B(\theta)$ in (12.6) is greater than $2H_B(\theta)$ for $\theta \neq \frac{1}{2}$. Thus, the random i.i.d. binning solution for the two-help-one problem is suboptimal, and inferior to (random) structured coding.

Does this mean that *any* random coding scheme (i.e., single-letter solution) would be suboptimal for the two-help-one problem? We shall return to this question at the end of the next section, which focuses on the quadratic-Gaussian variant of the two-help-one problem and a lattice coding solution.

---

[1] The distortion measure can also be written as $d(X_1, X_2, \hat{T}) = X_1 \oplus X_2 \oplus \hat{T}$.

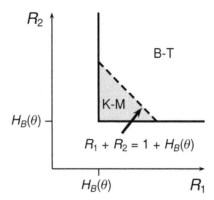

Figure 12.3 Comparison of the rate regions achievable for the Körner–Marton problem via random linear binning (K-M) and random i.i.d. binning via the Berger–Tung scheme (B-T).

12.1.2 Linear functions of two Gaussian sources

The Körner–Marton problem can be generalized naturally to the quadratic-Gaussian setting. Consider two encoders that observe jointly Gaussian sources X_1 and X_2, respectively. Each source has zero mean and variance σ^2. Let ρ denote the correlation coefficient, $E[X_1 X_2] = \rho \sigma^2$. As in the binary case, the goal of the receiver is to reconstruct a linear function,

$$T = X_1 + aX_2, \tag{12.7}$$

for some real-valued coefficient a. The quality of the reconstruction $\hat{T}$ is measured by the quadratic distortion measure, $D = E[(T - \hat{T})^2]$. Let

$$\sigma_T^2 = (1 + 2\rho a + a^2)\sigma^2 \tag{12.8}$$

denote the variance of T.

The conventional approach to this problem employs random i.i.d. quantization and binning. Its rate region follows by plugging in Gaussian auxiliary random variables and the quadratic distortion measure into the Berger–Tung region (10.25). It is important to note that even though we are only interested in reconstructing a function of the sources, this approach implicitly gives the decoder access to good estimates of the individual sources. If the product ρa is positive, then we cannot improve on this approach, i.e., the Berger–Tung region is optimal [264]. However, if ρa is negative, then lattice-based compression can sometimes yield lower rates.

To understand why this is the case, let us for now assume that the sources are positively correlated, $\rho > 0$. The sources can thus be expressed as $X_1 = Y_1 + Y_C$ and $X_2 = Y_2 + Y_C$, where Y_1, Y_2 and Y_C are independent zero mean Gaussian random variables with variances $(1 - \rho)\sigma^2$, $(1 - \rho)\sigma^2$, and $\rho\sigma^2$, respectively. It follows that $\mathrm{Var}(X_1) = \mathrm{Var}(X_2) = \sigma^2$, $\mathrm{Var}(X_1|X_2) = (1 - \rho)\sigma^2$, and $\mathrm{Var}(X_1 - X_2) =$

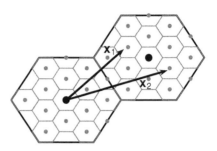

Figure 12.4 Nested lattice quantization for the distributed compression of a linear function of correlated Gaussian sources. The resolution of the coarse lattice is chosen so that, with high probability, the difference of the sources lies in its fundamental Voronoi region. The resolution of the fine lattice is chosen to meet the distortion constraint.

$2(1 - \rho)\sigma^2$. If the decoder wants the *sum* $T = X_1 + X_2 = Y_1 + Y_2 + 2Y_C$ (i.e., $a = 1$), then random quantization and binning is a natural solution. For example, the first encoder can send a quantized version of X_1 and the second encoder can send the innovation provided by the quantized X_2 with respect to X_1 via binning. This scheme can be mimicked via nested lattice quantization, as illustrated in Figure 12.4. That is, the first encoder can send the index obtained by quantizing X_1 onto the fine lattice. The second encoder can then send its innovation by quantizing X_2 onto the fine lattice and taking modulo the coarse lattice (corresponding to lattice Wyner–Ziv encoding from Section 10.4). Thus, distributed compression of the sum is an equivalent problem to the distributed compression of the individual sources.

Now, consider the scenario where the decoder wants the *difference* $T = X_1 - X_2 = Y_1 - Y_2$ (i.e., $a = -1$). Ideally, we would like to send *only the innovation* provided by each source versus the other. That is, there is no need to send any information regarding the common part Y_C. In this sense, compressing the difference corresponds to compressing the modulo sum in the binary case. The conventional approach, random i.i.d. quantization and binning via the Berger–Tung scheme (10.25) with Gaussian auxiliary random variables, implicitly encodes information about Y_C.[2] However, it is possible to create a nested lattice version of the Körner–Marton scheme. Each encoder quantizes its source onto the fine lattice but only sends the innovation with respect to its coarse lattice point. Owing to the fact that the encoders employ the same coarse lattice, the decoder can estimate T by computing the difference of the codewords modulo the coarse lattice. As we will show, the second moment of the coarse lattice should be chosen so that it contains the quantized difference T within its fundamental Voronoi region with high

---

[2] As discussed at the end of this section, it is possible to recover some of the benefits of lattice coding within the Berger–Tung framework by departing from Gaussian auxiliaries.

probability, and the second moment of the fine lattice should be set according to the desired distortion. Again, see Figure 12.4 for an illustration. This is why lattice encoding is useful: the shared structure of the coarse lattice can be used for *innovation alignment*.

We now provide a formal description of the lattice scheme for distributed compression of the difference

$$T = X_1 - X_2. \tag{12.9}$$

Each source is scaled by α and then encoded by a Voronoi quantizer, where the coarse lattice is identical at both encoders and tuned to match the variance of the difference.[3] Here, we also use a common fine lattice Λ_1, meaning that the rates of the encoders are equal, $R_1 = R_2 = R(\Lambda_1/\Lambda_2)$. The decoder subtracts the two codewords, modulo the coarse lattice, to isolate the desired (quantized) difference signal, and uses a Wiener coefficient β to reduce the overall mean-squared distortion:

Encoder 1 $\quad \mathbf{v}_1 = Q_{\Lambda_1}(\alpha \mathbf{x}_1 + \mathbf{u}_1) \bmod \Lambda_2 \qquad\qquad$ (12.10a)

Encoder 2 $\quad \mathbf{v}_2 = Q_{\Lambda_1}(\alpha \mathbf{x}_2 + \mathbf{u}_2) \bmod \Lambda_2 \qquad\qquad$ (12.10b)

Decoding $\quad \hat{\mathbf{t}} = \beta \cdot ([(\mathbf{v}_1 - \mathbf{u}_1) - (\mathbf{v}_2 - \mathbf{u}_2)] \bmod_{\mathcal{V}_0} \Lambda_2), \qquad$ (12.10c)

where $\mathbf{u}_1$ and $\mathbf{u}_2$ are dithers. Note that, as in the lattice Wyner–Ziv scheme (10.28), only the modulo operation at the decoder needs to be taken with respect to the fundamental Voronoi cell $\mathcal{V}_0$. We can also define an "ideal" reconstruction vector $\tilde{\mathbf{t}}$ without the modulo-Λ_2 operation at both the encoders and the decoder,

$$\tilde{\mathbf{t}} = \beta \cdot ((Q_{\Lambda_1}(\alpha \mathbf{x}_1 + \mathbf{u}_1) - \mathbf{u}_1) - (Q_{\Lambda_1}(\alpha \mathbf{x}_2 + \mathbf{u}_2) - \mathbf{u}_2)). \tag{12.11}$$

We also define $\mathcal{E} = \{\hat{\mathbf{T}} \neq \tilde{\mathbf{T}}\}$ to be the event that an overload occurs along with the probability of error

$$P_e = \Pr\{\mathcal{E}\}. \tag{12.12}$$

The end-to-end distortion can thus be written as $D_t \triangleq (1/n)E\|\hat{\mathbf{T}} - \mathbf{T}\|^2 = D_{cd} + D_o$ where the correct-decoding and overload distortion terms are defined following (10.32) and (10.33).

To attain the smallest distortion under this scheme, we set the scaling parameters α and β and the fine lattice second moment $\sigma^2(\Lambda_1)$ to

$$\alpha = \beta = \sqrt{1 - \frac{D}{\sigma_T^2}} \qquad\qquad \sigma^2(\Lambda_1) = \frac{D}{2}. \tag{12.13}$$

---

[3] Here, we have employed a symmetric pre/post-scaled dithered quantization. The same performance can be attained through post-scaled quantization, i.e., setting $\alpha = 1$ and adjusting the other parameters accordingly. See (10.35) in the case of lattice Wyner–Ziv coding.

Note that the factor of 2 appears in the denominator of the fine lattice second moment since there are *two quantization noise terms* due to the distributed encoding.

Theorem 12.1.1 (Performance characterization) *For the lattice Körner–Marton coding scheme* (12.10) *with parameters as specified in* (12.13), *and independent uniform (or modulo-uniform) dithers* $\mathbf{U}_1$ *and* $\mathbf{U}_2$ *with respect to* Λ_1, *the correct-decoding distortion satisfies*

$$D_{cd} \leq D, \tag{12.14a}$$

the coding rate satisfies

$$R(\Lambda_1/\Lambda_2) = \frac{1}{2} \log\left(\frac{2\sigma_T^2}{D}\right) + \frac{1}{2} \log(G(\Lambda_1) \cdot \mu_{\text{euclid}}(\Lambda_2, \mathbf{Z}_{\text{eq}}, P_e)), \tag{12.14b}$$

and the decoding error (i.e., overload) probability is equal to

$$P_e = \Pr\{\mathbf{Z}_{\text{eq}} \notin \mathcal{V}_0(\Lambda_2)\}, \tag{12.14c}$$

where

$$\mathbf{Z}_{\text{eq}} = \alpha\mathbf{T} + \mathbf{U}_{\text{eq},1} - \mathbf{U}_{\text{eq},2} \tag{12.14d}$$

is the equivalent source; and where $G(\Lambda_1)$ *is the NSM of the fine lattice;* $\mu_{\text{euclid}}(\Lambda_2, \mathbf{Z}_{\text{eq}}, P_e)$ *is the Euclidean mismatched NVNR (Definition 6.4.2) of the coarse lattice; and* $\mathbf{U}_{\text{eq},1}$ *and* $\mathbf{U}_{\text{eq},2}$ *are the equivalent dithers* (4.8), *which are uniform over* $\mathcal{V}_0(\Lambda_1)$ *and independent of each other and the source* $\mathbf{T}$.

Proof The analysis of this scheme closely follows that of Wyner–Ziv lattice coding in Sections 10.4.2 and 10.4.3, with the exception that there are now two quantization noise terms. As before, we define quantization error terms

$$\mathbf{e}_{q,1} = Q_{\Lambda_1}(\alpha\mathbf{x}_1 + \mathbf{u}_1) - (\alpha\mathbf{x}_1 + \mathbf{u}_1) \tag{12.15a}$$

$$\mathbf{e}_{q,2} = Q_{\Lambda_1}(\alpha\mathbf{x}_2 + \mathbf{u}_2) - (\alpha\mathbf{x}_2 + \mathbf{u}_2). \tag{12.15b}$$

We can express the final reconstruction as

$$\hat{\mathbf{t}} = \beta([(Q_{\Lambda_1}(\alpha\mathbf{x}_1 + \mathbf{u}_1) - \mathbf{u}_1) - (Q_{\Lambda_1}(\alpha\mathbf{x}_2 + \mathbf{u}_2) - \mathbf{u}_2)] \bmod_{\mathcal{V}_0} \Lambda_2) \tag{12.16a}$$

$$= \beta([(\alpha\mathbf{x}_1 + \mathbf{u}_1 + \mathbf{e}_{q,1} - \mathbf{u}_1) - (\alpha\mathbf{x}_2 + \mathbf{u}_2 + \mathbf{e}_{q,2} - \mathbf{u}_2)] \bmod_{\mathcal{V}_0} \Lambda_2) \tag{12.16b}$$

$$= \beta([\alpha\mathbf{t} + \mathbf{e}_{q,1} - \mathbf{e}_{q,2}] \bmod_{\mathcal{V}_0} \Lambda_2) \tag{12.16c}$$

$$\overset{\text{c.d.}}{=} \beta(\alpha\mathbf{t} + \mathbf{e}_{q,1} - \mathbf{e}_{q,2}) \tag{12.16d}$$

$$= \tilde{\mathbf{t}}, \tag{12.16e}$$

where (12.16a) follows by the distributive law of the modulo operation (8.21), $\overset{\text{c.d.}}{=}$ in (12.16d) denotes equality conditioned on correct decoding, and $\tilde{\mathbf{t}}$ is as defined in (12.11). Note that the overload probability (12.12) is

$$P_e = \Pr\{(\alpha\mathbf{T} + \mathbf{E}_{q,1} - \mathbf{E}_{q,2}) \notin \mathcal{V}_0(\Lambda_2)\}. \tag{12.17}$$

If decoding is successful, the reconstruction is equal in distribution to the channel $\hat{\mathbf{T}} = \beta(\alpha\mathbf{T} + \mathbf{U}_{\text{eq},1} - \mathbf{U}_{\text{eq2}})$; see the crypto lemma (Lemma 4.1.1) and Theorem 4.1.1. The correct-decoding distortion is thus upper bounded as

$$n D_{\text{cd}} \leq E\|\hat{\mathbf{T}} - \mathbf{T}\|^2 \tag{12.18a}$$
$$= E\|\beta\mathbf{U}_{\text{eq},1} - \beta\mathbf{U}_{\text{eq},2} - (1 - \alpha\beta)\mathbf{T}\|^2 \tag{12.18b}$$
$$= \beta^2 E\|\mathbf{U}_{\text{eq},1}\|^2 + \beta^2 E\|\mathbf{U}_{\text{eq},2}\|^2 + (1 - \alpha\beta)^2 E\|\mathbf{T}\|^2 \tag{12.18c}$$
$$= n\big(2\beta^2\sigma^2(\Lambda_1) + (1 - \alpha\beta)^2\sigma_T^2\big). \tag{12.18d}$$

Plugging in the choice of parameters from (12.13), we find that $D_{\text{cd}} \leq D$. $\qquad\square$

To keep the overload probability small, we should set the coarse lattice second moment $\sigma^2(\Lambda_2)$ to be greater than or equal to the second moment of the equivalent source $\mathbf{Z}_{\text{eq}}$, which is σ_T^2 for the parameter choices in (12.13). Following the proof of Theorem 10.4.2, it can be shown that the decoding error probability P_e and overload distortion D_{o} go to zero for an appropriate sequence of nested lattices. This leads us to the following corollary on the asymptotic performance of good nested lattice codes.

Corollary 12.1.1 (Optimality for good nested lattices) *For a sequence of nested lattices good for Voronoi quantization (Definition 10.4.1), the lattice coding scheme (12.10) achieves*

$$R_1 = \frac{1}{2}\log\left(\frac{2\sigma_T^2}{D_{\text{t}}}\right) \qquad R_2 = \frac{1}{2}\log\left(\frac{2\sigma_T^2}{D_{\text{t}}}\right) \tag{12.19}$$

with respect to the total distortion D_{t}.

Specifically, for the choice of parameters in (12.13) ($\alpha = \beta = \sqrt{1 - D/\sigma_T^2}$ and $\sigma^2(\Lambda_1^{(n)}) = D/2$ for all n) and a sequence of good nested lattices $\Lambda_2^{(n)} \subset \Lambda_1^{(n)}$, this coding scheme satisfies $\sigma^2(\Lambda_2^{(n)}) \to \sigma_T^2$, $P_e(n) \to 0$, $D_{\text{o}}(n) \to 0$, $D_{\text{t}}(n) \to D$, and $R(\Lambda_1^{(n)}/\Lambda_2^{(n)}) \to \frac{1}{2}\log(2\sigma_T^2/D_{\text{t}})$ as $n \to \infty$.

If X_2 (or X_1) is available perfectly at the decoder, then reconstructing $\mathbf{T}$ becomes equivalent to reconstructing X_1 (or X_2), i.e., to the Wyner–Ziv problem from Section 10.3. This leads to the "genie-aided" lower bound

$$R_1 \geq \frac{1}{2}\log\left(\frac{\sigma_T^2}{D_{\text{t}}}\right) \qquad R_2 \geq \frac{1}{2}\log\left(\frac{\sigma_T^2}{D_{\text{t}}}\right). \tag{12.20}$$

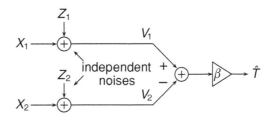

Figure 12.5 Standard Berger–Tung test channel with additive noise. The sum rate is equal to $I(X_1, X_2; V_1, V_2)$.

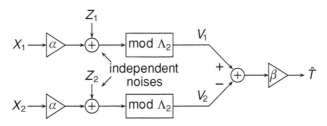

Figure 12.6 Lattice Körner–Marton test channel. The sum rate is equal to $I(X_1, X_2; V_1, V_2)$. See footnote 3 above for a post-scaling only ($\alpha = 1$) solution.

Observe that (12.19) is 1 bit (in sum rate) above the genie-aided lower bound (12.20), corresponding to a 3 dB loss in distortion at high-resolution quantization. Intuitively, this loss is due to the accumulation of the two independent[4] quantization noises in (12.18). Note also that the distortion $D = \sigma_T^2$ is trivially achievable using zero rate, $R_1 = R_2 = 0$, simply by setting $\hat{T} = 0$. Interestingly, this rate point is not captured by the expression in (12.19). This can be easily corrected by time-sharing between (12.19) and the zero-rate point, although it would be preferable for the coding scheme to attain this performance on its own.[5]

Overall, the lattice Körner–Marton scheme often outperforms the conventional random i.i.d. binning solution à la Berger–Tung (10.25), which (implicitly) encodes both sources X_1 and X_2 just to transmit their difference $T = X_1 - X_2$. In Figures 12.5 and 12.6, we have drawn the test channels corresponding to Berger–Tung coding (with Gaussian auxiliary random variables) and lattice binning. It can be shown that the minimum sum rate attainable via Berger–Tung coding with Gaussian auxiliaries is

$$R_{\text{BT,sum}} = \text{l.c.e.} \left\{ \left[\frac{1}{2} \log \left(\left(\frac{2\sigma^2}{D} \right)^2 \left(1 - \frac{\rho D}{\sigma^2} - \rho^2 \right) \right) \right]^+ \right\} \quad (12.21)$$

---

[4] A common (or correlated) dithering scheme is complicated to analyze, but may reduce this loss.
[5] This zero-rate point can be achieved directly through a modified coding scheme that includes a second linear binning stage. See [263] for more details.

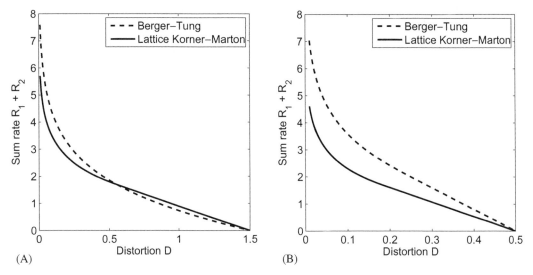

Figure 12.7 Comparison of the sum rate needed for compressing the difference $T = X_1 - X_2$ by the Berger–Tung scheme (with Gaussian auxiliary random variables) and the lattice Körner–Marton scheme: (A) $\rho = 0.25$, $\sigma^2 = 1$, $\sigma_T^2 = 1.5$; (B) $\rho = 0.75$, $\sigma^2 = 1$, $\sigma_T^2 = 0.5$.

where l.c.e. represents the lower convex envelope with respect to time sharing with $R_1 + R_2 = 0$ and $D = \sigma_T^2$ (as defined in (12.8)) and, as defined earlier, σ^2 refers to the variance of X_1 and X_2. See [142] for a derivation. Following Corollary 12.1.1, the sum rate attainable via lattice quantization and binning is

$$R_{\text{KM,sum}} = \text{l.c.e.}\left\{ \left[\log\left(\frac{4(1-\rho)\sigma^2}{D} \right) \right]^+ \right\}. \qquad (12.22)$$

In Figure 12.7, we have plotted the sum rate used by these two schemes. Notice that the performance gap widens as the sources become more correlated. It can also be shown that if we let the variance of the common signal Y_C tend to infinity while holding the variances of the innovations Y_1 and Y_2 fixed, the sum rate used by this Berger–Tung scheme tends to infinity as well. However, since the lattice Körner–Marton scheme manages to eliminate Y_C and send only the innovations Y_1 and Y_2, its sum rate will remain constant as the variance increases. We conjecture that in this scenario of a strong common part, the sum rate obtained by lattice Körner–Marton is optimal, i.e., the 3 dB loss in high-resolution quantization is unavoidable.

Essential structure Although we evaluated the lattice Körner–Marton scheme using both fine and coarse lattice quantization steps, only the coarse quantizer needs to be lattice structured and common to both encoders. The same performance can be attained by replacing the fine lattice with any good vector quantizer. For instance, take a lattice Λ_2 which is a good channel code for the difference signal T and select a "good" set of 2^{nR} vectors $\mathbf{v}(i)$ inside some fundamental cell $\mathcal{P}_0(\Lambda_2)$, for example, the vectors can be drawn independently and uniformly over the cell. Form 2^{nR} bins as the cosets

$$\mathbf{v}(i) + \Lambda_2, \; i = 1, 2, \ldots, 2^{nR}. \tag{12.23}$$

Now, the fine lattice quantization encoding step can be replaced by quantizing to the nearest coset, while the mod Λ_2 steps (at the encoders and decoder) remain unchanged. Note that the coding rate is R, regardless of the structure of the set $\mathbf{v}(1), \ldots, \mathbf{v}(2^{nR})$. Furthermore, each encoder can use a different set of $\mathbf{v}(i)$.

Beyond Gaussian It is also possible to design lattice schemes for non-Gaussian sources X_1 and X_2. In particular, the proposed scheme is *universal* with respect to the statistics of the common signal Y_C; in fact, we could use exactly the same scheme if Y_C were arbitrarily varying and Y_1 and Y_2 remained Gaussian. For non-Gaussian innovations Y_1 and Y_2, we should take the modulo operation with respect to an appropriately *noise-matched* (non-Voronoi) cell of Λ_2.

Loss in single-letter characterization The choice of Gaussian auxiliary random variables (U and V) for evaluating the Berger–Tung rate region (10.25) is quite natural since the sources are jointly Gaussian and the distortion measure is quadratic. Surprisingly, for compressing the difference, this choice is suboptimal in general. In fact, it is better to choose auxiliary random variables involving *scalar* modulo-lattice operations with respect to a *shared scalar lattice* (Figure 12.6 with $\Lambda_2 = \mathbb{Z}$). The resulting performance is that of the lattice Körner–Marton scheme, with a *cubic* coarse lattice. This implies a single-letter rate loss equal to the "coding loss": $\frac{1}{2} \log \left(\mu_{\text{euclid}}(\mathbb{Z}, \mathbf{Z}_{\text{eq}}, P_e)/(2\pi e) \right)$ bits per user, that can be assessed from the last row of Table 7.2.

12.1.3 Linear functions of K Gaussian sources

We now turn to the general case where K encoders observe jointly Gaussian sources $X_1, \ldots, X_K$ and a single receiver wishes to recover a linear function

$$T = a_1 X_1 + \cdots + a_K X_K. \tag{12.24}$$

As before, we assume that each source has zero mean and variance σ^2. Let $\mathbf{X} = [X_1 \cdots X_K]^T$ denote the vector of sources, $\Sigma_X = E[\mathbf{XX}^T]$ denote their covariance matrix, and $\mathbf{a} = [a_1 \cdots a_K]^T$ denote the vector of coefficients. The desired linear function has variance

$$\sigma_T^2 = \mathbf{a}^T \Sigma_X \mathbf{a} \tag{12.25}$$

and we assess the quality of the reconstruction $\hat{T}$ with a quadratic distortion measure $D = E[(\hat{T} - T)^2]$.

The lattice scheme developed above for the symmetric case can be generalized to this setting in a straightforward fashion. The key difference is that each encoder should be able to set its own resolution in order to optimize the sum rate. To this end, we provide each encoder ℓ with its own fine lattice $\Lambda_{1,\ell}$ and a common coarse lattice Λ_2 (with second moment chosen to match σ_T^2) to form a Voronoi quantizer with rate $R_\ell = R(\Lambda_{1,\ell}/\Lambda_2)$. Each encoder sends its own contribution to the linear function (12.24) by scaling its source $\mathbf{x}_\ell$ by a_ℓ and sending the dithered quantization modulo Λ_2. The decoder makes an estimate by removing the dithers, taking modulo Λ_2, and scaling [6] by β:

Encoder ℓ $\quad \mathbf{v}_\ell = Q_{\Lambda_{1,\ell}}(a_\ell \mathbf{x}_\ell + \mathbf{u}_\ell) \bmod \Lambda_2 \qquad \ell \in \{1, 2, \ldots, K\}$ (12.26a)

Decoding $\quad \hat{\mathbf{t}} = \beta \left(\left[\sum_{\ell=1}^{K} (\mathbf{v}_\ell - \mathbf{u}_\ell) \right] \bmod_{\mathcal{V}_0} \Lambda_2 \right).$ (12.26b)

Given sequences of nested lattices that are good for Voronoi quantization (Definition 10.4.1), one can optimize β, $\sigma^2(\Lambda_{1,\ell})$, and $\sigma^2(\Lambda_2)$ to show that it suffices to choose rates $R_1, \ldots, R_K$ satisfying

$$\sum_{\ell=1}^{K} 2^{-2R_\ell} \leq \frac{D}{\sigma_T^2} \qquad (12.27)$$

in order to achieve distortion D. See [142, Theorem 3.1] for a detailed proof. The minimum sum rate of this scheme is

$$R_{\mathrm{KP,sum}} = \frac{K}{2} \log \left(\frac{K \sigma_T^2}{D} \right) \qquad (12.28)$$

and is attained by setting $R_1 = R_2 = \cdots = R_K$. Note that better rates are sometimes achievable by first reconstructing intermediate functions and using these as side information for recovering T.

We now seek to compare the sum rate achievable via lattice coding to that achievable via the Berger–Tung scheme with Gaussian auxiliary random variables. The latter is difficult to characterize in closed form in general. However, the following expression is an excellent approximation to the sum rate in the high-resolution regime (i.e., for small D),

$$R_{\mathrm{BT,sum}} = \frac{K}{2} \log \left(\frac{K}{D} \left(\det(\Sigma_X) \prod_{\ell=1}^{K} a_\ell^2 \right)^{1/K} \right). \qquad (12.29)$$

---

[6] Notice that here we have used a post-scaled quantizer although the same performance is available via pre/post-scaling.

See [142] for more details. It follows that, in the high-resolution regime, the difference in sum rates is

$$R_{\text{BT,sum}} - R_{\text{KP,sum}} = \frac{1}{2} \log \left(\frac{\det(\Sigma_X) \prod_{\ell=1}^{K} a_\ell^2}{\left(\mathbf{a}^T \Sigma_X \mathbf{a} \right)^K} \right). \tag{12.30}$$

For the two-user setting from (12.7) where $T = X_1 + aX_2$, $E[X_1^2] = E[X_2^2] = \sigma^2$, and $E[X_1 X_2] = \rho\sigma^2$, this specializes to

$$\log \left(\frac{|a|\sqrt{1 - \rho^2}}{1 + 2\rho a + a^2} \right). \tag{12.31}$$

From this expression, we can see that the lattice-based scheme provides the largest benefits when $a = 1$ or $a = -1$ and, for these special cases, the gap tends to infinity as $\rho \to -1$ or $\rho \to 1$, respectively. Intuitively, this is due to the fact that the Berger–Tung scheme with Gaussian auxiliary random variables implicitly encodes both sources in their entirety, whereas the lattice Körner–Marton scheme only encodes the difference of the sources. In other words, since the sources are correlated, they will tend to agree on the most significant bits, so sending them via Berger–Tung is inherently wasteful.

For $|a| \neq 1$, the relative gain of lattice-based coding is lower, owing to the fact that the lattice is no longer perfectly matched to the function. A similar phenomenon occurs in the lattice network coding when the channel and desired function are mismatched, as we will explore in Section 12.3.

12.2   Dirty multiple-access channel

We next consider what seems to be the "dual" of the Körner–Marton problem: a generalization of the Gaussian dirty-paper problem of Section 10.5 to a multiple-access channel (MAC) setup.

The MAC is often compared to a cocktail party, where a curious guest wishes to listen to several conversations simultaneously. See Figure 12.8 for an illustration of the two-user MAC. Let us add another complication: suppose that hidden in each conversation is a secret message which is embedded into the speech signal (say, similar to hiding a binary message in an integer sequence, as described in the beginning of Chapter 10). Even if the guest knew how to extract the message that is embedded in a *single* (isolated) speech signal, can he/she extract all of the embedded messages in the cocktail party?

Recall that in the point-to-point Gaussian dirty-paper setting, there is a single transmitter with input X, an interferer whose signal S is known non-causally to the transmitter, and a receiver that observes $Y = X + S + Z$, where Z is independent Gaussian noise. In terms of the cocktail party scenario above, X carries the secret message embedded into the speech signal S. As discussed in Section 10.6,

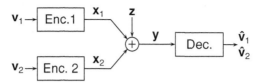

Figure 12.8 Two-user multiple-access channel.

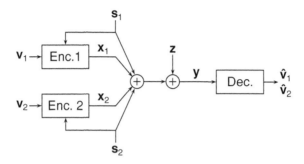

Figure 12.9 Two-user dirty multiple-access channel. Two transmitters communicate with a single receiver. Each transmitter knows part of the interference non-causally.

it is possible to communicate reliably at the AWGN capacity, regardless of the interference power. This can be accomplished via either random or lattice binning. In the multiple-access Gaussian dirty-paper setting, there are several transmitters each with input X_ℓ, an interferer with signal S, and a single receiver that observes $Y = \sum_\ell X_\ell + S + Z$. The key issue is how the knowledge of S is distributed across the transmitters. If S is completely known to all transmitters, then classical random binning techniques suffice (as does lattice binning). However, if the knowledge of S is *decentralized*, then structured coding seems to be essential.

Let us assume that the interference can be decomposed into the sum $S = \sum_\ell S_\ell$ of independent component interference terms S_ℓ. The ℓth transmitter has access to S_ℓ non-causally. See Figure 12.9 for an illustration of the two-user case. In this scenario, the natural generalization of Costa's random binning scheme will yield rates that *vanish* as the interference power tends to infinity. We will argue that lattice binning allows the transmitters to communicate reliably with the receiver at rates that are nearly unaffected by the interference. This is a special case of the general state-dependent multiple-access problem, which we describe below.

Consider two transmitters with independent inputs X_1 and X_2 as well as independent states S_1 and S_2. The channel output Y is the result of passing X_1 and X_2 through the memoryless channel defined by $p(y|x_1, x_2, s_1, s_2)$ where S_1 is i.i.d. according to $p(s_1)$ and S_2 is i.i.d. according to $p(s_2)$. As above, the first transmitter knows S_1 non-causally and the second knows S_2 non-causally. The best known rate region that can be achieved via random i.i.d. binning is the convex hull of all rate pairs

(R_1, R_2) satisfying

$$R_1 \leq I(U_1; Y|U_2) - I(U_1; S_1) \tag{12.32a}$$

$$R_2 \leq I(U_2; Y|U_1) - I(U_2; S_2) \tag{12.32b}$$

$$R_1 + R_2 \leq I(U_1, U_2; Y) - I(U_1; S_1) - I(U_2; S_2), \tag{12.32c}$$

for some distribution of the form $p(u_1, u_2, x_1, x_2|s_1, s_2) = p(u_1, x_1|s_1)p(u_2, x_2|s_2)$. The proof follows along similar lines as the point-to-point case as described in Section 10.5 (i.e., the Gelfand–Pinsker solution (10.54) for capacity with side information at the transmitter), with each user generating its bins independently of the other. See [126] for more details. If the channel $p(y|x_1, x_2)$ is independent of the state (S_1, S_2) (or the state is unknown to both the transmitters and the receiver), then the channel reduces to the well-known "clean" MAC shown in Figure 12.8, and the rate region (12.32) becomes the set of all rate pairs satisfying (see [53]):

$$R_1 \leq I(X_1; Y|X_2) \tag{12.33a}$$

$$R_2 \leq I(X_2; Y|X_1) \tag{12.33b}$$

$$R_1 + R_2 \leq I(X_1, X_2; Y), \tag{12.33c}$$

for some distribution of the form $p(x_1, x_2) = p(x_1)p(x_2)$.

We now demonstrate that classical random binning can be significantly outperformed by linear binning through two special cases: the noiseless binary and the Gaussian dirty multiple-access channels.

12.2.1 The noiseless binary dirty MAC

Consider a two-user, dirty-paper multiple-access channel (MAC) with binary-valued inputs, interferences, and output. That is, let $\mathbf{x}_1$ and $\mathbf{x}_2$ denote the length-n input sequences of transmitters 1 and 2, respectively. These inputs are subject to average Hamming-weight constraints, $w_H(\mathbf{x}_\ell) \leq q_\ell n$ for some $0 \leq q_1, q_2 \leq \frac{1}{2}$. Define $q = \min(q_1, q_2)$. Also, let $\mathbf{s}_1$ and $\mathbf{s}_2$ denote the binary interference sequences. The noise-free channel output is

$$\mathbf{y} = \mathbf{x}_1 \oplus \mathbf{x}_2 \oplus \mathbf{s}_1 \oplus \mathbf{s}_2. \tag{12.34}$$

As discussed above, the knowledge of the interference is decentralized: transmitter 1 only knows $\mathbf{s}_1$ and transmitter 2 only knows $\mathbf{s}_2$.

To remove the effects of the interference in a distributed fashion, we can employ a linear code and exploit the additive nature of the channel. More precisely, the coding scheme depends on the availability of a linear code $\mathcal{C}$ that is good for source coding for Bernoulli(q) sources in the sense of (10.8). Let H denote the $(n - k) \times n$ parity-check matrix associated with this linear code. As in Section 10.1.3, the messages will be encoded as "syndromes" of this parity-check matrix. The lengths of the messages $\mathbf{w}_1 \in \{0, 1\}^{k_1}$ and $\mathbf{w}_2 \in \{0, 1\}^{k_2}$ are chosen to sum up to the syndrome length, $k_1 + k_2 = n - k$. These messages are then zero-padded to length $n - k$ and mapped to coset leaders as follows:

$$\mathbf{v}_1 = \text{leader}\left(\begin{bmatrix} \mathbf{w}_1 \\ \mathbf{0} \end{bmatrix}\right) \qquad \mathbf{v}_2 = \text{leader}\left(\begin{bmatrix} \mathbf{0} \\ \mathbf{w}_2 \end{bmatrix}\right). \qquad (12.35)$$

From here, the encoding and decoding process mirrors that of the point-to-point setting,

$$\begin{aligned} \textbf{Encoding} \quad \mathbf{x}_1 &= [\mathbf{v}_1 \oplus \mathbf{s}_1] \bmod \mathcal{C} & (12.36a) \\ \mathbf{x}_2 &= [\mathbf{v}_1 \oplus \mathbf{s}_2] \bmod \mathcal{C} & (12.36b) \\ \textbf{Decoding} \quad \hat{\mathbf{t}} &= \mathbf{y} \bmod \mathcal{C} & (12.36c) \end{aligned}$$

where $\mathbf{t} = \mathbf{v}_1 \oplus \mathbf{v}_2$ is the sum of the coset leaders. Note that the interference is completely canceled out by this scheme,

$$\begin{aligned} \mathbf{y} \bmod \mathcal{C} &= [[\mathbf{v}_1 \oplus \mathbf{s}_1] \bmod \mathcal{C} \oplus [\mathbf{v}_2 \oplus \mathbf{s}_2] \bmod \mathcal{C} \oplus \mathbf{s}_1 \oplus \mathbf{s}_2] \bmod \mathcal{C} & (12.37) \\ &= [\mathbf{v}_1 \oplus \mathbf{v}_2] \bmod \mathcal{C} & (12.38) \\ &= \mathbf{t} & (12.39) \end{aligned}$$

where the second-to-last step is due to the distributive law of the modulo operation. Owing to the zero-padded structure of $\mathbf{v}_1$ and $\mathbf{v}_2$, the messages $\mathbf{w}_1$ and $\mathbf{w}_2$ can be directly inferred from $\mathbf{t}$.

Since $\mathcal{C}$ is assumed to be good for source coding, we have by (10.8) that $\frac{n-k}{n} \approx H_B(q)$. Moreover, if the interferences are i.i.d. Bernoulli(1/2), then we also have that $E\{w_H(\mathbf{x}_1)\} = E\{w_H(\mathbf{x}_2)\} = qn$ as desired. Overall, this means that any rates $R_1 = k_1/n$ and $R_2 = k_2/n$ satisfying

$$R_1 + R_2 \leq H_B(q), \qquad (12.40)$$

are achievable. This turns out to be the capacity region for the noiseless binary dirty MAC. The converse can be shown by giving the transmitters knowledge of each other's messages [215, Lemma 1].

Let us now focus on the special case where $q_1 = q_2 = q$ and compare the performance of the scheme above to that of random i.i.d. binning. Using (12.32), it can be shown that the rate region for random i.i.d. binning is equal to the set of all rate pairs satisfying

$$R_1 + R_2 \leq \text{u.c.e.}\{[2H_B(q) - 1]^+\} \qquad (12.41)$$

where u.c.e. stands for the upper convex envelope taken with respect to $R_1 = R_2 = 0$ and $q = 0$. The achievability part follows by setting the distributions in (12.32) according to $X_\ell \sim \text{Bernoulli}(q)$ and $U_\ell = X_\ell \oplus S_\ell$ for $\ell = 1, 2$. Showing that this choice yields the largest rate region is more involved and the proof can be found in [215, Appendix B].

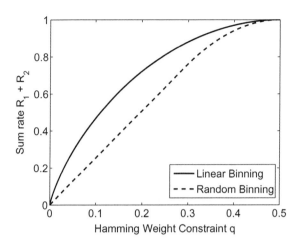

Figure 12.10 Comparison of sum rates for random linear and random i.i.d. binning over the binary dirty multiple-access channel.

In Figure 12.10, we have plotted the sum rates associated with linear binning from (12.40) and random i.i.d. binning from (12.41). Notice that the linear binning scheme strictly outperforms the random i.i.d. binning scheme. We discuss the loss of single-letter characterization in more detail at the end of the next section.

Arbitrary interference Even if the interferences $\mathbf{s}_1$ and $\mathbf{s}_2$ are not i.i.d. Bernoulli$(1/2)$, the same rate region is achievable by dithering the codewords prior to transmission. This parallels the universality property of the lattice Körner–Marton scheme in Section 12.1.2 towards the common signal Y_C.

Noisy binary dirty MAC One can obtain the capacity region of the *noisy* binary dirty MAC via linear binning by replacing the linear codes in the argument above with the *nested* linear codes from Section 8.3. See [216] for more details.

12.2.2 The Gaussian dirty MAC

For the two-user Gaussian dirty MAC, the channel output is

$$Y = X_1 + X_2 + S_1 + S_2 + Z \tag{12.42}$$

where the interference terms S_1 and S_2 are arbitrary and, as usual, the noise Z is i.i.d. Gaussian with variance N. See Figure 12.9 for an illustration. Encoder 1 generates the input X_1, subject to a power constraint P_1, and has non-causal knowledge of S_1. Encoder 2 generates the input X_2, subject to a power constraint P_2, and has non-causal knowledge of S_2.

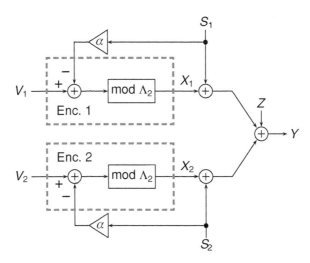

Figure 12.11 Lattice dirty multiple-access coding scheme.

The gains of lattice binning over random binning can be quite dramatic. For example, consider the scenario where the interferences S_1 and S_2 are i.i.d. Gaussian with variance Γ. As Γ tends to infinity, we will show that lattice-based dirty-paper multiple-access scheme can operate very close to the interference-free multiple-access capacity, whereas the rate of the random i.i.d. binning scheme (12.32) with Gaussian auxiliary random variables tends to zero.

The encoding and decoding process for the lattice dirty-paper multiple-access scheme is quite similar to that for the single-user scheme from (10.58), except that now each user removes a separate interference term. The essential ingredient is that both transmitters coordinate their interference removal through the use of a common coarse lattice Λ_2, as illustrated in Figure 12.11. The equivalent effect is that of *interference concentration* over the points of Λ_2, as we have seen in Figure 10.20 for the single-interference case. The second moment of Λ_2 is chosen to be the minimum[7] of the powers of the transmitters, $\sigma^2(\Lambda_2) = P \triangleq \min(P_1, P_2)$. We also employ a common fine lattice Λ_1 for coding:

Encoding $\quad \mathbf{x}_1 = [\mathbf{v}_1 + \mathbf{u}_1 - \alpha\mathbf{s}_1] \bmod_{\mathcal{V}_0} \Lambda_2$ $\hspace{2em}$ (12.43a)

$\qquad\qquad \mathbf{x}_2 = [\mathbf{v}_2 + \mathbf{u}_2 - \alpha\mathbf{s}_2] \bmod_{\mathcal{V}_0} \Lambda_2$ $\hspace{2em}$ (12.43b)

Decoding $\quad \hat{\mathbf{t}} = \left[\arg\min_{\lambda \in \Lambda_1} \|\alpha\mathbf{Y} - \mathbf{u}_1 - \mathbf{u}_2 - \lambda\| \right] \bmod \Lambda_2$ $\hspace{2em}$ (12.43c)

where $\mathbf{v}_1, \mathbf{v}_2 \in \mathcal{C}_{\Lambda_1, \mathcal{P}_0}$ are the message (coset) representatives, $\mathbf{u}_1$ and $\mathbf{u}_2$ are independent dithers, and $\hat{\mathbf{t}}$ is the decoded estimate of the modulo sum $\mathbf{t} = [\mathbf{v}_1 + \mathbf{v}_2] \bmod \Lambda_2$.

---

[7] Although this choice may seem arbitrary, it also appears naturally in the outer bound.

Our coding scheme will be designed so that the message representatives can be inferred directly from **t**. Note that only the encoding modulo operation needs to be taken with respect to the Voronoi region.

As in the single-user dirty-paper channel, the resulting equivalent channel will be completely interference free. However, owing to the distributed encoding process, the decoder will face two self-noise terms, one from each transmitter. This in turn results in a bounded rate loss from the point-to-point AWGN capacity. As we will see, this is a small price to pay to remove the potentially unbounded interference.

Lemma 12.2.1 (Equivalent modulo-Λ channel) *Consider the dirty Gaussian multiple-access channel* (12.42) *and the lattice coding scheme* (12.43) *with arbitrary input vectors* $\mathbf{v}_1$ *and* $\mathbf{v}_2$. *For independent uniform (or modulo-uniform) dithers* $\mathbf{U}_1$ *and* $\mathbf{U}_2$ *with respect to* Λ_2, *and any interference vectors* $\mathbf{s}_1$ *and* $\mathbf{s}_2$, *the channel from* $\mathbf{v}_1$ *and* $\mathbf{v}_2$ *to the decision vector* $\tilde{\mathbf{Y}} = [\alpha \mathbf{Y} - \mathbf{U}_1 - \mathbf{U}_2] \bmod \Lambda_2$ *is equivalent to the (zero-interference) modulo-Λ channel*

$$\tilde{\mathbf{Y}} = [\mathbf{v}_1 + \mathbf{v}_2 + \mathbf{Z}_{eq}] \bmod \Lambda_2, \tag{12.44a}$$

where the equivalent noise is defined as

$$\mathbf{Z}_{eq} = \underbrace{\left[\alpha \mathbf{Z} + (\alpha - 1)(\mathbf{U}_{eq,1} + \mathbf{U}_{eq,2}) \right]}_{\mathbf{Z}_{mix}} \bmod \Lambda_2, \tag{12.44b}$$

and $\mathbf{U}_{eq,1}$, $\mathbf{U}_{eq,2}$ *are equivalent dithers* (4.8) *which are uniform over* $\mathcal{V}_0(\Lambda_2)$ *and independent of each other and* $\mathbf{v}_1$, $\mathbf{v}_2$.

Proof See Problem 12.5. □

To simplify our analysis, we begin with the special case where the rate of the first user is $R_1 = R(\Lambda_1/\Lambda_2)$ and the rate of the second user is $R_2 = 0$ (i.e., $\mathbf{v}_2 = \mathbf{0}$ so that $\mathbf{t} = \mathbf{v}_1$).

Theorem 12.2.1 (System performance) *Assuming that* $R_2 = 0$, *the coding rate* $R_1 = R(\Lambda_1/\Lambda_2) = (1/n) \log |\Lambda_1/\Lambda_2|$ *of the lattice dirty MAC coding scheme* (12.43) *with a Wiener estimation coefficient* $\alpha^* = 2P/(2P + N)$ *satisfies*

$$R_1 \geq \left[\frac{1}{2} \log \left(\frac{1}{2} + \frac{P}{N} \right) - \mathcal{L}(\mathbf{Z}_{mix}) \right]^+ \tag{12.45}$$

where $\mathbf{Z}_{mix} = \alpha \mathbf{Z} + (\alpha - 1)(\mathbf{U}_{eq,1} + \mathbf{U}_{eq,2})$ *is the linear estimation error (equal to the mixture noise) in the modulo-Λ channel* (12.44a), *and*

$$\mathcal{L}(\mathbf{Z}_{mix}) \triangleq \frac{1}{2} \log(G(\Lambda_2)\mu_{euclid}(\Lambda_1, \mathbf{Z}_{mix}, P_e)) \tag{12.46}$$

is the total (shaping plus coding) rate loss. For nested lattice pairs that are good for Voronoi modulation over an AWGN channel, the scheme achieves $R_1 = [\frac{1}{2} \log(\frac{1}{2} + P/N)]^+$ *asymptotically in* n.

Proof We follow the analysis of Voronoi modulation with general estimation in Section 9.6.2. The second moment of the mixture noise from (12.44b) is equal to

$$\frac{1}{n}E\|\mathbf{Z}_{\text{mix}}\|^2 = \frac{1}{n}E\|\alpha\mathbf{Z} + (\alpha - 1)(\mathbf{U}_{\text{eq},1} + \mathbf{U}_{\text{eq},2})\|^2 \tag{12.47a}$$

$$= \frac{1}{n}\alpha^2 E\|\mathbf{Z}\|^2 + \frac{1}{n}(1 - \alpha)^2\left(E\|\mathbf{U}_{\text{eq},1}\|^2 + E\|\mathbf{U}_{\text{eq},1}\|^2\right) \tag{12.47b}$$

$$= \alpha^2 N + (1 - \alpha)^2 2P. \tag{12.47c}$$

This quantity is minimized by the Wiener coefficient $\alpha^* = 2P/(2P + N)$ to yield $(1/n)E\|\mathbf{Z}_{\text{mix}}\|^2 = 2PN/(2P + N)$. From here, (12.45) follows from the equivalent mod Λ channel (Lemma 12.2.1) and Lemma 9.6.1 by noting that

$$\frac{1}{2}\log\left(\frac{P}{\frac{1}{n}E\|\mathbf{Z}_{\text{mix}}\|^2}\right) = \frac{1}{2}\log\left(\frac{P(2P + N)}{2PN}\right) = \frac{1}{2}\log\left(\frac{1}{2} + \frac{P}{N}\right). \tag{12.48}$$

Now, recall that good Voronoi codes result in equivalent dithers $\mathbf{U}_{\text{eq},1}$ and $\mathbf{U}_{\text{eq},1}$ that correspond to semi-spherical noise (Definition 7.8.2 and the subsequent discussion) and, by Problem 7.12, the sum of semi-spherical noise and AWGN noise is itself semi-spherical noise. Thus, the equivalent noise meets the conditions of Definition 9.6.2 and the existence of asymptotically good codes is guaranteed by Theorem 8.5.1. $\qquad\square$

Thus, with lattice dirty-paper coding, we can nearly reach the AWGN capacity, up to the loss of the "one plus" term inside the log. The resulting rate loss vanishes as the SNR increases. Exchanging the roles of user 1 and user 2, we can also asymptotically achieve $R_1 = 0$ and $R_2 = \left[\frac{1}{2}\log(\frac{1}{2} + P/N)\right]^+$. We can also trivially achieve $R_1 = R_2 = 0$. By time sharing [8] between these three rate pairs, the following rate region is achievable:

$$R_1 + R_2 \leq \text{u.c.e.}\left\{\left[\frac{1}{2}\log\left(\frac{1}{2} + \frac{\min(P_1, P_2)}{N}\right)\right]^+\right\} \tag{12.49}$$

since $P = \min(P_1, P_2)$.

While the rate region above is achievable for arbitrary interference, we will need to specify the interference statistics to find an outer bound. Let us assume that $\mathbf{S}_1$ and $\mathbf{S}_2$ are independent of each other and i.i.d. Gaussian with zero mean and variance Γ. As Γ tends to infinity, the rate region of the dirty-paper multiple-access channel must satisfy the following outer bound:

$$R_1 + R_2 \leq \frac{1}{2}\log\left(1 + \frac{\min(P_1, P_2)}{N}\right). \tag{12.50}$$

---

[8] Alternatively, we could enable both transmitters to send information simultaneously by partitioning the fine lattice in a similar fashion to the binary scheme in (12.35).

See [217, Corollary 2] for more details. This nearly corresponds to the inner bound (12.49) established by the lattice scheme, with the addition of the "1 +" term (instead of "$\frac{1}{2}$ +") inside the logarithm. Interestingly, unlike the point-to-point case, there is a rate loss for the dirty multiple-access channel compared to the clean multiple-access channel (12.33), which has a sum rate of $\frac{1}{2}\log(1 + (P_1 + P_2)/N)$. We shall discuss the slight gap between (12.49) and (12.50) in the next section, where it reappears in the context of the two-way relay.

To find the rates available to the conventional random binning scheme (i.e., the natural extension of Costa's binning scheme), we plug Gaussian auxiliary random variables, $U_1 = \alpha S_1 + X_1$ and $U_2 = \alpha S_2 + X_2$, into the rate region from (12.32), as illustrated in Figure 12.12. It can be shown [217] that the resulting sum rate is upper bounded by

$$
R_1 + R_2 \leq \left[h(S_1 + S_2) - h(S_1) - h(S_2) + \frac{1}{2}\log\left(2\pi e \frac{P_1 P_2}{N}\right) + o(1) \right]^+,
$$

(12.51)

where $o(1) \to 0$ as $\Gamma \to \infty$. Since $h(S_1 + S_2) = \frac{1}{2}\log(4\pi e \Gamma)$ and $h(S_1) + h(S_2) = \log(2\pi e \Gamma)$, we find that by taking the interference power Γ to infinity, $R_1 + R_2 \leq 0$. In other words, conventional random binning *cannot achieve positive rates* as the interference power tends to infinity.

Essential structure Note that only the coarse lattices of the transmitters need to be aligned in order to perform distributed interference cancelation. For instance, a "semi-lattice" scheme as in the Körner–Marton problem (12.23) can achieve the same rates by replacing the fine lattice with a random code generated according to a uniform distribution over $\mathcal{V}_0$.

Beyond Gaussian As in the binary case, it is possible to handle arbitrary interferences $\mathbf{s}_1$ and $\mathbf{s}_2$ while attaining the same rate region, owing to the use of the dithering operation. We focused on the case of i.i.d. Gaussian interference above in order to make a clear comparison with classical random binning strategies as well as to develop the upper bound. For non-Gaussian noise, we can employ noise-matched (ML) decoding like in Voronoi demodulation; see Section 9.3.6 and the remark at the end of Section 12.3.2 below.

Loss in single-letter characterization Costa's binning scheme is derived from a Gaussian single-letter formula. It fails on the dirty MAC because, unlike for lattice binning, the sum of two independent bins (from the two users) results in a "bad" codebook. Mathematically, this failure is reflected by the vanishing of the right-hand side of (12.51) in the limit of a strong interference.

It turns out that we can improve upon this performance by evaluating (12.32) using auxiliary random variables that involve modulo operations with respect to a *shared scalar lattice*, instead of Gaussian random variables. This enables the

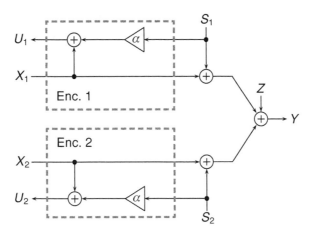

Figure 12.12 Costa dirty multiple-access test channel.

random i.i.d. binning scheme to mimic the lattice binning scheme proposed above with a cubic coarse lattice, i.e., with the exception of the shaping provided by a good coarse lattice. Compare Figures 12.12 and 12.11. We conjecture that the best single-letter formula for the dirty MAC in the limit of strong interference and high SNR is given in terms of a one-dimensional lattice. The resulting rate loss is thus the "shaping gain" $\frac{1}{2}\log(2\pi e/12) \approx 0.254$ bits per user, i.e., the divergence from Gaussianity of a scalar dither (see Example 9.3.1).

12.3   Lattice network coding

In a standard packet switching network, nodes act as routers – they wish to find the best route for a packet under the current conditions. If the inflow to a node is higher than its output capacity, then some of the packets will be discarded. The idea of network coding is that a bottleneck node can "combine" together packets rather than choose which one to pass on and which one to discard. If the final destination gets enough such "combinations" (from different routes), then it can resolve the ambiguity and decode all the transmitted packets reliably.

If the nodes are linked together by a network of lossless bit pipes, then the destination can decode from any end-to-end information-preserving mapping, i.e., a network code. Random binning at the nodes results in an information-preserving mapping with high probability and was used to establish the multicast capacity of such networks [4]. Much of the subsequent research on network coding has focused on *linear* schemes [138, 155], in part due to their reduced implementation complexity. However, when extending the network coding idea to noisy interference networks (i.e., networks in which the transmissions of the nodes interfere with one

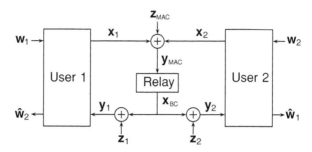

Figure 12.13 Two-way relay channel. There are two users who wish to exchange messages through a relay. The relay observes the noisy sum of the users' signals and communicates back to them via a broadcast channel.

another and are corrupted with noise), employing a network code with algebraic structure is essential to avoid rate loss.

This phenomenon is of particular interest in wireless networks, where each receiver naturally observes a noisy linear combination of all transmitted codewords (or packets). The separation-based approach suggests that each receiver should first decode a subset of the codewords, and only then perform linear network coding to create the packet to be transmitted over the next hop. However, using appropriately chosen lattice codes, it is possible for each receiver to *decode a linear combination of the codewords directly*, often at significantly higher rates. This technique falls into the family of *physical-layer network coding* strategies [157, 195], which aim to exploit the linear combinations taken by the channel as part of an end-to-end network code. Note that while the analog combination taken by the channel can be used directly as a form of network code, this suffers from the problem of *noise accumulation*, i.e., the destinations will have to decode from the sum of the noise terms of all the relays.

A compelling scenario for physical-layer network coding is provided by the linear two-way relay channel, as illustrated in Figure 12.13. User 1 has a message $\mathbf{w}_1$ which it wishes to send to user 2. Similarly, user 2 wishes to send $\mathbf{w}_2$ to user 1. The challenge is that the users can only communicate through a relay that observes the noisy sum of the transmitted signals. The separation-based approach to this problem combines classical multiple-access techniques (at the channel coding layer) with network coding (at the source coding layer): the relay first decodes both messages, then computes the modulo sum, and finally broadcasts to the users to solve for their desired messages. Interestingly, if both users employ the same linear or lattice codebook, then the relay can decode the sum directly without inferring the individual messages, which roughly doubles the achievable sum rate. Note that random i.i.d. codebooks are good for classical multiple-access and bad for computation for the same reason: they make each pair of codewords discernible from their sum. Below, we will explore this "compute-and-forward"

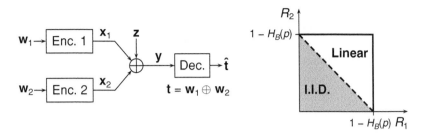

Figure 12.14 Computation over a multiple-access channel. The receiver wishes to recover the modulo sum of the messages. By using the same linear code at each transmitter, we can attain the capacity region, which is much larger than the rate region achievable via random i.i.d. codes.

technique in the context of binary and Gaussian channels, and compare its performance to that of random i.i.d. coding as well as to the theoretical performance limits.

12.3.1 Binary case, single receiver

We set the stage with a coding technique for computing the modulo sum of messages over a binary multiple-access channel. We will then use this as a building block for a compute-and-forward strategy for the binary two-way relay channel.

Computation over multiple-access channels

Consider a simple multiple-access channel with binary-valued messages, channel inputs, noise, and output (see Figure 12.14). Specifically, let $\mathbf{w}_1$ and $\mathbf{w}_2$ be the length-k messages of transmitters 1 and 2, respectively. Also, let $\mathbf{x}_1$ and $\mathbf{x}_2$ denote the transmitters' length-n channel inputs. The receiver observes a noisy sum of these inputs,

$$\mathbf{y} = \mathbf{x}_1 \oplus \mathbf{x}_2 \oplus \mathbf{z} \tag{12.52}$$

where $\mathbf{z}$ is i.i.d. Bernoulli(p) noise. The goal is for the receiver to recover the modulo sum

$$\mathbf{t} = \mathbf{w}_1 \oplus \mathbf{w}_2 \tag{12.53}$$

with low probability of error at the highest possible rate [9] $R = k/n$.

---

[9] The rate is defined in terms of the average number of bits per channel use sent by *each* transmitter, rather than the average number of bits obtained by the receiver. This distinction will be useful later on in the two-way relay, since we are ultimately interested in communicating messages between users.

If the channel were noiseless (e.g., $p = 0$), then communicating the sum would be trivial. The transmitters could send their messages in an uncoded fashion, $\mathbf{x}_1 = \mathbf{w}_1$ and $\mathbf{x}_2 = \mathbf{w}_2$, to give the receiver the sum $\mathbf{y} = \mathbf{w}_1 \oplus \mathbf{w}_2$ at the maximum rate of $R = 1$. However, in the noisy setting, some form of channel coding is required to attain a low probability of error.

The classical approach to proving achievability results for the multiple-access channel is to employ an independent random codebook at each transmitter whose codewords are generated element-wise i.i.d. This implicitly ensures that each possible pair of messages $(\mathbf{w}_1, \mathbf{w}_2)$ is assigned to a pair of codewords with a unique sum (assuming the rates are set appropriately). While this is a desirable property in the case where the receiver desires both messages, it leads to a significant rate loss when it only wants the sum. Specifically, this strategy only achieves a rate of $R = (1 - H_B(p))/2$ per user, which is what is required to send both messages in their entirety.

Now, consider employing the same linear code at each transmitter in the spirit of the Körner–Marton scheme describe in Section 12.1. Specifically, let G be an $n \times k$ generator matrix for a linear code that is good for coding over a BSC in the sense discussed in Section 10.1.1. The encoding and decoding procedure is

$$\textbf{Encoding} \quad \mathbf{x}_1 = G\mathbf{w}_1 \tag{12.54a}$$

$$\mathbf{x}_2 = G\mathbf{w}_2 \tag{12.54b}$$

$$\textbf{Decoding} \quad \hat{\mathbf{t}} = \mathbf{y} \oplus \Big(\underbrace{[\mathbf{y}] \bmod \mathcal{C}}_{\hat{\mathbf{z}}} \Big). \tag{12.54c}$$

Notice that, from the receiver's perspective, it appears as if the sum was directly encoded with G,

$$\mathbf{y} = \mathbf{x}_1 \oplus \mathbf{x}_2 \oplus \mathbf{z} \tag{12.55a}$$

$$= G\mathbf{w}_1 \oplus G\mathbf{w}_2 \oplus \mathbf{z} \tag{12.55b}$$

$$= G(\mathbf{w}_1 \oplus \mathbf{w}_2) \oplus \mathbf{z} \tag{12.55c}$$

$$= G\mathbf{t} \oplus \mathbf{z}. \tag{12.55d}$$

It follows that, since G is good for coding over a BSC, the second term in (12.54c), $\hat{\mathbf{z}} = [\mathbf{y}] \bmod \mathcal{C}$, is equal with high probability to the noise $\mathbf{z}$, so $\hat{\mathbf{t}} = \mathbf{t}$ with high probability as well. Thus, any rate up to

$$R = 1 - H_B(p) \tag{12.56}$$

is achievable. This turns out to be the capacity for sending the binary sum, as the transmitters could do no better if they were permitted to cooperate fully.

We can also consider the case where the transmitters have unequal rates. Specifically, let transmitter ℓ have k_ℓ bits and rate $R_\ell = k_\ell / n$. The messages $\mathbf{w}_1$ and $\mathbf{w}_2$ comprise these bit sequences zero-padded to the longer of the two lengths

$k = \max(k_1, k_2)$ so that the sum $\mathbf{w}_1 \oplus \mathbf{w}_2$ is well defined. It follows that any rate pair (R_1, R_2) satisfying

$$\max(R_1, R_2) \le 1 - H_B(p) \tag{12.57}$$

is achievable whereas random i.i.d. codes only achieve the multiple-access rate region

$$R_1 + R_2 \le 1 - H_B(p). \tag{12.58}$$

See Figure 12.14 for a comparison of these rate regions.

Two-way relaying

We can use this computation result as part of a compute-and-forward relaying strategy for the two-way relay channel depicted in Figure 12.13. Consider the scenario where all of the channels are binary

$$\mathbf{y}_{\text{MAC}} = \mathbf{x}_1 \oplus \mathbf{x}_2 \oplus \mathbf{z}_{\text{MAC}} \tag{12.59a}$$

$$\mathbf{y}_1 = \mathbf{x}_{\text{BC}} \oplus \mathbf{z}_1 \tag{12.59b}$$

$$\mathbf{y}_2 = \mathbf{x}_{\text{BC}} \oplus \mathbf{z}_2 \tag{12.59c}$$

and the independent noise sequences $\mathbf{z}_{\text{MAC}}$, $\mathbf{z}_1$, and $\mathbf{z}_2$ are i.i.d. Bernoulli(p). Each user's message is written as a length-k binary vector, $\mathbf{w}_1, \mathbf{w}_2 \in \{0, 1\}^k$. Thus, the users have equal rates $R_1 = R_2 = k/n$ where n is the number of channel uses. [10] The computation scheme developed above can be used to send the sum $\mathbf{t} = \mathbf{w}_1 \oplus \mathbf{w}_2$ to the relay at rates $R_1 = R_2 = 1 - H_B(p)$. Afterwards, the relay re-encodes its estimate of the sum using the same linear code [11] and conveys it to both users, $\mathbf{x}_{\text{BC}} = G\hat{\mathbf{t}}$. The users then make estimates $\hat{\mathbf{t}}_1$ and $\hat{\mathbf{t}}_2$ of the sum and combine these with their own messages to estimate their desired messages, $\hat{\mathbf{w}}_2 = \hat{\mathbf{t}}_1 \oplus \mathbf{w}_1$ and $\hat{\mathbf{w}}_1 = \hat{\mathbf{t}}_2 \oplus \mathbf{w}_2$. Overall, this shows that $R_1 = R_2 = 1 - H_B(p)$ is achievable. This corresponds to the sum capacity of the binary two-way relay channel since each user simultaneously attains the rate it could achieve if the other user were silent. Classical relaying strategies fall short of this performance. For instance, the "decode-and-forward" strategy would have the relay recover both messages $\mathbf{w}_1$ and $\mathbf{w}_2$ prior to forming the modulo sum $\mathbf{t}$ for the broadcast phase. We know from the multiple-access capacity region that this is possible only if $R_1 + R_2 \le 1 - H_B(p)$, meaning that this strategy cannot reach the capacity. Another possibility, known as "compress-and-forward," would have the relay compress its observation of the noisy sum $\mathbf{y}$ so that it can be conveyed over the broadcast segment. Each user can then receive a

[10] Here, we have implicitly assumed that the relay is half-duplex, meaning that it cannot transmit and receive simultaneously. Thus, it will have received all n channel outputs from the multiple-access stage prior to generating its n channel inputs for the broadcast stage.

[11] While the use of a linear code seems essential for the multiple-access phase, we are free to choose a standard capacity-achieving i.i.d. random code for the broadcast phase. Here, we used a linear code for the broadcast phase for ease of exposition.

noisy observation of its desired codeword by combining its estimate of $\mathbf{y}$ with its own codeword. Similarly, the relay could follow an "amplify-and-forward" strategy and simply repeat its observation uncoded over the broadcast segment. However, owing to the accumulation of noise, neither of these strategies will attain the capacity.

Modulo channels The binary framework developed above can be generalized to any modulo-additive multiple-access channel or relay with no input constraints. In all such cases, the compute-and-forward approach is "perfect," i.e., it achieves the cut-set bound on capacity. For real addition, the story is more involved, as we will see below in the Gaussian case.

12.3.2   Gaussian case, single receiver

We now consider the problem of computing over a Gaussian multiple-access channel and its application to two-way relaying. Ideally, we would like the receiver to decode the modulo sum of the codewords as this is well suited for two-way relaying. That is, the modulo sum provides enough information to each user to recover its desired message, without the need for any excess rate (which would be required for sending the real sum without binning). Note that there is mismatch between this desired function and the channel, which only returns a noisy version of the real sum. As we will see, nested lattice codes provide a means to bridge this gap.

Computation over multiple-access channels
Consider the setting where the receiver observes a linear combination

$$Y = X_1 + X_2 + Z \tag{12.60}$$

of the users' real-valued inputs X_1 and X_2, corrupted by i.i.d. Gaussian noise Z with variance N. As in the binary case, we would like the receiver to recover the sum of the codewords. In other words, we need to guarantee that *any sum of codewords is protected against Gaussian noise*. Lattice-based codes are an excellent choice for this problem due to the fact that lattices are closed under addition. Moreover, nested lattice codes provide a natural way to compute a modulo sum of the codewords.

For now, let us assume that both users have equal power constraints $P_1 = P_2 = P$. This allows each transmitter to use the same Voronoi code, created from nested lattices $\Lambda_2 \subset \Lambda_1$. The receiver decodes to the closest lattice point, just as in the point-to-point case, except that here the estimate will correspond to the modulo-lattice sum. The encoding and decoding operations are as follows:

$$\textbf{Encoding} \quad \mathbf{x}_1 = [\mathbf{v}_1 + \mathbf{u}_1] \bmod_{\mathcal{V}_0} \Lambda_2 \tag{12.61a}$$

$$\mathbf{x}_2 = [\mathbf{v}_2 + \mathbf{u}_2] \bmod_{\mathcal{V}_0} \Lambda_2 \tag{12.61b}$$

$$\textbf{Decoding} \quad \hat{\mathbf{t}} = \left[\underset{\lambda \in \Lambda_1}{\arg\min} \|\alpha Y - \mathbf{u}_1 - \mathbf{u}_2 - \lambda\| \right] \bmod \Lambda_2 \tag{12.61c}$$

where $\mathbf{v}_1, \mathbf{v}_2 \in \mathcal{C}_{\Lambda_1, \mathcal{P}_0}$ (for some nested lattice pair $\Lambda_2 \subset \Lambda_1$) are the message (coset) representatives, $\mathbf{u}_1$ and $\mathbf{u}_2$ are independent dithers, and $\hat{\mathbf{t}}$ is the decoded estimate of the modulo-lattice sum

$$\mathbf{t} = [\mathbf{v}_1 + \mathbf{v}_2] \bmod \Lambda_2. \tag{12.62}$$

The coding rate is defined as $R(\Lambda_1/\Lambda_2) = \frac{1}{n} \log |\Lambda_1/\Lambda_2|$, which, as above, is the number of bits per channel use sent by *each* transmitter; see footnote 9.

Note that the modulo operation need only be taken with respect to the Voronoi cell at the encoders, as is the case for point-to-point channel coding.

From the receiver's perspective, the effective channel closely resembles the point-to-point AWGN channel under Voronoi modulation studied in Chapter 9. The main difference is the distribution of the equivalent noise, which consists of two *self-noise* terms, one from each transmitter. This in turn results in a small rate loss from the point-to-point AWGN capacity, which closely resembles the penalty encountered in the the dirty MAC setting of Section 12.2.2.

Lemma 12.3.1 (Equivalent modulo-Λ channel) *Consider the Gaussian multiple-access channel* (12.60) *and the lattice coding scheme* (12.61) *with arbitrary input vectors* $\mathbf{v}_1$ *and* $\mathbf{v}_2$. *For independent uniform (or modulo-uniform) dithers* $\mathbf{U}_1$ *and* $\mathbf{U}_2$ *with respect to* Λ_2, *the channel from* $\mathbf{v}_1$ *and* $\mathbf{v}_2$ *to the decision vector* $\tilde{\mathbf{Y}} = [\alpha \mathbf{Y} - \mathbf{U}_1 - \mathbf{U}_2] \bmod \Lambda_2$ *is equivalent to the additive modulo-Λ_2 channel:*

$$\tilde{\mathbf{Y}} = [\mathbf{v}_1 + \mathbf{v}_2 + \mathbf{Z}_{\text{eq}}] \bmod \Lambda_2 \tag{12.63}$$

with equivalent noise

$$\mathbf{Z}_{\text{eq}} = \big[\underbrace{\alpha \mathbf{Z} + (\alpha - 1)(\mathbf{U}_{\text{eq},1} + \mathbf{U}_{\text{eq},2})}_{\mathbf{Z}_{\text{mix}}} \big] \bmod \Lambda_2, \tag{12.64}$$

where $\mathbf{U}_{\text{eq},1}$ *and* $\mathbf{U}_{\text{eq},2}$ *are the equivalent dithers* (4.8), *which are uniform over* $\mathcal{V}_0(\Lambda_2)$ *and independent of each other and* $\mathbf{v}_1, \mathbf{v}_2$.

Proof See Problem 12.5. □

Using this lemma, we now characterize the performance of our scheme.

Theorem 12.3.1 (System performance) *The coding rate* $R(\Lambda_1/\Lambda_2) = (1/n) \log |\Lambda_1/\Lambda_2|$ *of the lattice MAC computation scheme* (12.61) *with a Wiener estimation coefficient*

$$\alpha^* = \frac{2P}{2P + N} \tag{12.65}$$

satisfies

$$R(\Lambda_1/\Lambda_2) \geq \left[\frac{1}{2} \log \left(\frac{1}{2} + \frac{P}{N} \right) - \mathcal{L}(\mathbf{Z}_{\text{mix}}) \right]^{+} \tag{12.66}$$

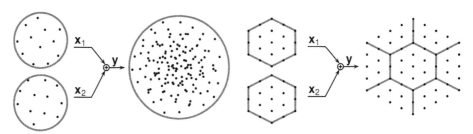

Figure 12.15 Conventional random i.i.d. coding ensures that sums of codewords are unique, hence they are denser and less resilient to noise. Lattice coding guarantees that sums of codewords are themselves codewords.

where $\mathbf{Z}_{\text{mix}} = \alpha\mathbf{Z} + (\alpha - 1)(\mathbf{U}_{\text{eq},1} + \mathbf{U}_{\text{eq},2})$ is the mixture noise in the modulo-Λ channel (12.64), and $\mathcal{L}(\mathbf{Z}_{\text{mix}})$ is the rate loss (12.46). For nested lattice pairs that are good for Voronoi modulation over an AWGN channel (in the sense of Definition 9.6.2), the scheme achieves $R(\Lambda_1/\Lambda_2) = \left[\frac{1}{2}\log(\frac{1}{2} + P/N)\right]^+$ asymptotically in n.

The proof is nearly identical to that of Theorem 12.2.1 and is omitted.

The rate $\left[\frac{1}{2}\log(\frac{1}{2} + P/N)\right]^+$ can be thought of as the "computation rate," meaning the effective rate of the codebook induced by the modulo sum of the codebooks of the two users. Since we have constrained the users to have equal rates, this computation rate is also a constraint on the rate of each user. One peculiar feature of this rate expression is the absence of the usual "$1 +$" term inside the logarithm. From one perspective, the loss seems to be due to a mismatch between the transmitters and the receiver. That is, the receiver observes the sum at power $2P$ as well as Gaussian noise at power N. The resulting effective SNR (after the modulo operation) imposes a constraint on the density of the fine lattice. This fine lattice, if combined with a coarse shaping lattice of second moment $2P$, would yield a Voronoi codebook with rate of $\frac{1}{2}\log(1 + 2P/N)$. However, each transmitter only has power P, meaning that the coarse lattice will have second moment P and the resulting Voronoi codebook will only have rate $\left[\frac{1}{2}\log(\frac{1}{2} + P/N)\right]^+$. (This is due to the fact that the volume of the Voronoi region is decreased by a factor of $2^{n/2}$.)

This performance is much higher than what is available via classical random i.i.d. coding for decoding the sum. Specifically, random i.i.d. coding ensures that, with high probability, each pair of codewords is mapped to a unique sum. Thus, in decoding the sum, the receiver will end up implicitly decoding the individual messages and will be constrained by multiple-access capacity bounds to a computation rate of $\frac{1}{4}\log(1 + 2P/N)$. See Figure 12.15 for a visual comparison of random i.i.d. and lattice coding.

If there are K transmitters (each with power P and equal rates) and a single transmitter that desires the modulo sum of their messages, Voronoi modulation can attain a computation rate of $[\frac{1}{2}\log(1/K + P/N)]^+$. This is a special case of the compute-and-forward framework that will be presented in Section 12.3.3.

Two-way relaying

We now turn back to the Gaussian two-way relay channel and propose a compute-and-forward relaying strategy. Specifically, assume that the channels in Figure 12.13 are given by

$$\mathbf{y}_{\text{MAC}} = \mathbf{x}_1 + \mathbf{x}_2 + \mathbf{z}_{\text{MAC}} \tag{12.67a}$$

$$\mathbf{y}_1 = \mathbf{x}_{\text{BC}} + \mathbf{z}_1 \tag{12.67b}$$

$$\mathbf{y}_2 = \mathbf{x}_{\text{BC}} + \mathbf{z}_2 \tag{12.67c}$$

where the real-valued inputs $\mathbf{x}_1, \mathbf{x}_2, \mathbf{x}_{\text{BC}}$ are power limited by P and the independent noise sequences $\mathbf{z}_{\text{MAC}}, \mathbf{z}_1$, and $\mathbf{z}_2$ are i.i.d. Gaussian with mean zero and variance N. Each user's message is mapped to a coset representative, $\mathbf{v}_1, \mathbf{v}_2 \in \mathcal{C}_{\Lambda_1, \mathcal{P}_0}$, which implies that the rates are equal $R_1 = R_2 = R(\Lambda_1/\Lambda_2)$. The lattice scheme developed above can be used as part of a compute-and-forward relaying strategy. The relay can decode the modulo sum $\mathbf{t} = [\mathbf{v}_1 + \mathbf{v}_2] \bmod \Lambda_2$ at rates $R_1 = R_2 = [\frac{1}{2}\log(\frac{1}{2} + P/N)]^+$. It can then simply transmit its estimate $\hat{\mathbf{t}}$ of the modulo sum to the users which can also decode reliably. Now, using their own messages as side information, the users can solve for their desired messages. For example, user 1 can form the estimate $\hat{\mathbf{v}}_2 = [\hat{\mathbf{t}}_1 - \mathbf{v}_1] \bmod \Lambda_2$ using its message representative $\mathbf{v}_1$ and its estimate $\hat{\mathbf{t}}_1$ of the modulo sum $\mathbf{t}$. Overall, this yields a sum rate of

$$R_{\text{compute, sum}} = \left[\log\left(\frac{1}{2} + \frac{P}{N}\right)\right]^+ \tag{12.68}$$

which nearly matches the upper bound on the sum rate

$$R_{\text{upper, sum}} = \log\left(1 + \frac{P}{N}\right) \tag{12.69}$$

that would be attainable if each user had a dedicated (i.e., interference-free) AWGN channel to the other. The only difference is the absence of the "1+" term inside the logarithm, as discussed above. It is an open problem as to whether the upper or lower bound is loose [12] (or both).

Comparison with other relaying strategies We now briefly review the performance of classical relaying strategies. For instance, consider a *decode-and-forward* strategy where the relay first decodes the individual messages $\mathbf{w}_1$ and $\mathbf{w}_2$ at the symmetric multiple-access capacity $R_1 = R_2 = \frac{1}{4}\log(1 + 2P/N)$ (via standard random i.i.d. coding). It then transmits the modulo sum $\mathbf{w}_1 \oplus \mathbf{w}_2$ to the users to attain a sum

---

[12] See [270] for an alternative proof of this lower bound based on spherically shaped lattice codes.

rate of

$$R_{\text{decode,sum}} = \frac{1}{2} \log \left(1 + \frac{2P}{N} \right). \tag{12.70}$$

Notice that the penalty for not taking advantage of the natural sum computation provided by the multiple-access channel is nearly a factor of $1/2$ in the limit of high SNR.

Under an *amplify-and-forward* strategy, the relay makes no attempt to decode. Instead, it simply repeats its channel observation $\mathbf{y}_{\text{MAC}}$ on the broadcast phase $\mathbf{x}_{BC} = \sqrt{\frac{P}{2P+N}} \mathbf{y}_{\text{MAC}}$ (scaled to meet the power constraint). Each user can cancel out the effect of its own codeword to get an effective point-to-point channel from the other user. If the users employ i.i.d. Gaussian codebooks, the resulting sum rate is

$$R_{\text{amplify,sum}} = \log \left(1 + \frac{P}{N} \frac{P}{3P + N} \right). \tag{12.71}$$

Note that this strategy matches the slope of the upper bound, owing to the fact that it exploits the sum taken by the multiple-access stage. However, it has a lower end-to-end effective SNR as compared to compute-and-forward, due to the fact that it does not remove noise at the relay.

Finally, under a *compress-and-forward* strategy, the relay quantizes its channel observation $\mathbf{y}_{\text{MAC}}$ and broadcasts the resulting bits to both users. Each user then forms an estimate of the relay's observation and again subtracts its own codeword to obtain an effective point-to-point channel from the other user. It can be shown that, for i.i.d. Gaussian codebooks, the sum rate is equal to (12.71), i.e, this strategy suffers from *noise accumulation* as in amplify-and-forward.

In Figure 12.16, we have plotted the performance of these strategies versus SNR. Notice that the compute-and-forward strategy is dominant except in the low SNR range (where the absence of the "1+" term has a more pronounced effect).

Asymmetric powers

It may be the case that the two users have different power constraints or that their signals travel across channels with unequal strengths. For instance, consider the Gaussian two-way relay channel (12.67) and assume that the input power constraints are

$$\frac{1}{n} E \|\mathbf{X}_1\|^2 \le P_1 \qquad \frac{1}{n} E \|\mathbf{X}_2\|^2 \le P_2 \qquad \frac{1}{n} E \|\mathbf{X}_{\text{BC}}\|^2 \le P_{\text{BC}}, \tag{12.72}$$

where, without loss of generality, we assume that $P_1 \ge P_2$. The i.i.d. zero mean Gaussian noise sequences $\mathbf{z}_{\text{MAC}}$, $\mathbf{z}_1$, and $\mathbf{z}_2$ have variances N_{MAC}, N_1, and N_2, respectively.

We could directly apply the lattice coding scheme developed for the symmetric powers by setting the second moment of the coarse lattice to the minimum of these

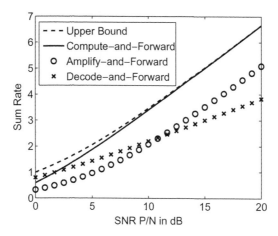

Figure 12.16 Comparison of the sum rates of various relaying strategies over the Gaussian two-way relay channel.

powers. However, this may be quite wasteful if one user has significantly more power than the other. Interestingly, it is possible to utilize fully the power at each transmitter while still reliably decoding the sum at the relay. The main idea is to use the same fine lattice at both users but different coarse lattices, each sized to meet its user's power constraint.

To this end, we will select a *triple* of nested lattices

$$\Lambda_3 \subset \Lambda_2 \subset \Lambda_1$$

to form two Voronoi codebooks $\mathcal{C}_1 = \Lambda_1 \cap \mathcal{P}_0(\Lambda_2)$ and $\mathcal{C}_2 = \Lambda_1 \cap \mathcal{P}_0(\Lambda_3)$. These codebooks share a common fine lattice, which ensures that the sum of codewords is itself an element of the fine lattice and thus afforded protection against noise. The second moments of the coarse lattices are set to meet the power constraints of the users $\sigma^2(\Lambda_2) = P_1$ and $\sigma^2(\Lambda_3) = P_2$, which allows each user to use a different rate, $R_1 = R(\Lambda_1/\Lambda_2)$ and $R_2 = R(\Lambda_1/\Lambda_3)$. The encoding process at the users and the decoding process at the relay is summarized as follows:

Encoding $\quad \mathbf{x}_1 = [\mathbf{v}_1 + \mathbf{u}_1] \bmod_{\mathcal{V}_0} \Lambda_2 \qquad\qquad\qquad (12.73\text{a})$

$\qquad\qquad\quad \mathbf{x}_2 = [\mathbf{v}_2 + \mathbf{u}_2] \bmod_{\mathcal{V}_0} \Lambda_3 \qquad\qquad\qquad (12.73\text{b})$

Decoding $\quad \hat{\mathbf{t}} = \left[\arg\min_{\lambda \in \Lambda_1} \|\alpha \mathbf{Y}_{\text{MAC}} - \mathbf{u}_1 - \mathbf{u}_2 - \lambda\| \right] \bmod \Lambda_3 \qquad (12.73\text{c})$

where $\mathbf{v}_1 \in \mathcal{C}_1$ and $\mathbf{v}_2 \in \mathcal{C}_2$ are the message (coset) representatives, $\mathbf{u}_1$ and $\mathbf{u}_2$ are independent dithers, and $\hat{\mathbf{t}}$ is the decoded estimate of the modulo-lattice sum $\mathbf{t} = \left[\mathbf{v}_1 + \mathbf{v}_2 - Q_{\Lambda_2}(\mathbf{v}_2 + \mathbf{u}_2) \right] \bmod \Lambda_3$. It can be shown that there exist sequences of lattices $\Lambda_1^{(n)}$, $\Lambda_2^{(n)}$, and $\Lambda_3^{(n)}$ such that both codebooks $\mathcal{C}_1$ and $\mathcal{C}_2$ are good for Voronoi modulation (Definition 9.6.2). By combining this fact with the steps in the proof of Lemma 12.3.1 and Theorem 12.3.1, it can be shown that for a Wiener coefficient

$\alpha^* = (P_1 + P_2)/(P_1 + P_2 + N)$, any rate pair (R_1, R_2) satisfying

$$R_1 \leq \left[\frac{1}{2} \log \left(\frac{P_1}{P_1 + P_2} + \frac{P_1}{N_{\text{MAC}}} \right) \right]^+ \quad R_2 \leq \left[\frac{1}{2} \log \left(\frac{P_2}{P_1 + P_2} + \frac{P_2}{N_{\text{MAC}}} \right) \right]^+$$

$$(12.74)$$

is achievable asymptotically in n. Note that for symmetric powers $(P_1 = P_2)$, (12.74) reduces to Theorem 12.3.1. See [192] for more details.

The final step is for the relay to communicate back to the users. One possibility is for the relay to randomly bin its estimate $\hat{t}$ and send the resulting index to the users via a capacity-achieving channel code. The users can then combine their channel observations with the side information provided by their own messages to infer their desired messages. This is possible so long as

$$R_1 \leq \frac{1}{2} \log \left(1 + \frac{P_{\text{BC}}}{N_1} \right) \quad R_2 \leq \frac{1}{2} \log \left(1 + \frac{P_{\text{BC}}}{N_2} \right). \quad (12.75)$$

Overall, the achievable rate region for the two-way relay channel with unequal powers is described by the intersection of the regions (12.74) and (12.75). This region turns out to be within half a bit of the outer bound that can be established by assuming that each user has exclusive access to the channel through the relay to the other user.

Notice that the compute-and-forward strategy would work just as well if the broadcast phase were not Gaussian. That is, classical random i.i.d. binning and channel coding suffices to convey the modulo sum back to the users. See [204] for more details.

Remarks

Essential structure Clearly, the fine lattice is essential to guarantee that the sum of two codewords is afforded protection against noise. However, the rate regions described above for the Gaussian two-way channel do not require the shaping code to be identical at both users or even to be a lattice. Specifically, we could generate the codebooks by intersecting the fine lattice with an n-dimensional ball to enforce the power constraint, i.e., spherical shaping as discussed in Section 9.3.3. Note that the broadcast stage will have to use a binning step to avoid sending the excess rate associated with the real sum of the codewords. See [270] for more details. A simpler, but suboptimal, scheme is to use a cubic shaping lattice, which results in a rate loss of $\frac{1}{2} \log(2\pi e/12) \approx 0.254$ bits per user (see Example 9.3.1).

Beyond Gaussian It is also possible to design compute-and-forward schemes for non-Gaussian noise. In particular, the fine lattice decoder should be designed according to the noise-matched NVNR criterion from Definition 6.4.1. Since from the receiver's perspective, the equivalent channel looks like a point-to-point

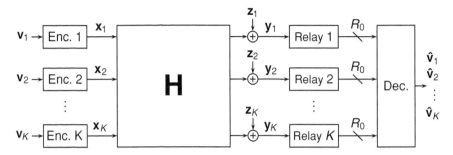

Figure 12.17 K users communicate to a single destination through K relays, each of which observes a noisy linear combination of the transmitted signals.

channel, we can turn to the ideas from Section 9.3.6 for analyzing the performance under non-Gaussian noise. Note that by Theorem 7.8.1, we can obtain a sequence of good fine lattices in the sense that the noise-matched NVNR tends to $2\pi e$.

Loss in single-letter characterization It does not seem that a random i.i.d. ensemble can attain similar performance for computation over a multiple-access channel. That is, standard approaches to derive single-letter expressions seem implicitly to decode both messages. Nevertheless, for the Gaussian two-way relay channel, it is possible to match nearly the lattice-based performance of compute-and-forward via a random (non-lattice) i.i.d. coding. Specifically, a more sophisticated version of the compress-and-forward relaying strategy, known as *noisy network coding* [159], comes within 1 bit (per user) of the capacity region. This rate loss is partly due to the fact that this strategy does not attempt to remove noise at the relay (i.e., it does not decode a linear combination), which may carry a larger cost in relay networks with many hops.

12.3.3   Gaussian case, multiple receivers

We now turn to consider a broader class of relay networks. In Section 12.3.2, we introduced the idea of having the relay decode the sum of the transmitted codewords, rather than the codewords themselves. Here, we expand the class of decodable functions to include *integer linear combinations of codewords*.

    To illustrate the key principles at work, consider the relay network depicted in Figure 12.17 which consists of K transmitters, K relays, and a single destination. The ℓth transmitter generates a channel input X_ℓ which is constrained to have power less than or equal to P. The kth relay observes a noisy linear combination of all transmitted signals,

$$Y_k = \sum_{\ell=1}^{K} h_{k\ell} X_\ell + Z_k \tag{12.76}$$

where the $h_{k\ell}$ are real-valued channel coefficients and the Z_k are i.i.d. Gaussian noise of variance N. Each relay has access to a noise-free bit-pipe that can carry R_0 bits per channel use to the destination, which must recover all of the transmitted messages.

One possible strategy would be to designate each relay responsible for the message from a particular transmitter. The relay would then aim to decode this transmitter's codeword while ignoring the rest (although it might use knowledge of their codebooks in its decoding scheme). This decode-and-forward strategy is *interference limited* as each relay must overcome $K - 1$ interfering signals to decode its targeted message. Of course, the relays are not required to decode anything: they could simply quantize their channel observations and send them to the destination for decoding. While this compress-and-forward strategy alleviates the problem of interference, it is *noise limited* as the relays end up wasting some of their rate on quantizing the noise. While this may not seem to be an issue in the high SNR regime, it will be exacerbated in relay networks with multiple hops between the transmitters and the destination(s).

Computation over interfering channels

Ideally, we would like the relays to "denoise" their observed linear combinations and let the destination disentangle the interference between users. It turns out that some version of this compute-and-forward strategy is possible via lattice coding. Specifically, consider the following lattice encoder (for each transmitter) and decoder (for each relay):

Encoding $\mathbf{x}_\ell = [\mathbf{v}_\ell + \mathbf{u}_\ell] \bmod_{\mathcal{V}_0} \Lambda_2$ $\tag{12.77a}$

Decoding $\hat{\mathbf{t}}_k = \left[\arg\min_{\lambda \in \Lambda_1} \left\| \alpha_k \mathbf{Y}_k - \sum_{\ell=1}^{K} a_{k\ell} \mathbf{u}_\ell - \lambda \right\| \right] \bmod \Lambda_2$ $\tag{12.77b}$

where the $\mathbf{v}_\ell \in \mathcal{C}_{\Lambda_1, \mathcal{P}_0}$ are message (coset) representatives, the $\mathbf{u}_\ell$ are independent dithers, and $\hat{\mathbf{t}}_k$ is the decoded estimate of the integer linear combination

$$\mathbf{t}_k = \left[\sum_{\ell=1}^{K} a_{k\ell} \mathbf{v}_\ell \right] \bmod \Lambda_2, \tag{12.78}$$

where the $a_{k\ell}$ are integer-valued [13] coefficients that can be freely chosen by each receiver. These coefficients should be chosen to ensure that the destination can

---

[13] The constraint that the $a_{k\ell}$ are integer valued ensures that $\mathbf{t}_k$ is an element of the nested lattice codebook $\mathcal{C}_{\Lambda_1, \mathcal{P}_0}$.

recover the messages, i.e., the matrix $\mathbf{A} = \{a_{k\ell}\}$ should be full rank. As we will see, the effective noise encountered in decoding is governed by how well the integer-valued $a_{k\ell}$ approximate the real-valued $h_{k\ell}$. In general, the channel will not be "perfectly matched" to any integer linear combination, resulting in a rate loss. This resembles the phenomenon encountered in the Gaussian many-help-one source coding problem when the desired function is not well matched to the lattice structure, i.e., it has non-integer coefficients (see Section 12.1.3).

The lemma below characterizes the equivalent channel to each relay.

Lemma 12.3.2 (Equivalent modulo-Λ channels) *Consider the Gaussian relay network* (12.76) *and the lattice coding scheme* (12.77) *with arbitrary input vectors* $\mathbf{v}_1, \ldots, \mathbf{v}_K$. *For independent uniform (or modulo-uniform) dithers* $\mathbf{U}_1, \ldots, \mathbf{U}_K$ *with respect to* Λ_2, *the channels from* $\mathbf{v}_1, \ldots, \mathbf{v}_K$ *to the decision vectors*

$$\tilde{\mathbf{Y}}_k = \left[\alpha_k \mathbf{Y}_k - \sum_{\ell=1}^{K} a_{k\ell} \mathbf{U}_\ell \right] \bmod \Lambda_2, \tag{12.79}$$

$k = 1, \ldots, K$, *are equivalent to the modulo-Λ_2 channels*

$$\tilde{\mathbf{Y}}_k = [\mathbf{t}_k + \mathbf{Z}_{\mathrm{eq},k}] \bmod \Lambda_2, \tag{12.80}$$

where $\mathbf{t}_k$ *is the integer linear combination of the inputs defined in* (12.78)*, and the equivalent noise is*

$$\mathbf{Z}_{\mathrm{eq},k} = \underbrace{\left[\alpha_k \mathbf{Z}_k + \sum_{\ell=1}^{K} (\alpha_k h_{k\ell} - a_{k\ell}) \mathbf{U}_{\mathrm{eq},\ell} \right]}_{\mathbf{Z}_{\mathrm{mix},k}} \bmod \Lambda_2 \tag{12.81}$$

where the $\mathbf{U}_{\mathrm{eq},\ell}$ *are equivalent dithers* (4.8)*, which are uniform over* $\mathcal{V}_0(\Lambda_2)$ *and independent of each other and the* $\mathbf{t}_k$.

Proof See Problem 12.5. □

In the next theorem, we will establish achievable rates for the compute-and-forward strategy. First, it is useful to define the *computation rate*

$$\mathcal{R}_{\mathrm{comp}}(\mathbf{h}_k, \mathbf{a}_k) = \frac{1}{2} \log^+ \left(\frac{N + P \|\mathbf{h}_k\|^2}{N \|\mathbf{a}_k\|^2 + P \left(\|\mathbf{h}_k\|^2 \|\mathbf{a}_k\|^2 - (\mathbf{h}_k^T \mathbf{a}_k)^2 \right)} \right) \tag{12.82}$$

where $\mathbf{h}_k^T = [h_{k1} \cdots h_{kK}]$ represents the channel vector to the kth relay and $\mathbf{a}_k^T = [a_{k1} \cdots a_{kK}]$ represents the vector of integer-valued equation coefficients desired by the kth relay. At a high level, the theorem states that a relay can reliably decode any [14] integer linear combination whose coefficients $\mathbf{a}_k$ yield a computation rate that exceeds the coding rate $R(\Lambda_1/\Lambda_2)$. Notice that the transmitters only emit codewords

---

[14] In fact, if several linear combinations satisfy this condition, the relay can decode all of them concurrently.

at a fixed coding rate and are oblivious to the channel and equation coefficients: the choice of equation coefficients is determined completely by the relays.

Theorem 12.3.2 (System performance) *The coding rate* $R(\Lambda_1/\Lambda_2) = (1/n)\log|\Lambda_1/\Lambda_2|$ *of the lattice compute-and-forward scheme* (12.77), *with random dithers and Wiener estimation coefficients*

$$\alpha_k^* = \frac{P\,\mathbf{a}_k^T\mathbf{h}_k}{N + P\|\mathbf{h}_k\|^2}, \tag{12.83}$$

satisfies

$$R(\Lambda_1/\Lambda_2) \geq \min_k[\mathcal{R}_{\mathrm{comp}}(\mathbf{h}_k, \mathbf{a}_k) - \mathcal{L}(\mathbf{Z}_{\mathrm{mix},k})]^+ \tag{12.84}$$

where $\mathbf{Z}_{\mathrm{mix},k}$ *is the kth estimation error in the modulo-*Λ *channel* (12.81) *and* $\mathcal{L}(\mathbf{Z}_{\mathrm{mix},k})$ *is the rate loss from* (12.46). *For nested lattice pairs that are good for Voronoi modulation over an AWGN channel, the scheme achieves* $R = \min_k \mathcal{R}_{\mathrm{comp}}(\mathbf{h}_k, \mathbf{a}_k)$ *asymptotically in n.*

Proof The second moment of the kth estimation error from (12.81) is equal to

$$\frac{1}{n}E\|\mathbf{Z}_{\mathrm{mix},k}\|^2 = \frac{1}{n}E\left\|\alpha_k\mathbf{Z}_k + \sum_{\ell=1}^{K}(\alpha_k h_{k\ell} - a_{k\ell})\mathbf{U}_{\mathrm{eq},\ell}\right\|^2 \tag{12.85a}$$

$$= \frac{1}{n}\alpha_k^2 E\|\mathbf{Z}_k\|^2 + \frac{1}{n}\sum_{\ell=1}^{K}(\alpha_k h_{k\ell} - a_{k\ell})^2 E\|\mathbf{U}_{\mathrm{eq},\ell}\|^2 \tag{12.85b}$$

$$= \alpha_k^2 N + P\|\alpha_k\mathbf{h}_k - \mathbf{a}_k\|^2. \tag{12.85c}$$

This quantity is minimized by the Wiener coefficient $\alpha_k^* = P\mathbf{a}_k^T\mathbf{h}_k/(N + P\|\mathbf{h}_k\|^2)$ to yield $(1/n)E\|\mathbf{Z}_{\mathrm{mix},k}\|^2 = P\|\mathbf{a}_k\|^2 - P^2(\mathbf{a}_k^T\mathbf{h}_k)^2/(N + P\|\mathbf{h}_k\|^2)$. From here, (12.84) follows from Lemma 9.6.1 by noting that

$$\frac{1}{2}\log\left(\frac{P}{\frac{1}{n}E\|\mathbf{Z}_{\mathrm{mix},k}\|^2}\right) = \frac{1}{2}\log\left(\frac{P}{P\|\mathbf{a}_k\|^2 - P^2\frac{(\mathbf{a}_k^T\mathbf{h}_k)^2}{N+P\|\mathbf{h}_k\|^2}}\right) = \mathcal{R}_{\mathrm{comp}}(\mathbf{h}_k, \mathbf{a}_k). \tag{12.86}$$

Note that the minimization with respect to k is to ensure that all relays can decode their linear combinations. Following the proof of Theorem 12.3.1, it can be argued that the equivalent noises are semi-spherical and thus good Voronoi codes exist that can achieve $\min_k \mathcal{R}_{\mathrm{comp}}(\mathbf{h}_k, \mathbf{a}_k)$ asymptotically in n. □

Linear combinations over a finite field

Interestingly, it is possible to make an explicit connection between this lattice-based strategy and linear network coding over a prime-sized finite field. Recall from Section 8.5.2 that the Voronoi codebook $\mathcal{C}_{\Lambda_1,\mathcal{P}_0}$ is generated from a pair of nested linear codes using construction A. In fact, the Voronoi codebook is isomorphic to a linear code over a prime-sized finite field $\mathbb{F}_p$. It can be shown that there exists a function $\phi : \mathbb{F}_p^k \to \mathcal{C}_{\Lambda_1, \mathcal{P}_0}$ that maps finite field messages $\mathbf{w}_\ell$ to lattice message representatives $\mathbf{v}_\ell = \phi(\mathbf{w}_\ell)$ such that

$$\phi^{-1}\left(\left[\sum_{\ell=1}^{K} a_{k\ell}\mathbf{v}_\ell\right] \bmod \Lambda_2\right) = \bigoplus_{\ell=1}^{K} q_{k\ell}\mathbf{w}_\ell \tag{12.87}$$

where $q_{k\ell}$ is the finite field element corresponding to $[a_{k\ell}] \bmod p$ under a natural mapping. In other words, this mapping preserves linearity. The function $\phi(\cdot)$ can be derived from nested construction A (Section 8.3), and is related to *coset enumeration* by *diagonal nesting* (Section 8.2). See [78] and [196] for more details.

Given this mapping and the fact that the underlying prime p tends to infinity asymptotically in n, it can be argued that the original message representatives $\mathbf{v}_1, \ldots, \mathbf{v}_K$ can be recovered from their integer linear combinations $\mathbf{t}_1, \ldots, \mathbf{t}_K$ provided that the matrix $\mathbf{A} = \{a_{k\ell}\}$ is full rank. Turning back to the relay network of Figure 12.17, we find that each relay can convey its estimated function $\hat{\mathbf{t}}_k$ to the destination if the rate of the Voronoi codebook does not exceed R_0 and the destination can decode if $\mathbf{A}$ is full rank over the reals. Overall, we find that the following coding rate (per user) is achievable via compute-and-forward (asymptotically in n),

$$R(\Lambda_1/\Lambda_2) = \min\left(\left(\max_{\substack{\mathbf{A} \in \mathbb{Z}^{K \times K} \\ \text{rank}(\mathbf{A})=K}} \min_{k=1,\ldots,K} \mathcal{R}_{\text{comp}}(\mathbf{h}_k, \mathbf{a}_k)\right), R_0\right). \tag{12.88}$$

Finding the best integers

We now examine the issue of finding the optimal integer coefficients for a single relay. As shown in Theorem 12.3.2, the achievable rate is determined by how well the integer coefficients approximate the real-valued channel gains. It turns out that determining the quality of this integer approximation is closely linked to the problem of *Diophantine approximation*, which, at a high level, studies how well the reals can be approximated with the rationals. See [201] for more details. Below, we discuss simple algorithms for selecting the integers.

As a starting point, consider an exhaustive search strategy, in which the relay plugs in all viable integer vectors into (12.82) to find the coefficients that yield the highest computation rate. From (12.82), it can be seen that only integer vectors satisfying

$$\|\mathbf{a}_k\|^2 \leq 1 + \frac{P\|\mathbf{h}_k\|^2}{N} \tag{12.89}$$

will result in a positive computation rate, i.e., it suffices to search over a bounded set.

To develop a deeper understanding of this issue, the following equivalent form of the computation rate is quite useful:

$$\mathcal{R}_{\text{comp}}(\mathbf{h}_k, \mathbf{a}_k) = \left[\frac{1}{2} \log \left(\frac{P}{\|\mathbf{F}_k \mathbf{a}_k\|^2} \right) \right]^+ \tag{12.90a}$$

$$\mathbf{F}_k = \left(\frac{1}{P} \mathbf{I} + \frac{1}{N} \mathbf{h}_k \mathbf{h}_k^T \right)^{-1/2}. \tag{12.90b}$$

From here, we observe that the search for the integer vector $\mathbf{a}_k$ that maximizes $\mathcal{R}_{\text{comp}}(\mathbf{h}_k, \mathbf{a}_k)$ is equivalent to the search for the shortest vector in the lattice $\mathbf{F}_k \mathbb{Z}^K$. Although finding this vector is challenging (in terms of computational complexity), efficient approximation algorithms, such as the LLL algorithm [154], can be used to find near-optimal solutions.

12.3.4   Integer-forcing MIMO equalization

So far, we have examined network scenarios where lattice codes can attain performance beyond what has been available for classical random i.i.d. codes. Here, we investigate a scenario for which the capacity can be attained via a random i.i.d. ensemble, and lattice codes can be used to reduce the decoding complexity significantly. Specifically, take a multiple-input multiple-output (MIMO) channel with a single transmitter and receiver, each equipped with K antennas. The channel output can be written as

$$\mathbf{Y} = \mathbf{HX} + \mathbf{Z} \tag{12.91}$$

where $\mathbf{H} \in \mathbb{R}^{K \times K}$ is the channel matrix, $\mathbf{X} \in \mathbb{R}^{K \times n}$ is the channel input, and the noise $\mathbf{Z} \in \mathbb{R}^{K \times n}$ is element-wise i.i.d. $\mathcal{N}(0, N)$. We will assume that each transmit antenna encodes an independent message and that the associated (independent) codewords $\mathbf{x}_1, \ldots, \mathbf{x}_K$ make up the rows of the channel input, $\mathbf{X} = [\mathbf{x}_1 \cdots \mathbf{x}_K]^T$.

The capacity of this point-to-point MIMO channel can be attained by drawing each codeword from an i.i.d. Gaussian codebook and employing a maximum likelihood (or joint-typicality) decoder at the receiver. Note that since we have restricted each transmit antenna to encoding independent information, the capacity is

$$R_{\text{MIMO}}(\mathbf{H}) = \min_{\mathcal{S} \subseteq \{1, \ldots, K\}} \frac{K}{2|\mathcal{S}|} \log \det \left(\mathbf{I} + \frac{P}{N} \mathbf{H}_{\mathcal{S}} \mathbf{H}_{\mathcal{S}}^T \right) \tag{12.92}$$

where $\mathbf{H}_{\mathcal{S}}$ denotes the submatrix of $\mathbf{H}$ consisting of the columns with indices in the subset $\mathcal{S}$. (Without this restriction, the capacity is simply $\frac{1}{2} \log \det(\mathbf{I} + \frac{P}{N} \mathbf{HH}^T)$.)

The issue with this capacity-achieving approach is that the complexity of the maximum likelihood decoder scales rapidly with the number of antennas. Therefore, it is often of interest to find a "single-user" decoding approach that decomposes the problem into K effective point-to-point channels for which single-antenna decoding suffices. The conventional approach, known as *zero-forcing*, tries to isolate the codewords from each other via a linear projection,

$$\tilde{\mathbf{Y}} = \mathbf{BY} = \mathbf{BHX} + \mathbf{BZ}, \tag{12.93}$$

and then tries to decode the individual codewords from the rows of $\tilde{\mathbf{Y}}$. If the channel matrix is full rank, setting $\mathbf{B} = \mathbf{H}^{-1}$ will completely eliminate interference between codewords. Unfortunately, this comes at the cost of noise amplification. Even the optimal choice of LMMSE projection, $\mathbf{B} = P\mathbf{H}^T(N\mathbf{I} + P\mathbf{HH}^T)^{-1}$, does not fully close the gap to capacity (unless combined with successive decoding). Furthermore, both methods have the unintended effect of spreading the noise power unequally across the resulting point-to-point channels. If the transmitter does not know the channel state, then it cannot counter this effect through rate allocation across its antennas, resulting in a significant rate loss.

Most of this rate loss can be eliminated via a twist on conventional single-user decoding known as *integer-forcing*. Instead of trying to decode the codewords directly, the receiver first recovers integer linear combinations, and only afterwards solves these for its desired messages. The linear projection (12.93) can be used to induce an effective integer-valued channel matrix $\mathbf{A} \in \mathbb{Z}^{K \times K}$. For instance, if the channel matrix is full rank, setting $\mathbf{B} = \mathbf{AH}^{-1}$ will perfectly match the channel to the desired equations, although the optimal choice is again the LMMSE projection $\mathbf{B} = P\mathbf{AH}^T(N\mathbf{I} + P\mathbf{HH}^T)^{-1}$. Overall, this closely resembles the compute-and-forward scheme (12.77), except that the receiver has full access to all channel observations. We summarize the encoding and decoding operations below:

Encoding $\quad \mathbf{x}_\ell = [\mathbf{v}_\ell + \mathbf{u}_\ell] \bmod_{\nu_0} \Lambda_2 \tag{12.94a}$

Decoding $\quad \hat{\mathbf{t}}_k = \left[\arg\min_{\lambda \in \Lambda_1} \left\| \mathbf{b}_k^T \mathbf{Y} - \sum_{\ell=1}^{K} a_{k\ell} \mathbf{u}_\ell - \lambda \right\| \right] \bmod \Lambda_2 \tag{12.94b}$

$$\hat{\mathbf{v}}_k = \left[\sum_{\ell=1}^{K} a_{k\ell}^{\text{inv}} \hat{\mathbf{t}}_\ell \right] \bmod \Lambda_2 \tag{12.94c}$$

where the $\mathbf{v}_\ell \in \mathcal{C}_{\Lambda_1, \mathcal{P}_0}$ are message (coset) representatives, the $\mathbf{u}_\ell$ are independent dithers, the $\hat{\mathbf{t}}_k$ are the decoded estimates of the integer linear combinations $\mathbf{t}_k = \left[\sum_{\ell=1}^{K} a_{k\ell} \mathbf{v}_\ell \right] \bmod \Lambda_2$ for some $a_{k\ell} \in \mathbb{Z}$, the $\mathbf{b}_k^T$ are the rows of the projection matrix $\mathbf{B}$, and the $a_{k\ell}^{\text{inv}}$ are the elements of an integer-valued matrix $\mathbf{A}^{\text{inv}}$ satisfying $[\mathbf{A}^{\text{inv}} \mathbf{A}] \bmod p = \mathbf{I}$ with respect to the matrix $\mathbf{A} = \{a_{k\ell}\}$.

It can be shown that if the coding rate $R(\Lambda_1/\Lambda_2)$ is less than the MIMO computation rate

$$\mathcal{R}_{\text{comp}}(\mathbf{H}, \mathbf{A}) = \min_{k=1,\dots,K} \left[\frac{1}{2} \log\left(\frac{P}{\|\mathbf{Fa}_k\|^2} \right) \right]^+ \tag{12.95a}$$

$$\mathbf{F} = \left(\frac{1}{P}\mathbf{I} + \frac{1}{N}\mathbf{H}^T\mathbf{H} \right)^{-1/2} \tag{12.95b}$$

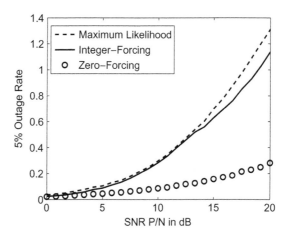

Figure 12.18 Comparison of 5% outage rates for maximum likelihood, integer-forcing, and zero-forcing decoding over a MIMO channel with two transmit and two receive antennas and i.i.d. $\mathcal{N}(0, 1)$ fading.

then the receiver can decode all desired integer linear combinations with vanishing error probability in n (for good Voronoi codes). It follows that, after optimizing over $\mathbf{A} \in \mathbb{Z}^{K \times K}$, the following rate is achievable via integer-forcing,

$$R_{\text{IF}}(\mathbf{H}) = K \cdot \max_{\substack{\mathbf{A} \in \mathbb{Z}^{K \times K} \\ \text{rank}(\mathbf{A})=K}} \mathcal{R}_{\text{comp}}(\mathbf{H}, \mathbf{A}). \tag{12.96}$$

This strictly improves on the rate available via zero-forcing, which can be expressed as

$$R_{\text{ZF}}(\mathbf{H}) = K \cdot \mathcal{R}_{\text{comp}}(\mathbf{H}, \mathbf{I}). \tag{12.97}$$

That is, zero-forcing can be viewed as a special case of integer-forcing that always sets $\mathbf{A} = \mathbf{I}$.

The optimization over the integer matrix $\mathbf{A}$ in (12.96) is closely related to the problem of finding a *good basis* for a lattice discussed in Section 2.1.1. In particular, near-optimal solutions can be obtained by applying the LLL algorithm to the lattice $\Lambda(\mathbf{F})$. (The difference between this optimization and finding a good basis is that we do not restrict $\mathbf{A}$ to be unimodular; see [294].)

To compare these approaches, we consider, as an example, a MIMO channel with $K = 2$ transmit and receive antennas. The channel matrix $\mathbf{H}$ is generated element-wise i.i.d. $\mathcal{N}(0, 1)$ and is only revealed to the receiver. To deal with the fact that the transmitter lacks channel state information and cannot set its coding rate appropriately, we allow for a $1/20$ probability of outage. For example, for the integer-forcing approach, we set the coding rate so that

$$\Pr\{K \cdot R(\Lambda_1/\Lambda_2) \geq R_{\text{IF}}(\mathbf{H})\} = \frac{1}{20}, \tag{12.98}$$

meaning that with probability $19/20$, decoding is successful. In Figure 12.18, we have plotted the resulting 5% outage rates $K \cdot R(\Lambda_1/\Lambda_2)$. Notice that integer-forcing

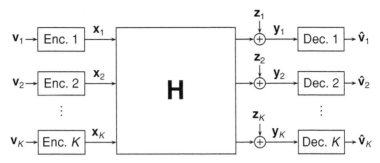

Figure 12.19 K-user Gaussian interference channel. There are K transmitter-receiver pairs that share a common wireless channel.

nearly matches the performance of maximum likelihood decoding, whereas zero-forcing falls far short of it. It can also be shown that integer-forcing outperforms successive cancelation decoding (i.e., after recovering a codeword, the receiver is able to subtract its contribution to the effective noise encountered in decoding subsequent codewords). See [294] for more details.

12.4 Interference alignment

We now explore the phenomenon of interference alignment, which is one of the most surprising consequences of structured signaling. To demonstrate its potential, we consider K transmitters that share a common wireless channel. Each transmitter wishes to communicate with a unique receiver. From the perspective of each receiver, one of the transmitters emits a useful signal while the other $K - 1$ transmitters emit interference. The goal is to design a scheme that maximizes the total throughput of this network, known as the K-user interference channel. See Figure 12.19 for an illustration of the Gaussian case.

At a first glance, this problem may seem simple. That is, the transmitters can eliminate the interference issue via time or frequency division (or some other form of orthogonalization). For example, by dividing up time slots equally, each transmitter can attain $1/K$ its interference-free throughput (i.e., the throughput that would be available in the absence of other users). This corresponds to the throughput scaling encountered in a K-user Gaussian multiple-access or broadcast channel. As it turns out, this is quite far from the optimal throughput scaling, which enables each user to attain $1/2$ its interference-free throughput *regardless of the total number of users*.

This striking behavior is due to the application of interference alignment. At a high level, the aim of this technique is to make it appear to each receiver that its observed signal comprises only its desired signal and a single effective interfering signal. Inducing this effect is the main challenge underlying interference alignment. Initial work demonstrated the feasibility of alignment for time-varying channels in the high SNR regime through the selection of appropriate beamforming vectors [26].

Subsequent work has focused on uncovering settings in which alignment is feasible as well as novel techniques for inducing it.

We now take a closer look at a specific alignment scheme for time-varying channels. Let us say that, at some time t_1, the kth transmitter emits a symbol X_k with power P and that the kth receiver observes

$$Y_k[t_1] = h_{kk}[t_1]X_k + \sum_{\ell \neq k} h_{k\ell}[t_1]X_\ell + Z_k[t_1] \qquad (12.99)$$

where $h_{k\ell}[t_1]$ are the channel gains and $Z_k[t_1]$ is independent noise of variance N. Now, assume that, at some later time t_2, the channel gains from each transmitter to its intended receiver are identical, $h_{kk}[t_2] = h_{kk}[t_1]$, and the channel gains to all unintended receivers are flipped, $h_{k\ell}[t_2] = -h_{kk}[t_1]$ for all $\ell \neq k$. If this is indeed the case and this fact is known to all transmitters and receivers, the following strategy induces alignment. Each transmitter repeats its symbol X_k from time t_1 so that each receiver observes

$$Y_k[t_2] = h_{kk}[t_1]X_k - \sum_{\ell \neq k} h_{k\ell}[t_1]X_\ell + Z_k[t_2] \qquad (12.100)$$

where $Z_k[t_2]$ is again independent noise of variance N. Each receiver can now simply combine its observations from time t_1 and t_2 to obtain an interference-free look at its desired symbol

$$Y_k[t_1] + Y_k[t_2] = 2h_{kk}[t_1]X_k + Z_k[t_1] + Z_k[t_2]. \qquad (12.101)$$

The simple idea described above can be developed into a full-fledged communication strategy by pairing up channel gains that are "close enough" rather than waiting indefinitely for a perfect match. It can be shown that if the time sequences of channel phases are independent of one another, marginally uniform, and ergodic, then this strategy, known as ergodic alignment, permits each user to operate at $1/2$ its interference-free throughput *at any SNR*. See [197] for more details.

Although this scheme attains a very high throughput, it comes at the cost of extremely high latency or delay. Specifically, the expected delay between channel pairings scales at least like $(KP/N)^{K^2}$ since K^2 channel gains must be matched to within precision $N/(KP)$ in order to remove most of the effects of the interference. This leaves us with the following question: is alignment possible with low latency? Going further, is it possible to induce alignment when the channel gains are *static*? It is here that lattice codes enter the picture.

The capacity region of the static K-user Gaussian interference channel lies beyond our current understanding. Below, we discuss a class of symmetric channels where alignment is easy to induce in order to gain a better understanding of the problem in the finite SNR regime. For a broader view of the interference alignment phenomenon, see [127].

12.4.1   Symmetric K-user Gaussian interference channel

Consider a symmetric interference channel in which each receiver sees a noisy version of its desired signal plus the sum of the interfering signal scaled by g,

$$Y_k = X_k + g \sum_{\ell \neq k} X_\ell + Z_k. \tag{12.102}$$

As usual, we assume that X_k is power constrained to P and Z_k is i.i.d. Gaussian with variance N.

If all of the codewords are drawn from the same lattice codebook, then the sum of the interfering codewords will itself be a codeword. Thus, each receiver will see an effective two-user multiple-access channel,

$$Y_k = X_k + g X_{\text{int},k} + Z_k \tag{12.103}$$

where $X_{\text{int},k} = \sum_{\ell \neq k} X_\ell$. It might seem that the problem is now easy to solve as the capacity region of the multiple-access channel is well understood. However, the fact that the transmitters employ the same lattice codebook introduces additional complications.

To understand why this is the case, consider two transmitters that employ the same PAM constellations $X_1, X_2 \in \{0, 1, 2, \ldots, q - 1\}$, which can be viewed as a simple Voronoi code. The sum $X_1 + X_2$ takes values over $2q - 1$ points, which is not enough to distinguish the q^2 possible (X_1, X_2) pairs. However, the linear combination $X_1 + \sqrt{2}X_2$ does take values over q^2 points, meaning that it is possible to recover X_1 and X_2 for sufficiently large SNR. Turning back to the symmetric interference channel, it follows that the receiver will have much more difficulty discerning its desired codeword from the interference when $g = 1$ than when $g = \sqrt{2}$. This is because when $g = 1$, the desired codeword is aligned with the interfering ones. In fact, in this case the channel is equivalent to a compound multiple-access channel for which the sum capacity is only $\frac{1}{2} \log(1 + KP/N)$.

It can be argued that the sensitivity of the rate to the channel gains is a fundamental property of the interference channel with $K \geq 3$ users, owing to the possibility of alignment. For instance, the number of degrees of freedom is $K/2$ for irrational g and strictly less for rational g [73, 190]. From an engineering perspective, extreme sensitivity to the channel parameters is quite undesirable. The lattice coding approaches explored below will enable us to develop a better understanding of this phenomenon in the finite SNR regime.

12.4.2   The very strong regime and successive cancelation

If the interference is much stronger than the desired signal, a natural decoding strategy is first to decode the sum of the interfering codewords while treating the desired codeword as noise. Then, the interference can be canceled to yield an interference-free effective channel. This completely circumvents the issue of alignment between the desired and interfering signals.

The compute-and-forward strategy from Section 12.3.3 can be used at each receiver to decode the modulo sum of the interfering lattice codewords. However, in order to cancel the interfering signals, we need to recover the real sum of the dithered codewords $\mathbf{x}_{\text{int},k}$. This is possible via an additional decoding step that estimates the closest coarse lattice point to $\mathbf{x}_{\text{int},k}$.

We focus on the case of symmetric [15] rates $R_1 = \cdots = R_K = R_{\text{SYM}}$. The following encoding and decoding operations encapsulate the successive interference cancelation strategy for lattice-based alignment:

$$\textbf{Encoding} \quad \mathbf{x}_\ell = [\mathbf{v}_\ell + \mathbf{u}_\ell] \bmod_{\mathcal{V}_0} \Lambda_2 \tag{12.104a}$$

$$\textbf{Decoding} \quad \hat{\mathbf{t}}_{\text{int},k} = \left[\underset{\lambda \in \Lambda_1}{\arg\min} \left\| \alpha \mathbf{Y}_k - \sum_{\ell \neq k} \mathbf{u}_\ell - \lambda \right\| \right] \bmod \Lambda_2 \tag{12.104b}$$

$$\hat{\mathbf{r}}_{\text{int},k} = \left[\hat{\mathbf{t}}_{\text{int},k} + \sum_{\ell \neq k} \mathbf{u}_\ell \right] \bmod_{\mathcal{V}_0} \Lambda_2 \tag{12.104c}$$

$$\hat{\mathbf{x}}_{\text{int},k} = Q_{\Lambda_2}(\alpha \mathbf{Y}_k - \hat{\mathbf{r}}_{\text{int},k}) + \hat{\mathbf{r}}_{\text{int},k} \tag{12.104d}$$

$$\hat{\mathbf{v}}_k = \left[\underset{\lambda \in \Lambda_1}{\arg\min} \left\| \beta \mathbf{Y}_k - \beta g \hat{\mathbf{x}}_{\text{int},k} - \mathbf{u}_k - \lambda \right\| \right] \bmod \Lambda_2 \tag{12.104e}$$

where the $\mathbf{v}_\ell \in \mathcal{C}_{\Lambda_1, \mathcal{P}_0}$ are message (coset) representatives, the $\mathbf{u}_\ell$ are independent dithers, $\hat{\mathbf{t}}_{\text{int},k}$ is the decoded estimate of the modulo sum of the message representatives $\mathbf{t}_{\text{int},k} = \left[\sum_{\ell \neq k} \mathbf{v}_k \right] \bmod \Lambda_2$, $\hat{\mathbf{r}}_{\text{int},k}$ is the decoded estimate of the modulo sum of the dithered codewords $[\mathbf{x}_{\text{int},k}] \bmod_{\mathcal{V}_0} \Lambda_2$, $\hat{\mathbf{x}}_{\text{int},k}$ is the decoded estimate of the real sum of the dithered codewords $\mathbf{x}_{\text{int},k}$, and $\hat{\mathbf{v}}_k$ is the decoded estimate of the desired message.

Theorem 12.4.1 (System performance) *For nested lattice pairs that are good for Voronoi modulation over an AWGN channel, the coding rate $R_{\text{sym}} = (1/n) \log |\Lambda_1/\Lambda_2|$ of the lattice interference-alignment scheme* (12.104) *with Wiener estimation coefficients*

$$\alpha^* = \frac{P(K-1)g}{N + P(1 + (K-1)g^2)} \qquad \beta^* = \frac{P}{N+P} \tag{12.105}$$

achieves

$$R_{\text{SYM}} = \min\left(\frac{1}{2} \log\left(\frac{1}{K-1} + \frac{g^2 P}{N+P} \right), \frac{1}{2} \log\left(1 + \frac{P}{N} \right) \right) \tag{12.106}$$

asymptotically in n.

---

[15] It can be argued that the sum capacity of the symmetric interference channel is equal to K times the symmetric capacity.

Proof The decoded estimate $\hat{\mathbf{t}}_{\text{int},k}$ is identical to the compute-and-forward decoding operation for a linear combination with integer coefficient vector

$$\mathbf{a}_k^T = [\underbrace{1 \cdots 1}_{k-1 \text{ users}} \, 0 \, \underbrace{1 \cdots 1}_{K-k \text{ users}}] \tag{12.107}$$

over the channel vector

$$\mathbf{h}_k^T = [\underbrace{g \cdots g}_{k-1 \text{ users}} \, 1 \, \underbrace{g \cdots g}_{K-k \text{ users}}]. \tag{12.108}$$

It follows from Theorem 12.3.2 that with Wiener coefficient α^* (12.83), each receiver can reliably estimate its modulo sum $\mathbf{t}_{\text{int},k}$ at achievable rate (12.106). For the remainder of the proof, we condition on the event that $\hat{\mathbf{t}}_{\text{int},k} = \mathbf{t}_{\text{int},k}$. This implies that $\hat{\mathbf{r}}_{\text{int},k} = [\mathbf{x}_{\text{int},k}] \bmod_{\mathcal{V}_0} \Lambda_2$.

Using the fact that $[\mathbf{x}_{\text{int},k}] \bmod_{\mathcal{V}_0} \Lambda_2 = \mathbf{x}_{\text{int},k} - Q_{\Lambda_2}(\mathbf{x}_{\text{int},k})$, the term inside the quantizer in (12.104d) can be written as

$$\alpha \mathbf{Y}_k - \hat{\mathbf{r}}_{\text{int},k} = Q_{\Lambda_2}(\mathbf{x}_{\text{int},k}) + \alpha \mathbf{x}_k + (\alpha g - 1)\mathbf{x}_{\text{int},k} + \alpha \mathbf{z}_k. \tag{12.109}$$

By assumption, R_{SYM} is positive, which via (12.106) implies that $\frac{1}{n} E \|\alpha \mathbf{x}_k + (\alpha g - 1)\mathbf{x}_{\text{int},k} + \alpha \mathbf{z}_k\|^2 < P$. Since Λ_2 is a good quantizer, we have that $Q_{\Lambda_2}(\alpha \mathbf{Y}_k - \hat{\mathbf{r}}_{\text{int},k}) = Q_{\Lambda_2}(\mathbf{x}_{\text{int},k})$ and $\hat{\mathbf{x}}_{\text{int},k} = \mathbf{x}_{\text{int},k}$ with high probability (asymptotically in n).

Finally, the receiver cancels the interference by subtracting $g \hat{\mathbf{x}}_{\text{int},k}$ from $\mathbf{Y}_k$. Conditioned on correct decoding, the receiver now has a point-to-point AWGN channel between $\mathbf{x}_k$ and $\mathbf{Y}_k - g \hat{\mathbf{x}}_{\text{int},k}$. It follows from Theorem 9.6.2 that reliable decoding is possible if $R_{\text{SYM}} < \frac{1}{2} \log(1 + P/N)$. $\qquad\square$

Clearly, the symmetric capacity is upper bounded by $\frac{1}{2} \log(1 + P/N)$. Therefore, in the regime where

$$\frac{1}{2} \log \left(\frac{1}{K-1} + \frac{g^2 P}{N + P} \right) > \frac{1}{2} \log \left(1 + \frac{P}{N} \right) \tag{12.110}$$

we have determined the symmetric capacity. The *very strong interference regime* is usually taken to mean the regime where successive interference cancelation suffices to achieve the (interference-free) capacity $\frac{1}{2} \log(1 + P/N)$. From (12.110), this certainly includes all channel gains satisfying

$$g^2 \geq \frac{(N + P)^2}{NP} - \frac{N + P}{(K - 1)P}. \tag{12.111}$$

As a comparison point, consider a scheme that employs i.i.d. Gaussian coding and decodes *all* of the interfering signals individually prior to successive cancelation. It can be shown that this only attains the interference-free capacity when

$$g^2 \geq \frac{\left(\left(1 + \frac{P}{N}\right)^{K-1} - 1 \right)(N + P)}{(K - 1)P}. \tag{12.112}$$

That is, the threshold for the "decode all" successive cancelation strategy increases exponentially with the number of users K, whereas the threshold for lattice-aligned successive cancelation decreases slightly.

12.4.3   The strong regime and joint decoding

For the two-user symmetric interference channel (i.e., $K = 2$), it is well known that successive interference cancelation is suboptimal outside the very strong regime. For instance, in the *strong regime*, $1 \leq g^2 \leq 1 + \frac{P}{N}$, both receivers should jointly decode the desired and interfering codewords. The symmetric capacity in this regime is

$$R_{\text{SYM}} = \frac{1}{4} \log \left(1 + (1 + g^2)\frac{P}{N} \right). \tag{12.113}$$

See [116, 233] for more details.

Thus, for $K > 2$ users, it seems that each transmitter should encode its message $\mathbf{v}_k$ using a lattice codebook as in (12.104a) and each receiver should jointly decode its desired message $\mathbf{v}_k$ and the sum of the interfering messages $\sum_{\ell \neq k} \mathbf{v}_\ell$. Unfortunately, a direct analysis of this strategy's achievable rates lies beyond our current understanding, owing to dependencies between competing codeword pairs. See [208] for more details. Another possibility is to decode indirectly by first *recovering two linear combinations*

$$\mathbf{t}_{k1} = \left[b_{11}\mathbf{v}_k + b_{12} \sum_{\ell \neq k} \mathbf{v}_\ell \right] \bmod \Lambda_2 \tag{12.114}$$

$$\mathbf{t}_{k2} = \left[b_{21}\mathbf{v}_k + b_{22} \sum_{\ell \neq k} \mathbf{v}_\ell \right] \bmod \Lambda_2. \tag{12.115}$$

If the integer vectors $[b_{11}\ b_{12}]$ and $[b_{21}\ b_{22}]$ are linearly independent, then each receiver can solve for its desired codeword $\mathbf{v}_k$.

From Theorem 12.3.2, it follows that, asymptotically in n, each receiver can decode its linear combinations if

$$R_{\text{SYM,align}} = \max_{\substack{\mathbf{B} \in \mathbb{Z}^{2 \times 2} \\ \text{rank}(\mathbf{B}) = 2}} \min(\mathcal{R}_{\text{comp}}(\mathbf{h}, \mathbf{a}_1), \ \mathcal{R}_{\text{comp}}(\mathbf{h}, \mathbf{a}_2)) \tag{12.116}$$

where

$$\mathbf{B} = \begin{bmatrix} b_{11} & b_{12} \\ b_{21} & b_{22} \end{bmatrix} \qquad\qquad \mathbf{h}^T = \begin{bmatrix} 1 \underbrace{g \ \cdots \ g}_{K-1 \text{ users}} \end{bmatrix} \tag{12.117}$$

$$\mathbf{a}_1^T = \begin{bmatrix} b_{11} \underbrace{b_{12} \ \cdots \ b_{12}}_{K-1 \text{ users}} \end{bmatrix} \qquad\qquad \mathbf{a}_2^T = \begin{bmatrix} b_{21} \underbrace{b_{22} \ \cdots \ b_{22}}_{K-1 \text{ users}} \end{bmatrix}. \tag{12.118}$$

In Figure 12.20, we have plotted the sum rate of this strategy for $K = 3$ users. In the very strong regime (corresponding to about 20 dB onwards), we applied the

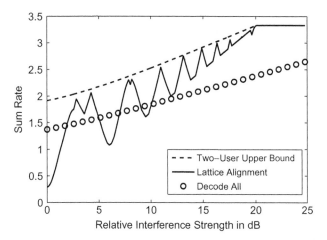

Figure 12.20 Comparison of the sum rates of the upper bound, lattice alignment, and the "decode all" strategy over the symmetric Gaussian three-user interference channel.

successive cancelation scheme from Theorem 12.4.1. For comparison, we have also plotted a "two-user" upper bound (i.e., the rate that would be achievable if each receiver had one interferer's message as side information),

$$R_{\text{SYM,upper}} = \min\left(\frac{1}{4}\log\left(1 + (1 + g^2)\frac{P}{N}\right), \frac{1}{2}\log\left(1 + \frac{P}{N}\right)\right), \quad (12.119)$$

as well as the rate for decoding all of the messages at each receiver,

$$R_{\text{SYM,all}} = \min\left(\frac{1}{6}\log\left(1 + (1 + g^2)\frac{P}{N}\right), \frac{1}{2}\log\left(1 + \frac{P}{N}\right)\right). \quad (12.120)$$

Notice that the local maxima of the lattice alignment rate nearly reach the two-user upper bound. The minima occur when the channel vector lies very close to one of the integer vectors and is thus far from the other. Since both equations must be decoded to recover the message, this leads to a decrease in performance. Interestingly, this phenomenon captures the channel sensitivity encountered earlier.

The gap between the lattice alignment rate (12.116) and the two-user upper bound (12.119) can be lower bounded analytically. In particular, for $g \in [1, \sqrt{P/N}]$, it can be shown that, for any $c \geq 0$, the lattice-alignment rate is lower bounded by

$$R_{\text{SYM,align}} \geq \frac{1}{4}\log\left(\frac{g^2 P}{N}\right) - \frac{c}{2} - 3 \quad (12.121)$$

up to an "outage set" of channel gains with measure at most 2^{-c}. The proof relies on two interesting facts. The first is that the sum of the computation rates $\mathcal{R}_{\text{comp}}(\mathbf{h}, \mathbf{a}_1) + \mathcal{R}_{\text{comp}}(\mathbf{h}, \mathbf{a}_2)$ is always at least equal to the multiple-access sum capacity minus a constant gap (assuming that $\mathbf{a}_1$ and $\mathbf{a}_2$ maximize (12.116)). The second is that the rate of the best linear combination can be upper bounded using the theory of Diophantine approximation. Combining these two facts enables us to lower bound the rate of the worst of the two linear combinations, corresponding to a lower bound on (12.116). See [209] for more details.

Asymmetric channels For general channel matrices, it is still unclear whether alignment results in throughput gains outside the very high SNR regime. The difficulty lies with inducing lattice alignment between users. If each transmitter emits a single lattice codeword, aligning the interference at all receivers corresponds to an overconstrained problem. This barrier can be circumvented by employing many layers of lattice codewords at each transmitter and inducing alignment between layers. This approach, sometimes referred to as "real interference alignment," is able to achieve $K/2$ degrees of freedom up to a set of channel matrices of measure 0 [190]. Unfortunately, each additional layer incurs an additional rate loss, owing to the loss of the "$1 +$" term in the computation rate for decoding the sum (12.68). Overall, there is still much work to be done in terms of developing new alignment techniques for general channels and characterizing their performance.

Essential structure For interference alignment, the key point is that the sum of the interfering codewords must itself be a codeword. As argued in Section 12.3.2 for lattice network coding, this does not require the use of a coarse shaping lattice, only a fine coding lattice.

Beyond Gaussian If the noise has non-Gaussian statistics, then the fine lattice decoder can be tuned as discussed at the end of Section 12.3.2.

Loss in single-letter characterization Unlike the scenarios considered earlier, we are unaware of a random i.i.d. coding scheme whose performance approximates that of our lattice-based scheme. However, a *multi-letter* characterization of the capacity region is known, which can be used to obtain directly the degrees of freedom for general $\mathbf{H}$ [273].

12.5   Summary and outlook

In this chapter, we have examined several network scenarios where linear or lattice coding can offer a performance advantage over the conventional random coding approach. In each of these scenarios, the (explicit or implicit) goal was to send functions of the transmitters' information to the receivers. Random i.i.d. ensembles are not well suited for this task as they ensure that all of the codewords can be individually distinguished (with high probability). For linear or lattice ensembles,

Table 12.1 Essential structure for the coding schemes considered in this chapter

Coding scheme	Coarse (shaping) lattice	Fine lattice
Lattice Körner–Marton	aligned	–
Dirty multiple-access	aligned	–
Lattice network coding	–	aligned
Lattice interference alignment	–	aligned

the resulting codebook is closed under linear operations, meaning that the codeword corresponding to a linear function can be generated by the transmitters in a distributed fashion.

It is important to note that while linear or lattice codes outperform the best known random i.i.d. code constructions in the scenarios considered above, this does not itself imply that structured coding is *necessary*. That is, even though it seems natural to insist on linear or lattice codebooks when the goal is to communicate a linear function, we do not yet have a proof that using such a codebook is required to reach the capacity. In fact, for most of the Gaussian scenarios we have considered, the capacity itself is unknown. We believe that part of this gap can be closed by developing novel converse arguments that account for the algebraic structure inherent to the problem. These arguments may in turn reveal what form of codebook structure is required to approach the capacity limits.

As discussed in the "essential structure" remark in each section, it is often unnecessary to employ a nested linear or lattice code: depending on the scenario, it may suffice to replace either the fine or coarse code with a conventional random i.i.d. codebook. In Table 12.1, we have summarized, for each of the Gaussian network scenarios, whether the fine or the coarse codebook should be aligned across the transmitters. By aligned, we mean that the specified lattice code is either the same across all transmitters or that it is taken from a chain of nested lattices. If no requirement is placed on the coding (fine) lattice, each transmitter can either replace it with a random i.i.d. code or employ a lattice that has no nesting relationship with any other fine or coarse code. If the shaping (coarse) lattice is unconstrained, it can be replaced with a spherical shaping region (i.e., the codebook is formed by taking all fine lattice points that fall within the sphere). In some cases, it may also be desirable to align both the fine and coarse lattices across transmitters. For instance, this is needed to establish the link between the lattice-based compute-and-forward scheme and linear network coding over a finite field (Section 12.3.3).

The scenarios considered in this chapter should be viewed as examples of the promise of structured coding in networks, rather than as representatives of a general theory. Much work is needed to establish a structured random coding framework for network information theory that is on a par with the existing random i.i.d. coding framework. (See the book of El Gamal and Kim [64] for an in-depth treatment of the random i.i.d. coding framework.) These efforts will certainly include taking the coding techniques studied here beyond the Gaussian (or binary) setting. More broadly, part of the power of the random i.i.d. coding framework lies in the availability of sophisticated multi-user encoding and decoding techniques, such as binning and joint-typicality decoding. While structured counterparts of these techniques are sometimes available, more effort is needed to link the examples presented here into a robust framework.

On the practical side, the code design criteria for these scenarios are quite similar to the criteria for point-to-point Voronoi modulation and quantization, as discussed in Chapter 9. Specifically, good lattice codes for Gaussian channel and source coding can be directly employed to implement the coding techniques proposed in this chapter. In some cases, there are additional benefits to paying close attention to the algebraic structure of the lattice code. For example, in the compute-and-forward technique, it is possible to create a direct correspondence with a finite field if the lattice code is generated over the same finite field via construction A. (More generally, one can link compute-and-forward to a finite ring, see [78] for more details.)

Summary of Chapter 12

Linear Körner–Marton coding To encode the modulo sum $T = X_1 \oplus X_2$ of distributed binary sources X_1 and X_2, send the syndromes

$$\mathbf{S}_1 = H \cdot \mathbf{X}_1 \qquad \mathbf{S}_2 = H \cdot \mathbf{X}_2$$

and reconstruct $\mathbf{T}$ as

$$\hat{\mathbf{T}} = \text{leader}(\mathbf{S}_1 \oplus \mathbf{S}_2),$$

where the code $\mathcal{C}$ (with parity-check matrix H) is good for a BSC with noise T.

Lattice Körner–Marton coding To encode the difference $T = X_1 - X_2$ of distributed Gaussian sources X_1 and X_2, send the relative cosets

$$\mathbf{V}_1 = [Q_{\Lambda_1}(\alpha^* \mathbf{X}_1 + \mathbf{U}_1)] \bmod \Lambda_2 \qquad \mathbf{V}_2 = [Q_{\Lambda_1}(\alpha^* \mathbf{X}_2 + \mathbf{U}_2)] \bmod \Lambda_2$$

and reconstruct $\mathbf{T}$ as

$$\hat{\mathbf{T}} = \beta^*([(\mathbf{V}_1 - \mathbf{U}_1) - (\mathbf{V}_2 - \mathbf{U}_2)] \bmod_{\mathcal{V}_0} \Lambda_2)$$

where $\mathbf{U}_1$ and $\mathbf{U}_2$ are random dithers and, for targeted distortion D, the fine lattice Λ_1 is a good quantizer matched to distortion $D/2$, the coarse lattice Λ_2 is a good channel code for noise T, and $\alpha^* = \beta^* = \sqrt{1 - D/\text{Var}(T)}$.

Lattice dirty multiple-access coding To send coset information $\mathbf{v}_1, \mathbf{v}_2 \in \mathcal{C}_{\Lambda_1, \mathcal{P}_0}$ over the multiple-access channel $Y = X_1 + X_2 + S_1 + S_2 + Z$ where the interference S_1 is available at the first encoder and S_2 at the second, transmit the vectors

$$\mathbf{X}_1 = [\mathbf{v}_1 + \mathbf{U}_1 - \alpha^* \mathbf{S}_1] \bmod_{\mathcal{V}_0} \Lambda_2 \qquad \mathbf{X}_2 = [\mathbf{v}_2 + \mathbf{U}_2 - \alpha^* \mathbf{S}_2] \bmod_{\mathcal{V}_0} \Lambda_2$$

and infer $\mathbf{v}_1$ and $\mathbf{v}_2$ from $\mathbf{t} = [\mathbf{v}_1 + \mathbf{v}_2] \bmod \Lambda_2$, which is decoded as

$$\hat{\mathbf{t}} = [Q_{\Lambda_1}(\alpha^* \mathbf{Y} - \mathbf{U}_1 - \mathbf{U}_2)] \bmod \Lambda_2$$

where the fine lattice Λ_1 is a good channel code for the noise Z, the coarse lattice Λ_2 is a good quantizer (matched to the target transmit power), and α^* is the Wiener coefficient for estimating $\mathbf{X}_1 + \mathbf{X}_2$ from $\mathbf{Y}$.

Equivalence to an interference-free channel The channel from $\mathbf{v}_1$ and $\mathbf{v}_2$ to $[\alpha^* \mathbf{Y} - \mathbf{U}_1 - \mathbf{U}_2] \bmod \Lambda_2$ is independent of the statistics of $\mathbf{S}_1$ and $\mathbf{S}_2$.

Dirty multiple-access rate If Λ_1 and Λ_2 are good for coding and shaping, respectively, the coding rate $R(\Lambda_1/\Lambda_2)$ can approach

$$\left[\tfrac{1}{2} \log \left(\tfrac{1}{2} + P/N \right) \right]^+$$

while driving the error probability to zero.

Computation over a binary MAC To send the modulo sum $\mathbf{t} = \mathbf{w}_1 \oplus \mathbf{w}_2$ of information vectors $\mathbf{w}_1$ and $\mathbf{w}_2$ over a binary MAC $Y = X_1 \oplus X_2 \oplus Z$, transmit

$$\mathbf{x}_1 = G \cdot \mathbf{w}_1 \qquad \mathbf{x}_2 = G \cdot \mathbf{w}_2,$$

and decode $\mathbf{t}$ as

$$\hat{\mathbf{t}} = \mathbf{Y} \oplus ([\mathbf{Y}] \bmod \mathcal{C}),$$

where the code $\mathcal{C}$ (with generator matrix G) is good for a BSC with noise Z.

Compute-and-forward To send integer linear combinations $\mathbf{t}_k = [\sum_\ell a_{k\ell} \mathbf{v}_\ell] \bmod \Lambda_2$ of the message (coset) representatives $\mathbf{v}_\ell \in \mathcal{C}_{\Lambda_1, \mathcal{P}_0}$ to K relays over interfering channels $Y_k = h_{k1} X_1 + \cdots + h_{kK} X_K + Z_k$, transmit

$$\mathbf{X}_\ell = [\mathbf{v}_\ell + \mathbf{U}_\ell] \bmod_{\mathcal{V}_0} \Lambda_2$$

and reconstruct $\mathbf{t}_k$ as

$$\hat{\mathbf{t}}_k = \left[Q_{\Lambda_1} \left(\alpha_k^* \mathbf{Y}_k - \sum_{\ell=1}^K a_{k\ell} \mathbf{U}_\ell \right) \right] \bmod \Lambda_2$$

where the fine lattice Λ_1 is a good channel code for the effective noises $\mathbf{Z}_{\text{eff},k} = \alpha_k^* \mathbf{Z}_k + \sum_\ell (\alpha_k^* h_{k\ell} - a_{k\ell}) \mathbf{X}_\ell$, the coarse lattice Λ_2 is a good quantizer (matched to the target transmit power), and α_k^* is the Wiener coefficient for estimating $\sum_\ell a_{k\ell} \mathbf{X}_\ell$ from $\mathbf{Y}_k$.

Computation rate If Λ_1 and Λ_2 are good for coding and shaping, respectively, the coding rate $R(\Lambda_1/\Lambda_2)$ for compute-and-forward can approach $\min_k \mathcal{R}_{\mathrm{comp}}(\mathbf{h}_k, \mathbf{a}_k)$ while driving the error probability to zero where

$$
\mathcal{R}_{\mathrm{comp}}(\mathbf{h}_k, \mathbf{a}_k) = \left[\frac{1}{2} \log \left(\frac{N + P\|\mathbf{h}_k\|^2}{N\|\mathbf{a}_k\|^2 + P\left(\|\mathbf{h}_k\|^2\|\mathbf{a}_k\|^2 - (\mathbf{h}_k^T \mathbf{a}_k)^2\right)} \right) \right]^+ .
$$

Alignment via two equations To send coset information $\mathbf{v}_1, \ldots, \mathbf{v}_K \in \mathcal{C}_{\Lambda_1, \mathcal{P}_0}$ over the symmetric interference channel $Y_k = X_k + g \sum_{\ell \neq k} X_\ell + Z_k$, transmit

$$
\mathbf{X}_\ell = [\mathbf{v}_\ell + \mathbf{U}_\ell] \bmod_{\mathbf{v}_0} \Lambda_2 ,
$$

and decode two independent linear combinations

$$
\mathbf{t}_{k1} = \left[b_{11}\mathbf{v}_k + b_{12} \sum_{\ell \neq k} \mathbf{v}_\ell \right] \bmod \Lambda_2 \qquad \mathbf{t}_{k2} = \left[b_{21}\mathbf{v}_k + b_{22} \sum_{\ell \neq k} \mathbf{v}_\ell \right] \bmod \Lambda_2
$$

at each receiver via the compute-and-forward framework, and solve for $\mathbf{v}_k$.

Problems

P.12.1 (Asymmetric two-help-one) Consider two dependent binary sources X_1 and X_2 with probability distribution $p_{ij} = \mathrm{Pr}(X_1 = i, X_2 = j)$ for $i, j \in \{0, 1\}$. In Section 12.1, we argued that the Körner–Marton scheme (12.2) is optimal for the distributed compression of the modulo sum for the symmetric case where $p_{00} = p_{11} = 1 - \theta$ and $p_{10} = p_{01} = \theta$. Find p_{ij} such that the scheme (12.2) is suboptimal (as compared to the rate region in (12.6)).

P.12.2 (Quantization and innovation) Show that the sum rate for the Berger–Tung scheme in (12.21) can be equivalently written as

$$
R_{\mathrm{BT,sum}} = \frac{1}{2} \log \left(1 + \frac{\sigma^2}{D_1} \right) + \frac{1}{2} \log \left(1 + \frac{\mathrm{Var}\,(X_2|X_1 + Z_1)}{D_2} \right) \tag{12.122}
$$

where $D_1 = D_2 = \tilde{D}/2$, $Z_1 \sim \mathcal{N}(0, D_1)$, and $\tilde{D}$ satisfies $1/D = (1/\tilde{D}) + (1/\sigma_T^2)$. The first term is the rate needed to send X_1 at distortion D_1 and the second term is the rate needed to send the innovations to describe X_2 at distortion D_2.

P.12.3 (Noisy binary dirty MAC) Consider the following twist on the binary dirty MAC studied in Section 12.2.1: the channel output is now $\mathbf{y} = \mathbf{x}_1 \oplus \mathbf{x}_2 \oplus \mathbf{s}_1 \oplus \mathbf{s}_2 \oplus \mathbf{z}$ where $\mathbf{z}$ is an i.i.d. Bernoulli(δ) noise sequence. Develop a nested linear coding scheme to achieve the capacity region

$$
\mathcal{C} = \left\{ (R_1, R_2) : R_1 + R_2 \leq \mathrm{u.c.e.}\{[H_B(q) - H_B(\delta)]^+\} \right\} \tag{12.123}
$$

where the upper convex envelope is taken with respect to the weight constraints q_1 and q_2.

P.12.4 (Computation with dependent sources) Consider the problem of computing over a binary multiple-access channel from Section 12.3.1. Develop an optimal coding scheme for the scenario where the messages are replaced with the dependent sources from Problem 12.1 (and the decoder still wants their modulo sum). Write down the associated rate region.

P.12.5 (Equivalent modulo-Λ_2 channels) Provide a proof for Lemmas 12.2.1 and 12.3.1, 12.3.2. (Hint: see Lemma 9.5.1.)

P.12.6 (Computation rate) Show that the two expressions for the computation rate in (12.82) and (12.90) are equivalent.

P.12.7 (Amplify-and-forward) Prove that the rate (12.71) is achievable via an amplify-and-forward strategy. Specifically, calculate the effective SNR of the point-to-point channel that is established once user 1 has subtracted its own codeword from its observation $\mathbf{y}_1 = \sqrt{\frac{P}{2P+N}}(\mathbf{x}_1 + \mathbf{x}_2 + \mathbf{z}_{\mathrm{MAC}}) + \mathbf{z}_1$. (The analysis for user 2 follows by symmetry.)

P.12.8 (Integer-forcing) Provide a full proof that the integer-forcing rate (12.96) is achievable. (Hint: see the proof of Theorem 12.3.2.) Show that the zero-forcing rate (which is attained using i.i.d. Gaussian codebooks and the MMSE projection $\mathbf{B} = P\mathbf{H}^T(N\mathbf{I} + P\mathbf{H}\mathbf{H}^T)^{-1}$) can be expressed as (12.97).

P.12.9 (Sum of the computation rates is nearly the MAC sum capacity) Consider a K-user Gaussian multiple-access channel with output $Y = h_1 X_1 + \cdots + h_K X_K + Z$ where $Z \sim \mathcal{N}(0, N)$ and $\mathbf{h} = [h_1 \ \cdots \ h_K]^T$ denotes the channel vector. Let $\mathbf{a}_1, \ldots, \mathbf{a}_K \in \mathbb{Z}^K$ denote the K linearly independent integer coefficient vectors with the highest computation rates $\mathcal{R}_{\mathrm{comp}}(\mathbf{h}, \mathbf{a}_1), \ldots, \mathcal{R}_{\mathrm{comp}}(\mathbf{h}, \mathbf{a}_K)$ over this channel. Show that the sum of these computation rates is lower bounded by the multiple-access sum capacity minus a constant gap,

$$\sum_{k=1}^{K} \mathcal{R}_{\mathrm{comp}}(\mathbf{h}, \mathbf{a}_k) \geq \frac{1}{2} \log \left(1 + \frac{P\|\mathbf{h}\|^2}{N} \right) - \frac{K}{2} \log K. \qquad (12.124)$$

(Hint: use the equivalent form of the computation rate from (12.90).)

P.12.10 (Decoding the real sum) Show that, within the context of the compute-and-forward scheme in Theorem 12.3.2, if a receiver has successfully decoded an integer linear combination of the message representatives modulo the coarse lattice, $\left[\sum_{\ell=1}^{K} a_\ell \mathbf{v}_\ell \right] \bmod \Lambda_2$, then it can also decode an integer linear combination of the dithered codewords, $\sum_{\ell=1}^{K} a_\ell \mathbf{x}_\ell$, with vanishing probability of error (as n tends to infinity).

P.12.11 (Successive compute-and-forward) Consider the scenario from Problem 12.9 where a single receiver wishes to decode K linear combinations with integer coefficient vectors $\mathbf{a}_1, \ldots, \mathbf{a}_K \in \mathbb{Z}^K$ over a MAC with channel vector $\mathbf{h} \in \mathbb{R}^K$. Assume that, after decoding each equation, the receiver also decodes the corresponding integer linear combination of the dithered codewords as in Problem 12.10. Thus, when decoding the equation with integer coefficient vector $\mathbf{a}_m$, the receiver has access to $\mathbf{y}, \sum_{\ell=1}^{K} a_{1\ell}\mathbf{x}_\ell, \ldots, \sum_{\ell=1}^{K} a_{(m-1)\ell}\mathbf{x}_\ell$, from which it can form a new effective channel observation (via an MMSE projection). Determine the successive computation rate $\mathcal{R}_{\mathrm{comp}}(\mathbf{h}, \mathbf{a}_m|\mathbf{a}_{m-1}, \ldots, \mathbf{a}_1)$, i.e., the highest coding rate $R(\Lambda_1/\Lambda_2)$ at which $\left[\sum_{\ell=1}^{K} a_{m\ell}\mathbf{v}_\ell\right]$ mod Λ_2 can be decoded.

P.12.12 (Sum rate optimality) Show that the sum of the successive computation rates derived in Problem 12.11 is equal to the multiple-access sum capacity for any unimodular matrix $\mathbf{A} = [\mathbf{a}_1 \cdots \mathbf{a}_K]^T$ (i.e., an integer-valued matrix with determinant equal to one),

$$\sum_{k=1}^{K} \mathcal{R}_{\mathrm{comp}}(\mathbf{h}, \mathbf{a}_k|\mathbf{a}_{k-1}, \ldots, \mathbf{a}_1) = \frac{1}{2}\log\left(1 + \frac{P\|\mathbf{h}\|^2}{N}\right). \tag{12.125}$$

Historical notes

The first instance of a network scenario (the two-help-one problem from Section 12.1) in which random structured coding outperforms random i.i.d. coding was discovered by Körner and Marton in 1979 [139]. With the exception of a 1983 paper by Ahlswede and Han [5] which proposed an achievable region that includes the Körner–Marton and Slepian–Wolf regions as special cases, this research thread appears to have lain dormant until the 2000s. Motivated in part by renewed interest in network information theory, several groups independently and concurrently uncovered new scenarios where structured coding helps. Specifically, Krithivasan and Pradhan [142] generalized the Körner–Marton scheme to the case of correlated Gaussian sources, Nazer and Gastpar [193] proposed the problem of computation over a MAC and its application to network coding, and Philosof et al. [217] introduced the dirty Gaussian multiple-access problem. Soon thereafter, lattice coding was recognized as a promising approach for relaying and interference alignment as well. Below, we survey [16] the literature on structured coding within the context of each of these scenarios. Note that we will not attempt to give a full account of the vast literature on random i.i.d. coding strategies associated to these networks.

For distributed source coding, the Körner–Marton scheme serves as an example that random i.i.d. quantization and binning (e.g., the Berger–Tung scheme [20])

---

[16] Since these topics were the focus of active research when this chapter was written, this list is necessarily incomplete.

is not always optimal. As mentioned earlier, in the quadratic-Gaussian setting, Wagner et al. [264] established that, for two users, the Berger–Tung scheme is optimal (optimality for general sources at the high-resolution regime was shown in [285]). They also demonstrated that, for the many-help-one problem, Berger–Tung is optimal if the Gaussian sources can be expressed in terms of a Gauss–Markov (binary) tree [253]. If this condition does not hold, the lattice Körner–Marton scheme proposed by Krithivasan and Pradhan [142] yields achievable rate tuples outside the Berger–Tung region. Subsequent work by Wagner [263] showed that this scheme can be further improved (in the low rate regime) through the use of a second linear binning stage. Additionally, Wagner developed new lower bounds for the distributed compression of a linear function of correlated Gaussian sources (as described in Section 12.1.2). Building on these results, Tse and Maddah-Ali approximated the rate-distortion region to within a constant gap [258]. The upper and lower bounds were further refined by Yang and Xiong [278]. For the more general setting of distributed source coding for discrete memoryless sources, Krithivasan and Pradhan have also proposed a framework based on Abelian group codes which strictly contains the Berger–Tung and Körner–Marton regions [143].

In the context of computing over a MAC, Nazer and Gastpar proposed linear and lattice coding strategies for sending linear functions of discrete and Gaussian sources, respectively [193]. They also showed that these techniques are useful for (linear) network coding over noisy, interfering links by characterizing the multicast capacity for a class of finite field multiple-access networks. Independently and concurrently, Zhang et al. [295] as well as Popovski and Yomo [223] noticed that the physical layer can be exploited for network coding (with the former coining the phrase "physical-layer network coding"). See [157, 195] for surveys of the physical-layer network coding literature. Subsequent work led to the discovery of the compute-and-forward framework for Gaussian relay networks, starting with the work of Nazer and Gastpar [194] and Wilson et al. [270] for the single receiver case with equal transmit powers, followed by that of Nam et al. [192] for the single receiver case with unequal transmit powers, and finally that of Nazer and Gastpar [196] for the multiple receiver case with equal transmit powers. Later work by Ntranos et al. [203] extended the multiple receiver case to unequal transmit powers. Building on these ideas, Zhan et al. [294] proposed integer-forcing for MIMO channels. Ordentlich and Erez [207] showed that integer-forcing, when combined with linear "space-time coding" at the transmitter, achieves the MIMO capacity to within a constant gap. Hong and Caire [122] proposed "reverse compute-and-forward" for a single transmitter communicating to many receivers with the help of relays: the transmitter pre-inverts (over the finite field) the linear combinations of the messages targeted by the relays. Although compute-and-forward offers rate gains over decoding the messages in their entirety, this advantage may disappear in the high SNR regime if the transmitters do not know the channel realization, as argued by Niesen and Whiting [201] for the scheme of [196]. Their bounds were further refined by Ordentlich et al. [209]. Feng et al. [78] proposed a generalized algebraic framework for compute-and-forward that links the underlying nested lattice code to a finite ring and provides guidance for practical code design.

For the dirty Gaussian two-user MAC considered in Section 12.2.2, it was conjectured by Gel'fand and Pinsker [101] that the capacity region is equal to that of the Gaussian MAC with no interference. While this is true if the interference terms are available at both transmitters (as shown by Gel'fand and Pinsker [101] as well as Kim *et al.* [134]), this is not the case when knowledge of the interference is distributed. In particular, Philosof *et al.* [217] developed the lattice coding scheme and upper bounds discussed in Section 12.2.2, which established that for i.i.d. Gaussian interferences (whose power is taken to infinity), the sum capacity is determined by the minimum of the transmitters' powers, as opposed to their sum. Philosof and Zamir [215] then derived the capacity region for the noiseless binary setting in Section 12.2.1. Subsequent work by Philosof *et al.* [216] generalized this result to the noisy setting using nested linear codes. The best known random i.i.d. coding region for the general dirty MAC is due to Jafar [126]. For the dirty Gaussian K-user multiple-access channel, Wang [265] characterized the capacity region to within a constant gap.

The phenomenon of interference alignment has been thoroughly studied since its introduction as a technique for wireless communication in the papers of Maddah-Ali *et al.* [177], Jafar and Shamai [128], and Cadambe and Jafar [26]. In particular, [26] made the surprising discovery that the (time-varying) K-user Gaussian interference channel has $K/2$ degrees of freedom. See the survey of Jafar [127] for an in-depth account and further references. For the two-user case, Etkin *et al.* [74] characterized the capacity region to within one bit per user. Although their coding scheme does not rely on alignment, their capacity bounds are often used as a benchmark for $K > 2$ users. Bresler *et al.* [25] were the first to note the potential of lattice codes for alignment over K-user static channels, in the context of a "many-to-one" Gaussian interference channel where only one receiver observes interference. Afterwards, Sridharan *et al.* [251] determined the capacity of the K-user symmetric Gaussian interference channel in the very strong regime via successive cancelation. For the same channel, Jafar and Vishwanath [129] characterized the generalized degrees of freedom. For general channel gains, Etkin and Ordentlich [73] showed that the degrees of freedom is discontinuous at rational-valued channel matrices. Subsequently, Motahari *et al.* [190] showed that $K/2$ degrees of freedom are available (up to a set of channel matrices of measure zero) through the use of carefully layered codebooks comprising lattice symbols. Wu *et al.* found an alternative perspective of this result based on the multi-letter characterization of the capacity region [273]. Later efforts sought to characterize the sensitivity of alignment to the channel gains. In particular, Niesen and Maddah-Ali [200] approximated the sum capacity of the two-user Gaussian X channel up to an outage set through the aid of a deterministic model. Ordentlich *et al.* [209] established a similar characterization of the K-user symmetric Gaussian interference channel via the compute-and-forward framework (as discussed in Section 12.4.3). They also showed that the sum of the computation rates (for K independent linear combinations) is nearly equal to the MAC sum capacity (see Problem 12.9). In subsequent work [210], they established that successive compute-and-forward can attain the exact MAC sum capacity (see Problem 12.12). Finally, we mention that the ergodic alignment technique is due to Nazer *et al.* [197].

We also note that there are many other network scenarios where structured coding can improve performance (which we did not manage to discuss in this chapter). For instance, Song and Devroye [250] as well as Nokleby and Aazhang [202] have combined lattice coding with classical relaying techniques, He and Yener [118] as well as Shashank and Kashyap [245] have proposed lattice techniques for physical-layer secrecy, and Haim *et al.* [115] have improved upon existing multiple-access error exponents using linear codes.

13 Error exponents

The Shannon capacity C is the most important theoretic performance figure of a communication channel. But it is only one extreme point. Information theory provides a more elaborate characterization of a channel, in terms of its *error exponent* – that is, the exponent governing the decay of the decoding error probability as the dimension of the code increases. This exponent depends on the gap to capacity: it is large for a coding rate R which is far below C, and it vanishes for $R \geq C$.

In a similar manner, we can assess the asymptotic behavior of good lattice constellations by considering their error exponent over *un*constrained additive-noise channels. Such exponential behavior occurs under various circumstances. For example, in the AWGN channel, the error probability of good modulation lattices decays exponentially if we keep the VNR μ (3.30) fixed at some value above the ideal VNR of $2\pi e$ – the equivalent of capacity in the unconstrained channel setup.

As in Chapter 7, a simple and effective way to assess the error exponent is provided by the *sphere bound* (7.14), i.e., the probability $P_e(\mathcal{B}_{r_{\text{eff}}(\Lambda)}, \sigma^2)$ that the noise leaves a ball with the same volume as the lattice cell. The sphere bound gives rise to an exponential *lower* bound of the form [1]

$$P_e(\Lambda, \sigma^2) \overset{\cdot}{\geq} e^{-n E_{\text{sp}}\left(\frac{r_{\text{eff}}^2(\Lambda)}{r_{\text{noise}}^2}\right)}, \tag{13.2}$$

where $P_e(\Lambda, \sigma^2)$ is the probability (3.32) that an AWGN of variance σ^2 leaves the fundamental Voronoi cell of a lattice Λ (corresponding to the error probability in NN decoding), n is the lattice dimension, $r_{\text{noise}} = \sigma\sqrt{n}$ is the *typical noise radius* (7.27), and the function $E_{\text{sp}}(x)$ – known as the *Chernoff "sphere packing" exponent* – is given by $\frac{1}{2}[x - 1 - \ln(x)]$, for $x \geq 1$, and zero elsewhere.

The argument of $E_{\text{sp}}(\cdot)$ in (13.2) is the square of the effective *radius to noise ratio* $r_{\text{eff}}(\Lambda)/r_{\text{noise}}$. Like the VNR μ, it measures the lattice density with respect to AWGN of variance σ^2. (For a large lattice dimension, $(r_{\text{eff}}(\Lambda)/r_{\text{noise}})^2 \approx \mu/2\pi e$.)

---

[1] An *exponential inequality* $a_n \overset{\cdot}{\geq} b_n$ means that

$$\liminf_{n\to\infty} \frac{1}{n} \log\left(\frac{a_n}{b_n}\right) \geq 0. \tag{13.1}$$

The sphere packing exponent vanishes, as expected, when this radius ratio is smaller than or equal to 1, corresponding to VNR below $2\pi e$. As we shall see further, the lower bound (13.2) is *exponentially tight*, i.e.,

$$P_e(\Lambda_n, \sigma^2) \doteq e^{-n E_{\mathrm{sp}}\left(\frac{r_{\mathrm{eff}}^2(\Lambda_n)}{r_{\mathrm{noise}}^2}\right)},$$

for a sequence of good modulation lattices Λ_n, operating at a fixed radius ratio in the range $(1, \sqrt{2}]$ (or a fixed VNR in the range $(2\pi e, 4\pi e]$).

Another way to assess the error exponent in the presence of AWGN is via the minimum-distance bound $P_e(\Lambda, \sigma^2) \geq Q(d_{\min}/2\sigma)$ of (3.34). This bound, combined with the conjecture that the best asymptotic packing efficiency $\rho_{\mathrm{pack}}(\Lambda)$ is one-half (see the discussion following Corollary 7.6.1), gives rise to an exponential lower bound of the form

$$P_e(\Lambda, \sigma^2) \gtrdot e^{-\frac{n}{8}\left(\frac{r_{\mathrm{eff}}(\Lambda)}{r_{\mathrm{noise}}}\right)^2}. \tag{13.3}$$

The bound (13.3) is better than (13.2) for a *large* radius to noise ratio, i.e., large VNR. And in fact, it is asymptotically *tight* for VNR greater than or equal to $8\pi e$ (radius ratio greater than 2), for lattices which are good for packing.

Achievability of the exponential lower bound (13.2) is shown using the MHS ensemble, and of (13.3) using an expurgated version of the MHS ensemble. The derivation uses the same "averaging argument" as in proving the achievability of NVNR of $2\pi e$ (Section 7.7): there must exist a lattice in the ensemble which is at least as good as the ensemble average. (An alternative derivation follows from the Gallager error exponent of a modulo-additive-noise channel; see Problem 13.6.) The combination of (13.2) and (13.3) is known as the *Poltyrev error exponent*.

In the end of the chapter we shall apply these results to *finite* lattice constellations, and derive the error exponent of a lattice-shaped (Voronoi) codebook.

13.1   Sphere packing exponent

A useful tool to estimate the exponent of the probability that i.i.d. noise leaves a ball (and more general events associated with threshold crossing) is provided by the *Chernoff bound*.

Proposition 13.1.1 (Chernoff bound) *Let the vector $\mathbf{Z}$ consist of n i.i.d. copies of a random variable Z, and let $g(s) = E\{e^{sZ^2}\}$ denote the characteristic function of Z^2. Then, the probability that $\mathbf{Z}$ leaves a ball of radius $r = t\sqrt{n}$ satisfies*

$$\Pr\{\|\mathbf{Z}\| > r\} \leq \left(\frac{g(s)}{e^{st^2}}\right)^n, \quad \forall s > 0. \tag{13.4}$$

The Chernoff bound is meaningful when the radius r is larger than the root mean square of $\|\mathbf{Z}\|$, i.e., $r > \sqrt{E\|\mathbf{Z}\|^2}$ (or $t > \sqrt{EZ^2}$); see Problem 13.1. For example, for $Z \sim N(0, \sigma^2)$, the interesting range is $r > \sigma \sqrt{n}$. Otherwise the upper bound can be replaced by 1. The proof of this bound is based on a well-known probabilistic inequality.

Proposition 13.1.2 (Markov inequality) *If X is a non-negative random variable, then*

$$\Pr\{X > t\} \leq \frac{E\{X\}}{t}.$$

Proof $E\{X\} = \Pr\{X \leq t\} \cdot E\{X \mid X \leq t\} + \Pr\{X > t\} \cdot E\{X \mid X > t\} \geq 0 + \Pr\{X > t\} \cdot t.$ □

Proof of Proposition 13.1.1 The event $\|\mathbf{Z}\| > r$ is equivalent to $e^{s\|\mathbf{Z}\|^2} > e^{sr^2}$, for all $s > 0$. Thus, by Markov's inequality,

$$\Pr\{\|\mathbf{Z}\| > r\} = \Pr\left\{e^{s\|\mathbf{Z}\|^2} > e^{sr^2}\right\} \leq \frac{E\left\{e^{s\|\mathbf{Z}\|^2}\right\}}{e^{sr^2}}. \tag{13.5}$$

Since $e^{s\|\mathbf{Z}\|^2} = e^{s \sum_{i=1}^{n} Z_i^2} = \prod_{i=1}^{n} e^{sZ_i^2}$, and since the Z_i are i.i.d., the expectation in the numerator breaks into $\prod_{i=1}^{n} E\{e^{sZ_i^2}\} = g^n(s)$, and the bound follows. □

The Chernoff bound is *exponentially tight* under mild conditions on the distribution of Z. These conditions hold in the Gaussian case.

Proposition 13.1.3 (Gaussian sphere crossing) *If $\mathbf{Z}$ is a zero-mean white-Gaussian vector of variance σ^2, then $g(s) = 1/\sqrt{1 - 2\sigma^2 s}$, and the best (minimizing) value of the parameter s in (13.4) is $s^* = \frac{1}{2\sigma^2}(1 - \sigma^2/t^2)$, for $t > \sigma$. The resulting Chernoff bound is*

$$\Pr\{\|\mathbf{Z}\| > r\} \leq e^{-n E_{\text{sp}}\left(\frac{r^2}{r_{\text{noise}}^2}\right)}, \tag{13.6}$$

where $r_{\text{noise}} = \sigma \sqrt{n}$ is the typical noise radius (7.27), and $E_{\text{sp}}(\cdot)$ is the sphere packing exponent:

$$E_{\text{sp}}(x) = \begin{cases} \frac{1}{2}[x - 1 - \ln(x)], & \text{if } x > 1 \\ 0, & \text{if } x \leq 1. \end{cases} \tag{13.7}$$

Furthermore, asymptotically as $n \to \infty$, the bound is tight for $r < r_{\text{noise}}$ (where $\Pr\{\|\mathbf{Z}\| > r\}$ goes to 1), and is exponentially tight for $r > r_{\text{noise}}$, i.e.,

$$\Pr\{\|\mathbf{Z}\| > \alpha r_{\text{noise}}\} \doteq e^{-n E_{\text{sp}}(\alpha^2)} \tag{13.8}$$

for $\alpha > 1$, where $\doteq$ denotes exponential equality as $n \to \infty$; see (7.26).

Proof The first part of the proposition follows by analyzing the (simple) characteristic function $g(s)$ in the Gaussian case. The second part (the exponential tightness (13.8)) can be shown, for example, using the *Laplace method of integration*, presented later in this chapter. See Problem 13.2. □

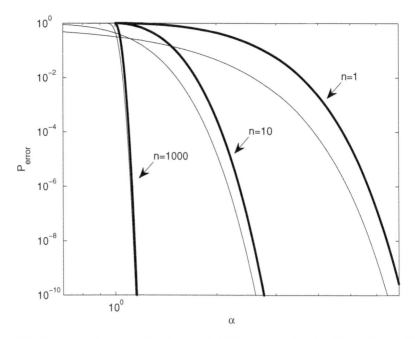

Figure 13.1 Gaussian sphere crossing: true probability compared to the Chernoff bound for several block lengths n, as a function of $\alpha =$ the sphere radius divided by the noise radius $r/\sigma \sqrt{n}$.

As can be seen in Figure 13.1, the bound (13.6) is not necessarily tight for small n. For example, for $n = 1$, the probability that $Z > \alpha \sigma$ is given by the Q-function calculated at α, which is upper bounded by [62, section 1; 272, p. 83],

$$Q(\alpha) \leq \frac{1}{2} e^{-\alpha^2/2}. \tag{13.9}$$

The exponent $\alpha^2/2$ in (13.9) is indeed slightly stronger (higher) than the sphere packing exponent (13.7).

In light of the sphere bound (7.14), the Gaussian sphere crossing exponent implies the exponential lower bound (13.2), which we now state formally. See Figure 13.2.

Theorem 13.1.1 (Sphere packing exponent) *For any sequence of lattices* Λ_n *of increasing dimension* n, *with a fixed effective radius to noise ratio* $r_{\mathrm{eff}}(\Lambda_n)/r_{\mathrm{noise}} = \alpha$ *for all* n, *the error probability* (3.32) *in nearest-neighbor decoding in the presence of AWGN with variance* σ^2, *is lower bounded by*

$$P_e(\Lambda_n, \sigma^2) \geq P_e(\mathcal{B}_{r_{\mathrm{eff}}(\Lambda_n)}, \sigma^2) \doteq e^{-n E_{\mathrm{sp}}(\alpha^2)}. \tag{13.10}$$

Table 13.1 Transition points in error exponents

	Lattice to noise measure			
	Δ [bit]	μ	$\alpha = \dfrac{r_{\text{eff}}(\Lambda)}{r_{\text{noise}}}$	α [dB]
Capacity	0	$2\pi e$	1	0
Cutoff rate	$\frac{1}{2}\log(4/e) \approx 0.28$	8π	$2/\sqrt{e} \approx 1.21$	1.7
Critical rate	1/2	$4\pi e$	$\sqrt{2}$	3
Expurgated rate	1	$8\pi e$	2	6

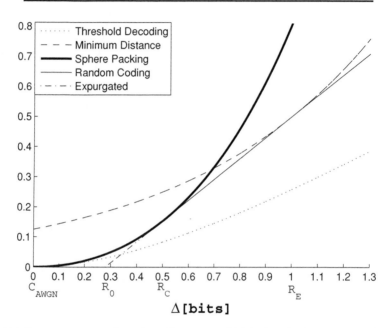

Figure 13.2 Error exponents of good (uniform-ensemble) lattices in the presence of AWGN, as a function of the gap to capacity $\Delta = C_\infty - R_\infty$: (i) sphere packing exponent (outer bound); (ii) threshold-decoding exponent; (iii) ML decoding random-coding exponent; (iv) minimum-distance (conjectured) exponent (outer bound); and (v) expurgated exponent. The transition points in the graphs are summarized in Table 13.1.

13.2   Measures of lattice to noise density

The effective *radius to noise ratio* (ERNR)

$$\frac{r_{\text{eff}}(\Lambda)}{r_{\text{noise}}} = \frac{r_{\text{eff}}(\Lambda)}{\sqrt{n\sigma^2}} \tag{13.11}$$

in the exponents above is a measure for the density of the lattice Λ with respect to AWGN with variance σ^2. It is tightly connected to the *gap to capacity* (6.26):

$$\Delta = C_\infty - R(\Lambda) = \frac{1}{2}\log\left(\frac{\mu}{2\pi e}\right) \tag{13.12}$$

and to the VNR μ. Indeed, using the definitions of the effective radius (3.10) and the VNR (3.30), we obtain

$$\log\left(\frac{r_{\text{eff}}(\Lambda)}{r_{\text{noise}}}\right) = \Delta + \frac{1}{2}\log\left(\frac{2\pi e}{n V_n^{2/n}}\right) \to \Delta \qquad (13.13)$$

as $n \to \infty$, where the limit follows from (7.28). Hence, for large n we have

$$\frac{r_{\text{eff}}(\Lambda)}{r_{\text{noise}}} \approx 2^\Delta = \sqrt{\frac{\mu}{2\pi e}}. \qquad (13.14)$$

As Table 13.1 shows, *at capacity* we have $\Delta = 0$, $\mu = 2\pi e$, and the ERNR is equal to 1.

Since the error exponent is, by definition, asymptotic in n, we shall see essentially the same exponential behavior of the error probability either by fixing the ERNR, or by fixing the VNR (equivalently the gap to capacity) as n goes to infinity. In particular, using (13.14) in (13.7), the sphere packing exponent for a fixed VNR μ is given by

$$E_{\text{sp}}\left(\frac{\mu}{2\pi e}\right) = \begin{cases} \frac{1}{2}[\frac{\mu}{2\pi e} - 1 - \ln(\frac{\mu}{2\pi e})], & \text{if } \mu > 2\pi e \\ 0, & \text{if } \mu \leq 2\pi e. \end{cases} \qquad (13.15)$$

For the sake of consistency, we shall stick from now on to the VNR as a measure for the lattice density with respect to the noise. Keeping the VNR $\mu = V^{2/n}(\Lambda)/\sigma^2$ fixed for a given noise variance σ^2, amounts to keeping the rate per unit volume $R(\Lambda) = \frac{1}{n}\log(1/V(\Lambda))$, and the gap to capacity $\Delta = C_\infty - R(\Lambda)$ fixed; hence the lattice point density $\gamma = 1/V(\Lambda)$ scales exponentially with the dimension:

$$\gamma = \frac{1}{(\sigma\sqrt{\mu})^n}.$$

Furthermore, from (13.14), the effective radius scales roughly as the square root of the dimension, i.e., $r_{\text{eff}}(\Lambda) \approx \sqrt{n\mu/2\pi e}$. (The correction factor is $\sqrt{2\pi e/n}V_n^{1/n}$, which goes to 1 as $n \to \infty$.)

13.3   Threshold-decoding exponent

Our next goal is to try to approach the sphere packing exponential lower bound of Theorem 13.1.1 by a high-dimensional lattice from the MHS ensemble. We saw in Section 7.7 that a random lattice from this ensemble can achieve, with *threshold decoding*, an arbitrary small error probability, for VNR as close as desired to $2\pi e$. For AWGN, this corresponds to achieving the capacity per unit volume, i.e., $R(\Lambda, P_e) \to C_\infty$ for all P_e, or $\Delta \to 0$.

In this section we examine the exponential behavior of the threshold decoder of Section 7.7, which will turn out to be *weaker* than the sphere packing exponent

(13.10). In the next section we will examine an NN decoder, whose exponential behavior is better, and meets (13.10) at a range of rates near capacity.

Recall that the MHS ensemble threshold-decoding bound of Lemma 7.7.2 is parametric in the search radius $r_{\text{th}} = r$:

$$P_e^{\text{TB}}(r, n) = \underbrace{\Pr\{\|\mathbf{Z}\| > r\}}_{\text{sphere crossing}} + \underbrace{\gamma V_n r^n}_{\text{competing codewords}}, \quad \text{for } r > 0, \quad (13.16)$$

where γ is the MHS ensemble point density. It is given by the sum of two terms: a sphere crossing term $\Pr\{\|\mathbf{Z}\| > r\}$ which decreases with r, and a competing codewords term $\gamma V_n r^n = (r/r_{\text{eff}})^n$ which increases with r. Interestingly, for a white-Gaussian noise $\mathbf{Z}$ the optimum r that minimizes the sum has a simple closed form expression.

Proposition 13.3.1 (Optimal search radius) *For AWGN at VNR $\mu \geq 2\pi$ (i.e., ensemble point density $\gamma \leq (2\pi\sigma^2)^{-n/2}$), the optimal search radius in the threshold-decoding bound $P_e^{\text{TB}}(r, n)$ of Lemma 7.7.2 is given by*

$$r_{\text{th}}^* = r_{\text{noise}} \cdot \sqrt{1 + \ln(\mu/2\pi e)}$$
$$= r_{\text{noise}} \cdot \sqrt{1 + 2\ln(e)\Delta}, \quad (13.17)$$

where $r_{\text{noise}} = \sigma\sqrt{n}$.

Proof Taking the derivative of $P_e^{\text{TB}}(r, n)$ with respect to r and equating to zero, we obtain

$$\frac{d P_e^{\text{TB}}(r, n)}{dr} = \gamma V_n n r^{n-1} - f_{\|\mathbf{Z}\|}^*(r; n) = 0$$

where $f_{\|\mathbf{Z}\|}^*(r; n)$ denotes the density function of the norm $\|\mathbf{Z}\|$ of an n-dimensional AWGN vector with variance σ^2. Due to the isotropy of the AWGN distribution, the latter density is given by the density of $\mathbf{Z}$ (at $\|\mathbf{Z}\| = r$) multiplied by the area of a spherical shell of radius r (see (7.11c)): [2]

$$f_{\|\mathbf{Z}\|}^*(r; n) = V_n n r^{n-1} \left(\frac{1}{\sqrt{2\pi\sigma^2}}\right)^n e^{-r^2/2\sigma^2}. \quad (13.18)$$

Substituting in (13.17), we obtain $\gamma = (2\pi\sigma^2)^{-n/2} \exp(-r^2/2\sigma^2)$. This equation is solved by $r = r_{\text{th}}^*$ of (13.17), provided $\gamma \leq (2\pi\sigma^2)^{-n/2}$. □

As we see in the proof above, the optimum search radius r_{th}^* equalizes the *derivatives* of the two terms in the threshold bound (13.16). It turns out, not surprisingly, that it also equalizes their exponents; see Figure 13.3. Thus, the exponent of the sum at $r = r_{\text{th}}^*$ is equal to the exponent of each of the terms.

---

[2] It can also be written in terms of the nth-order *chi-square* density function:

$$f_{\|\mathbf{Z}\|}^*(r; n) = \frac{1}{\sigma} \chi_{\text{PDF}}^2(r/\sigma; n).$$

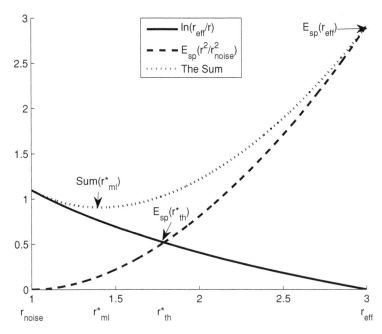

Figure 13.3 The sub-exponents $\ln(r_{\text{eff}}/r)$ and $E_{\text{sp}}(r^2/r_{\text{noise}}^2)$ of the threshold-decoding bound, for r in the range $r_{\text{noise}} < r < r_{\text{eff}}$, where $r_{\text{eff}}/r_{\text{noise}} = 3$. Also shown is their sum for the NN decoding bound below.

Proposition 13.3.2 (Threshold-decoding exponent) *For AWGN and a fixed VNR* $\mu \geq 2\pi$, *the optimal threshold-decoding bound* (13.16) *is given exponentially by*

$$P_e^{\text{TB}}(r_{\text{th}}^*, n) \doteq e^{-n E_{\text{sp}}\left(\frac{r_{\text{th}}^{*2}}{r_{\text{noise}}^2}\right)} = e^{-n E_{\text{sp}}\left(1 + \ln\left(\frac{\mu}{2\pi e}\right)\right)}. \tag{13.19}$$

Proof Writing the competing codeword term as $\gamma V_n r^n = (r/r_{\text{eff}})^n$, the threshold-decoding bound (13.16) is given exponentially by

$$P_e^{\text{TB}}(r, n) \doteq e^{-n \ln\left(\frac{r_{\text{eff}}}{r}\right)} + e^{-n E_{\text{sp}}\left(\frac{r^2}{r_{\text{noise}}^2}\right)} \tag{13.20}$$

and it is dominated by the *smaller* of the two exponents. Figure 13.3 shows these two exponents for $r_{\text{noise}} < r < r_{\text{eff}}$. We see that while the competing codeword exponent decreases (from $\ln(r_{\text{eff}}/r_{\text{noise}})$ to zero), the sphere crossing exponent increases (from zero to $E_{\text{sp}}(r_{\text{eff}}^2/r_{\text{noise}}^2)$), and the exponents meet at

$$r = r_{\text{noise}} \cdot \sqrt{1 + \ln(r_{\text{eff}}/r_{\text{noise}})^2} \approx r_{\text{th}}^*$$

where the last approximation holds for large n (see (13.14)); hence we can plug it in the right term of (13.20) to obtain the desired exponent. $\qquad\square$

We see that near capacity (i.e., a small gap to capacity $\Delta \to 0$, or $\mu \to 2\pi e$ and $r_{\text{eff}}/r_{\text{noise}} \to 1$), the optimum search radius r_{th}^* is approximately equal to the typical noise radius $r_{\text{noise}} = \sqrt{n\sigma^2}$. This is in line with the proof of the existence of lattices which are good for channel coding (Theorem 7.7.1), where we want the lattice to be the densest possible with respect to the noise.

Channel dispersion

A careful analysis of the threshold-decoding bound (13.16) of the MHS ensemble reveals that near capacity $P_e^{\text{TB}}(r_{\text{th}}^*, n)$ goes approximately like $e^{-n\Delta^2}$; more precisely, the gap to capacity decreases with the dimension n as

$$\Delta_{\text{MHS}}(n, P_e) = \sqrt{\frac{1}{2n}} Q^{-1}(P_e) + O\left(\frac{\ln n}{n}\right), \qquad (13.21)$$

for a fixed error probability P_e (where here Δ is measured in nats). The factor $1/2$ in the first term is known as the *channel dispersion*. This term confirms the $O(1/\sqrt{n})$ convergence of the NVNR to the ideal value of $2\pi e$ in (7.58). See [125, 222] and Problem 13.3.

When deviating from capacity, however, r_{th}^* increases much slower than r_{eff} (the square radius increases *logarithmically* with the VNR instead of linearly). Hence, as can be seen in Figure 13.2, the threshold-decoding exponent is much worse than the sphere packing exponent (13.15).

13.4   Nearest-neighbor decoding exponent

Setting a *fixed* search radius around the received vector, as in threshold decoding, may reduce the search complexity. However, it is clearly *not optimal*.

As we have seen in the discussion above, if the rate is far from capacity, then the optimal search radius r_{th}^* is much smaller than the effective radius r_{eff} of the lattice cells. On the other hand, quantization goodness (Section 7.9.4) implies that the distance of a random point to the nearest lattice point of a "good" lattice is $\approx r_{\text{eff}}$ with a high probability. It follows that even if the noise $\mathbf{Z}$ exceeds r_{th}^*, implying a threshold-decoding failure, it is still unlikely that there are competing codewords within a distance $\|\mathbf{Z}\|$ around the received point.[3]

As illustrated in Figure 13.4, a more efficient decoder strategy is to use an *adaptive threshold*: enlarge the search radius until it hits the first codeword. Thus, the search radius is usually small, around the noise typical radius r_{noise}, and it is large only if the noise is atypically large. This strategy amounts to *nearest-neighbor*

---

[3] The reason for this conservative choice of r_{th}^* is that it avoids ambiguity in the rare event (in the MHS ensemble) of a lattice with an atypically short minimum distance. See more on that in Section 13.7.

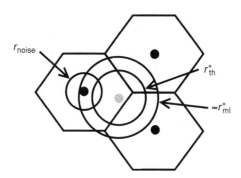

Figure 13.4 A situation where the typicality decoder declares an error ("no codeword inside search radius"), while ML decoding is successful.

(NN) decoding, which for AWGN amounts to ML decoding. We shall next modify the derivation of Lemma 7.7.2 to account for NN decoding.

Given that the zero lattice point was transmitted, and that the noise vector is $\mathbf{z}$, the NN decoder will make an error if and only if there is another lattice point within a radius $\|\mathbf{z}\|$ around the point $\mathbf{z}$. This amounts to the event $N_{\mathcal{B}(\mathbf{z},\|\mathbf{z}\|)}(\Lambda) \geq 1$. The error probability of the NN decoder is thus

$$P_e^{NN}(\Lambda) = \int_0^\infty f_{\|\mathbf{Z}\|}(r,n) \cdot \Pr\{N_{\mathcal{B}(\mathbf{Z},r)}(\Lambda) \geq 1 \mid \|\mathbf{Z}\| = r\}dr, \qquad (13.22)$$

where $f_{\|\mathbf{Z}\|}(r,n)$ is the radius density function of the noise $\mathbf{Z}$.

For a random lattice from the MHS ensemble, the probability of the event $N_{\mathcal{B}(\mathbf{z},\|\mathbf{z}\|)}(\Lambda) \geq 1$ is bounded from above by (see Lemma 7.7.2)

$$\gamma \, V_n \|\mathbf{z}\|^n = \left(\frac{\|\mathbf{z}\|}{r_{\text{eff}}}\right)^n. \qquad (13.23)$$

When $\|\mathbf{z}\| = r_{\text{eff}}$, this bound is equal to 1. Thus, for larger values of $\|\mathbf{z}\|$ the bound is useless, and we can replace it by 1. Substituting (13.23) in (13.22) (with a switch to 1 at $r = r_{\text{eff}}$), we obtain the following bound on the average NN decoding error probability over the MHS ensemble $\mathbb{L}$:

$$P_e^{NN}(\mathbb{L}) \leq \int_0^\infty f_{\|\mathbf{Z}\|}(r,n) \cdot \min\left\{\left(\frac{r}{r_{\text{eff}}}\right)^n, 1\right\}dr \qquad (13.24\text{a})$$

which is simply

$$P_e^{NN}(\mathbb{L}) = \int_0^{r_{\text{eff}}} f_{\|\mathbf{Z}\|}(r,n) \cdot \left(\frac{r}{r_{\text{eff}}}\right)^n dr + \Pr\left\{\|\mathbf{Z}\| \geq r_{\text{eff}}\right\}, \qquad (13.24\text{b})$$

where the second term amounts to $\int_{r_{\text{eff}}}^\infty f_{\|\mathbf{Z}\|}(r,n)dr$. Using integration by parts for the integral in (13.24b), the second term in (13.24b) cancels, and we obtain

$$P_e^{NN}(\mathbb{L}) \leq \int_0^{r_{\text{eff}}} \Pr\{\|\mathbf{Z}\| \geq r\} \cdot \frac{n}{r} \cdot \left(\frac{r}{r_{\text{eff}}}\right)^n dr. \qquad (13.24\text{c})$$

The bound (13.24) holds for a general noise distribution. In the case of AWGN, NN decoding amounts to ML decoding, and we denote the upper bound of (13.24) by $P_e^{\text{MLB}}(n)$. Neglecting sub-exponential terms in (13.24c), we conclude that the bound on the average ML decoding error probability of the MHS ensemble in the presence of AWGN is given exponentially by

$$P_e^{\text{MLB}}(n) \doteq \int_0^{r_{\text{eff}}} e^{-n E_{\text{sp}}\left(\frac{r^2}{r_{\text{noise}}^2}\right)} \cdot e^{-n \ln\left(\frac{r_{\text{eff}}}{r}\right)} dr. \tag{13.25}$$

Interestingly, the integrand in (13.25) is a product of the two exponential terms in the *threshold-decoding bound* (13.20). Hence, the exponent of the integrand is the *sum* of the two sub-exponents, shown in Figure 13.3. To assess the integral of an exponential function, we shall use the *Laplace method of integration*.

Proposition 13.4.1 (Laplace's method of integration) *If $g(x, n) \doteq e^{-nf(x)}$ is Riemann integrable, and $f(x)$ is a continuous function in the interval $[a, b]$, then*

$$\int_a^b g(x, n)dx \doteq e^{-n f_{\min}}, \tag{13.26}$$

where

$$f_{\min} = \min_{a \leq x \leq b} f(x).$$

In other words, the integral is dominated exponentially by the smallest exponent of the integrand.

Proof See Problem 13.5. □

It follows from Proposition 13.4.1, that the exponent of the integral in (13.25) is given by the minimum exponent of the integrand in the range $(0, r_{\text{eff}})$. Hence, the exponent E_r of $P_e^{\text{MLB}}(n)$ is given by

$$E_r = \min_{0 \leq r \leq r_{\text{eff}}} \left\{ E_{\text{sp}}\left(\frac{r^2}{r_{\text{noise}}^2}\right) + \ln\left(\frac{r_{\text{eff}}}{r}\right) \right\}. \tag{13.27}$$

Note that at the edge $r = r_{\text{eff}}$, the minimum is equal to the sphere packing exponent $E_{\text{sp}}(r_{\text{eff}}^2/r_{\text{noise}}^2)$. A closer inspection reveals that for $r_{\text{eff}} \leq \sqrt{2} r_{\text{noise}}$, this edge is also the minimizing point in the entire interval $0 \leq r \leq r_{\text{eff}}$ (see Figure 13.3); hence, in this case E_r coincides with the sphere packing exponent. In general, the minimum occurs at

$$r_{\text{ml}}^* = \begin{cases} r_{\text{eff}}, & \text{for } r_{\text{noise}} \leq r_{\text{eff}} \leq \sqrt{2} r_{\text{noise}} \\ \sqrt{2} r_{\text{noise}}, & \text{for } r_{\text{eff}} \geq \sqrt{2} r_{\text{noise}}, \end{cases} \tag{13.28}$$

hence for $r_{\text{eff}} > \sqrt{2} r_{\text{noise}}$ the minimum is strictly smaller than the sphere packing exponent.

Replacing the square radius to noise ratio $r_{\text{eff}}^2/r_{\text{noise}}^2$ by $\mu/2\pi e$ (which, by (13.14), holds for large n), we thus obtain the following exponential upper bound on the error probability of the MHS ensemble, under ML decoding in the presence of AWGN. See Figure 13.2.

Theorem 13.4.1 (Random coding exponent) *The ML decoding bound of* (13.25), *on the average error probability of the MHS ensemble in the presence of AWGN at VNR* μ, *is given exponentially by*

$$P_e^{\mathrm{MLB}}(n) \doteq e^{-nE_r\left(\frac{\mu}{2\pi e}\right)} \tag{13.29}$$

where $E_r(\cdot)$ *is the* random coding error exponent, *defined as*

$$E_r(x) = \begin{cases} \frac{1}{2}[\ln(x) + \ln(e/4)], & if\ x \geq 2 \\ \frac{1}{2}[x - 1 - \ln(x)], & if\ 2 \geq x \geq 1 \\ 0, & if\ x \leq 1. \end{cases} \tag{13.30}$$

Since in the mid range of VNRs between $2\pi e$ and $4\pi e$ the exponent $E_r(\cdot)$ meets the sphere packing exponent (13.15) – which is an exponential *lower* bound on the error probability for *any* lattice – we conclude that in this range $E_r(\cdot)$ cannot be improved. A stronger analysis of the average performance of a random (lattice) ensemble shows that the upper bound of Theorem 13.4.1 is, in fact, exponentially *tight* for *all* VNR; see [12, 13, 60]. Hence, $E_r(\mu/2\pi e)$ is the *true* error exponent of the random ensemble. Is this also the best possible error exponent of *any lattice* for $\mu > 4\pi e$?

It turns out that for rates per unit volume which are significantly smaller than capacity (i.e., $\mu \gg 2\pi e$) the *minimum distance* dominates the error behavior. In this regime, the MHS ensemble average suffers from a small subset of lattices which have a "bad" minimum distance. By excluding these lattices from the ensemble – in a process called *expurgation* – the exponent of the remaining ensemble exceeds that of the full ensemble (Theorem 13.4.1).

But in order to capture the improvement due to expurgation, we must deviate from our previous line of analysis. We shall develop bounds on the error probability based on the minimum distance, as well as on the more elaborate *distance spectrum*.

13.5 Distance spectrum and pairwise errors

The threshold-based analysis of Sections 13.3 and 13.4 took an a posteriori (or "reverse channel") viewpoint: given the received vector $\mathbf{Y}$, examine the presence of competing codewords within the search range. We now center our viewpoint around the transmitted codeword λ, and ask: what is the chance that a neighboring codeword λ' will be more likely?

A *pairwise error* between a transmitted codeword λ and a competing codeword λ' is an error event for a codebook that contains only these two codewords. The error probability in this case is simple to compute.

Proposition 13.5.1 (Pairwise error) *For AWGN with variance σ^2 and NN decoding, the pairwise error probability between two codewords λ and λ' is given by $Q(\|\lambda - \lambda'\|/2\sigma)$, where $Q(\cdot)$ denotes the Q-function.*

Proof Since the AWGN distribution is rotation invariant, it can be rotated so that the first noise component would be parallel to the line connecting λ and λ'. All the other noise components are then orthogonal to this line. It follows that an error will occur if and only if the first noise component (after rotation) is larger than half the distance between λ and λ'. □

The simple dependence on the pairwise distance in Proposition 13.5.1 gives rise to the definition of a *distance spectrum*.

Definition 13.5.1 *The distance spectrum $N_\Lambda(d)$, for $d \geq d_{\min}$, is defined as*

$$N_\Lambda(d) = \textit{number of lattice points at a distance d from the origin.}$$

Note that discreteness of the lattice implies that $N_\Lambda(d)$ is non-zero for a *countable* set of values of d. Also, the *geometric uniformity* of the lattice implies that each lattice point sees the same distance spectrum; that is, $N_\Lambda(d)$ is, in fact, the number of lattice points at distance d from *any* lattice point. [4]

Consider now a decomposition of the error event $\{\mathbf{Z} \notin \mathcal{V}_0\}$ into a *union* of pairwise errors to the Voronoi cell's neighbors. Assuming nearest-neighbor (Euclidean) partition, we have

$$\Pr\{\mathbf{Z} \notin \mathcal{V}_0\} = \Pr\{\mathbf{Z} \text{ is closer to } \lambda \text{ than to } \mathbf{0}, \text{ for some } \lambda \neq \mathbf{0} \}. \quad (13.31)$$

Clearly, only points λ that determine the faces of the fundamental Voronoi cell affect this probability. This includes lattice points whose distance from the origin is between $d_{\min} = 2r_{\text{pack}}(\Lambda)$ and $2r_{\text{cov}}(\Lambda)$.

The probability of the union of events in (13.31) is, on the one hand, lower bounded by the probability of each one of them; in particular, by the most likely one, i.e., the error to the non-zero lattice point closest to the origin:

$$\Pr\{\mathbf{Z} \notin \mathcal{V}_0\} \geq \max_{\lambda \in \Lambda, \lambda \neq 0} \Pr\{\mathbf{Z} \text{ is closer to } \lambda \text{ than to } \mathbf{0}\}$$

$$= Q\left(\frac{d_{\min}/2}{\sigma}\right),$$

where in the second line we assumed AWGN and used Proposition 13.5.1. This gives us the *minimum-distance bound* (3.34).

On the other hand, by the union bound, (13.31) is upper bounded by summing the pairwise error probabilities over all non-zero lattice points.

---

[4] The Laplace transform of the function $N_\Lambda(d)$ is related to the *theta series* of the lattice; see [49].

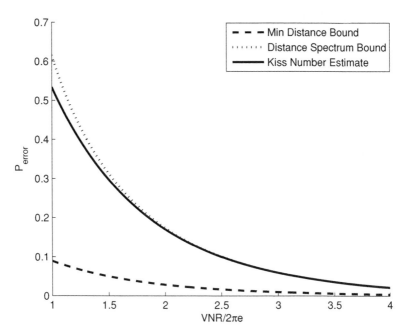

Figure 13.5 Pairwise approximation for error probability for the hexagonal lattice.

Proposition 13.5.2 (Distance-spectrum bound)

$$\Pr\{\mathbf{Z} \notin \mathcal{V}_0\} \leq \sum_{d \geq d_{\min}} N_\Lambda(d) \cdot Q\left(\frac{d/2}{\sigma}\right). \tag{13.32}$$

Note that the sum above can be limited to distances in the range $2r_{\text{pack}}(\Lambda) \leq d \leq 2r_{\text{cov}}(\Lambda)$.

A "compromise" between the minimum-distance lower bound (3.34) and the distance-spectrum upper bound (13.32), is given by the first term in the sum in (13.32):

$$P_e(\Lambda, \sigma^2) \approx N_\Lambda(d_{\min}) \cdot Q\left(\frac{d_{\min}/2}{\sigma}\right), \tag{13.33}$$

where $N_\Lambda(d_{\min})$ is the *kissing number* of the lattice (see Section 3.1). As can be seen from Figure 13.5, this approximation is useful at high VNR ($\mu \gg 2\pi e$), when the lattice is sparse with respect to the noise. As it turns out, the exact value of the coefficient $N_\Lambda(d_{\min})$ does not affect the error exponent in this regime.

13.6 Minimum-distance exponent

We know that the packing efficiency $\rho_{\text{pack}}(\Lambda)$ is always smaller than or equal to 1, and it is equal to one-half for "Minkowski good" lattices. Furthermore, it is conjectured that for large-dimensional lattices, $\rho_{\text{pack}}(\Lambda)$ cannot exceed one-half;

see the discussion after Corollary 7.6.1. That is, for any sequence of lattices Λ_n of increasing dimension n,

$$\limsup_{n \to \infty} \rho_{\text{pack}}(\Lambda_n) \leq 1/2. \tag{13.34}$$

Using the minimum-distance bound (3.34), the relation between minimum distance and packing efficiency, and using the fact that the Q-function satisfies $Q(\sqrt{nt}) \doteq e^{-nt/2}$, as $n \to \infty$ (see (13.9)), we obtain the following.

Proposition 13.6.1 (Minimum-distance exponent) *For a sequence of lattices Λ_n of increasing dimension with a bounded packing efficiency $\rho_{\text{pack}}(\Lambda_n) \leq \rho$ for all n, the decoding error probability in the presence of AWGN at VNR μ is exponentially lower bounded by*

$$P_e(\Lambda_n, \sigma^2) \stackrel{\cdot}{\geq} e^{-n\rho^2 \frac{\mu}{4\pi e}}.$$

As a consequence, if the conjecture (13.34) is also true, then

$$P_e(\Lambda_n, \sigma^2) \stackrel{\cdot}{\geq} e^{-n \frac{\mu}{16\pi e}} \tag{13.35}$$

for any sequence of lattices of increasing dimension. (This is the same as (13.3) with ERNR replaced by $\mu/2\pi e$.)

Proof By the definition of packing efficiency, $\rho_{\text{pack}}(\Lambda)r_{\text{eff}}(\Lambda) = r_{\text{pack}}(\Lambda)$. Thus, for any lattice Λ_n in the sequence $d_{\min}(\Lambda_n)/2 = r_{\text{pack}}(\Lambda_n) \leq \rho r_{\text{eff}}(\Lambda_n)$, so the minimum-distance lower bound (3.34) becomes

$$P_e(\Lambda_n, \sigma^2) \geq Q\left(\sqrt{n\rho^2 r_{\text{eff}}(\Lambda_n)/r_{\text{noise}}}\right)$$

where $r_{\text{noise}} = \sigma\sqrt{n}$. The proof follows since the Q-function satisfies $Q(\sqrt{nt}) \doteq e^{-nt/2}$, as $n \to \infty$ (see (13.9)), and by the asymptotic ERNR–VNR relation (13.14). □

13.7   The expurgated MHS ensemble

The exponent (13.35) ignores the multiplicity of code words at distance d, and it is too optimistic near capacity; see Figure 13.2. Nevertheless, we shall now show that far from capacity it is, in fact, achievable by a slight modification of the MHS ensemble.

The volume of a spherical shell about the origin of radius d and thickness Δd is approximately $nV_n d^{n-1}\Delta d$. Thus, for an MHS ensemble $\mathbb{L}$ of point density γ, the average number of lattice points inside this shell is approximately $\gamma n V_n d^{n-1}\Delta d$. Substituting in the distance-spectrum bound (13.32), and taking the limit of infinitely small increments, we obtain the bound

$$P_e(\mathbb{L}_n, \sigma^2) \leq \gamma n V_n \int_{d_{\min}}^{\infty} t^{n-1} Q\left(\frac{t/2}{\sigma}\right) dt. \tag{13.36}$$

If we take the MHS ensemble as is, it contains lattices with arbitrarily small minimum distance. Hence, the integral above must start at zero, making the upper bound poor. Nevertheless, the probability of such "bad" lattices is small. Minkowski's theorem (see Corollary 7.6.1) assures us that in each dimension $n \geq 2$, there exists a lattice whose minimum distance is greater than or equal to its effective radius. As evident from the following lemma, this property holds not only for one lattice but for *most* lattices in the MHS ensemble as the dimension goes to infinity.

Proposition 13.7.1 (Bad minimum distance) *For a random n-dimensional lattice Λ in the MHS ensemble $\mathbb{L}_n$,*

$$\Pr\{d_{\min}(\Lambda) < r_{\text{eff}}\} \to 0 \quad \text{as } n \to \infty, \tag{13.37}$$

where r_{eff} is the effective radius of the ensemble (i.e., $r_{\text{eff}} = (\gamma V_n)^{-1/n}$ where γ is the ensemble point density). Hence, $d_{\min}(\Lambda)$ is asymptotically greater than or equal to r_{eff} in probability.

Proof Note that $d_{\min}(\Lambda) \leq r$ amounts to $N_{\mathcal{B}(0,r)}(\Lambda) \geq 1$. Thus, for a random lattice $\Lambda \in \mathbb{L}_n$,

$$\Pr\{d_{\min}(\Lambda) \leq r\} = \Pr\left\{N_{\mathcal{B}(0,r)}(\Lambda) \geq 1\right\} \tag{13.38}$$

$$\leq E_{\mathbb{L}}\left\{N_{\mathcal{B}(0,r)}(\Lambda)\right\} \tag{13.39}$$

$$= \left(\frac{r}{r_{\text{eff}}}\right)^n, \tag{13.40}$$

where in (13.39) we followed the proof of Lemma 7.7.1. It follows that

$$\Pr\left\{d_{\min}(\Lambda) < \frac{r_{\text{eff}}}{1+\epsilon}\right\} \leq \frac{1}{(1+\epsilon)^n}, \tag{13.41}$$

which goes to zero as n goes to infinity, for all $\epsilon > 0$. □

We can therefore exclude the lattices with minimum distance smaller than r_{eff} from the MHS ensemble, almost (as $n \to \infty$) without affecting the distribution of the remaining lattices. Specifically, for $0 < \alpha < 1$, let $\mathbb{L}_{\exp}(\alpha)$ denote the conditional MHS ensemble given $d_{\min}(\Lambda) \geq \alpha r_{\text{eff}}$. Then, by (13.41) (setting $\alpha = 1/(1+\epsilon)$), the integral (13.36) becomes

$$P_e(\mathbb{L}_{\exp}(\alpha), \sigma^2) \leq (1 - \alpha^n) \int_{\alpha r_{\text{eff}}}^{\infty} \gamma n V_n t^{n-1} Q\left(\frac{t/2}{\sigma}\right) dt,$$

so

$$P_e(\mathbb{L}, \sigma^2) \dot{\leq} \int_{r_{\text{eff}}}^{\infty} \left(\frac{t}{r_{\text{eff}}}\right)^n e^{-\frac{n}{2}\left(\frac{t/2}{r_{\text{noise}}}\right)^2} dt, \tag{13.42}$$

where the factor before the first integral follows from Bayes' law (noting that $p(A|B) \leq P(A)/P(B)$, where B is the event that $d_{\min}(\Lambda) \geq \alpha r_{\text{eff}}$), and the asymptotic inequality follows by taking $n \to \infty$ and then $\alpha \to 1$, and using the bound (13.9) on the Q-function. See Problem 13.7.

We can now apply the Laplace method of integration to investigate the exponential behavior of this integral. It turns out that for $r_{\text{eff}}/r_{\text{noise}} \geq 2$, i.e., VNR greater than or equal to $\mu_x = 8\pi e$, the integral (13.42) is dominated by its lower boundary r_{eff}. In this case, its exponent is strictly better than the random coding error exponent (13.30), and it coincides with the conjectured minimum-distance exponent (13.35). For lower values of VNRs, the integral (13.42) is dominated by $t^* = 2r_{\text{noise}}$, and the new exponent coincides with the middle section of the random coding error exponent.

Theorem 13.7.1 (Expurgated exponent) *The average error probability of the expurgated MHS ensemble is exponentially upper bounded by*

$$P_e(\mathbb{L}_{\text{exp}}, \sigma^2) \stackrel{.}{\leq} e^{-nE_x(\mu/2\pi e)}$$

where $E_x(\cdot)$ is the expurgated error exponent, *defined as*

$$E_x(x) = \begin{cases} x/8, & \text{if } x \geq 4 \\ \frac{1}{2}[\ln(x) + \ln(e/4)], & \text{if } 4/e \leq x \leq 4 \\ 0, & \text{if } x \leq 4/e. \end{cases} \qquad (13.43)$$

We see that in the range $1 \leq x \leq 2$, the expurgated exponent (13.43) is worse than the random coding exponent (13.30). The best achievable exponent is the combination of both exponents. This combination is the full *Poltyrev exponent*, shown in Figure 13.2.

13.8   Error exponents of Voronoi codes

So far we have dealt with infinite (unbounded) lattice constellations. In this section, we turn to examine the error exponent in Voronoi modulation, where the receiver employs estimation and lattice decoding; see Chapter 9. The analysis applies also for lattice dirty-paper coding, because it has the same equivalent channel as in the zero-interference case; see Section 10.6.

For a linear estimator $g(\mathbf{Y}) = \alpha \mathbf{Y}$, the estimation error vector is a linear *mixture* of AWGN and dither (9.58). The (Euclidean) lattice decoding error probability is bounded by the probability (9.67) that this mixture noise falls outside the fundamental Voronoi cell of the fine lattice:

$$P_e \leq \Pr\{\alpha \mathbf{Z} + (\alpha - 1)\mathbf{U} \notin \mathcal{V}_0(\Lambda_1)\}, \qquad (13.44)$$

where the dither $\mathbf{U}$ is uniform over the fundamental Voronoi cell $\mathcal{V}_0(\Lambda_2)$ of the coarse lattice. Since the dither is not Gaussian, this bound may have a different exponential behavior than that characterized in Theorems 13.1.1–13.4.1.

We shall first investigate nested lattice pairs, where the fine lattice is good while the coarse lattice is arbitrary (e.g., $\Lambda_2 = \mathbb{Z}^n$). We then show that if *both* fine and coarse lattices are good (the fine for AWGN-channel coding, and the coarse for covering, i.e., Rogers good), and the estimation coefficient is the *Wiener* coefficient α^*, then the error exponent is at least as good as the Poltyrev random coding exponent (13.30) at the same gap to capacity. Finally, we observe that if we reduce α below α^*, then the error exponent can improve beyond that of unbounded lattices; in fact, with an optimal $\alpha(\text{SNR}, R)$ it achieves the optimum ML decoding error exponent of the power-constrained AWGN channel for all rates R near capacity. Thus, for a good Voronoi modulation system, the linear estimation lattice decoder is error-exponent-wise *optimal*.

13.8.1 Sphere crossing exponent for mixture noise

We start our analysis with modifying the Gaussian sphere crossing bound of Proposition 13.1.3 to the case of a mixture noise.

Theorem 13.8.1 (Mixture sphere crossing) *If* $\mathbf{Z}$ *is an n-dimensional AWGN with variance N, and* $\mathbf{U} \sim \text{Unif}(\mathcal{V}_0(\Lambda))$ *for some lattice* Λ *in* $\mathbb{R}^n$*, then*

$$\Pr\{\|\alpha\mathbf{Z} + (\alpha - 1)\mathbf{U}\| > r\} \le e^{-nE_{\text{sp}}(r^2/r_{\text{mix}}^2)}, \tag{13.45}$$

where $E_{\text{sp}}(\cdot)$ *is the* sphere packing exponent (13.7),

$$r_{\text{mix}}^2 \triangleq n\alpha^2 N + (1 - \alpha)^2 r_{\text{cov}}^2(\Lambda), \tag{13.46}$$

and $r_{\text{cov}}(\Lambda)$ *is the lattice-covering radius* (3.16).

In view of the pure AWGN case (Proposition 13.1.3), we conclude the following.

Corollary 13.8.1 (A Gaussian bound) *The sphere crossing exponent of a mixture noise* (13.45) *is the same as the sphere crossing exponent of a pure AWGN with power*

$$N_{\text{eq}} = \frac{r_{\text{mix}}^2}{n} = \alpha^2 N + (1 - \alpha)^2 \frac{r_{\text{cov}}^2(\Lambda)}{n}. \tag{13.47}$$

In particular, setting $\sigma^2(\Lambda) = P$, if Λ is a high-dimensional "Rogers-good" lattice, then $r_{\text{cov}}^2(\Lambda)/n \approx \sigma^2(\Lambda) = P$; thus $N_{\text{eq}} \approx \alpha^2 N + (1 - \alpha)^2 P = \text{MSE}(\alpha)$, which is the variance of the mixture noise (9.61). The exponent in (13.45) then becomes that of an AWGN with the *same variance*. At the other extreme, if Λ is a cubic lattice, then $r_{\text{cov}}(a\mathbb{Z}^n) = \sqrt{n}a/2$, while $\sigma^2(a\mathbb{Z}^n) = a^2/12$, so $r_{\text{cov}}^2(a\mathbb{Z}^n)/n = 3\sigma^2(a\mathbb{Z}^n) = 3P$; thus, N_{eq} becomes $\alpha^2 N + 3(1 - \alpha)^2 P$, i.e., the self-noise component of the mixture noise is 5 dB higher than in (9.61).

The proof of Theorem 13.8.1 follows from the Chernoff bound (Proposition 13.1.1) and the following three lemmas. The first lemma bounds the mixture sphere crossing probability by conditioning on a specific dither value.

Lemma 13.8.1 (Conditioning on a large dither) *If $\mathbf{Z}$ is AWGN and $\|\mathbf{U}\| \leq u_{\max}$, then*

$$\Pr\{\|\alpha\mathbf{Z} + (\alpha - 1)\mathbf{U}\| > r\} \leq \Pr\{\|\alpha\mathbf{Z} + (\alpha - 1)\mathbf{u}_0\| > r\} \tag{13.48}$$

for any vector $\mathbf{u}_0$ with norm $u_{\max}$.

Proof The right-hand side amounts to conditioning the sphere crossing event on $\mathbf{U} = \mathbf{u}_0$. The unconditional sphere crossing probability is the expectation of the right-hand side over the random dither $\mathbf{U}$. Now, due to the isotropy of the AWGN $\mathbf{Z}$, the conditional sphere crossing probability depends only on the norm of $\mathbf{u}_0$, and increases monotonically with it; see Problem 9.11 and Appendix A.4. Thus, the conditional probability is maximum for the largest possible norm of the dither, which is $r_{\text{cov}}(\Lambda)$. $\qquad\square$

The vector $\alpha\mathbf{Z} + (\alpha - 1)\mathbf{u}_0$ in (13.48) is a *non-zero-mean* Gaussian vector. The second lemma gives the characteristic function of its squared norm.

Lemma 13.8.2 (Characteristic function of a squared non-zero-mean Gaussian) *If $\tilde{\mathbf{Z}} \sim N(\eta, \sigma^2 I)$ is Gaussian with mean η and white covariance matrix $\sigma^2 I$, then the characteristic function of $\|\tilde{\mathbf{Z}}\|^2$ is given by*

$$E\left\{e^{s\|\tilde{\mathbf{Z}}\|^2}\right\} = \frac{e^{s\|\eta\|^2/(1-2s\sigma^2)}}{(1 - 2s\sigma^2)^{n/2}}, \quad \text{for } 0 \leq s < 1/2\sigma^2. \tag{13.49}$$

Proof The proof is based on a straightforward integration, and is left as an exercise; see Problem 13.9. $\qquad\square$

Note that (13.49) depends on the mean vector η only through its norm. For $\eta = \mathbf{0}$, it reduces to the squared AWGN characteristic function of Proposition 13.1.3.

To complete the proof of Theorem 13.8.1, the last lemma shows that the characteristic function associated with a *non-zero*-mean Gaussian is upper bounded by that of a *zero*-mean Gaussian with the *same second moment*.

Lemma 13.8.3 (Zero mean maximizes Chernoff) *For all $0 \leq s < 1/2\sigma^2$,*

$$\frac{e^{\frac{2}{n}s\|\eta\|^2/(1-2s\sigma^2)}}{1 - 2s\sigma^2} \leq \frac{1}{1 - 2s(\sigma^2 + \|\eta\|^2/n)} \tag{13.50}$$

with equality (for $\eta \neq \mathbf{0}$) if and only if $s = 0$.

Proof Inequality (13.50) is equivalent to

$$e^{-\frac{2}{n}s\|\eta\|^2/1-2s\sigma^2} \geq 1 - \frac{2s\|\eta\|^2/n}{1 - 2s\sigma^2},$$

which follows from the well-known inequality $e^{-x} \geq 1 - x$, with equality if and only if $x = 0$. $\qquad\square$

Proof of Theorem 13.8.1 Lemma 13.8.1 implies that the sphere crossing probability of the mixture noise (left-hand side of (13.45)) is upper bounded by that of a biased Gaussian $\tilde{\mathbf{Z}} = (\alpha - 1)u_{\max} + N(0, \alpha^2 N \cdot I)$. Lemmas 13.8.2 and 13.8.3 then imply that the characteristic function of $\|\tilde{\mathbf{Z}}\|^2$ is upper bounded by the characteristic function of a squared AWGN with variance r_{mix}^2/n (see (13.47)). The desired result now follows from the Chernoff bound (Proposition 13.1.1), with the optimum parameter s^* for AWGN with variance r_{mix}^2/n. □

The equivalence to Gaussian noise in Corollary 13.8.1 hides the fact that the mixture noise may have a *lighter tail* – hence a *smaller error probability* – than that of a white-Gaussian noise. See the discussion near Figure 9.13. This weakness of the exponential bound (13.45) is due to the zero mean bound of Lemma 13.8.3, and we shall reconsider it in Section 13.8.3.

13.8.2 Random coding exponent for mixture noise

The mixture sphere crossing bound of Theorem 13.8.1 is a useful tool for deriving exponential bounds on the error probability (9.66) in NN lattice decoding.

Following our earlier analysis of an unbounded lattice, we use a random coding argument with respect to a fine lattice satisfying the *ensemble uniformity property* – (7.52) for the MHS ensemble or (7.92) for the $\mathbb{L}_{n,k,p,\gamma}$ (Loeliger) ensemble. The shaping (coarse) lattice can be arbitrary, although eventually we shall take a Rogers-good lattice; see Definition 7.3.1. See Section 8.5 for the existence of good nested lattice pairs.

We obtain the following exponential upper bound on P_e.

Theorem 13.8.2 (Wiener estimation achieves the Poltyrev random-coding exponent) *Consider a rate-R/block-length-n Voronoi modulation system, with a random fine lattice Λ_1 satisfying the ensemble uniformity property, and a shaping (coarse) lattice Λ_2 with a covering efficiency upper bounded by $\bar{\rho}_{\mathrm{cov}}(\Lambda_2)$ (3.17). The system transmits over an AWGN channel at some SNR, and the receiver employs Wiener estimation (with $\alpha = \alpha^* = \mathrm{SNR}/(1 + \mathrm{SNR})$) and NN lattice decoding. As n goes to infinity, the error probability (13.44) is exponentially upper bounded by*

$$P_e \lesssim e^{-nE_r\left(2^{2\Delta} / \bar{\rho}_{\mathrm{cov}}^2(\Lambda_2)\right)}, \tag{13.51}$$

where $E_r(\cdot)$ is the Poltyrev random coding exponent (13.30), and

$$\Delta = C_{\mathrm{AWGN}} - R = \frac{1}{2}\log(1 + \mathrm{SNR}) - R \tag{13.52}$$

is the gap to capacity. At high SNR conditions (where $\alpha^* \to 1$), the bound (13.51) can be tightened by replacing $\bar{\rho}_{\mathrm{cov}}^2(\Lambda_2)$ by the shaping loss $2\pi e\, \bar{G}(\Lambda_2)$, where $\bar{G}(\Lambda_2)$ is an upper bound on the NSM of the coarse lattice.

It follows that for a Rogers-good shaping lattice (where $\rho_{\text{cov}}(\Lambda_2) \approx 1$), the error exponent is at least as good as the Poltyrev random coding exponent, at the same gap to capacity Δ:

$$P_e \stackrel{\cdot}{\leq} e^{-nE_r(2^{2\Delta})}. \tag{13.53}$$

In particular, Theorem 13.8.2 shows that a good Voronoi modulation system can approach the AWGN channel capacity with an arbitrarily small error probability; thus, it provides an alternative proof for Theorem 9.6.1 (with a slightly stronger condition on the shaping lattice, replacing quantization goodness by covering goodness).

Proof of Theorem 13.8.2   We observe from Section 13.4, that for any additive noise, the NN decoding error exponent of a uniform lattice ensemble is governed *only* by the sphere crossing exponent of the noise. Furthermore, by Theorem 13.8.1, the mixture sphere crossing probability is exponentially upper bounded by that of a corresponding AWGN. It thus follows from Theorems 13.4.1 and 13.8.1 that

$$P_e \stackrel{\cdot}{\leq} e^{-nE_r\left(\frac{r_{\text{eff}}^2(\Lambda_1)}{r_{\text{mix}}^2}\right)}, \tag{13.54}$$

where $r_{\text{mix}}^2 = n\alpha^2 N + (1-\alpha)^2 r_{\text{cov}}^2(\Lambda_2)$; see (13.46). We can write the inverse of the argument of this exponent as

$$\frac{r_{\text{mix}}^2}{r_{\text{eff}}^2(\Lambda_1)} = \frac{n\alpha^2 N + (1-\alpha)^2 \cdot \rho_{\text{cov}}^2(\Lambda_2) \cdot r_{\text{eff}}^2(\Lambda_2)}{r_{\text{eff}}^2(\Lambda_1)} \tag{13.55a}$$

$$= 2^{2R} \cdot \rho_{\text{cov}}^2(\Lambda_2) \cdot \left[\frac{nN}{\rho_{\text{cov}}^2(\Lambda_2) \cdot r_{\text{eff}}^2(\Lambda_2)} \cdot \alpha^2 + (1-\alpha)^2 \right] \tag{13.55b}$$

$$= 2^{2R} \cdot \rho_{\text{cov}}^2(\Lambda_2) \cdot \left[\frac{nV_n^{2/n} \cdot G(\Lambda_2)}{\rho_{\text{cov}}^2(\Lambda_2)} \cdot \frac{N}{P} \cdot \alpha^2 + (1-\alpha)^2 \right] \tag{13.55c}$$

$$\leq 2^{2R} \cdot \bar{\rho}_{\text{cov}}^2(\Lambda_2) \cdot \left[\frac{\alpha^2}{\text{SNR}} + (1-\alpha)^2 \right], \tag{13.55d}$$

where $P = \sigma^2(\Lambda_2)$ and N are the transmitter and noise power, respectively; and where (13.55a) follows from the relation $r_{\text{cov}}(\Lambda_2) = r_{\text{eff}}(\Lambda_2) \cdot \rho_{\text{cov}}(\Lambda_2)$; (13.55b) follows since the coding rate is given by $R = \log(r_{\text{eff}}(\Lambda_2)/r_{\text{eff}}(\Lambda_1))$; (13.55c) follows from the definition of the NSM; and finally (13.55d) follows from the assumed bound $\rho_{\text{cov}}(\Lambda_2) \leq \bar{\rho}_{\text{cov}}(\Lambda_2)$, and using the sphere bound for the NSM (7.14)–(7.15):

$$\frac{G(\Lambda_2)}{\rho_{\text{cov}}^2(\Lambda_2)} \leq G_n^* = \frac{1}{(n+2)V_n^{2/n}} \leq \frac{1}{nV_n^{2/n}}.$$

The theorem now follows by plugging $\alpha = \alpha^*$ in (13.55d).   □

13.8.3 Rate-optimized linear estimation

It turns out that at high SNR conditions, the Poltyrev random coding exponent $E_r(2^{2\Delta})$ of (13.53), with $\Delta = C_{\text{AWGN}} - R$, coincides with the *best achievable* random coding exponent of the AWGN channel, denoted $E_{\text{AWGN}}(R, \text{SNR})$ [97]. In this regime the linear coefficient α is close to 1; thus, the simple (no-estimation) lattice decoder of Section 9.3 is error-exponent optimal at a range of rates near capacity (specifically, for $C_{\text{AWGN}} - 1/2 \le R \le C_{\text{AWGN}}$), in line with the discussion in Section 9.3.4.

For non-high SNR conditions, however, $E_{\text{AWGN}}(R, \text{SNR})$ is strictly *better* than the Poltyrev random coding exponent at the same gap to capacity. The exponent loss in Theorem 13.8.2 has two sources: (i) the weakness of our analysis, in particular, the mixture sphere crossing bound of Theorem 13.8.1; and (ii) the choice of the linear estimation coefficient α. We shall improve our analysis by giving up the (loose) bound for a non-zero-mean Gaussian of Lemma 13.8.3. This will allow us to optimize the linear coefficient α, depending on *both* the SNR and rate R. And luckily, the resulting exponent meets $E_{\text{AWGN}}(R, \text{SNR})$.

A direct incorporation of Lemmas 13.8.1 and 13.8.2 into the Chernoff bound (Proposition 13.1.1) implies the following parametric sphere crossing bound depending on s.

Corollary 13.8.2 (Parametric bound) *Let* $\|\eta\| = (1 - \alpha)r_{\text{cov}}(\Lambda)$, *and* $\sigma^2 = \alpha^2 N$. *Then, for* $0 \le s < 1/2\sigma^2$,

$$\Pr\{\|\alpha\mathbf{Z} + (\alpha - 1)\mathbf{U}\| > r\} \le \left(\frac{e^{\frac{s}{n}\|\eta\|^2/1 - 2s\sigma^2}/\sqrt{1 - 2s\sigma^2}}{e^{\frac{s}{n}r^2}} \right)^n \tag{13.56}$$

$$\overset{\Delta}{=} e^{-nE_{\text{mix}}(r,s,\alpha)}. \tag{13.57}$$

In view of the derivation of the NN decoding bound in Section 13.4, we can replace the Gaussian sphere crossing exponent in (13.27) by $E_{\text{mix}}(r, s, \alpha)$ from (13.57). This will give us an exponential upper bound on the average error probability of a fine lattice Λ_1 from a uniform ensemble (the covering efficiency of the coarse lattice is assumed to be bounded for all n as in Theorem 13.8.2):

$$E_r = \max_{0 < s, \, 0 \le \alpha \le 1} \min_{0 \le r \le r_{\text{eff}}(\Lambda_1)} \left\{ E_{\text{mix}}(r, s, \alpha) + \ln\left(\frac{r_{\text{eff}}(\Lambda_1)}{r} \right) \right\}, \tag{13.58}$$

where the effective radius $r_{\text{eff}}(\Lambda_1)$ is a parameter of the ensemble of fine lattices, which is determined by the rate, power constraint and NSM of the coarse lattice (see (13.55)):

$$r_{\text{eff}}^2(\Lambda_1) = r_{\text{eff}}^2(\Lambda_2) \cdot 2^{-2R} \tag{13.59a}$$

$$= \frac{P}{V_n^{2/n} \cdot G(\Lambda_2)} \cdot 2^{-2R} \tag{13.59b}$$

$$\approx nP \cdot 2^{-2R}, \tag{13.59c}$$

where the approximation holds for a large n and good shaping lattice. As in Theorem 13.4.1, for some range of rates near capacity the minimum is achieved at the edge $r = r_{\text{eff}}(\Lambda_1)$; hence the resulting exponential bound meets the mixture sphere crossing exponent E_{mix} for the best choice of s and α.

Theorem 13.8.3 (Rate-optimized estimation) *For rate-R Voronoi modulation with lattice decoding in the presence of AWGN as in Theorem 13.8.2, where the fine lattice is drawn from a uniform ensemble, and the shaping (coarse) lattice is taken from a sequence of Rogers-good lattices, if the estimation coefficient α is equal to*

$$\alpha_{\text{opt}} = -\frac{\gamma}{2} + \sqrt{\frac{\gamma^2}{4} + \gamma}, \tag{13.60}$$

where

$$\gamma \overset{\Delta}{=} (1 - 2^{-2R}) \cdot \text{SNR}, \tag{13.61}$$

then the error probability is exponentially upper bounded by

$$P_e \doteq e^{-nE_{\text{AWGN}}(R, \text{SNR})}, \tag{13.62}$$

for all rates $R_{\text{cr}} \leq R \leq C_{\text{AWGN}}$, where E_{AWGN} is the optimal error exponent of the power-constrained AWGN channel, given by

$$2E_{\text{AWGN}}(R, \text{SNR}) = 2^{-2R} \cdot \text{SNR} - \alpha_{\text{opt}} + \ln\left[(1 - \alpha_{\text{opt}}) \cdot 2^{2R}\right], \tag{13.63}$$

and R_{cr} is the critical rate, *given by*

$$R_{\text{cr}} = \frac{1}{2} \log\left(\frac{1}{2} + \frac{1}{4}\text{SNR} + \frac{1}{2}\sqrt{1 + \frac{\text{SNR}^2}{4}}\right). \tag{13.64}$$

Proof The proof is based on optimizing (13.58) over s and α, and is left as an exercise; see Problem 13.11. □

Note that as R varies from zero to C_{AWGN}, γ increases from zero to $\text{SNR}^2/(1 + \text{SNR})$, while α_{opt} increases from zero to the Wiener coefficient $\alpha^* = \text{SNR}/(1 + \text{SNR})$. As a by-product of the proof, we get that the error exponent for a specific α (i.e., omitting the maximization over α in (13.58)) is given by

$$E_{\text{AWGN}}(R, \text{SNR}, \alpha) = E\left(\frac{1}{\alpha^2}\text{SNR} \cdot 2^{-2R}, \frac{(1 - \alpha)^2}{\alpha^2}\text{SNR}\right) \tag{13.65}$$

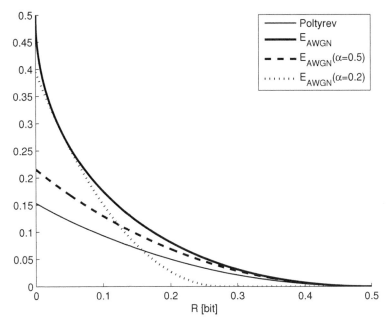

Figure 13.6 The error exponent $E_{\text{AWGN}}(R, \text{SNR}, \alpha)$ for Voronoi modulation with a linear estimation lattice decoder. The graph is drawn for $\text{SNR} = 1$, for three cases of the estimation coefficient: (i) $\alpha = \alpha^* = 1/2$, (ii) $\alpha = 0.2$, and (iii) the rate-optimized value $\alpha_{\text{opt}} = \alpha(R, \text{SNR})$ of (13.60). For comparison, the Poltyrev random-coding exponent $E_{\text{sp}}\left(2^{2(C-R)}\right)$ is also shown; see (13.68).

for

$$0 \leq R \leq C(\text{SNR}, \alpha) \triangleq \frac{1}{2} \log \left(\frac{\text{SNR}}{(1-\alpha)^2 \text{SNR} + \alpha^2} \right) \qquad (13.66)$$

where $C(\text{SNR}, \alpha)$ is the capacity of the mod Λ channel for a general (non-Wiener) estimation coefficient α (see (9.62b)), and where $E(x, y)$ is given by

$$2E(x, y) = x + y - \sqrt{1 + 4xy} + \ln \left[\frac{1}{2x} \left(1 + \sqrt{1 + 4xy} \right) \right]. \qquad (13.67)$$

It is easy to verify that within half a bit from capacity, the error exponent for the Wiener coefficient $\alpha = \alpha^*(\text{SNR})$ is lower and upper bounded as

$$E_{\text{AWGN}}(R, \text{SNR}) > E_{\text{AWGN}}(R, \text{SNR}, \alpha^*) > E_{\text{sp}}\left(2^{2(C_{\text{AWGN}} - R)}\right), \qquad (13.68)$$

and the gap between the bounds vanishes as the SNR goes to infinity. Figure 13.6 shows the dependence of the error exponent $E_{\text{AWGN}}(R, \text{SNR} = 1, \alpha)$ on the estimation coefficient α.

Like for unconstrained constellations, the error exponent can be improved for low rates by expurgation; see [171] and [252].

Summary of Chapter 13

Gap to capacity of an unbounded lattice constellation

$$\Delta \triangleq C_\infty - R(\Lambda) = \frac{1}{2} \log\left(\frac{\mu}{2\pi e}\right) \approx \log\left(\frac{r_{\text{eff}}(\Lambda)}{r_{\text{noise}}}\right),$$

where $\mu = V^{2/n}(\Lambda)/\sigma^2$ is the VNR, and $r_{\text{noise}} = \sqrt{n\sigma^2}$ is the typical noise radius.

Sphere crossing exponent For AWGN $\mathbf{Z} = (Z_1, \ldots, Z_n)$ with power σ^2,

$$\Pr\{\|\mathbf{Z}\| > r\} \doteq e^{-nE_{\text{sp}}\left(\frac{r^2}{r^2_{\text{noise}}}\right)},$$

where $E_{\text{sp}}(x) = \frac{1}{2}[x - 1 - \ln x]$ for $x \geq 1$, and zero elsewhere.

Threshold-decoding exponent (for good unbounded lattices) For the optimal search radius, $r^*_{\text{th}} = r_{\text{noise}} \cdot \sqrt{1 + 2\ln(2)\Delta}$,

$$P_e^{\text{TB}}(n, r^*_{\text{th}}) \doteq e^{-nE_{\text{sp}}(1+2\ln(2)\Delta)}.$$

NN decoding exponent (for good unbounded lattices) For a gap to capacity $0 \leq \Delta \leq 1/2$ bit,

$$P_e^{\text{MLB}}(n) \doteq e^{-nE_{\text{sp}}(2^{2\Delta})}.$$

Power-constrained lattice decoding exponent (for a good Voronoi constellation) The decoding error probability

- is exponentially upper bounded by $e^{-nE_{\text{sp}}(2^{2\Delta})}$, for a Wiener estimation coefficient $\alpha^* = \text{SNR}/(1 + \text{SNR})$, and a gap to capacity $\Delta \triangleq \frac{1}{2}\log(1 + \text{SNR}) - R$ smaller than half a bit; and
- is exponentially given by the *optimal* error exponent of the power-constrained AWGN channel, for a rate-R-optimized linear estimation coefficient $\alpha = \alpha_{\text{opt}}(R, \text{SNR})$ at all rates above the critical rate.

Both exponents coincide in the limit of high SNR.

Problems

P.13.1 (Chernoff bound) Prove that the Chernoff bound (Proposition 13.1.1) is informative (i.e., smaller than 1 for some s) for $r > \sqrt{nE\{Z^2\}}$, and otherwise it is greater than or equal to 1. Guidance: compute the first and second derivatives of the function $g(s)e^{-st^2}$ at $s = 0$.

P.13.2 (Tightness of sphere crossing exponent) Prove the tightness (13.8) of the Gaussian sphere crossing exponent for $r > r_{\text{noise}}$, using the Laplace method of integration (Proposition 13.4.1).

P.13.3 (Dispersion) Prove that near capacity, the threshold-decoding bound (13.16) can be approximated as $P_e^{\text{TB}}(r_{\text{th}}^*, n) = e^{-n(\Delta^2 + O(\Delta^3))}$, where Δ is the gap to capacity in nats. Use the approximation $Q(x) \approx e^{-x^2/2}$, to show that it agrees with the first term ("dispersion") in (13.21).

P.13.4 (Exponent of a white-Gaussian norm distribution) Show that the tail probability (1-CDF) and the density (13.18) of the norm of a white-Gaussian vector have the same exponential behavior, given by the Chernoff bound (13.6):

$$f_{\|\mathbf{Z}\|}^*(r, n) \doteq e^{-n E_{\text{sp}}\left(\frac{r^2}{r_{\text{noise}}^2}\right)}. \tag{13.69}$$

P.13.5 (Laplace's method of integration) Prove Proposition 13.4.1. Guidance: write the integral as a Riemann sum, bound from below and from above in terms of the dominating bin, take the limit as n goes to infinity (where the bounds meet), and then the limit of infinitely small bins. Extension: show the pre-factor of the exponent in the case of a smooth maximum, relative to a "peaky" maximum.

P.13.6 (Relating the Gallager and Poltyrev exponents [97, 221]) The Gallager error exponent of a modulo-additive-noise channel, $Y = X + Z \mod a$, is achieved by a uniform input, and it can be written in the following parametric form:

$$E_r(\Delta, \rho, a) = \rho \left[\Delta + h(Z) - h_{\tilde{\rho}}(Z)\right], \quad 0 \le \rho \le 1,$$

where $\Delta = C_a - R = \log(a) - h(Z) - R$ is the gap to capacity, $h(Z)$ is the Shannon entropy of Z, $h_\alpha(Z) = 1/(1 - \alpha) \cdot \log\left(\int f_Z^\alpha(\mathbf{x})d\mathbf{x}\right)$ is the *Rényi entropy* of order α of Z, and $\tilde{\rho} = 1/(1 + \rho)$ (see Csiszár [54] and Erez and Zamir [70]). Show that $E_r(\Delta, \rho, a)$ (for the best value of the parameter $\rho = \rho(\Delta)$) coincides with the random coding exponent of the MHS ensemble (13.30) at the same gap to capacity Δ.

P.13.7 (Distance-spectrum exponent) Prove the asymptotic inequality in (13.42).

P.13.8 (An alternative expurgated ensemble exponent) Consider the exponent (13.27) of the ML bound. Show that for the expurgated ensemble, the range of r in the inner minimization of (13.27) narrows, and becomes $r_{\text{eff}}/2 \le r \le r_{\text{eff}}$ (instead of $0 \le r \le r_{\text{eff}}$). Show that as a result, the dominating radius r^* for $r_{\text{eff}} > 2\sqrt{2}r_{\text{noise}}$ becomes $r_{\text{eff}}/2$ (instead of $\sqrt{2}r_{\text{noise}}$); furthermore, that this improves the random coding exponent E_r (Theorem 13.4.1) for low rates (find the range), but the exponent is still worse than the expurgated exponent of Theorem 13.7.1.

P.13.9 (Characteristic function for a non-zero-mean Gaussian) Prove the formula in (13.49) for the characteristic function of a squared non-zero-mean Gaussian.

P.13.10 (Chernoff bound for a scalar mixture noise) Calculate the characteristic function $E\{e^{s(\alpha Z+(1-\alpha)U)^2}\}$ for a scalar dither U. (Note that the cross-term does not allow one to break the expression, in spite of the independence of Z and U.) Find the corresponding Chernoff bound, and compare with the mixture sphere crossing bound of Theorem 13.8.1.

P.13.11 (Rate-optimized estimation) Prove Theorem 13.8.3, under the assumption that the minimizing radius in (13.58) is the effective radius of the lattice, i.e., $r = r_{\text{eff}}(\Lambda_1)$. (This assumption holds for the range of rates $R_{\text{cr}} \leq R \leq C_{\text{AWGN}}$.) Guidance: denote the right-hand side of the parametric bound of Corollary 13.8.2 as $A(s)$. Show that $A(0) = 1$ and $\lim_{s \to 1/2\sigma^2}\{A(s)\} = +\infty$, and that $A'(0) \leq 0$ for $r^2 \geq n\sigma^2 + \|\eta\|^2$. Conclude that $A(s)$ has a unique minimum in the interval $0 \leq s < 1/2\sigma^2$, given by

$$s^* = \frac{2r^2 - n\sigma^2 - \sqrt{(n\sigma^2)^2 + 4\|\eta\|^2 r^2}}{4\sigma^2 r^2}, \tag{13.70}$$

for $r^2 \geq n\sigma^2 + \|\eta\|^2$. Next, plug this value back into (13.56), to obtain

$$-\frac{1}{n}\ln\left(\Pr\left\{\|\alpha Z + (\alpha - 1)U\| > r\right\}\right) \geq \begin{cases} E\left(\frac{r^2}{n\sigma^2}, \frac{\|\eta\|^2}{n\sigma^2}\right), & \frac{r^2}{n\sigma^2 + \|\eta\|^2} \geq 1 \\ 0, & 0 < \frac{r^2}{n\sigma^2 + \|\eta\|^2} \leq 1, \end{cases} \tag{13.71}$$

where the exponent function $E(x, y)$ is defined in (13.67). Now, use the relation (13.59) between $r_{\text{eff}}(\Lambda_1)$ and the system parameters R, P and $G(\Lambda_2)$, and assume a Rogers-good shaping lattice ($G(\Lambda_2) \approx 2\pi e$ and $\rho_{\text{cov}}(\Lambda_2) \approx 1$) to obtain the formula (13.65) for the α-dependent exponent $E_{\text{AWGN}}(R, \text{SNR}, \alpha)$. Differentiate (13.65) with respect to α, equate to zero, and obtain the following polynomial equation of degree 4:

$$(\alpha^2 + \gamma\alpha - \gamma)((1 + \text{SNR})\alpha^2 - 2\text{SNR}\alpha + \gamma) = 0. \tag{13.72}$$

Finally, show that α_{opt} of (13.60) is the only zero of (13.72) satisfying the condition $R < C(\text{SNR}, \alpha)$, which maximizes $E_{\text{AWGN}}(R, \text{SNR}, \alpha)$.

Historical notes

The error exponent of an infinite constellation in the presence of AWGN was studied by Poltyrev [221] in 1994. His achievability proof assumes a sequence of good lattices, and it follows de Buda's analysis from 1975 and 1989 of spherical lattice codes using the Minkowski–Hlawka theorem [57, 58]. Loeliger [175] re-derived the Poltyrev exponent using averaging bounds on construction A lattices. Not too surprisingly, this exponent coincides with the Gallager reliability function of the power-constrained AWGN channel [97], in the limit of high SNR, for the same gap to capacity. (Although for a general SNR it is strictly lower.)

The analysis of infinite lattice constellations in this chapter follows Ingber *et al.* [125] (for the random coding error exponent) and Ingber and Zamir [124] (for the expurgated exponent). Like Poltyrev's analysis, it is based on the uniformity property of the MHS ensemble (the Siegel version of the Minkowski–Hlawka theorem [36]). But our spherical error bounds – used both for the typicality (threshold) exponent and for the ML decoding exponent – are simpler. (The latter is reminiscent of "Gallager's type II bound"; see, for example, Divsalar [59].) The derivation of Ingber *et al.* [124, 125] also contains tighter estimates for the error probability of finite-dimensional lattices (lower bounds for any lattice, upper bound for random lattices), as well as comparison to some state-of-the-art lattices, and a relation to the dispersion analysis of Polyanskiy *et al.* [222].

The error exponent of finite (bounded) lattice constellations was studied by de Buda [57] – for spherical shaping with ML decoding – and by Forney *et al.* [92] – in the high SNR regime. Erez and Zamir [71] showed that, with Wiener estimation at the receiver, Voronoi modulation with nearest-neighbor lattice decoding can achieve the Poltyrev error exponent at the same gap to capacity. Thus, this modulation and demodulation scheme is error-exponent-wise optimal at high SNR, but possibly suboptimal for a general SNR (although it achieves capacity). Liu *et al.* [171] strengthened this result using a large-deviation approach; they showed that with a rate-optimized (non-MMSE) linear estimator, the error exponent of the mod Λ channel with nearest-neighbor decoding coincides with the Gallager reliability function at all SNR. Swannack *et al.* [252] provide a geometric explanation for this increase in the error exponent. The derivation in Section 13.8 is based on simple parametric characterization and bounds for the Chernoff exponent, due to Yeredor [279] and Tridenski [257].

Appendix

A.1   Entropy and mutual information

This appendix summarizes formulas and properties of information measures: chain rule, discrete and continuous distributions, maximum entropy and data-processing inequalities. For a comprehensive (and elegant) survey, see the book of Cover and Thomas [53].

1. The *entropy* of a discrete random variable $X \sim p(x)$:

$$H(X) = -\sum_x p(x) \log_2 p(x) \text{ [bit]}. \tag{A.1}$$

H is non-negative and upper bounded by the logarithm of the size of the alphabet:

$$0 \leq H(X) \leq \log |\mathcal{A}|. \tag{A.2}$$

2. The *conditional entropy*:

$$H(X|Y) = \sum_y p(y) \cdot H(X|Y=y) = \sum_{x,y} p(x,y) \log p(x|y). \tag{A.3}$$

3. Conditioning reduces entropy:

$$H(X|Y) \leq H(X). \tag{A.4}$$

4. The *joint entropy* (the entropy of the joint distribution) satisfies the *chain rule*:

$$H(X,Y) = H(X) + H(Y|X) = H(Y) + H(X|Y). \tag{A.5}$$

5. The *mutual information* between two random variables is the reduction in the entropy of one of them when the other becomes available:

$$I(X;Y) = H(X) - H(X|Y) \tag{A.6a}$$
$$= H(Y) - H(Y|X) \tag{A.6b}$$
$$= H(X) + H(Y) - H(X,Y). \tag{A.6c}$$

Futhermore, the conditional mutual information is given by $I(X;Y|Z) = H(X|Z) - H(X|Y,Z)$.

6. The *chain rule* for mutual information:

$$I(X; Y, Z) = I(X; Y) + I(X; Z|Y). \tag{A.7}$$

7. The *differential entropy* of a continuous random variable $X \sim f(x)$:

$$h(X) = - \int f(x) \log f(x) dx. \tag{A.8}$$

8. *Regular and differential entropy* If $Q_\Delta(\cdot)$ is a step-Δ uniform scalar quantizer, and $U_\Delta \sim \text{Unif}(0, \Delta)$ is independent of X, then

$$H\left(Q_\Delta(X)\right) + \log(\Delta) = h\left(Q_\Delta(X) + U_\Delta\right). \tag{A.9}$$

9. *Differential entropy via fine lattice quantization* If X has a density, then

$$\lim_{\Delta \to 0} \left[H(Q_\Delta(X)) + \log(\Delta) \right] = h(X). \tag{A.10}$$

10. If $Y = X + Z$ is an additive-noise channel, then

$$I(X; Y) = h(X + Z) - h(Z). \tag{A.11}$$

11. The *maximum entropy* under a peak-amplitude constraint is achieved by a uniform distribution,

$$\max_{X \in (a,b)} h(X) = h\left(\text{Unif}(a, b)\right) = \log |b - a|, \tag{A.12}$$

and under a power constraint it is achieved by a Gaussian distribution,

$$\max_{X : EX^2 \le P} h(X) = h\left(N(0, P)\right) = \tfrac{1}{2} \log(2\pi e P). \tag{A.13}$$

12. The *capacity* of a modulo-additive-noise channel $Y = [X + Z] \bmod a$ is achieved by a uniform input $X \sim \text{Unif}(0, a)$:

$$C = I\left(X; [X + Z] \bmod a\right) = \log(a) - h(Z). \tag{A.14}$$

13. The *entropy power* of a random variable X is the variance of a Gaussian having entropy $h(X)$:

$$P_E(X) = \frac{2^{2h(X)}}{2\pi e}. \tag{A.15}$$

14. The *entropy power inequality* (EPI) for a sum of independent random variables:

$$P_E(X + Y) \ge P_E(X) + P_E(Y), \tag{A.16}$$

with equality if and only if X and Y are Gaussians.

15. The *data-processing inequality* for a Markov triplet $X \leftrightarrow Y \leftrightarrow Z$:

$$I(X; Y) \ge I(X; Z). \tag{A.17}$$

16. The *divergence* (known also as *relative entropy* or Kullback–Leibler distance) between two density functions:

$$D(f\|g) = \int f(x) \log \left(\frac{f(x)}{g(x)} \right) dx. \tag{A.18}$$

17. *Divergence data-processing inequality ("second law of thermodynamics")* If $\tilde{f}$ is the output of a channel with an input $\sim f$, and $\tilde{g}$ is the output of the same channel with an input $\sim g$, then

$$D(\tilde{f}\|\tilde{g}) \le D(f\|g). \tag{A.19}$$

A.2   Success-threshold exponent

An exponential number of trials with an exponentially small "success" probability obeys a simple asymptotic "0–1 law." This law is very useful in random coding arguments.

Lemma A.2.1 (Success-threshold exponent) *Suppose we conduct 2^{nR} independent trails where the "success" probability in each trial is 2^{-nr}. Then*

$$\lim_{n \to \infty} \Pr\{\text{success in at least one trial}\} \to \begin{cases} 0, & R < r \\ 1, & R > r. \end{cases}$$

Moreover, the first case holds even if the trials are dependent *(with the same success probability in each trial).*

Proof By the union bound, this probability is upper bounded (even if the trails are dependent) by

$$\sum_{i=1}^{2^{nR}} \Pr\{\text{success in } i\text{th trial}\} = 2^{nR} 2^{-nr},$$

which vanishes as n goes to infinity if $R < r$, thus proving the first case. As for the second case, the probability that all trials fail (which is the complementary event) is given, due to independence, by

$$\prod_{i=1}^{2^{nR}} \Pr\{i\text{th trial fails}\} = (1 - 2^{-nr})^{2^{nR}} \le (\exp[-2^{-nr}])^{2^{nR}} = \exp[-2^{-nr} 2^{nR}],$$

where the inequality follows since $\exp(-x) \ge 1 - x$ for all x. The last term vanishes as n goes to infinity if $R > r$, thus proving the second case.   □

A.3   Coset density and entropy

Since the operation $\mathbf{x} \bmod \Lambda$ maps a whole coset of Λ to its representative point in some fundamental cell $\mathcal{P}_0$, we can think of the density of $\mathbf{X} \bmod \Lambda$ as the *coset density* of Λ with respect to the random variable $\mathbf{X}$ and the cell $\mathcal{P}_0$.

Lemma A.3.1 (Lemma 4.2.2) *Let* $\mathbf{X}$ *be a continuous random variable with a density* $f_{\mathbf{X}}(\mathbf{x})$. *The density of* $\mathbf{X}$, *after modulo-Λ reduction with respect to some fundamental cell* $\mathcal{P}_0$, *is given by*

$$f_{\mathbf{X} \bmod \Lambda}(\mathbf{x}) = \begin{cases} f_{\mathbf{X} \text{ rep } \Lambda}(\mathbf{x}), & \mathbf{x} \in \mathcal{P}_0 \\ 0, & \mathbf{x} \notin \mathcal{P}_0, \end{cases} \tag{A.20}$$

where

$$f_{\mathbf{X} \text{ rep } \Lambda}(\mathbf{x}) = \sum_{\lambda \in \Lambda} f_{\mathbf{X}}(\mathbf{x} + \lambda) \tag{A.21}$$

is the periodic replication *of* $f_{\mathbf{X}}$ *by the lattice* Λ.

Proof (due to Tamas Linder [162])   The lemma amounts to

$$\Pr\{[\mathbf{X} \in B] \bmod \Lambda\} = \int_B \sum_{\lambda \in \Lambda} f_{\mathbf{X}}(\mathbf{x} + \lambda) d\mathbf{x}, \tag{A.22}$$

for any measurable set B that can be packed by Λ (i.e., $\Lambda + B$ contains no overlaps). But $[\mathbf{X} \in B] \bmod \Lambda$ amounts to $\mathbf{X} \in [\Lambda + B]$, so the probability of this event is $\sum_{\lambda} \int_{B+\lambda} f(\mathbf{x}) d\mathbf{x}$, which amounts to (A.22). $\square$

Since $f_{\mathbf{X} \text{ rep } \Lambda}$ is invariant of the choice of the fundamental cell $\mathcal{P}_0$, and since any fundamental cell is a period of the (Λ-periodic) function $f_{\mathbf{X} \text{ rep } \Lambda}$, it follows from Lemma A.3.1 that the *coset entropy*

$$h(\mathbf{X} \bmod \Lambda) = -\int_{\mathcal{P}_0} f_{\mathbf{X} \text{ rep } \Lambda}(\mathbf{x}) \cdot \log\left(f_{\mathbf{X} \text{ rep } \Lambda}(\mathbf{x})\right) d\mathbf{x} \tag{A.23}$$

is invariant of the choice of $\mathcal{P}_0$.

Let us denote $Q = Q_\Lambda(\mathbf{X})$ for some quantization cell $\mathcal{P}_0$. We have the following properties:

$$H(Q|\mathbf{X}) = 0 \tag{A.24}$$

$$h(\mathbf{X} \bmod \Lambda|Q) = h(\mathbf{X} - Q|Q) = h(\mathbf{X}|Q) \tag{A.25}$$

(since entropy is invariant to a shift). Thus,

$$I(Q; \mathbf{X} \bmod \Lambda) = h(\mathbf{X} \bmod \Lambda) - h(\mathbf{X}|Q). \tag{A.26}$$

By the data-processing inequality (A.17) (since $Q \leftrightarrow \mathbf{X} \leftrightarrow [\mathbf{X} \bmod \Lambda]$)

$$I(Q; \mathbf{X} \bmod \Lambda) \le I(Q; \mathbf{X}), \tag{A.27}$$

implying, by (A.25), $h(\mathbf{X} \bmod \Lambda) \le h(\mathbf{X})$.

We can also find the difference term. Specifically, using (A.24),

$$H(Q) = I(Q; \mathbf{X}) = I(Q; \mathbf{X} \bmod \Lambda) + H(Q|\mathbf{X} \bmod \Lambda), \qquad (A.28)$$

so by (A.26),

$$h(\mathbf{X}) = h(\mathbf{X} \bmod \Lambda) + H(Q|\mathbf{X} \bmod \Lambda). \qquad (A.29)$$

Due to the non-negativity of H, we conclude the following.

Lemma A.3.2 (Modulo reduces entropy) *For a random variable X with a differential entropy $h(X)$, and a lattice Λ:*

$$h(\mathbf{X}) \geq h(\mathbf{X} \bmod \Lambda),$$

and the difference is $H(Q|X \bmod \Lambda)$. See Problem 9.7. [1]

A.4   Convolution of log-concave functions

We say that a non-negative function $f : \mathbb{R}^n \to \mathbb{R}^+$ is log-concave if

$$f(\theta \mathbf{x} + (1 - \theta)\mathbf{y}) \geq f(\mathbf{x})^\theta f(\mathbf{y})^{1-\theta}$$

for all $\mathbf{x}, \mathbf{y} \in \mathbb{R}^n$ and $0 < \theta < 1$. If $f(\mathbf{x})$ is strictly positive, then this amounts to the function $\log[f(\mathbf{x})]$ being concave $\cap$.

Lemma A.4.1 (Convolution of log-concave functions) *If f and g are two log-concave functions, then so is their convolution $f * g(\mathbf{x}) = \int f(\mathbf{y})g(\mathbf{x} - \mathbf{y})d\mathbf{y}$.*

Proof See [56, 225].   □

This lemma is used in Problem 9.11 and in the proof of Lemma 13.8.1 to show that the error probability in biased decoding is larger than in unbiased decoding, and is monotonically non-decreasing with the bias.

---

[1] For *discrete* random variables, this lemma follows easily from the properties of regular (non-differential) entropy: since there is a one-to-one mapping between X and the error-quantization pair $(X \bmod \Lambda, Q)$, the chain rule for regular entropy implies

$$H(X) = H(X \bmod \Lambda, Q) = H(X \bmod \Lambda) + H(Q|X \bmod \Lambda).$$

A.5   Mixture versus Gaussian noise

This section confirms that the tail of a mixture noise distribution becomes heavier if we replace the uniform component by another Gaussian. This implies that the Gaussian NVNR upper bounds the mixture NVNR for a sufficiently small P_e; see Section 9.6.1.

Lemma A.5.1 (Mixture versus Gaussian tail) *Let* $\mathbf{Z}$ *denote an n-dimensional AWGN vector. Let* $\mathbf{U}$ *denote an independent random variable which is uniform over a compact set in* $\mathbb{R}^n$, *and let* $\mathbf{U}^*$ *be AWGN with the same variance as* $\mathbf{U}$. *Let* Λ *be a lattice in* $\mathbb{R}^n$. *Then,*

$$\Pr\{\mathbf{Z} + \mathbf{U} \notin \mathcal{V}_0(a \cdot \Lambda)\} < \Pr\{\mathbf{Z} + \mathbf{U}^* \notin \mathcal{V}_0(a \cdot \Lambda)\} \qquad (A.30)$$

for a sufficiently large value of the scaling parameter a.

Proof Let $r = \sup_{\mathbf{x} \in \mathcal{S}} \|\mathbf{x}\|$ denote the radius of the set $\mathbf{S}$ (the support of $\mathbf{U}$) about the origin. Since the density $f_{\mathbf{Z}+\mathbf{U}}(\mathbf{x})$ is the average of of $f_{\mathbf{Z}}(\mathbf{x})$ over $\mathbf{x}$ in $\mathcal{S}$, $f_{\mathbf{Z}+\mathbf{U}}(\mathbf{x}) < K_1 \exp\left((\|\mathbf{x}\| - r)^2 / 2\sigma_z^2\right)$, for all $\mathbf{x}$ with $\|\mathbf{x}\| > r$. On the other hand, $f_{\mathbf{Z}+\mathbf{U}^*}(\mathbf{x}) = K_2 \exp\left((\|\mathbf{x}\|)^2 / 2(\sigma_z^2 + \sigma_u^2)\right)$. It follows that there exists some threshold T (which is a function of r, σ_z^2 and σ_u^2), such that

$$f_{\mathbf{Z}+\mathbf{U}}(\mathbf{x}) < f_{\mathbf{Z}+\mathbf{U}^*}(\mathbf{x}), \qquad (A.31)$$

for all $\mathbf{x}$ with $\|\mathbf{x}\| > T$. This implies inequality (A.30) for all a such that $r_{\text{pack}}(a \cdot \Lambda) > T$. □

Since the NVNR (6.33) is monotonic in the error probability P_e and invariant to scaling, it follows that

$$\mu_{\text{euclid}}(\Lambda, \mathbf{Z} + \mathbf{U}, P_e) < \mu_{\text{euclid}}(\Lambda, \mathbf{Z} + \mathbf{U}^*, P_e) = \mu(\Lambda, P_e) \qquad (A.32)$$

for a sufficiently small P_e. In particular, this behavior holds in the case where $\mathbf{U}$ is uniform over the Voronoi cell of another lattice.

The application of this result to *nested lattices* (where the uniform component $\mathbf{U}$ is induced by the quantization lattice, while the error event is measured with respect to the coding lattice) must be done with care. This is because the coupling between the lattices does not allow scaling of one lattice while keeping the other lattice fixed. In particular, the mixture NVNR (9.69), $\mu_{\text{mix}}(\Lambda_1, P_e; \Lambda_2, \alpha)$, obeys (A.32) for α close enough to one. It is an open question whether this relation holds eventually for a sequence of good nested lattices of increasing dimension.

A.6   Lattice-distributed noise

Corrupting the channel output by a discrete noise which is nearly uniform over a lattice, is equivalent to reducing the output modulo the lattice. The following lemma – which states this fact formally – is used in Section 10.6.1 to prove that for a strong interference, the modulo-lattice output of the dirty-paper channel is a sufficient statistic for the decoder. (The proof of the lemma is due to Tamas Linder [162].)

Lemma A.6.1 (Strong lattice-distributed noise) *If* $\mathbf{Y}$ *and* $\mathbf{Y}'$ *have finite second moments, then for any* $\mathbf{S}$ *which is independent of* $(\mathbf{X}, \mathbf{Y}, \mathbf{Y}')$, *has a p.d.f. and a finite second moment,*

$$\lim_{a \to \infty} I(\mathbf{X}; \mathbf{Y} + Q_\Lambda[a\mathbf{S} + \mathbf{Y}']) = I(\mathbf{X}; \mathbf{Y} \bmod \Lambda). \tag{A.33}$$

Note that the variable $Q_\Lambda[a\mathbf{S} + \mathbf{Y}']$ is nearly uniform over the points of Λ in a region that increases with a, and is asymptotically independent of $\mathbf{Y}'$.

Proof To simplify the exposition, we shall assume $\mathbf{Y}' \equiv 0$. Let us use the simplified notation $\tilde{\mathbf{S}} = a\mathbf{S}$, $\mathbf{Y}_q = Q_\Lambda[\mathbf{Y}]$ and $\tilde{\mathbf{S}}_q = Q_\Lambda[\tilde{\mathbf{S}}]$. We have the following identities:

$$I(\mathbf{X}; \mathbf{Y} + \tilde{\mathbf{S}}_q) = I(\mathbf{X}; Q_\Lambda[\mathbf{Y} + \tilde{\mathbf{S}}_q], [\mathbf{Y} + \tilde{\mathbf{S}}_q] \bmod \Lambda) \tag{A.34a}$$
$$= I(\mathbf{X}; \mathbf{Y}_q + \tilde{\mathbf{S}}_q, \mathbf{Y} \bmod \Lambda) \tag{A.34b}$$
$$= I(\mathbf{X}; \mathbf{Y} \bmod \Lambda) + I(\mathbf{X}; \mathbf{Y}_q + \tilde{\mathbf{S}}_q \mid \mathbf{Y} \bmod \Lambda), \tag{A.34c}$$

where (A.34a) follows from the one-to-one decomposition (2.12); (A.34b) follows since a translation by a lattice point does not affect the modulo operation (2.22); and (A.34c) follows from the chain rule (A.7). Proving (A.33) thus amounts to showing that the second term of (A.34c) vanishes as a goes to infinity. We write this term as

$$I(\mathbf{X}; \mathbf{Y}_q + \tilde{\mathbf{S}}_q \mid \mathbf{Y} \bmod \Lambda) = H(\mathbf{Y}_q + \tilde{\mathbf{S}}_q \mid \mathbf{Y} \bmod \Lambda)$$
$$\qquad - H(\mathbf{Y}_q + \tilde{\mathbf{S}}_q \mid \mathbf{Y} \bmod \Lambda, \mathbf{X}) \tag{A.35a}$$
$$\leq H(\mathbf{Y}_q + \tilde{\mathbf{S}}_q)$$
$$\qquad - H(\mathbf{Y}_q + \tilde{\mathbf{S}}_q \mid \mathbf{Y} \bmod \Lambda, \mathbf{X}, \mathbf{Y}_q) \tag{A.35b}$$
$$= H(\mathbf{Y}_q + \tilde{\mathbf{S}}_q) - H(\tilde{\mathbf{S}}_q), \tag{A.35c}$$

where (A.35a) follows from the definition of the (conditional) mutual information (A.6); (A.35b) follows since conditioning reduces entropy (A.4); and (A.35c) follows since $\mathbf{S}$ is independent of $(\mathbf{X}, \mathbf{Y})$. Thus, to prove that the second term of (A.34c) indeed vanishes, we need to show that $\mathbf{Y}_q$ does not much affect the entropy of the first term in (A.35c) for a large a. To this end, observe that $\frac{1}{a} \cdot Q_\Lambda(a\mathbf{S}) = Q_{\Lambda/a}[\mathbf{S}]$, by the lattice-scaling law (2.43). We can thus write (A.35c) as

$$H\left(Q_{\Lambda/a}[\mathbf{S}] + Q_{\Lambda/a}[\mathbf{Y}/a]\right) - H\left(Q_{\Lambda/a}[\mathbf{S}]\right) \tag{A.36a}$$

$$= [H\left(Q_{\Lambda/a}[\mathbf{S}] + Q_{\Lambda/a}[\mathbf{Y}/a]\right) - \log(a^n)] - [H\left(Q_{\Lambda/a}[\mathbf{S}]\right) - \log(a^n)]. \tag{A.36b}$$

By the "fine-quantization property" of the differential entropy (A.10), the second square bracket [·] in (A.36b) goes to $h(\mathbf{S})$, as a goes to infinity (where for simplicity we assume a unit lattice volume $V(\Lambda) = 1$). We next show that the first square bracket in (A.36b) is asymptotically equal to $h(\mathbf{S})$. For that, we use (A.9) to rewrite it as a differential entropy:

$$H\left(Q_{\Lambda/a}[\mathbf{S}] + Q_{\Lambda/a}[\mathbf{Y}/a]\right) - \log(a) = h\left(Q_{\Lambda/a}[\mathbf{S}] + Q_{\Lambda/a}[\mathbf{Y}/a] + \mathbf{U}_a\right),$$
$$\tag{A.37}$$

where $\mathbf{U}_a$ is uniform over the cell of $Q_{\Lambda/a}$. Then, since $Q_{\Lambda/a}[\mathbf{S}]$ converges to $\mathbf{S}$ in distribution, and $Q_{\Lambda/a}[\mathbf{Y}/a]$ and $\mathbf{U}_a$ converge to zero in distribution, we have

$$\limsup_{a\to\infty} h\left(Q_{\Lambda/a}[\mathbf{S}] + Q_{\Lambda/a}[\mathbf{Y}/a] + \mathbf{U}_a\right) \leq h(\mathbf{S}) \tag{A.38}$$

by the lower semi-continuity of the divergence; see [164, proof of theorem 1]. The reverse inequality follows since the independent term $Q_{\Lambda/a}[\mathbf{Y}/a]$ cannot reduce the entropy [53]. It follows that the two square brackets in (A.36b) are asymptotically equal. Hence (A.35c) goes to zero, as desired. $\square$

References

[1] J. Abrahams. Variable-rate unequal cost parsing and coding for shaping. *IEEE Transactions on Information Theory*, **44**:1648–1650, July 1998.

[2] Stella Achtenberg and Dan Raphaeli. Theoretic shaping bounds for single letter constraints and mismatched decoding. arXiv:1308.5938 [cs.IT], August 2013.

[3] E. Agrell, T. Eriksson, A. Vardy, and K. Zeger. Closet point search in lattices. *IEEE Transactions on Information Theory*, **48**:2201–2214, August 2002.

[4] R. Ahlswede, N. Cai, S.-Y. R. Li, and R. W. Yeung. Network information flow. *IEEE Transactions on Information Theory*, **46**(4):1204–1216, July 2000.

[5] Rudolf Ahlswede and Te Sun Han. On source coding with side information via a multiple-access channel and related problems in multi-user information theory. *IEEE Transactions on Information Theory*, **29**(3):396–412, May 1983.

[6] M. Ajtai. The shortest vector problem in l_2 is np-hard for randomized reductions. In *Proc. 30th ACM Symposium on the Theory of Computing (STOC)*, pages 10–19, ACM, 1998.

[7] E. Akyol and K. Rose. Nonuniform dithered quantization. In *Proc. Data Compression Conference*, pages 435–445, 2009.

[8] N. Alon and A. Orlitsky. Source coding and graph entropies. *IEEE Transactions on Information Theory*, **42**:1329–1339, 1996.

[9] V. Anantharam and F. Baccelli. A palm theory approach to error exponents. *Proc. IEEE Int. Symposium on Information Theory (ISIT)*, pages 1768–1772, July 2008.

[10] S. Arimoto. An algorithm for computing the capacity of arbitrary discrete memoryless channels. *IEEE Transactions on Information Theory*, **18**:14–20, January 1972.

[11] N. W. Ashcroft and N. D. Mermin. *Solid State Physics*. Cengage Learning, 1976.

[12] A. Barg. On the asymptotic accuracy of the union bound. Available at: http://arxiv.org/abs/cs/0412111, 2004.

[13] A. Barg and G. D. Forney, Jr. Random codes: minimum distances and error exponents. *IEEE Transactions on Information Theory*, **48**:2568–2573, September 2002.

[14] E. S. Barnes and N. J. A. Sloane. The optimal lattice quantizer in three dimensions. *SIAM Journal on Algebraic Discrete Methods*, **4**:30–41, March 1983.

[15] R. Barron, B. Chen, and G. Wornell. The duality between information embedding and source coding with side information, and some applications. *IEEE Transactions on Information Theory*, **49**:1159–1180, May 2003.

[16] J.-C. Belfiore. Lattice codes for the compute-and-forward protocol: the flatness factor. In *Proc. IEEE Information Theory Workshop*, pages 1–4, 2011.

[17] W. R. Bennett. Spectra of quantized signals. *Bell System Technical Journal*, **27**:446–472, July 1948.

[18] T. Berger. *Rate Distortion Theory: A Mathematical Basis for Data Compression*. Prentice-Hall, Englewood Cliffs, NJ, 1971.

[19] T. Berger. *Multiterminal Source Coding*. In G. Longo, editor, *The Information Theory Approach to Communications*, Springer-Verlag, New York, 1977.

[20] T. Berger and S. Y. Tung. Encoding of correlated analog sources. In *Proc. IEEE–USSR Joint Workshop on Information Theory*, pages 7–10, 1975.

[21] E. Berlekamp. *Key Papers in the Development of Coding Theory*. IEEE Press, New York, 1974.

[22] R. E. Blahut. Computation of channel capacity and rate-distortion functions. *IEEE Transactions on Information Theory*, **18**:460–473, 1972.

[23] R. E. Blahut. *Theory and Practice of Error Control Codes*. Addison Wesley, Reading, MA, 1983.

[24] S. Borade and L. Zheng. Writing on fading paper, dirty tape with little ink: wideband limits for causal transmitter csi. *IEEE Transactions on Information Theory*, **58**:5388–5397, August 2012.

[25] Guy Bresler, Abhay Parekh, and David Tse. The approximate capacity of the many-to-one and one-to-many Gaussian interference channels. *IEEE Transactions on Information Theory*, **56**(9):4566–4592, September 2010.

[26] Viveck R. Cadambe and Syed A. Jafar. Interference alignment and the degrees of freedom for the K-user interference channel. *IEEE Transactions on Information Theory*, **54**(8):3425–3441, August 2008.

[27] G. Caire and S. Shamai. On the achievable throughput of a multi-antenna Gaussian broadcast channel. *IEEE Transactions on Information Theory*, **49**:1649–1706, July 2003.

[28] A. R. Calderbank. *Class notes*.

[29] A. R. Calderbank. Multilevel codes and multistage decoding. *IEEE Transactions on Communications*, **37**:222–229, March 1989.

[30] A. R. Calderbank. The art of signaling: fifty years of coding theory. *IEEE Transactions on Information Theory*, **44**:2561–2595, October 1998.

[31] A. R. Calderbank and P. C. Fishburn. The normalized second moment of the binary lattice determined by a convolutional code. *IEEE Transactions on Information Theory*, **40**:166–174, January 1994.

[32] A. R. Calderbank, P. C. Fishburn, and A. Rabinovich. Covering properties of convolutional codes and associated lattices. *IEEE Transactions on Information Theory*, **41**:732–746, May 1995.

[33] A. R. Calderbank and L. H. Ozarow. Nonequiprobable signaling on the Gaussian channel. *IEEE Transactions on Information Theory*, **36**:726–740, July 1990.

[34] A. R. Calderbank and N. J. A Sloane. New trellis codes based on lattices and cosets. *IEEE Transactions on Information Theory*, **33**:726–740, March 1987.

[35] J. C. Candy and G. C. Temes. *Oversampling Delta-Sigma Data Converters*. IEEE Press, New York, 1992.

[36] J. W. H. Cassels. *An Introduction to the Geometry of Numbers*. Springer, 1971, 1991.

[37] B. Chen and G. Wornell. Analog error-correcting codes based on chaotic dynamical systems. *IEEE Transactions on Communications*, **46**:881–890, July 1998.

[38] B. Chen and G. W. Wornell. Quantization index modulation: a class of provably good methods for digital watermarking and information embedding. *IEEE Transactions on Information Theory*, **47**:1423–1443, May 2001.

[39] J. Chen, C. Tian, T. Berger, and S. Hemami. Multiple description quantization via Gram–Schmidt orthogonalization. *IEEE Transactions on Information Theory*, **52**:5197–5217, December 2006.

[40] P. A. Chou, T. Lookabaugh, and R. M. Gray. Entropy constrained vector quantization. *IEEE Transactions on Acoustics, Speech, and Signal Processing*, **37**:31–42, January 1989.

[41] J. M. Cioffi, G. P. Dudevoir, M. V. Eyuboglu, and G. D. Forney, Jr. MMSE decision-feedback equalizers and coding – Part I: Equalization results. *IEEE Transactions on Communications*, **43**:2582–2594, October 1995.

[42] M. Cioffi and G. D. Forney, Jr. Generalized decision-feedback equalization for packet transmission with ISI and Gaussian noise. In A. Paulraj, V. Roychowdhury, and C. D. Schaper, editors, *Communications, Computation, Control and Signal Processing, a tribute to Thomas Kailath*, pages 79–129, Kluwer Academic Publishers, Boston, MA, 1997.

[43] A. S. Cohen and A. Lapidoth. The Gaussian watermarking game. *IEEE Transactions on Information Theory*, **48**:1639–1667, June, 2002. See also, *On the Gaussian Watermarking Game, Proc. IEEE Int. Symposium on Information Theory (ISIT), Sorrento, Italy*, page 48, June 2000.

[44] P. M. Cohn. *Classic Algebra*. Wiley, 2000.

[45] J. H. Conway and N. J. A. Sloane. Voronoi regions of lattices, second moments of polytops, and quantization. *IEEE Transactions on Information Theory*, **28**:211–226, March 1982.

[46] J. H. Conway and N. J. A. Sloane. Fast quantizing and decoding algorithms for lattice quantizers and codes. *IEEE Transactions on Information Theory*, **28**:227–231, March 1982.

[47] J. H. Conway and N. J. A. Sloane. A fast encoding method for lattice codes and quantizers. *IEEE Transactions on Information Theory*, **29**:820–824, November 1983.

[48] J. H. Conway and N. J. A. Sloane. On the Voronoi region of certain lattices. *SIAM Journal on Algebraic Discrete Methods*, **5**:294–305, September 1984.

[49] J. H. Conway and N. J. A. Sloane. *Sphere Packings, Lattices and Groups*. Springer-Verlag, New York, 1988.

[50] M. H. M. Costa. Writing on dirty paper. *IEEE Transactions on Information Theory*, **29**:439–441, May 1983.

[51] T. M. Cover. A proof of the data compression theorem of Slepian and Wolf for ergodic sources. *IEEE Transactions on Information Theory*, **21**:226–228, March 1975.

[52] T. M. Cover and M. Chiang. Duality of channel capacity and rate distortion with two sided state information. *IEEE Transactions on Information Theory*, **48**:1629–1638, June 2002.

[53] T. M. Cover and J. A. Thomas. *Elements of Information Theory*. Wiley, New York, 1991.

[54] I. Csiszár. Generalized cutoff rates and Rényi's information measures. *IEEE Transactions on Information Theory*, **41**:26–34, January 1995.

[55] I. Csiszár and P. Narayan. Channel capacity for a given decoding metric. *IEEE Transactions on Information Theory*, **41**:35–43, January 1995.

[56] J. S. Davidovic, B. I. Korenbljum, and B. I. Hacet. A certain property of logarithmically concave functions. *Doklady Akademii Nauk SSSR*, **185**:1215–1218, 1969. English translation: *Sov. Math. Dokl.*, **10**:477–480, 1969.

[57] R. de Buda. The upper error bound of a new near-optimal code. *IEEE Transactions on Information Theory*, **21**:441–445, July 1975.

[58] R. de Buda. Some optimal codes have structure. *IEEE Journal on Selected Areas in Communications*, **7**:893–899, August 1989.

[59] D. Divsalar. A simple tight bound on error probability of block codes with application to turbo codes. JPL, TMO Progress Report, pp. 42139, November 1999.

[60] Y. Domb, R. Zamir, and M. Feder. The random coding bound is tight for the average linear code or lattice. arXiv:1307.5524, 2013.

[61] J. J. Duistermaat and J. A. C. Kolk. *Multidimensional Real Analysis II: Integration*. Cambridge University Press, 2004.

[62] R. Durrett. *Probability Theory and Examples*. Wadsworth and Brooks, Pacific Grove, CA, 1989.

[63] J. J. Eggers, R. Bäuml, R. Tzschoppe, and B. Girod. Scalar Costa scheme for information embedding. *IEEE Transactions on Signal Processing*, **51**:1003–1019, April 2002.

[64] A. El Gamal and T.-H. Kim. *Network Information Theory*. Cambridge University Press, 2011.

[65] P. Elias. Coding for noisy channels. *IRE Convention Record*, Part 4, pp. 37–46, 1955. Also appears in D. Slepian, editor, *Key Papers in the Development of Information Theory*, pages 102–111, IEEE Press, 1974.

[66] U. Erez, S. Litsyn, and R. Zamir. Lattices which are good for (almost) everything. *IEEE Transactions on Information Theory*, **51**:3401–3416, October 2005.

[67] U. Erez, S. Shamai, and R. Zamir. Additive noise channels with side information at the transmitter. In *Proc. IEEE 21st Convention of Electrical and Electronics Engineers in Israel (IEEEI)*, pages 373–376, April 2000.

[68] U. Erez, S. Shamai (Shitz), and R. Zamir. Capacity and lattice strategies for cancelling known interference. *IEEE Transactions on Information Theory*, **51**:3820–3833, November 2005.

[69] U. Erez and S. ten Brink. A close-to-capacity dirty paper coding scheme. *IEEE Transactions on Information Theory*, **51**:3417–3432, October 2005.

[70] U. Erez and R. Zamir. Error exponents of modulo-additive noise channels with side-information at the transmitter. *IEEE Transactions on Information Theory*, **47**:210–218, January 2001.

[71] U. Erez and R. Zamir. Achieving 1/2 log(1+SNR) on the AWGN channel with lattice encoding and decoding. *IEEE Transactions on Information Theory*, **50**:2293–2314, October 2004.

[72] U. Erez and R. Zamir. A modulo-lattice transformation for multiple-access channels. In *Proc. IEEE 25th Convention of Electrical and Electronics Engineers in Israel (IEEEI)*, pages 836–840, December 2008 (presented also at ITA 2009, UCSD).

[73] Raul Etkin and Erik Ordentlich. The degrees-of-freedom of the K-user Gaussian inter-ference channel is discontinuous at rational channel coefficients. *IEEE Transactions on Information Theory*, **55**(11):4932–4946, November 2009.

[74] Raul H. Etkin, David N. C. Tse, and Hua Wang. Gaussian interference channel capac-ity to within one bit. *IEEE Transactions on Information Theory*, **54**(12):5534–5562, December 2008.

[75] M. V. Eyuboglu and G. D. Forney, Jr. Trellis precoding: combined coding, precoding and shaping for intersymbol interference channels. *IEEE Transactions on Information Theory*, **38**:301–314, March 1992.

[76] M. V. Eyuboglu and G. D. Forney, Jr. Lattice and trellis quantization with lattice- and trellis-bounded codebooks: high-rate theory for memoryless sources. *IEEE Transac-tions on Information Theory*, **39**:46–59, January 1993.

[77] A. Feinstein. A new basic theorem of information theory. *IRE Transactions on Infor-mation Theory*, **4**:2–22, 1954.

[78] Chen Feng, Danilo Silva, and Frank Kschischang. An algebraic approach to physical-layer network coding. *IEEE Transactions on Information Theory*, **59**:7576–7596, November 2013.

[79] W. A. Finamore and W. A. Pearlman. Optimal encoding of discrete-time continuous-amplitude memoryless sources with finite output alphabets. *IEEE Transactions on Information Theory*, **26**:144–155, March 1980.

[80] Robert F. H. Fischer. The modulo-lattice channel: the key feature in precoding schemes. *International Journal of Electronics and Communications*, **59**:244–253, June 2005.

[81] G. D. Forney, Jr. *Class notes at MIT*.

[82] G. D. Forney. The Viterbi algorithm. *Proceedings of the IEEE*, **61**:268–278, 1973.

[83] G. D. Forney, Jr. Coset codes I: introduction and geometrical classification. *IEEE Transactions on Information Theory*, **34**:1123–1151, September 1988.

[84] G. D. Forney, Jr. Coset codes II: binary lattices and related codes. *IEEE Transactions on Information Theory*, **34**:1152–1187, September 1988.

[85] G. D. Forney, Jr. Multidimensional constellations – part II: Voronoi constellations. *IEEE Journal of Selected Areas in Communications*, **7**:941–958, August 1989.

[86] G. D. Forney, Jr. Geometrically uniform codes. *IEEE Transactions on Information Theory*, **37**:1241–1260, September 1991.

[87] G. D. Forney, Jr. Trellis shaping. *IEEE Transactions on Information Theory*, **38**:281–300, March 1992.

[88] G. D. Forney. On the duality of coding and quantizing. In *DIMACS Series on Discrete Mathematics and Theoretical Computer Science*, Vol. 14, 1993.

[89] G. D. Forney. On the role of MMSE estimation in approaching the information-theoretic limits of linear gaussian channels: Shannon meets Wiener. In *Allerton Con-ference, Allerton House, Urbana, Illinois*, October 2003.

[90] G. D. Forney, Jr. Shannon meets Wiener II: on MMSE estimation in successive decod-ing schemes. In *42nd Annual Allerton Conference on Communication, Control, and Computing, Allerton House, Monticello, Illinois*, October 2004.

[91] G. D. Forney, R. G. Gallager, O. R. Lang, F. M. Longstaff, and S. U. Quereshi. Efficient modulation for band-limited channels. *IEEE Journal on Selected Areas in Communications*, **2**:632–647, September 1984.

[92] G. D. Forney, M. D. Trott, and S.-Y. Chung. Sphere-bound-achieving coset codes and multilevel coset codes. *IEEE Transactions on Information Theory*, **46**:820–850, May 2000.

[93] G. D. Forney and G. Ungerboeck. Modulation and coding for linear Gaussian channels. *IEEE Transactions on Information Theory*, **44**:2384–2415, October 1998.

[94] G. D. Forney and L.-F. Wei. Multidimensional constellations – part I: introduction, figures of merit, and generalized cross constellations. *IEEE Journal on Selected Areas in Communications*, **7**:877–892, August 1989.

[95] Y. Frank-Dayan and R. Zamir. Dithered lattice-based quantizers for multiple descriptions. *IEEE Transactions on Information Theory*, **48**:192–204, January 2002.

[96] R. G. Gallager. A simple derivation of the coding theorem and some applications. *IEEE Transactions on Information Theory*, **11**:3–18, 1965.

[97] R. G. Gallager. *Information Theory and Reliable Communication*. Wiley, New York, 1968.

[98] T. Gariby and U. Erez. On general lattice quantization noise. In *IEEE International Symposium on Information Theory*, pages 2717–2721, July 2008.

[99] M. Gastpar, B. Rimoldi, and M. Vetterli. To code, or not to code: lossy source-channel communication revisited. *IEEE Transactions on Information Theory*, **49**(5):1147–1158, 2003.

[100] S. I. Gelfand and M. S. Pinsker. Coding for channels with random parameters. *Problemy Peredachi Informatsii (Probl. Inform. Trans.)*, **9**(1):19–31, 1980.

[101] S. I. Gel'fand and M. S. Pinsker. On Gaussian channels with random parameters. In *IEEE International Symposium on Information Theory*, pages 247–250, Tashkent, USSR, September 1984.

[102] A. Gersho. Asymptotically optimal block quantization. *IEEE Transactions on Information Theory*, **25**:373–380, July 1979.

[103] A. Gersho and R. M. Gray. *Vector Quantization and Signal Compression*. Kluwer Academic, Boston, MA, 1992.

[104] G. D. Gibson, T. Berger, T. Lookabaugh, D. Lindbergh, and R. L. Baker. *Digital Compression for Multimedia: Principles and Standards*. Morgan Kaufmann, San Fansisco, CA, 1998.

[105] G. Ginis and J. M. Cioffi. Vectored-DMT: a FEXT canceling modulation scheme for coordinating users. In *ICC2001, Helsinki, Finland*, volume 1, pages 305–309, June 2001.

[106] H. Gish and N. J. Pierce. Asymptotically efficient quantization. *IEEE Transactions on Information Theory*, **14**:676–683, September 1968.

[107] T. J. Goblick. Theoretical limitations on the transmission of data from analog sources. *IEEE Transactions on Information Theory*, **11**:558–567, 1965.

[108] O. Goldreich, S. Goldwasser, and S. Halevi. Eliminating decryption errors in the Ajtai–Dwork cryptosystem. In *Advances in Cryptology*, volume 1294 of Lecture Notes in Computer Science, pages 105–111, Springer, 1997.

[109] R. M. Gray. Quantization noise spectra. *IEEE Transactions on Information Theory*, **36**:1220–1244, November 1990.

[110] R. M. Gray, T. Linder, and J. Li. A Lagrangian formulation of Zador's entropy-constrained quantization theorem. *IEEE Transactions on Information Theory*, **48**:695–707, March 2002.

[111] R. M. Gray and D. L. Neuhoff. Quantization. *IEEE Transactions on Information Theory*, **44**:2325–2383, October 1998.

[112] R. M. Gray and T. J. Stockham, Jr. Dithered quantizers. *IEEE Transactions on Information Theory*, **39**:805–812, May 1993.

[113] P. M. Gruber and C. G. Lekkerkerker. *Geometry of Numbers*. North-Holland Mathematical Library, Vol. 37, 1987.

[114] M. Gutman. On uniform quantization with various distortion measures. *IEEE Transactions on Information Theory*, **33**:169–171, January 1987.

[115] E. Haim, Y. Kochman, and U. Erez. Distributed structure: joint expurgation for the multiple-access channel. *IEEE Transactions on Information Theory*, submitted 2012. Available at: http://arxiv.org/abs/1207.1345.

[116] Te Sun Han and Kingo Kobayashi. A new achievable rate region for the interference channel. *IEEE Transactions on Information Theory*, **27**(1):49–60, January 1981.

[117] H. Harashima and H. Miyakawa. Matched-transmission technique for channels with intersymbol interference. *IEEE Transactions on Communications*, **20**:774–780, August 1972.

[118] X. He and A. Yener. Providing secrecy with structured codes: tools and applications to two-user Gaussian channels. *IEEE Transactions on Information Theory*, **60**:2121–2138, April 2014.

[119] C. Heegard. Partitioned linear block codes for computer memory with "stuck-at" defects. *IEEE Transactions on Information Theory*, **29**:831–842, November 1983.

[120] C. Heegard and A. El Gamal. On the capacity of computer memory with defects. *IEEE Transactions on Information Theory*, **29**:731–739, September 1983.

[121] J. Hoffstein, J. Pipher, and J. H. Silverman. NTRU: a ring based public key cryptosystem. In *Proceedings of ANTS-III*, volume 1423 of LNCS, pages 267–288, Springer, June 1998.

[122] S. N. Hong and G. Caire. Compute-and-forward strategies for cooperative distributed antenna systems. *IEEE Transactions on Information Theory*, **59**:5227–5243, September 2013.

[123] Hideki Imai and Shuji Hirakawa. A new multilevel coding method using error-correcting codes. *IEEE Transactions on Information Theory*, **23**:371–377, May 1977.

[124] Amir Ingber and Ram Zamir. Expurgated infinite constellations at finite dimensions. *Proc. IEEE Int. Symposium on Information Theory*, pages 130–134, July 2012.

[125] A. Ingber, R. Zamir, and M. Feder. Finite-dimensional infinite constellations. *IEEE Transactions on Information Theory*, **59**:1630–1656, March 2013.

[126] S. A. Jafar. Capacity with causal and noncausal side information: a unified view. *IEEE Transactions on Information Theory*, **52**(12):5468–5474, December 2006.

[127] Syed A. Jafar. Interference alignment – a new look at signal dimensions in a communication network. In *Foundations and Trends in Communications and Information Theory*, volume 7, NOW Publishers, 2011.

[128] S. A. Jafar and Shlomo Shamai (Shitz). Degrees of freedom region for the MIMO X channel. *IEEE Transactions on Information Theory*, **54**(1):151–170, January 2008.

[129] Syed A. Jafar and Sriram Vishwanath. Generalized degrees of freedom of the symmetric Gaussian K-user interference channel. *IEEE Transactions on Information Theory*, **56**(7):3297–3303, July 2010.

[130] N. S. Jayant and P. Noll. *Digital Coding of Waveform*. Prentice-Hall, Englewood Cliffs, NJ, 1984.

[131] N. S. Jayant and L. R. Rabiner. The application of dither to the quantization of speech signals. *Bell System Technical Journal*, **51**:1293–1304, July/August 1972.

[132] G. A. Kabatiansky and V. I. Levenshtein. Bounds for packings on a sphere and in space (in Russian). *Problemy Peredachi Informatsii*, **14**(1):3–25, 1978. English translation: *Probl. Inform. Trans.*, **14**(1):1–17, 1978.

[133] A. K. Khandani and P. Kabal. Shaping multidimensional signal spaces, part I: optimum shaping, shell mapping. *IEEE Transactions on Information Theory*, **39**:1799–1808, November 1993.

[134] Y.-H. Kim, A. Sutivong, and S. Sigurjónsson. Multiple user writing on dirty paper. In *IEEE International Symposium on Information Theory*, page 534, June 2004.

[135] Y. Kochman, A. Khina, U. Erez, and R. Zamir. Rematch-and-forward: joint source/channel coding for parallel relaying with spectral mismatch. *IEEE Transactions on Information Theory*, accepted for publication.

[136] Y. Kochman and R. Zamir. Joint Wyner–Ziv/dirty-paper coding by modulo-lattice modulation. *IEEE Transactions on Information Theory*, **55**(11):4878–4889, 2009.

[137] Y. Kochman and R. Zamir. Analog matching of colored sources to colored channels. *IEEE Transactions on Information Theory*, **57**(6):3180–3195, 2011.

[138] Ralf Koetter and Muriel Medard. An algebraic approach to network coding. *IEEE/ACM Transactions on Networking*, **11**:782–795, October 2003.

[139] J. Körner and K. Marton. How to encode the modulo-two sum of binary sources. *IEEE Transactions on Information Theory*, **25**:219–221, 1979.

[140] V. A. Kotel'nikov. *The Theory of Optimum Noise Immunity*. McGraw-Hill, 1959.

[141] P. Koulgi, E. Tuncel, S. L. Regunathan, and K. Rose. On zero-error source coding with decoder side information. *IEEE Transactions on Information Theory*, **49**:99–111, January 2003.

[142] D. Krithivasan and S. S. Pradhan. Lattices for distributed source coding: Jointly Gaussian sources and reconstruction of a linear function. *IEEE Transactions on Information Theory*, **55**:5628–5651, December 2009.

[143] Dinesh Krithivasan and Sandeep Pradhan. Distributed source coding using Abelian group codes. *IEEE Transactions on Information Theory*, **57**(3):1495–1519, March 2011.

[144] H. Krúger, B. Geiser, P. Vary, H. T. Li, and D. Zhang. Gosset lattice spherical vector quantization with low complexity. *Proc. IEEE Int. Conference on Acoustics, Speech and Signal Processing (ICASSP)*, pages 485–488, 2011.

[145] F. R. Kschischang and S. Pasupathy. Optimal nonuniform signaling for Gaussian channels. *IEEE Transactions on Information Theory*, **39**:913–929, May 1993.

[146] B. Kudryashov and K. Yurkov. Random quantization bounds for lattices over q-ary linear codes. In *Proc. IEEE Int. Symposium on Information Theory (ISIT), Nice, France*, June 2007.

[147] A. V. Kuznetsov and B. S. Tsybakov. Coding in a memory with defective cells. *Problemy Peredachi Informatsii*, **10**:52–60, April–June 1974.

[148] A. Lapidoth, N. Merhav, G. Kaplan, and S. Shamai. On information rates for mismatched decoders. *IEEE Transactions on Information Theory*, **40**:1953–1967, November 1994.

[149] R. Laroia. Coding for intersymbol interference channels, combined coding and pre-coding. *IEEE Transactions on Information Theory*, **42**:1053–1061, July 1996.

[150] R. Laroia, N. Farvardin, and S. Tretter. On optimal shaping of multidimensional constellations. *IEEE Transactions on Information Theory*, **40**:1044–1056, July 1994.

[151] R. Laroia, S. Tretter, and N. Farvardin. A simple and effective precoding scheme for noise whitening on intersymbol interference channels. *IEEE Transactions on Communications*, **41**:1460–1463, October 1993.

[152] L. Lastras and T. Berger. All sources are nearly successively refinable. *IEEE Transactions on Information Theory*, **47**:918–926, March 2001.

[153] J. Leech and N. J. A. Sloane. Sphere packing and error-correcting codes. *Canadian Journal of Mathematics*, **23**:718–745, 1971.

[154] A. K. Lenstra, H. W. Lenstra, and L. Lovasz. Factoring polynomials with rational coefficients. *Mathematische Annalen*, **261**(4):515–534, 1982.

[155] S.-Y. R. Li, R. W. Yeung, and N. Cai. Linear network coding. *IEEE Transactions on Information Theory*, **49**(2):371–381, February 2003.

[156] Y. Liang, H. V. Poor, and S. Shamai. *Information Theoretic Security*. Foundations and Trends in Communications and NOW Publishers, Hanover, MA, 2009.

[157] Soung Chang Liew, Shengli Zhang, and Lu Lu. Physical-layer network coding: tutorial, survey, and beyond. *Physical Communication*, **6**:4–42, 2013.

[158] R. Lifshitz. What is a crystal? *Zeitschrift für Kristallographie*, **222**:313–317, 2007.

[159] Sung Hoon Lim, Young-Han Kim, Abbas El Gamal, and Sae-Young Chung. Noisy network coding. *IEEE Transactions on Information Theory*, **57**(5):3132–3152, May 2011.

[160] J. O. Limb. Design of dithered waveforms for quantized visual signals. *Bell System Technical Journal*, **48**:2555–2582, September 1968.

[161] Y. Linde, A. Buzo, and R. M. Gray. An algorithm for vector quantizer design. *IEEE Transactions on Communications*, **28**:84–95, January 1980.

[162] T. Linder. Private communication.

[163] T. Linder, Ch. Schlegel, and K. Zeger. Corrected proof of de Buda's theorem. *IEEE Transactions on Information Theory*, **39**:1735–1737, September 1993.

[164] T. Linder and R. Zamir. On the asymptotic tightness of the Shannon lower bound. *IEEE Transactions on Information Theory*, **40**:2026–2031, November 1994.

[165] T. Linder and K. Zeger. Asymptotic entropy constrained performance of tesselating and universal randomized lattice quantization. *IEEE Transactions on Information Theory*, **40**:575–579, March 1994.

[166] C. Ling and J. C. Belfiore. Achieving the AWGN channel capacity with lattice Gaussian distribution. In *Proc. Int. Symposium on Information Theory*, July 2013.

[167] C. Ling, L. Luzzi, and J. C. Belfiore. Lattice codes achieving strong secrecy over the mod-λ Gaussian channel. In *Proc. Int. Symposium on Information Theory, Cambridge, MA*, June 2012.

[168] C. Ling, L. Luzzi, J. C. Belfiore, and D. Stehlé. Semantically secure lattice codes for the Gaussian wiretap channel. arXiv:1210.6673v3[cs.IT], October 2013.

[169] Yu. N. Linkov. Evaluation of epsilon entropy of random variables for small epsilon. *Problemy Peredachi Informatsii*, **1**:18–28, 1965. English translation: *Probl. Inform. Trans.*, **1**:12–18, 1965.

[170] Stanley P. Lipshitz, Robert A. Wannamaker, and John Vanderkooy. Quantization and dither: a theoretical survey. *Journal of the Audio Engineering Society*, 5:355–375, May 1992.

[171] T. Liu, P. Moulin, and R. Koetter. On error exponents of nested lattice codes for the AWGN channel. *IEEE Transactions on Information Theory*, 52:454–471, February 2006.

[172] T. Liu and P. Viswanath. Opportunistic orthogonal writing on dirty paper. *IEEE Transactions on Information Theory*, 52:1828–1846, May 2006.

[173] Z. Liu, S. Cheng, A. Liveris, and Z. Xiong. Slepian–Wolf coded nested lattice quantization for Wyner–Ziv coding: high-rate performance analysis and code design. *IEEE Transactions on Information Theory*, 52:4358–4379, October 2006.

[174] S. P. Lloyd. Least squares quantization in PCM. *IEEE Transactions on Information Theory*, 28:129–137, March 1982 (originally presented at the Institute of Mathematical Statistics Meeting 1957).

[175] H. A. Loeliger. Averaging bounds for lattices and linear codes. *IEEE Transactions on Information Theory*, 43:1767–1773, November 1997.

[176] T. Lookabaugh and R. M. Gray. High resolution quantization theory and the vector quantizer advantage. *IEEE Transactions on Information Theory*, 35:1020–1033, September 1989.

[177] Mohammad Ali Maddah-Ali, Abolfazl Seyed Motahari, and Amir Keyvan Khandani. Communication over MIMO X channels: interference alignment, decomposition, and performance analysis. *IEEE Transactions on Information Theory*, 54(8):3457–3470, August 2008.

[178] M. W. Marcellin and T. R. Fischer. Trellis coded quantization of memoryless and Gauss–Markov sources. *IEEE Transactions on Communications*, 38:82–93, January 1990.

[179] E. Martinian, G. W. Wornell, and R. Zamir. Source coding with distortion side-information. *IEEE Transactions on Information Theory*, 54:4638–4665, October 2008.

[180] K. Marton. A coding theorem for the discrete memoryless broadcast channel. *IEEE Transactions on Information Theory*, 22:374–377, May 1979.

[181] A. Mashiach, J. Østergaard, and R. Zamir. Multiple description delta-sigma quantization with individual and central receivers. In *Proc. IEEE 26th Convention of Electrical and Electronics Engineers in Israel (IEEEI), Eilat, Israel*, pages 942–946, November 2010.

[182] A. Mashiach and R. Zamir. Noise-shaped quantization for nonuniform sampling. In *Proc. IEEE Int. Symposium on Information Theory (ISIT) Istanbul, Turkey*, July 2013.

[183] J. Massey. Causality, feedback and directed information. In *Proc. IEEE Int. Symp. Information Theory and Its Applicatios*, pages 303–305.

[184] R. J. McAulay and D. J. Sakrison. A PPM/PM hybrid modulation system. *IEEE Transactions on Communication Technology*, 17(4):458–469, 1969.

[185] N. Merhav and S. Shamai. On joint source-channel coding for the Wyner–Ziv source and the Gel'fand–Pinsker channel. *IEEE Transactions on Information Theory*, 49(11):2844–2855, 2003.

[186] D. Micciancio and S. Goldwasser. *Complexity of Lattice Problems*. Kluwer, 2002.

[187] D. Micciancio and O. Regev. Worst-case to average-case reductions based on Gaussian measures. In *Proc. 45th Annual IEEE Symposium on Foundations of Computer Science, 2004*, pages 372–381, October 2004.

[188] D. Micciancio and O. Regev. Lattice-based cryptography. In D. J. Bernstein and J. Buchmann, editors, *Post-Quantum Cryptography*, Springer, 2008.

[189] U. Mittal and N. Phamdo. Hybrid digital-analog (HDA) joint source-channel codes for broadcasting and robust communications. *IEEE Transactions on Information Theory*, **48**(5):1082–1102, 2002.

[190] A. S. Motahari, S. O. Gharan, Mohammad-Ali Maddah-Ali, and Amir Keyvan Khandani. Real interference alignment: exploiting the potential of single antenna systems. *IEEE Transactions on Information Theory*, submitted 2009. Available at: http://arxiv.org/abs/0908.2282.

[191] P. Moulin J. A. O'Sullivan and J. M. Ettinger. Information-theoretic analysis of steganography. In *Proc. IEEE Symposium on Information Theory, Boston, MA*, page 297, August 1998.

[192] W. Nam, S. Y. Chung, and Y. H. Lee. Nested lattice codes for Gaussian relay networks with interference. *IEEE Transactions on Information Theory*, **57**(12):7733–7745, December 2011.

[193] B. Nazer and M. Gastpar. Computation over multiple-access channels. *IEEE Transactions on Information Theory*, **53**(10):3498–3516, 2007.

[194] Bobak Nazer and Michael Gastpar. Lattice coding increases multicast rates for Gaussian multiple-access networks. In *Allerton Conference on Communications, Control, and Computing, Monticello, IL*, September 2007.

[195] B. Nazer and M. Gastpar. Reliable physical layer network coding. *Proceedings of the IEEE*, **99**(3):438–460, March 2011.

[196] Bobak Nazer and Michael Gastpar. Compute-and-forward: harnessing interference through structured codes. *IEEE Transactions on Information Theory*, **57**(10):6463–6486, October 2011.

[197] Bobak Nazer, Michael Gastpar, Syed A. Jafar, and Sriram Vishwanath. Ergodic interference alignment. *IEEE Transactions on Information Theory*, **58**(10):6355–6371, October 2012.

[198] D. L. Neuhoff. Source coding strategies: simple quantizers vs. simple noiseless codes. *Proc. 1986 Conference on Information Sciences and Systems*, Vol. 20, pages 267–271, Princeton, NJ, March 1986.

[199] D. L. Neuhoff, R. M. Gray, and L. D. Davisson. Fixed rate universal block source coding with a fidelity criterion. *IEEE Transactions on Information Theory*, **21**:511–523, September 1975.

[200] Urs Niesen and M. A. Maddah-Ali. Interference alignment: from degrees of freedom to constant-gap capacity approximations. *IEEE Transactions on Information Theory*, **59**(8):4855–4888, August 2013.

[201] Urs Niesen and Phil Whiting. The degrees-of-freedom of compute-and-forward. *IEEE Transactions on Information Theory*, **58**(8):5214–5232, August 2012.

[202] M. Nokleby and B. Aazhang. Cooperative compute-and-forward. *IEEE Transactions on Information Theory*, submitted 2012. Available at: http://arxiv.org/abs/1203.0695.

[203] Vasileios Ntranos, Viveck R. Cadambe, Bobak Nazer, and Giuseppe Caire. Asymmetric compute-and-forward. In *Allerton Conference on Communication, Control, and Computing, Monticello, IL*, October 2013.

[204] T. J. Oechtering, C. Schnurr, I. Bjelakovic, and H. Boche. Broadcast capacity region of two-phase bidirectional relaying. *IEEE Transactions on Information Theory*, **54**(1):454–458, January 2008.

[205] B. M. Oliver, J. R. Pierce, and C. E. Shannon. The philosophy of PCM. *Proceedings of the IRE*, **36**:1324–1331, November 1948.

[206] O. Ordentlich and U. Erez. A simple proof for the existence of good pairs of nested lattices. In *Proc. 27th IEEE Convention of Electrical and Electronics Engineers in Israel (IEEEI)*, 2012.

[207] Or Ordentlich and Uri Erez. Precoded integer-forcing equalization universally achieves the MIMO capacity up to a constant gap. *IEEE Transactions on Information Theory*, submitted 2013.

[208] Or Ordentlich and Uri Erez. On the robustness of lattice interference alignment. *IEEE Transactions on Information Theory*, **59**(5):2735–2759, May 2013.

[209] Or Ordentlich, Uri Erez, and Bobak Nazer. The approximate sum capacity of the symmetric K-user Gaussian interference channel. *IEEE Transactions on Information Theory*, to appear 2014.

[210] Or Ordentlich, Uri Erez, and Bobak Nazer. Successive integer-forcing and its sum rate optimality. In *Allerton Conference on Communication, Control, and Computing, Monticello, IL*, October 2013.

[211] J. Østergaard and R. Zamir. Multiple descriptions by dithered delta-sigma quantization. *IEEE Transactions on Information Theory*, **55**:4661–4675, October 2009.

[212] M. Palgy, J. Østergaard, and R. Zamir. Multiple description image/video compression using oversampling and noise shaping in the dct-domain. In *Proc. 26th IEEE Convention of Electrical and Electronics Engineers in Israel (IEEEI)*, pages 965–969, November 2010.

[213] N. Palgy and R. Zamir. Dithered probabilistic shaping. In *Proc. 27th IEEE Convention of Electrical and Electronics Engineers in Israel (IEEEI)*, pages 1–5, November 2012.

[214] T. Philosof, U. Erez, and R. Zamir. Combined shaping and precoding for interference cancellation at low snr. In *Proc. IEEE Int. Symposium on Information Theory (ISIT), Yokohama, Japan*, page 68, July 2003.

[215] T. Philosof and R. Zamir. On the loss of single-letter characterization: the dirty multiple access channel. *IEEE Transactions on Information Theory*, **55**:2442–2454, June 2009.

[216] T. Philosof, R. Zamir, and U. Erez. The capacity region of the binary dirty MAC. In *IEEE Information Theory Workshop, Taormina, Italy*, pages 273–277, October 2009.

[217] T. Philosof, R. Zamir, U. Erez, and A. J. Khisti. Lattice strategies for the dirty multiple-access channel. *IEEE Transactions on Information Theory*, **57**:5006–5035, Aug. 2011.

[218] J. T. Pinkston. *Encoding Independent Sample Information Sources*. PhD Thesis, Massachusetts Institute of Technology, September 1967.

[219] M. S. Pinsker. *Information and Information Stability of Random Variables and Processes*, Holden Day, San Francisco, CA, 1964.

[220] M. S. Pinsker. Capacity of noiseless broadcast channels. *Problemy Peredachi Informat-sii*, **14**:28–34, 1978. English translation: *Probl. Inform. Trans.*, **14**:97–102. April–June 1978.

[221] G. Poltyrev. On coding without restrictions for the AWGN channel. *IEEE Transactions on Information Theory*, **40**:409–417, March 94.

[222] Y. Polyanskiy, H. V. Poor, and S. Verdú. Channel coding rate in the finite blocklength regime. *IEEE Transactions on Information Theory*, **56**(5):2307–2359, May 2010.

[223] Petar Popovski and Hiroyuki Yomo. The anti-packets can increase the achievable throughput of a wireless multi-hop network. In *IEEE International Conference on Communications, Istanbul, Turkey*, June 2006.

[224] S. S. Pradhan and K. Ramchandran. Distributed source coding using syndromes (DIS-CUS): design and constructions. In *Proc. Snowbird, UT, 1999 IEEE Data Compression Conference*, pages 158–167, March 1999.

[225] A. Prékopa. On logarithmic concave measures and functions. *Acta Scientiarum Mathematicarum*, **34**:335–343, 1973.

[226] R. Puri, A. Majumdar, and K. Ramchandran. Prism: a video coding paradigm with motion estimation at the decoder. *IEEE Transactions on Image Processing*, **16**:2436–2448, October 2007.

[227] Z. Reznic, M. Feder, and R. Zamir. Distortion bounds for broadcasting with bandwidth expansion. *IEEE Transactions on Information Theory*, **52**(8):3778–3788, 2006.

[228] L. G. Roberts. Picture coding using pseudo-random noise. *IRE Transactions on Information Theory*, **8**:145–154, 1962.

[229] C. A. Rogers. A note on coverings. *Mathematica*, **4**:1–6, 1957.

[230] C. A. Rogers. Lattice coverings of space. *Mathematica*, **6**:33–39, 1959.

[231] C. A. Rogers. *Packing and Covering*. Cambridge University Press, Cambridge, 1964.

[232] K. Rose. A mapping approach to rate-distortion computation and analysis. *IEEE Transactions on Information Theory*, **40**:1939–1952, November 1994.

[233] H. Sato. The capacity of the Gaussian interference channel under strong interference. *IEEE Transactions on Information Theory*, **27**(6):786–788, November 1981.

[234] L. Schuchman. Dither signals and their effects on quantization noise. *IEEE Transactions on Communications*, **12**:162–165, 1964.

[235] S. Servetto. Lattice quantization with side information. In *Proc. 2000 IEEE Data Compression Conference, Snowbird, UT*, pages 510–519, March 2000.

[236] O. Shalvi, N. Sommer, and M. Feder. Signal codes: convolutional lattice codes. *IEEE Transactions on Information Theory*, **57**:5203–5226, August 2011.

[237] S. Shamai and R. Laroia. The intersymbol interference channel: lower bounds on capacity and channel precoding loss. *IEEE Transactions on Information Theory*, **42**:1388–1404, September 1996.

[238] S. Shamai, S. Verdú, and R. Zamir. Digital broadcasting back-compatible with analog broadcasting: information theoretic limits. In *Proc. Int. Workshop on Digital Signal Processing Techniques Applied to Space Communication, Sitges, Barcelona, Spain*, pages 8.25–8.39, September 1996.

[239] S. Shamai, S. Verdú, and R. Zamir. Systematic lossy source/channel coding. *IEEE Transactions on Information Theory*, **44**:564–579, March 1998.

[240] C. E. Shannon. A mathematical theory of communication. *Bell System Technical Journal*, **27**:379–423, July 1948.

[241] C. E. Shannon. Communication in the presence of noise. *Proceedings of the IRE*, **37**:10–21, January 1949.

[242] C. E. Shannon. The lattice theory of information. *IRE Transactions on Information Theory*, **1**:105–107, February 1953.

[243] C. E. Shannon. Channels with side information at the transmitter. *IBM Journal of Research and Development*, **2**:289–293, October 1958.

[244] C. E. Shannon. Probability of error for optimal codes in a Gaussian channel. *Bell System Technical Journal*, **38**:611–656, 1959.

[245] V. Shashank and Naveen Kashyap. Lattice coding for strongly secure compute-and-forward in a bidirectional relay. In *Proc. IEEE Int. Symposium on Information Theory, Istanbul, Turkey (ISIT)*, pages 2775–2779, July 2013.

[246] M. Shemer. *A Korner Marton Approach to Low Complexity Video Encoding*. Master's Thesis, Tel Aviv University, February 2009.

[247] D. Slepian and J. K. Wolf. Noiseless coding of correlated information sources. *IEEE Transactions on Information Theory*, **19**:471–480, July 1973.

[248] N. J. A. Sloane. Shannon lecture, *Proc. IEEE Int. Symposium on Information Theory (ISIT)*, 1998.

[249] A. Somech-Baruch and N. Merhav. On the capacity game of public watermarking systems. *IEEE Transactions on Information Theory*, **50**(3):511–524, March 2004.

[250] Y. Song and N. Devroye. Lattice codes for the Gaussian relay channel: decode-and-forward and compress-and-forward. *IEEE Transactions on Information Theory*, **59**(8):4927–4948, August 2013.

[251] S. Sridharan, A. Jafarian, S. Vishwanath, and S. A. Jafar. Capacity of symmetric K-user Gaussian very strong interference channels. In *IEEE Global Communications Conference, New Orleans, LA*, December 2008.

[252] C. Swannack, U. Erez, and G. W. Wornell. Reflecting on the AWGN error exponent. In *43rd Annual Allerton Conference on Communication, Control, and Computing, Allerton House, Monticello, IL*, September 2005.

[253] S. Tavildar, P. Viswanath, and A. B. Wagner. The Gaussian many-help-one distributed source coding problem. *IEEE Transactions on Information Theory*, **56**(1):564–571, January 2010.

[254] M. Tomlinson. New automatic equalizer employing modulo arithmetic. *Electronics Letters*, **7**:138–139, March 1971.

[255] G. Fejes Tóth. Sur la representation d'une population infinie par un nombre fini d'elements. *Acta Mathematica Academiae Scientiarum Hungaricae*, **10**:299–304, 1959.

[256] S. Tretter. *Constellation Shaping, Non-linear Precoding, and Trellis Coding for Voice-band Telephone Channel Modems*. Kluwer Academic Publishers, 2002.

[257] S. Tridenski. Private communication.

[258] D. N. C. Tse and Mohammad Ali Maddah-Ali. Interference neutralization in distributed lossy source coding. In *Proc. IEEE Int. Symposium on Information Theory (ISIT), Austin, TX*, pages 166–170, June 2010.

[259] G. Ungerboeck. Channel coding with multilevel/phase signals. *IEEE Transactions on Information Theory*, **28**:55–67, 1982.

[260] G. Ungerboeck. Huffman shaping. In *Codes, Graphs, and Systems: A Celebration of the Life and Career of G. David Forney, Jr. on the Occasion of his Sixtieth Birthday*, Kluwer Academic Publishers, 2002.

[261] R. Urbanke and B. Rimoldi. Lattice codes can achieve capacity on the AWGN channel. *IEEE Transactions on Information Theory*, **44**:273–278, January 1998.

[262] S. Verdú. On channel capacity per unit cost. *IEEE Transactions on Information Theory*, **36**:1019–1030, September 1990.

[263] Aaron B. Wagner. On distributed compression of linear functions. *IEEE Transactions on Information Theory*, **57**(1):79–94, January 2011.

[264] A. B. Wagner, S. Tavildar, and P. Viswanath. Rate region of the quadratic Gaussian two-encoder source-coding problem. *IEEE Transactions on Information Theory*, **54**:1938–1961, May 2008.

[265] I.-Hsiang Wang. Approximate capacity of the dirty multiple-access channel with partial state information at the encoders. *IEEE Transactions on Information Theory*, **58**(5):2781–2787, May 2012.

[266] F. M. J. Willems. Computation of the Wyner–Ziv rate-distortion function. Technical Report 83-E-140, Eindhoven University, July 1983.

[267] F. M. J. Willems. On Gaussian channels with side information at the transmitter. In *Proc. 9th Symposium on Information Theory in the Benelux, Enschede, The Netherlands*, 1988.

[268] F. M. J. Willems. Signalling for the Gaussian channel with side information at the transmitter. In *Proc. IEEE Int. Symposium on Information Theory (ISIT), Sorrento, Italy*, page 348, June 2000.

[269] M. P. Wilson, K. Narayanan, and G. Caire. Joint source channel coding with side information using hybrid digital analog codes. *IEEE Transactions on Information Theory*, **56**(10):4922–4940, 2010.

[270] Makesh Pravin Wilson, Krishna Narayanan, Henry Pfister, and Alex Sprintson. Joint physical layer coding and network coding for bidirectional relaying. *IEEE Transactions on Information Theory*, **11**(56):5641–5654, November 2010.

[271] H. S. Witsenhausen and A. D. Wyner. Interframe coder for video signals. *US Patent 4191970*, March 1980.

[272] J. M. Wozencraft and I. M. Jacobs. *Principles of Communication Engineering*. Wiley, New York, 1967.

[273] Yihong Wu, Shlomo Shamai (Shitz), and Sergio Verdú. Degrees of freedom of the interference channel: a general formula. In *Proc. IEEE Int. Symposium on Information Theory (ISIT), St. Petersburg, Russia*, pages 1362–1366, August 2011.

[274] A. D. Wyner. Recent results in Shannon theory. *IEEE Transactions on Information Theory*, **20**:2–10, January 1974.

[275] A. D. Wyner. The rate-distortion function for source coding with side information at the decoder II: general sources. *Information and Control*, **38**:60–80, 1978.

[276] A. D. Wyner and J. Ziv. The rate-distortion function for source coding with side information at the decoder. *IEEE Transactions on Information Theory*, **22**:1–10, January 1976.

[277] Y. Yamada, S. Tazaki, and R. M. Gray. Asymptotic performance of block quantizers with difference distortion measures. *IEEE Transactions on Information Theory*, **26**:6–14, January 1980.

[278] Y. Yang and Z. Xiong. Distributed compression of linear functions: partial sum-rate tightness and gap to optimal sum-rate. *IEEE Transactions on Information Theory*, to appear 2013.

[279] A. Yeredor. Private communication.

[280] Y. Yona and E. Haim. Private communication.

[281] W. Yu and J. M. Cioffi. Trellis precoding for the broadcast channel. In *Proc. IEEE Global Telecommunications Conference, San Antonio, TX*, Vol. 2, pages 1344–1348, November 2001.

[282] P. Zador. *Development and Evaluation of Procedures for Quantizing Multivariate Distributions*. PhD Thesis, Stanford University, 1964.

[283] P. Zador. Asymptotic quantization error of continuous signals and their quantization dimension. *IEEE Transactions on Information Theory*, **28**:139–149, March 1982 (previously an unpublished Bell Laboratories memo, 1966).

[284] R. Zamir. The rate loss in the Wyner–Ziv problem. *IEEE Transactions on Information Theory*, **43**:2073–2084, November 1996.

[285] R. Zamir and T. Berger. Multiterminal source coding with high resolution. *IEEE Transactions on Information Theory*, **45**:106–117, January 1999.

[286] R. Zamir and U. Erez. A Gaussian input is not too bad. *IEEE Transactions on Information Theory*, **50**:1362–1367, June 2004.

[287] R. Zamir and M. Feder. On universal quantization by randomized uniform/lattice quantizer. *IEEE Transactions on Information Theory*, **38**:428–436, March 1992.

[288] R. Zamir and M. Feder. Rate distortion performance in coding band-limited sources by sampling and dithered quantization. *IEEE Transactions on Information Theory*, **41**:141–154, January 1995.

[289] R. Zamir and M. Feder. On lattice quantization noise. *IEEE Transactions on Information Theory*, **42**:1152–1159, July 1996.

[290] R. Zamir and M. Feder. Information rates of pre/post-filtered dithered quantizers. *IEEE Transactions on Information Theory*, **42**:1340–1353, September 1996.

[291] R. Zamir, Y. Kochman, and U. Erez. Achieving the Gaussian rate-distortion function by prediction. *IEEE Transactions on Information Theory*, **54**:3354–3364, July 2008.

[292] R. Zamir and S. Shamai. Nested linear/lattice codes for Wyner–Ziv encoding. In *Proc. Information Theory Workshop, Killarney, Ireland*, pages 92–93, June 1998.

[293] R. Zamir, S. Shamai, and U. Erez. Nested linear/lattice codes for structured multiterminal binning. *IEEE Transactions on Information Theory*, **48**:1250–1276, June 2002.

[294] Jiening Zhan, Bobak Nazer, Uri Erez, and Michael Gastpar. Integer-forcing linear receivers. *IEEE Transactions on Information Theory*, to appear 2014.

[295] S. Zhang, S.-C. Liew, and P. Lam. Hot topic: physical-layer network coding. In *ACM Int. Conference on Mobile Computing and Networking, Los Angeles, CA*, September 2006.

[296] Q. Zhao and M. Effros. Lossless and near-lossless source coding for multiple access networks. *IEEE Transactions on Information Theory*, **49**:112–128, January 2003.

[297] Q. Zhao, H. Feng, and M. Effros. Multiresolution source coding using entropy constrained dithered scalar quantization. In *Proc. 2004 IEEE Data Compression Conference, Snowbird, UT*, pages 22–31, March 2004.

[298] J. Ziv. The behavior of analog communication systems. *IEEE Transactions on Information Theory*, **16**:587–594, 1970.

[299] J. Ziv. Coding of sources with unknown statistics – part II: distortion relative to a fidelity criterion. *IEEE Transactions on Information Theory*, **18**:389–394, 1972.

[300] J. Ziv. On universal quantization. *IEEE Transactions on Information Theory*, **31**:344–347, May 1985.

Index